T0212119

Lecture Notes in Computer Science 11968

More information about this series at http://www.springer.com/series/7407

Nikolaos F. Matsatsinis · Yannis Marinakis ·
Panos Pardalos (Eds.)

Learning and Intelligent Optimization

13th International Conference, LION 13
Chania, Crete, Greece, May 27–31, 2019
Revised Selected Papers

 Springer

Editors
Nikolaos F. Matsatsinis (iD)
Technical University of Crete
Chania, Greece

Yannis Marinakis
Technical University of Crete
Chania, Greece

Panos Pardalos
Department of Industrial
and Systems Engineering
University of Florida
Gainesville, FL, USA

ISSN 0302-9743   ISSN 1611-3349 (electronic)
Lecture Notes in Computer Science
ISBN 978-3-030-38628-3   ISBN 978-3-030-38629-0 (eBook)
https://doi.org/10.1007/978-3-030-38629-0

LNCS Sublibrary: SL1 – Theoretical Computer Science and General Issues

This Springer imprint is published by the registered company Springer Nature Switzerland AG
The registered company address is: Gewerbestrasse 11, 6330 Cham, Switzerland

Preface

This volume, edited by Nikolaos Matsatsinis, Yannis Marinakis, and Panos Pardalos, contains peer-reviewed papers from the 13th Learning and Intelligent Optimization (LION-13) Conference, held in Chania, Crete, Greece, during May 27–31, 2019.

The LION-13 conference continued the successful series of the constantly expanding and worldwide recognized LION events (LION-1: Andalo, Italy, 2007; LION-2 and LION-3: Trento, Italy, 2008 and 2009; LION-4: Venice, Italy, 2010; LION-5: Rome, Italy, 2011; LION-6: Paris, France, 2012; LION-7: Catania, Italy, 2013; LION-8: Gainesville, USA, 2014; LION-9: Lille, France, 2015; LION-10: Ischia, Italy, 2016; LION-11: Nizhny Novgorod, Russia, 2017; and LION-12: Kalamata, Greece, 2018). This edition was organized by Nikolaos Matsatsinis and Yannis Marinakis from the Decision Support Systems Laboratory (ERGA.S.Y.A.) at the Technical University of Crete, Greece, and Panos Pardalos from the Center for Applied Optimization at the University of Florida, USA, who were conference general and Technical Program Committee chairs.

Like its predecessors, the LION-13 international meeting explored advanced research developments in interconnected fields such as mathematical programming, global optimization, machine learning, and artificial intelligence. The location of LION-13 in Chania, Crete, Greece, was an excellent occasion for researchers to meet and consolidate research and personal links.

A total of 38 papers were accepted for oral presentation and presented during the conference. The following three plenary lecturers shared their current research directions with the LION-13 participants:

- Xin-She Yang, Middlesex University London, UK: "Open Problems and Analysis of Nature-Inspired Algorithms"
- Panos Pardalos, Center for Applied Optimization, University of Florida, USA: "Sustainable Interdependent Networks"
- Marco Dorigo, IRIDIA Lab, Université Libre de Bruxelles, Belgium: "Controlling Robot Swarms"

A total of 31 papers were accepted for publication in this LNCS volume after thorough peer reviewing (up to three review rounds for some manuscripts) by the members of the LION-13 Program Committee and independent reviewers. These papers describe advanced ideas, technologies, methods, and applications in optimization and machine learning.

The editors thank all the participants for their dedication to the success of LION-13 and are grateful to the reviewers for their valuable work. The support of the Springer LNCS editorial staff is greatly appreciated.

The editors express their gratitude to the organizers and sponsors of the LION-13 international conference:

- Technical University of Crete
- Region of Crete
- Region of Crete, Regional Unit of Chania
- Cooperative Bank of Chania
- ARIAN Maritime S.A.
- Masters in Technology and Innovation Management
- Smart Tours
- Technical Chamber of Greece, Branch of Western Crete

Their support was essential for the success of this event.

November 2019

Nikolaos Matsatsinis
Yannis Marinakis
Panos Pardalos

Organization

General Chair

Nikolaos Matsatsinis   DSS Lab, Technical University of Crete, Greece

Technical Program Committee Chairs

Panos Pardalos   University of Florida, USA
Yannis Marinakis   DSS Lab, Technical University of Crete, Greece

Technical Program Committee

Adil Erzin	Sobolev Institute of Mathematics, Novosibirsk State University, Russia
Alexander Kelmanov	Sobolev Institute of Mathematics, Siberian Branch of the Russian Academy of Sciences, Russia
Alexander Strekalovskiy	Matrosov Institute for System Dynamics and Control Theory SB RAS, Russia
Andre Augusto Cire	University of Toronto Scarborough, Canada
Angelo Sifaleras	University of Macedonia, Greece
Bistra Dilkina	Georgia Institute of Technology, USA
Celso C. Ribeiro	Universidade Federal Fluminense, Brazil
Christian Blum	Spanish National Research Council (CSIC), Spain
Christos Tarantilis	Athens University of Economics and Business, Greece
Daniela Lera	University of Cagliari, Italy
Daniele Vigo	University of Bologna, Italy
Dario Landa-Silva	University of Nottingham, UK
Dmitri Kvasov	DIMES, University of Calabria, Italy
Eric D. Taillard	University of Applied Science Western Switzerland, HEIG-Vd, Switzerland
Francesca Rossi	IBM, University of Padova, Italy
Frank Hutter	University of Freiburg, Germany
Gerardo Toraldo	Universita di Napoli Federico II, Italy
Giovanni Fasano	University Ca'Foscari of Venice, Italy
Helena Ramalhinho	Universitat Pompeu Fabra, Spain
John Chinneck	Carleton University, Canada
John N. Tsitsiklis	Massachusetts Institute of Technology (MIT), USA
Julius Zilinskas	Vilnius University, Lithuania
Jun Pei	Hefei University of Technology, China
Konstantinos Parsopoulos	University of Ioannina, Greece
Lars Kotthoff	University of Wyoming, Canada
Luca Di Gaspero	DPIA, University of Udine, Italy

Magdalene Marinaki	Technical University of Crete, Greece
Marc Schoenauer	Projet TAO, Inria Saclay Ile-de-France, France
Marie-Eleonore Marmion	Université de Lille, Laboratoire CRIStAL, France
Martin Golumbic	University of Haifa, Israel
Massimo Roma	Università La Sapienza di Roma, Italy
Mauro Brunato	University of Trento, Italy
Michel Gendreau	Université de Montréal, Canada
Michael Khachay	Krasovsky Institute of Mathematics and Mechanics, Russia
Michael Trick	Carnegie Mellon University, USA
Mikhail Posypkin	Dorodnicyn Computing Centre, FRC CSC RAS, Russia
Nick Sahinidis	Carnegie Mellon University, USA
Nikolaos Samaras	University of Macedonia, Greece
Oleg Prokopyev	University of Pittsburgh, USA
Panos Pardalos	University of Florida, USA
Renato De Leone	Universita di Camerino, Italy
Roberto Battiti	Universita' di Trento, Italy
Stefan Boettcher	Emory University, USA
Stelios Tsafarakis	DSS Lab, Technical University of Crete, Greece
Sonia Cafieri	École Nationale de l'Aviation Civile, France
Sotiris Batsakis	DSS Lab, Technical University of Crete, Greece
Tatiana Tchemisova	University of Aveiro, Portugal
Tias Guns	Vrije Universiteit Brussel (VUB), Belgium
Vangelis Paschos	LAMSADE, University of Paris-Dauphine, France
Vladimir Grishagin	Nizhni Novgorod State University, Russia
Yaroslav Sergeyev	University of Calabria, Italy
Yury Kochetov	Sobolev Institute of Mathematics, Russia

Local Organizing Committee Chairs

Yannis Marinakis	DSS Lab, Technical University of Crete, Greece
Magdalene Marinaki	Technical University of Crete, Greece
Sotiris Batsakis	DSS Lab, Technical University of Crete, Greece
Evangelia Krassadaki	DSS Lab, Technical University of Crete, Greece

Local Organizing Committee

Katerina Zompanaki	Region of Crete, Regional Unit of Chania, Greece
Iakovos Roussos	Cooperative Bank of Chania, Greece
Nikos Spanoudakis	DSS Lab, Technical University of Crete, Greece
Sifis Kontaxakis	DSS Lab, Technical University of Crete, Greece
Alkaios Sakellaris	DSS Lab, Technical University of Crete, Greece
Tasos Kyriakidis	DSS Lab, Technical University of Crete, Greece

Konstantina Miteloudi   DSS Lab, Technical University of Crete, Greece
Garyfallia Matsatsini   DSS Lab, Mediterranean Agronomic Institute
                             of Chania, Greece
Yannis Karaparisis   DSS Lab, Technical University of Crete, Greece
Dimitra Trachanatzi   DSS Lab, Technical University of Crete, Greece

Contents

A Hybrid Immunological Search for the Weighted Feedback Vertex Set Problem

Vincenco Cutello$^{(\boxtimes)}$, Maria Oliva, Mario Pavone$^{(\boxtimes)}$, and Rocco A. Scollo

Department of Mathematics and Computer Science, University of Catania,
V.le A. Doria 6, 95125 Catania, Italy
cutello@unict.it, mpavone@dmi.unict.it

Abstract. In this paper we present a hybrid immunological inspired algorithm (HYBRID-IA) for solving the Minimum Weighted Feedback Vertex Set ($M\ W\ F\ V\ S$) problem. $MWFVS$ is one of the most interesting and challenging combinatorial optimization problem, which finds application in many fields and in many real life tasks. The proposed algorithm is inspired by the clonal selection principle, and therefore it takes advantage of the main strength characteristics of the operators of (i) cloning; (ii) hypermutation; and (iii) aging. Along with these operators, the algorithm uses a local search procedure, based on a deterministic approach, whose purpose is to refine the solutions found so far. In order to evaluate the efficiency and robustness of HYBRID-IA several experiments were performed on different instances, and for each instance it was compared to three different algorithms: (1) a memetic algorithm based on a genetic algorithm (MA); (2) a tabu search metaheuristic (XTS); and (3) an iterative tabu search (ITS). The obtained results prove the efficiency and reliability of HYBRID-IA on all instances in term of the best solutions found and also similar performances with all compared algorithms, which represent nowadays the state-of-the-art on for $MWFVS$ problem.

Keywords: Immunological algorithms · Immune-inspired computation · Metaheuristics · Combinatorial optimization · Feedback vertex set · Weighted feedback vertex set

1 Introduction

Immunological inspired computation represents an established and rich family of algorithms inspired by the dynamics and the information processing mechanisms that the Immune System uses to detect, recognise, learn and remember foreign entities to the living organism [11]. Thanks to these interesting features, immunological inspired algorithms represent successful computational methodologies in search and optimization tasks [4,20]. Although there exist several immune theories at the basis of immunological inspired computation that could characterize

© Springer Nature Switzerland AG 2020
N. F. Matsatsinis et al. (Eds.): LION 13 2019, LNCS 11968, pp. 1–16, 2020.
https://doi.org/10.1007/978-3-030-38629-0_1

their natural application to anomaly detection and classification tasks, one that
has been proven to be quite effective and robust is based on the clonal selection
principle. Algorithms based on the clonal selection principle work on a population of immunological cells, better known as *antibodies*, that proliferate, i.e. clone
themselves – the number of copies depends on the quality of their foreign entities
detection – and undergo a mutation process usually at a high rate. This process,
biologically called *Affinity Maturation*, makes these algorithms very suitable in
functions and combinatorial optimization problems. This statement is supported
by several experimental applications [6–8, 18], as well as by theoretical analyses
that prove their efficiency with respect to several Randomized Search Heuristics
[14, 15, 23, 25].

In light of the above, we have designed and developed an immune inspired
hypermutation algorithm - based precisely on the clonal selection principle -
in order to solve a classical combinatorial optimization problem, namely the
Weighted Feedback Vertex Set (WFVS).

In addition, we take into account what clearly emerges from the evolutionary computation literature: in order to obtain good performances and solve
hard a combinatorial optimization problem, it is necessary to combine together
metaheuristics and other classical optimization techniques, such as local search,
dynamic programming, exact methods, etc. For this reason, we have designed a
hybrid immunological algorithm, by including some deterministic criteria inside
in order to refine the found solutions and to help the convergence of the algorithm
towards the global optimal solution.

The hybrid immunological algorithm proposed in this paper, hereafter simply
called HYBRID-IA, takes advantage of the immunological operators of cloning,
hypermutation and aging, to carefully explore the search space and properly
exploit the information learned, but it also makes use of local search for improving the quality of the solutions, and deterministically trying to remove that
vertex that break off one or more cycles in the graph, and has the minor weight-
degree ratio. The many experiments performed have proved the fruitful impact
of such a greedy idea, since it was almost always able to improve the solutions,
leading the HYBRID-IA towards the global optimal solution. For evaluating the
reliability and efficiency of HYBRID-IA, many experiments on several different
instances have been performed, taken by [2] and following the same experimental
protocol proposed in it. HYBRID-IA was also compared with other three metaheuristics: *Iterated Tabu Search* (ITS) [3]; *eXploring Tabu Search* (XTS) [1, 9];
and *Memetic Algorithm* (MA) [2]. From all comparisons performed and analysed,
HYBRID-IA proved to be always competitive and comparable with the best of
the three metaheuristics. Furthermore, it is worth emphasizing that HYBRID-IA
outperforms the three algorithms on some instances, since, unlike them, it is
able to find the global best solution.

2   The Weighted Feedback Vertex Set Problem

Given a directed or undirected graph $G = (V, E)$, a feedback vertex set of G is a
subset $S \subset V$ of vertices whose removal makes G acyclic. More formally, if $S \subset V$

we can define the subgraph $G[S] = (V \backslash S, E_{V \backslash S})$ where $E_{V \backslash S} = \{(u, v) \in E : u, v \in V \backslash S\}$. If $G[S]$ is acyclic, then S is a feedback set. The *Minimum Feedback Vertex Set Problem* ($MFVS$) is the problem of finding a feedback vertex set of minimal cardinality. If S is a feedback set, we say that a vertex $v \in S$ is *redundant* if the induced subgraph $G[S \backslash \{v\}] = ((V \backslash S) \cup \{v\}, E_{(V \backslash S) \cup \{v\}}, w)$ is still an acyclic graph. It follows that S is a minimal FVS if it doesn't contain any redundant vertices. It is well known that the decisional version of the $MFVS$ is a $\mathcal{NP}$-complete problem for general graphs [12,17] and even for bipartite graphs [24].

If we associate a positive weight $w(v)$ to each vertex $v \in V$, let S be any subset of V, then its weight is the sum of the weights of its vertices, i.e. $\sum_{v \in S} w(v)$. The *Minimum Weighted Feedback Vertex Set Problem* ($MWFVS$) is the problem of finding a feedback vertex set of minimal weight.

The $MWFVS$ problem is an interesting and challenging task as it finds application in many fields and in many real-life tasks, such as in the context of (*i*) operating systems for preventing or removing deadlocks [22]; (*ii*) combinatorial circuit design [16]; (*iii*) information security [13]; and (*iv*) the study of monopolies in synchronous distributed systems [19]. Many heuristics and metaheuristics have been developed for the simple $MFVS$ problem, whilst very few, instead, have been developed for the weighted variant of the problem.

3 HYBRID-IA: A Hybrid Immunological Algorithm

HYBRID-IA is an immunological algorithm inspired by the clonal selection principle, which represents an excellent example of bottom up intelligent strategy where adaptation operates at local level, whilst a complex and useful behaviour emerges at global level. The basic idea of this metaphor is how the cells adapt for binding and eliminate foreign entities, better known as Antigene (*Ag*). The key features of HYBRID-IA are the cloning, hypermutation and aging operators, which, respectively, generate new populations centered on higher affinity values; explore the neighborhood of each solution into the search space; and eliminate the old cells and the less promising ones, in order to keep the algorithm from getting trapped into a local optimal. To these immunological operators, a local search strategy is also added, which, by restoring one or more cycles with the addition of a previously removed vertex, it tries again to break off the cycle, or the cycles, by conveniently removing a new vertex that improves the fitness of the solution. Usually the best choice is to select the node with the minimal weight-degree ratio. The existence of a cycle, or more cycles, is computed via the well-know procedure *Depth First Search* (DFS) [5]. The aim of the local search is then to repair the random choices done by the stochastic operators via more locally appropriate choices.

HYBRID-IA is based on two main entities, such as the antigen that represents the problem to tackle, and the B cell receptor that instead represents a point (configuration) of the search space. Each B cell, in particular, represents a permutation of vertices that determines the order of the nodes to be removed:

starting from the first position of the permutation, the relative node is selected and removed from the graph, and this is iteratively repeated, following the order in the permutation, until an acyclic graph is obtained, or, in general, there are no more vertices to be removed. Once an acyclic graph is obtained, all possible redundant vertices are restored in V. Now, all removed vertices represent the S set, i.e. a solution to the problem. Such process is outlined in simple way in Algorithm 1. Some clarification about Algorithm 1:

- Input to the Algorithm is an undirected graph $G = (V, E)$;
- given any $X \subseteq V$ by $d_X(v)$ we denote the degree of vertex v in the graph induced by X, i.e. the graph $G(X)$ obtained from G by removing all the vertices not in X.
- if, given any $X \subseteq V$ a vertex v has degree $d_X(v) \leq 1$, such a vertex cannot be involved in any cycle. So, when searching for a feedback vertex set, any such a vertex can simply be ignored, or removed from the graph.
- the algorithm associates deterministically a subset S of the graph vertices to any given permutation of the vertices in V. The sum of the weights of the vertices in S will be the fitness value associated to the permutation.

Algorithm 1. Create Solution

$X \leftarrow V$
$P \leftarrow permutation(V)$
$S \leftarrow \emptyset$
for $u \in P$ **do**
    if $u \in X$ and $G(X)$ not acyclic **then**
        $X \leftarrow X \setminus \{u\}$
        $S \leftarrow S \cup \{u\}$
        while $\exists v \in X : d_X(v) < 2$ **do**
            $X \leftarrow X \setminus \{u\}$
        end while
    end if
end for
Removing all redundant vertices from S
return S

At each time step t, we maintain a population of B cells of size d that we label $P^{(t)}$. Each permutation, i.e. B cell, is randomly generated during the initialization phase ($t = 0$) using a uniform distribution. A summary of the proposed algorithm is shown below. HYBRID-IA terminates its execution when the fixed termination criterion is satisfied. For our experiments and for all outcomes presented in this paper, a maximum number of generations has been considered.

Algorithm 2. HYBRID-IA (d, dup, ρ, τ_B)

$t \leftarrow 0;$
$P^{(t)} \leftarrow$ Initialize_Population$(d);$
Compute_Fitness$(P^{(t)});$
repeat
    $P^{(clo)} \leftarrow$ Cloning $(P^{(t)}, dup);$
    $P^{(hyp)} \leftarrow$ Hypermutation$(P^{(clo)}, \rho);$
    Compute_Fitness$(P^{(hyp)});$
    $(P_a^{(t)}, P_a^{(hyp)}) \leftarrow$ Aging$(P^{(t)}, P^{(hyp)}, \tau_B);$
    $P^{(select)} \leftarrow (\mu + \lambda)$-Selection$(P_a^{(t)}, P_a^{(hyp)});$
    $P^{(t+1)} \leftarrow$ Local_Search$(P^{(select)});$
    Compute_Fitness$(P^{(t+1)});$
    $t \leftarrow t + 1;$
until (termination criterion is satisfied)

Cloning Operator: It is the first immunological operator to be applied, and it has the aim to reproduce the proliferation mechanism of the immune system. Indeed it simply copies dup times each B cell producing an intermediate population $P^{(clo)}$ of size $d \times dup$. Once a B cell copy is created, i.e. the B cell is cloned, to this is assigned an age that determines its lifespan, during that it can mature, evolves and improves: from the assigned age it will evolve into the population for producing robust offspring until a maximum age reachable (a user-defined parameter), i.e. a prefixed maximum number of generations. It is important to highlight that the assignment of the age together to the aging operator play a key role on the performances of HYBRID-IA [10, 21], since their combination has the purpose to reduce premature convergences, and keep an appropriate diversity between the B cells.

The cloning operator, coupled with the hypermutation operator, performs a local search around the cloned solutions; indeed the introduction of a high number of blind mutations produces individuals with higher fitness function values, which will be after selected to form ever better progenies.

Inversely Hypermutation Operator: This operator has the aim to explore the neighborhood of each clone taking into account, however, the quality of the fitness function of the clone it is working on. It acts on each element of the population $P^{(clo)}$, performing M mutations on each clone, whose number M is determined by an *inversely proportional law* to the clone fitness value: the higher is the fitness function value, the lower is the number of mutations performed on the B cell. Unlike of any evolutionary algorithm, no mutation probability has been considered in HYBRID-IA.

Given a clone $\boldsymbol{x}$, the number of mutations M that it will be undergo is determined by the following *potential mutation*:

$$\alpha = e^{-\rho \hat{f}(\boldsymbol{x})}, \tag{1}$$

where α represents the *mutation rate*, and $\hat{f}(\boldsymbol{x})$ the fitness function value normalized in $[0,1]$. The number of mutations M is then given by

$$M = \lfloor (\alpha \times \ell) + 1 \rfloor, \tag{2}$$

where ℓ is the length of the B cell. From Eq. 2, is possible to note that at least one mutation occurs on each cloned B cell, and this happens to the solutions very close to the optimal one. Since any B cell is represented as a vertices permutation that determines their removing order, the position occupied by each vertex in the permutation becomes crucial for the achievement of the global best solution. Consequently, the hypermutation operator adopted is the well-known *Swap Mutations*, through which the right permutation, i.e. the right removing order, is searched: for any $\boldsymbol{x}$ B cell, choices two positions i and j, the vertices x_i and x_j are exchanged of position, becoming $x'_i = x_j$ and $x'_j = x_i$, respectively.

Aging Operator: This operator has the main goal to help the algorithm for jumping out from local optimal. Simply it eliminates the old B cells from the populations $P^{(t)}$ and $P^{(hyp)}$: once the age of a B cell exceeds the maximum age allowed (τ_B), it will be removed from the population of belonging independently from its fitness value. As written above, the parameter τ_B represents the maximum number of generations allowed so that every B cell can be considered to remain into the population. In this way, this operator is able to maintain a proper turnover between the B cells in the population, producing high diversity inside it, and this surely help the algorithm to avoid premature convergences and, consequently, to get trapped into local optima.

An exception about the removal may be done for the best current solution that is kept into the population even if its age is older than τ_B. This variant of the aging operator is called *elitist aging operator*.

$(\mu+\lambda)$**-Selection Operator:** Once the aging operator ended its work, the best d survivors from both populations $P_a^{(t)}$ and $P_a^{(hyp)}$ are selected for generating a temporary population $P^{(select}$ that, afterwards, will be undergo to the local search. For this process the classical $(\mu+\lambda)$-*Selection operator* has been considered, where, in our case, $\mu = d$ and $\lambda = (d \times dup)$. In a nutshell, this operator reduces the offspring B cell population $(P_a^{(hyp)})$ of size $\lambda \geq \mu$ to a new population $(P^{(select)})$ of size $\mu = d$. Since this selection mechanism identifies the d best elements between the offspring set and the old parent B cells, then it guarantees monotonicity in the evolution dynamics. Nevertheless, may happens that all survived B cells are less than the required population size (d), i.e. $d_a < d$. This can easily happens depending on the chosen age assignment, and fixed value of τ_B. In this case, the selection mechanism randomly generates $d - d_a$ new B cells.

Local Search: The main idea at the base of this local search is to repair in a proper and deterministic way the solutions produced by the stochastic mutation operator. Whilst the hypermutation operator determines the vertices to be removed in a blind and random way, i.e. choosing them independently of its weight, then the local search try to change one, or more of them with another node of lesser weight, improving thus the fitness function of that B cell. Given a solution x, all vertices in x are sorted in decreasing order with respect their weights. Then, starting from vertex u with largest weight, the local search procedure iteratively works as follows: the vertex u is inserted again in $V \setminus S$, generating then one or more cycles; via the classical DFS procedure, are computed the number of the cycles produced, and, thus one of these (if there are more) is taken into account, together with all vertices involved in it. These last vertices are now sorted in increasing way with respect their weight-degree ratio; thus, the vertex v with smaller ratio is selected to break off the cycle. Of course, may also happens that v break off also more cycles. Anyway, if from the removal of v the subgraph becomes acyclic, and this removal improves the fitness function then v is considered for the solution and it replaces u; otherwise the process is repeated again taking into account a new cycle. At the end of the iterations, if the sum of the new removed vertices (i.e. the fitness) is less to the previous one, then such vertices are inserted in the new solution in place of vertex u.

4   Results

In this section all analysis, experiments, and comparisons of HYBRID-IA are presented in order to measure the goodness of the proposed approach. The main goal of these experiments is obviously to prove the reliability and competitiveness of the HYBRID-IA with respect the state-of-the-art, but also to test the efficiency and the computational impact provided by the designed local search. Thus, for properly evaluating the performances of the proposed algorithm, a set of benchmark instances proposed in [3] have been considered, whose set includes grid, toroidal, hypercube, and random graphs. Each instance considered, besides to the topology, differs in the number of vertices; number of edges; and range of values for the vertices weights. In this way, as suggested in [2], is possible to inspect the computational performances of HYBRID-IA with respect to the density of the graph and weight ranges. Further, HYBRID-IA has been also compared with three different algorithms, which represent nowadays the state-of-the-art for the WFVS problem: *Iterated Tabu Search* (ITS) [3]; *eXploring Tabu Search* (XTS) [1,9]; and *Memetic Algorithm* (MA) [2]. All three algorithms, and their results, have been taken in [2].

Each experiment was performed by HYBRID-IA with population size $d = 100$, duplication parameter $dup = 2$, maximum age reachable $\tau_B = 20$, mutation rate parameter $\rho = 0.5$, and maximum number of generations $maxgen = 300$. Further, each experiment reported in the tables below is the average over five instances with the same characteristics but different assignment of the vertices weights. The experimental protocol used, and not included above, has been taken by [2]. It is important to highlight that in [2] a fixed maximum number of generators is not given, but the authors use a stop criterion based on a formula ($MaxIt$) that depends on the density of the graph: the algorithm will end its evolution process when it will reach $MaxIt$ consecutive iterations without improvements. Note that this threshold is reset every time an improvement occurs. From a simple calculus, it is possible to check how 300 generations used in this work are almost always lowest or equal to the minimum ones performed by the algorithms presented in [2], considering primarily the simplicity of fitness improvements in the first steps of the generations.

4.1   Dynamic Behaviour

Before to presents the experiments, and comparisons performed, the dynamic behavior and the learning ability of HYBRID-IA are presented. An analysis on the computational impact, and convergence advantages provided by the use of the local search developed has been conducted as well.

In Fig. 1, left plot, is shown the dynamic behavior of HYBRID-IA, where are displayed the curves of the (i) best fitness, (ii) average fitness of the population, and (iii) average fitness of the cloned hypermutated population over the generations. For this analysis the squared grid instance S_SG9 has been considered, whose features, and weights range are shown in Table 1. From this plot is possible to see how HYBRID-IA go down quickly in a very short generations to acceptable solutions for then oscillating between values close to each other. This oscillatory behavior, with main reference to the cloned hypermutated population curve, proves how the algorithm HYBRID-IA has good solutions diversity, which is surely helpful for the search space exploration. These fluctuations are instead less pronounced in the curve of the average fitness of the population, and this is due to the use of the local search that provides greater convergence stability. Last curve, the one of the best fitness, indicates how the algorithm takes the best advantage of the generations, managing to still improve in the last generations. Since this is one of few instances where HYBRID-IA didn't reach the optimal solution, inspecting this curve we think that likely increasing the number of generations (even just a little), HYBRID-IA will be able to find the global best. Right plot of Fig. 1 shows the comparison between the average fitness values of the survivors' population ($P^{(select)}$), and the one produced by the local search. Although they show a very similar parallel behavior, this plot proves the usefulness and benefits produced by the local search, which is always able to improve the solutions generated by the stochastic immune operators. Indeed the curve of the average fitness produced by the local search is always below to the survivors one.

Besides the convergence analysis, it becomes important also to understand the *learning ability* of the algorithm, i.e. how many information it is able to gain during all evolutionary process, which affects the performances of any evolutionary algorithm in general. For analyzing the learning process we have then used the well-known entropy function, *Information Gain*, which measures the quantity of information the system discovers during the learning phase [6,7,18]. Let B_m^t be the number of the B cells that at the timestep t have the fitness function value m; we define the candidate solutions distribution function $f_m^{(t)}$ as the ratio between the number B_m^t and the total number of candidate solutions:

$$f_m^{(t)} = \frac{B_m^t}{\sum_{m=0}^{h} B_m^t} = \frac{B_m^t}{d}. \qquad (3)$$

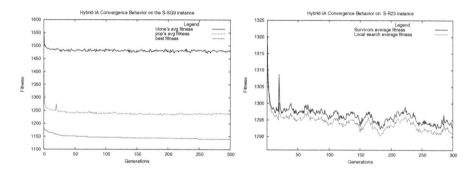

Fig. 1. Convergence behavior of the average fitness function values of $P^{(t)}$, $P^{(hyp)}$, and the best B cell versus generations on the grid S_SG9 instance (left plot). Average fitness function of $P^{(select)}$ vs. average fitness function $P^{(t)}$ on the random S_R23 instance (right plot).

It follows that the information gain $K(t, t_0)$ and entropy $E(t)$ can be defined as:

$$K(t, t_0) = \sum_m f_m^{(t)} \log(f_m^{(t)} / f_m^{(t_0)}), \qquad (4)$$

$$E(t) = \sum_m f_m^{(t)} \log f_m^{(t)}. \qquad (5)$$

The gain is the amount of information the system learned compared to the randomly generated initial population $P^{(t=0)}$. Once the learning process begins, the information gain increases until to reach a peak point.

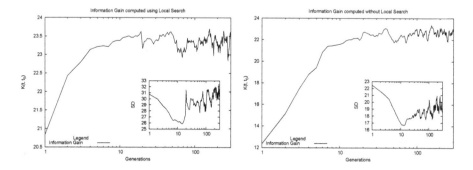

Fig. 2. Learning during the evolutionary process. Information gain curves of Hybrid-IA with local search strategy (left plot) and without it (right plot). The inset plots, in both figures, display the relative standard deviations.

Figure 2 shows the information gain obtained with the use of the local search mechanism (left plot), and without it (right plot) when HYBRID-IA is applied to the random S_R23 instance. In both plots we report also the standard deviation, showed as inset plot. Inspecting both plots is possible to note that with the using of the local search, the algorithm quickly gains high information until to reach the higher peak within the first generations, which exactly corresponds to the achievement of the optimal solution. It is important to highlight that the higher peak of the information gain corresponds to the lower point of the standard deviation, and then this confirm that HYBRID-IA reaches its maximum learning at the same time as it show less uncertainty. Due to the deterministic approach of the local search, once reached the highest peak, and found the global optimal solution, the algorithm begins to lost information and starts to show an oscillatory behavior. The same happens for the standard deviation. From the right plot, instead, without the use of the local search, HYBRID-IA gains information more slowly. Also for this version, the higher peak of information learned corresponds to the lowest uncertainty degree, and in this temporal window HYBRID-IA reaches the best solution. Unlike the curve produced with the use of the local search, once found the optimal solution and reached the highest information peak, the algorithm starts to lost information as well but its behavior seems to be more steady-state.

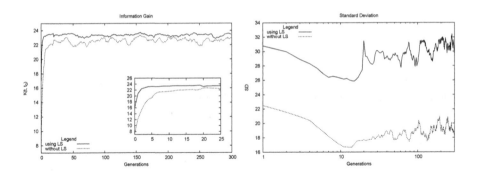

Fig. 3. Information Gain and standard deviation.

The two curves of the information gain have been compared and showed in Fig. 3 (left plot) over all generations. The relative inset plot is a zoom of the information gain behavior over the first 25 generations. From this plot is clear how the local search approach developed helps the algorithm to learn more information already from the first generations, and in a quickly way; the inset plot, in particular, highlight the important existing distance between the two curves. From an overall view over all 300 generations, it is possible to see how the local search gives more steady-state even after reaching the global optimal. In the right plot of the Fig. 3 is instead shown the comparison between the two standard deviations produced by HYBRID-IA with and without the use of the local search. Of course, as we expected, the curve of the standard deviation using the local search is taller as the algorithm gains higher amount of information than the version without local search.

Finally, from the analysis of all these figures – from convergence speed to learning ability – clearly emerges how the local search designed and developed helps HYBRID-IA in having an appropriate and correct convergence towards the global optimum, and a greater amount of information gained.

4.2 Experiments and Comparisons

In this subsection all experimental results and comparisons done with the state-of-the-art are presented in order to evaluate the efficiency, reliability and robustness of the HYBRID-IA performances. Many experiments have been performed, and the benchmark instances proposed in [2] have been considered for our tests and comparisons. In particular, HYBRID-IA has been compared with three different metaheuristics: ITS [3], XTS [1,9], and MA [2]. As written in Sect. 4, the same experimental protocol proposed in [2] has been used. Tables 1 and 2 show the results obtained by HYBRID-IA on different sets of instance: *squared grid graph*, *no squared grid graph*, and *toroidal graph* in Table 1; *hypercube graph* and *random graphs* in Table 2. In both tables are reported: the name of the instance in *1st* column; number of vertices in *2nd*; number of edges in *3rd*; lower and upper bounds of the vertex weights in *4th* and *5th*; and optimal solution K^* in *6th*. In the next columns are reported the results of HYBRID-IA (*7th*) and of the other three algorithms compared. It is important to clarify that for the grid graphs (squared and not squared) n and m indicate the number of the rows and columns of the grid. The last column of both tables shows the difference between the result obtained by HYBRID-IA and the best result among the three compared algorithms. Further, in each line of the tables the best results among all are reported in boldface.

On the *squared grid graph* instances (top of Table 1) is possible to see how HYBRID-IA is able to reach the optimal solution in 8 instances over 9, unlike of MA that instead reaches it in all instances. However, in this instance (S_SG7) the performances showed by HYBRID-IA are very close to the optimal solution (+0.2), and anyway better than the other two compared algorithms. On the *not square grid graphs* (middle of Table 1) instead HYBRID-IA reaches the optimal solution on all instances (9 over 9), outperforming all three algorithms on the instance S_NG7, where none of the three compared algorithms is able to find the optimal solution. Also on the *toroidal graphs* (bottom of Table 1) HYBRID-IA is able to reaches the global optimal solution in 9 over 9 instances. In the overall, inspecting all results in Table 1, is possible to see how HYBRID-IA shows competitive and comparable performances to MA algorithm, except in the instance S_SG7 where it shows slight worst results, whilst instead it is able to outperform MA in the instance S_NG7 where it reaches the optimal solution unlike of MA. Analysing the results with respect the other two algorithms, is clear how HYBRID-IA outperform ITS and XTS on all instances.

In Table 2 are presented the comparisons on the hypercube, and random graphs, which present larger problem dimensions with respect the previous ones. Analysing the results obtained on the *hypercube graph* instances, it is very clear how HYBRID-IA outperform all three algorithms in all instances (9 over 9),

Table 1. HYBRID-IA versus MA, ITS and XTS on the set of the instances: *squared grid; not squared grid* and *toroidal* graphs.

INSTANCE									
n	m	Low	Up	K^*	HYBRID-IA	ITS	XTS	MA	$\pm$
SQUARED GRID GRAPHS									
S_SG1	5	5	10	25	114.0	**114.0**	**114.0**	**114.0**	**114.0**	=
S_SG2	5	5	10	50	199.8	**199.8**	**199.8**	**199.8**	**199.8**	=
S_SG3	5	5	10	75	312.4	**312.4**	312.6	**312.4**	**312.4**	=
S_SG4	7	7	10	25	252.0	**252.0**	252.4	**252.0**	**252.0**	=
S_SG5	7	7	10	50	437.6	**437.6**	439.8	**437.6**	**437.6**	=
S_SG6	7	7	10	75	713.6	**713.6**	718.4	717.4	**713.6**	=
S_SG7	9	9	10	25	442.2	442.4	444.2	442.8	**442.2**	+0.2
S_SG8	9	9	10	50	752.2	**752.2**	754.6	753.0	**752.2**	=
S_SG9	9	9	10	75	1134.4	**1134.4**	1138.0	**1134.4**	**1134.4**	=
NOT SQUARED GRID GRAPHS									
S_NG1	8	3	10	25	96.8	**96.8**	**96.8**	**96.8**	**96.8**	=
S_NG2	8	3	10	50	157.4	**157.4**	**157.4**	**157.4**	**157.4**	=
S_NG3	8	3	10	75	220.0	**220.0**	**220.0**	**220.0**	**220.0**	=
S_NG4	9	6	10	25	295.6	**295.6**	295.8	295.8	**295.6**	=
S_NG5	9	6	10	50	488.6	**488.6**	489.4	**488.6**	**488.6**	=
S_NG6	9	6	10	75	755.0	**755.0**	**755.0**	755.2	**755.0**	=
S_NG7	12	6	10	25	398.2	**398.2**	399.8	398.8	398.4	−0.2
S_NG8	12	6	10	50	671.8	**671.8**	673.4	**671.8**	**671.8**	=
S_NG9	12	6	10	75	1015.2	**1015.2**	1017.4	1015.4	**1015.2**	=
TOROIDAL GRAPHS									
S_T1	5	5	10	25	101.4	**101.4**	**101.4**	**101.4**	**101.4**	=
S_T2	5	5	10	50	124.4	**124.4**	**124.4**	**124.4**	**124.4**	=
S_T3	5	5	10	75	157.8	**157.8**	**157.8**	158.8	**157.8**	=
S_T4	7	7	10	25	195.4	**195.4**	197.4	**195.4**	**195.4**	=
S_T5	7	7	10	50	234.2	**234.2**	**234.2**	**234.2**	**234.2**	=
S_T6	7	7	10	75	269.6	**269.6**	**269.6**	**269.6**	**269.6**	=
S_T7	9	9	10	25	309.6	**309.8**	310.4	**309.8**	**309.8**	=
S_T8	9	9	10	50	369.6	**369.6**	370.0	**369.6**	**369.6**	=
S_T9	9	9	10	75	431.8	**431.8**	432.2	432.2	**431.8**	=

reaching even the optimal solution on the S_H7 instance where instead the three algorithms fail. On the *random graphs* HYBRID-IA still shows comparable results to MA on all instances, even reaching the optimum on the instances S_R20 and S_R23 where instead MA, and the other two algorithms fail. In the overall, analyzing all results of this table is easy to assert that HYBRID-IA is comparable,

Table 2. HYBRID-IA versus MA, ITS and XTS on the set of the instances: *hypercube* and *random* graphs.

INSTANCE									
	n	m	Low	Up	K^*	HYBRID-IA	ITS	XTS	MA	$\pm$
HYPERCUBE GRAPHS										
S_H1	16	32	10	25	72.2	**72.2**	72.2	72.2	72.2	=
S_H2	16	32	10	50	93.8	**93.8**	93.8	93.8	93.8	=
S_H3	16	32	10	75	97.4	**97.4**	97.4	97.4	97.4	=
S_H4	32	80	10	25	170.0	**170.0**	170.0	170.0	170.0	=
S_H5	32	80	10	50	240.6	**240.6**	241.0	240.6	240.6	=
S_H6	32	80	10	75	277.6	**277.6**	277.6	277.6	277.6	=
S_H7	64	192	10	25	353.4	**353.4**	354.6	353.8	353.8	−0.4
S_H8	64	192	10	50	475.6	**475.6**	476.0	475.6	475.6	=
S_H9	64	192	10	75	503.8	**503.8**	503.8	504.8	503.8	=
RANDOM GRAPHS										
S_R1	25	33	10	25	63.8	**63.8**	63.8	63.8	63.8	=
S_R2	25	33	10	50	99.8	**99.8**	99.8	99.8	99.8	=
S_R3	25	33	10	75	125.2	**125.2**	125.2	125.2	125.2	=
S_R4	25	69	10	25	157.6	**157.6**	157.6	157.6	157.6	=
S_R5	25	69	10	50	272.2	**272.2**	272.2	272.2	272.2	=
S_R6	25	69	10	75	409.4	**409.4**	409.4	409.4	409.4	=
S_R7	25	204	10	25	273.4	**273.4**	273.4	273.4	273.4	=
S_R8	25	204	10	50	507.0	**507.0**	507.0	507.0	507.0	=
S_R9	25	204	10	75	785.8	**785.8**	785.8	785.8	785.8	=
S_R10	50	85	10	25	174.6	**174.6**	175.4	176.0	**174.6**	=
S_R11	50	85	10	50	280.8	**280.8**	280.8	281.6	**280.8**	=
S_R12	50	85	10	75	348.0	**348.0**	348.0	349.2	**348.0**	=
S_R13	50	232	10	25	386.2	**386.2**	389.4	386.8	**386.2**	=
S_R14	50	232	10	50	708.6	**708.6**	708.6	708.6	708.6	=
S_R15	50	232	10	75	951.6	**951.6**	951.6	951.6	951.6	=
S_R16	50	784	10	25	602.0	**602.0**	602.2	602.0	602.0	=
S_R17	50	784	10	50	1171.8	**1171.8**	1172.2	1172.0	**1171.8**	=
S_R18	50	784	10	75	1648.8	**1648.8**	1649.4	1648.8	1648.8	=
S_R19	75	157	10	25	318.2	**318.2**	321.0	320.0	**318.2**	=
S_R20	75	157	10	50	521.6	**522.2**	526.2	525.0	522.6	−0.4
S_R21	75	157	10	75	751.0	**751.0**	757.2	754.2	**751.0**	=
S_R22	75	490	10	25	635.8	**635.8**	638.6	**635.8**	635.8	=
S_R23	75	490	10	50	1226.6	**1226.6**	1230.6	1228.6	1227.6	−1.0
S_R24	75	490	10	75	1789.4	**1789.4**	1793.6	**1789.4**	1789.4	=
S_R25	75	1739	10	25	889.8	**889.8**	891.0	**889.8**	889.8	=
S_R26	75	1739	10	50	1664.2	**1664.2**	1664.8	1664.2	1664.2	=
S_R27	75	1739	10	75	2452.2	**2452.2**	2452.8	**2452.2**	2452.2	=

and sometime the best, with respect to MA also on this set of instances, winning even on three instances the comparison with it. Extending the analysis to the comparison with ITS and XTS algorithms, it is quite clear that HYBRID-IA outperforms them in all instances, finding always better solutions than these two compared algorithms.

Finally, from the analysis of the convergence behavior, learning ability, and comparisons performed it is possible to clearly assert how HYBRID-IA is comparable with the state-of-the-art for the $MWFVS$ problem, showing reliable and robustness performances, and good ability in information learning. These efficient performances are due to the combination of the immunological operators, which introduce enough diversity in the search phase, and the local search designed, which instead refine the solution via more appropriate and deterministic choices.

5   Conclusion

In this paper we introduce a hybrid immunological algorithm, simply called HYBRID-IA, which takes advantage by the immunological operators (cloning, hypermutation and aging) for carefully exploring the search space, introducing diversity, and avoiding to get trapped into a local optima, and by a Local Search whose aim is to refine the solutions found through appropriate choices and based on a deterministic approach.

The algorithm was designed and developed for solving one of the most challenging combinatorial optimization problems, such as the *Weighted Feedback Vertex Set*, which simply consists, given an undirected graph, in finding the subset of vertices of minimum weight such that their removal produce an acyclic graph. In order to evaluate the goodness and reliability of HYBRID-IA, many experiments have been performed and the results obtained have been compared with three different metaheuristics (ITS, XTS, and MA), which represent nowadays the state-of-the-art. For these experiments a set of graph benchmark instances has been considered, based on different topologies: *grid* (squared and no squared), *toroidal*, *hypercube*, and *random* graphs.

An analysis on the convergence speed and learning ability of HYBRID-IA has been performed in order to evaluate, of course, its efficiency but also the computational impact and reliability provided by the developed local search. From this analysis, clearly emerges how the developed local search helps the algorithm in having a correct convergence towards the global optima, as well as a high amount of information gained during the learning process.

Finally, from the results obtained, and the comparisons done, it is possible to assert how HYBRID-IA is competitive and comparable with the $WFVS$ state-of-the-art, showing efficient and robust performances. In all instances tested it was able to reach the optimal solutions, except in only one (S_SG7). It is important to point out how HYBRID-IA has been instead the only one to reach the global optimal solutions on 4 instances, unlike the three compared algorithms.

References

1. Brunetta, L., Maffioli, F., Trubian, M.: Solving the feedback vertex set problem on undirected graphs. Discret. Appl. Math. **101**, 37–51 (2000)
2. Carrabs, F., Cerrone, C., Cerulli, R.: A memetic algorithm for the weighted feedback vertex set problem. Networks **64**(4), 339–356 (2014)
3. Carrabs, F., Cerulli, R., Gentili, M., Parlato, G.: A tabu search heuristic based on k-diamonds for the weighted feedback vertex set problem. In: Pahl, J., Reiners, T., Voß, S. (eds.) INOC 2011. LNCS, vol. 6701, pp. 589–602. Springer, Heidelberg (2011). https://doi.org/10.1007/978-3-642-21527-8_66
4. Conca, P., Stracquadanio, G., Greco, O., Cutello, V., Pavone, M., Nicosia, G.: Packing equal disks in a unit square: an immunological optimization approach. In: Proceedings of the International Workshop on Artificial Immune Systems (AIS), pp. 1–5. IEEE Press (2015)
5. Cormen, T.H., Leiserson, C.E., Rivest, R.L., Stein, C.: Introduction to Algorithms, 3rd edn. MIT Press, Cambridge (2009)
6. Cutello, V., Nicosia, G., Pavone, M.: An immune algorithm with stochastic aging and kullback entropy for the chromatic number problem. J. Comb. Optim. **14**(1), 9–33 (2007)
7. Cutello, V., Nicosia, G., Pavone, M., Timmis, J.: An immune algorithm for protein structure prediction on lattice models. IEEE Trans. Evol. Comput. **11**(1), 101–117 (2007)
8. Cutello, V., Nicosia, G., Pavone, M., Prizzi, I.: Protein multiple sequence alignment by hybrid bio-inspired algorithms. Nucl. Acids Res., Oxf. J. **39**(6), 1980–1992 (2011)
9. Dell'amico, M., Lodi, A., Maffioli, F.: Solution of the cumulative assignment problem with a well-structured tabu search method. J. Heuristics **5**(2), 123–143 (1999)
10. Di Stefano, A., Vitale, A., Cutello, V., Pavone, M.: How long should offspring lifespan be in order to obtain a proper exploration? In: Proceedings of the IEEE Symposium Series on Computational Intelligence (IEEE SSCI), pp. 1–8. IEEE Press (2016)
11. Fouladvand, S., Osareh, A., Shadgar, B., Pavone, M., Sharafia, S.: DENSA: an effective negative selection algorithm with flexible boundaries for selfspace and dynamic number of detectors. Eng. Appl. Artif. Intell. **62**, 359–372 (2016)
12. Garey, M.R., Johnson, D.S.: Computers and Intractability: A Guide to Theory of NP-Completeness. Freeman, New York (1979)
13. Gusfield, D.: A graph theoretic approach to statistical data security. SIAM J. Comput. **17**(3), 552–571 (1988)
14. Jansen, T., Zarges, C.: Computing longest common subsequences with the B-cell algorithm. In: Coello Coello, C.A., Greensmith, J., Krasnogor, N., Liò, P., Nicosia, G., Pavone, M. (eds.) ICARIS 2012. LNCS, vol. 7597, pp. 111–124. Springer, Heidelberg (2012). https://doi.org/10.1007/978-3-642-33757-4_9
15. Jansen, T., Oliveto, P.S., Zarges, C.: On the analysis of the immune-inspired B-cell algorithm for the vertex cover problem. In: Liò, P., Nicosia, G., Stibor, T. (eds.) ICARIS 2011. LNCS, vol. 6825, pp. 117–131. Springer, Heidelberg (2011). https://doi.org/10.1007/978-3-642-22371-6_13
16. Johnson, D.B.: Finding all the elementary circuits of a directed graph. SIAM J. Comput. **4**, 77–84 (1975)

17. Karp, R.M.: Reducibility among combinatorial problems. In: Miller, R.E., Thatcher, J.W., Bohlinger, J.D. (eds.) Complexity of Computer Computations. The IBM Research Symposia Series, pp. 85–103. Springer, Heidelberg (1972). https://doi.org/10.1007/978-1-4684-2001-2_9

18. Pavone, M., Narzisi, G., Nicosia, G.: Clonal selection - an immunological algorithm for global optimization over continuous spaces. J. Glob. Optim. **53**(4), 769–808 (2012)

19. Peleg, D.: Size bounds for dynamic monopolies. Discret. Appl. Math. **86**, 263–273 (1998)

20. Tian, Y., Zhang, H.: Research on B cell algorithm for learning to rank method based on parallel strategy. PLoS ONE **11**(8), e0157994 (2016)

21. Vitale, A., Di Stefano, A., Cutello, V., Pavone, M.: The influence of age assignments on the performance of immune algorithms. In: Proceedings of the 18th Annual UK Workshop on Computational Intelligence (UKCI), Advances in Computational Intelligence Systems. Advances in Intelligent Systems and Computing series, vol. 840, pp. 16–28 (2018)

22. Wang, C.C., Lloyd, E.L., Soffa, M.L.: Feedback vertex set and cyclically reducible graphs. J. Assoc. Comput. Mach. (ACM) **32**(2), 296–313 (1985)

23. Xia, X., Yuren, Z.: On the effectiveness of immune inspired mutation operators in some discrete optimization problems. Inf. Sci. **426**, 87–100 (2018)

24. Yannakakis, M.: Node-deletion problem on bipartite graphs. SIAM J. Comput. **10**(2), 310–327 (1981)

25. Zarges, C.: On the utility of the population size for inversely fitness proportional mutation rates. In: Proceedings of the 10*th* ACM SIGEVO Workshop on Foundations of Genetic Algorithms (FOGA), pp. 39–46 (2009)

A Statistical Test of Heterogeneous Subgraph Densities to Assess Clusterability

Pierre Miasnikof[1]([⊠]), Liudmila Prokhorenkova[2,3], Alexander Y. Shestopaloff[4], and Andrei Raigorodskii[2,3]

[1] Department of Chemical Engineering and Applied Chemistry,
University of Toronto, Toronto, Canada
`p.miasnikof@mail.utoronto.ca`
[2] Moscow Institute of Physics and Technology, Dolgoprudny, Russia
[3] Yandex, Moscow, Russia
[4] The Alan Turing Institute, London, UK

Abstract. Determining if a graph displays a clustered structure prior to subjecting it to any cluster detection technique has recently gained attention in the literature. Attempts to group graph vertices into clusters when a graph does not have a clustered structure is not only a waste of time; it will also lead to misleading conclusions. To address this problem, we introduce a novel statistical test, the δ-test, which is based on comparisons of local and global densities. Our goal is to assess whether a given graph meets the necessary conditions to be meaningfully summarized by clusters of vertices. We empirically explore our test's behavior under a number of graph structures. We also compare it to other recently published tests. From a theoretical standpoint, our test is more general, versatile and transparent than recently published competing techniques. It is based on the examination of intuitive quantities, applies equally to weighted and unweighted graphs and allows comparisons across graphs. More importantly, it does not rely on any distributional assumptions, other than the universally accepted definition of a clustered graph. Empirically, our test is shown to be more responsive to graph structure than other competing tests.

1 Introduction

Graph clustering has been very well and very broadly covered in the literature (e.g., [2,8,14,15,29,30]). Nonetheless, objective measures of a graph's suitability for clustering are lacking. There are multiple graph clustering and community detection techniques in the literature and they all rely on the unmentioned assumption that the graph has a clustered (community) structure, to begin with. However, not all graphs have this type of structure. It is important to determine if a graph is a good candidate for clustering, before any vertex-grouping effort is

© Springer Nature Switzerland AG 2020
N. F. Matsatsinis et al. (Eds.): LION 13 2019, LNCS 11968, pp. 17–29, 2020.
https://doi.org/10.1007/978-3-030-38629-0_2

undertaken. Applying clustering techniques to a graph that does not have a clustered structure is not only a waste of time; it will inevitably lead to misleading conclusions.

In this article, our objective is to provide a test to determine if a graph meets the prerequisite conditions for it to have a meaningful cluster structure or if it displays no significant clustered structure. Here, our goal is not to identify graph clusters and their component vertices or assess the quality of the clusters identified by an algorithm. Our goal is to assess whether a given graph meets the necessary conditions to be meaningfully summarized by grouping its vertices into clusters; whether its vertices are concentrated within densely interconnected subgraphs. In other words, our aim is to provide a technique for accurately answering the question recently posed by Chiplunkar et al. [9], "(...) given access to a graph G = (V, E), can we quickly determine whether the graph can be partitioned into a few clusters with good inner conductance (...)?", using statistical sampling and hypothesis testing.

We want to assess whether a graph is more likely to be partition-able, like the graph shown in Fig. 1a, or if it has a cluster-less structure as in Fig. 1b.[1]

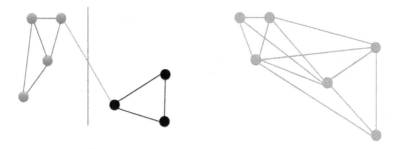

(a) Graph with Clustered Structure   (b) Graph without Clustered Structure

Fig. 1. Graphs displaying clustered and unclustered structure

It is important to note that we are not looking to determine if all vertices are grouped within clusters, but if a significant amount of them may belong to separate clusters. We want to determine if clusters might be a good model for the graph, if clusters might help us properly partition the graph in a meaningful way. For example, a graph may have a small number of clusters that contain a small fraction of vertices while the rest of the graph is randomly connected; or a graph may have most of its vertices contained in densely connected clusters with a very small number of vertices that do not belong to clusters. In the first instance, we would not want to conclude that the graph displays a clustered structure, since clusters are uninformative features of the graph and are not useful indicators for

---

[1] Also, note that in this article we assume undirected graphs with no self-loops.

its partitioning. Meanwhile, in the second instance, we would want to conclude that clusters are likely to offer a meaningful picture of a graph's structure.

With a view on large data sets, we develop a test that only relies on small samples from the graph and does not examine the graph in its entirety. Here, we follow the examples of many authors who have inferred various graph properties through sampling (e.g., [3,11,12,17–19]).

The rest of this article is organized as follows. We begin with a review of previous work in the field, describe and justify our methods, compare them to recent techniques from the literature and share results from our empirical tests. As a point of comparison, we also implement the techniques proposed by Gao and Lafferty [17,18] and empirically compare their performance to our method's performance.

2   Related Work

Our work is primarily inspired by the recent work of Gao and Lafferty [17,18], who use sampling and statistical testing to determine if a network has community structure. We are also guided by the question, "(...) given access to a graph $G = (V, E)$, can we quickly determine whether the graph can be partitioned into a few clusters with good inner conductance (...)?", which was recently posed by Chiplunkar et al. [9].

While a formal definition of "community structure" remains a topic of debate (e.g., [15,30]), virtually all authors agree a cluster (or community) is a subset of vertices that exhibit a high-level of interconnection between themselves and a low-level of connection to vertices in the rest of the graph [14,27,28,33] (we quote these authors, but their definition is very common across the literature).

The graph testing (through sampling) literature is very rich. Many authors have proposed tests for various graph properties other than the existence of clusters (e.g., [3,11,19,20,25]). Tests for clustering have also been studied in the past. For example, Arias-Castro and Verzelen [4,32] set up clustering as a hypothesis test. They use the Erdős-Rényi random (ER) graph [1,13] as a null model, but their alternative hypothesis is the existence of only one dense subgraph. Czumaj et al. [10] describe a test of clustering, but they impose a restrictive $\kappa - \phi$ model as their benchmark. Bikel and Sarkar develop tests for graph spectra, in order to determine if a graph follows the ER model [7].

As mentioned earlier, Gao and Lafferty [17,18] propose clustering tests, but their tests focus on vertex triplets and asymptotic properties. More recently, Jin et al. [23] raise the point that testing for clustering remains a non-trivial problem and propose potentially (computationally) costly tests based on the number of paths and cycles of fixed length. Also, very recently Chiplunkar et al. [9] propose a test of clustering based on a restrictive $\kappa - \phi$ model, while He et al. [22] highlight the problem as still being unresolved.

Unfortunately, most of these authors' approaches are restrictive and rely on rigid models for their hypotheses tests. Indeed, Ugander et al. [31] claim that incomplete non-empty $k-$node subgraphs are infrequent not just in social networks, but there exist mathematical reasons why such subgraphs are infrequent in general settings. They proved a theoretical upper bound on their probability of occurrence. Later, Yin and Leskovec discussed the need to look at more complex structures, in order to gain an understanding of complex network structures [34]. Finally, it is important to remember that closed triplet frequencies and clustering coefficients may be higher than under the ER hypothesis, while the network does not exhibit clustering [16].

The choice of null hypothesis is also a topic of debate in the literature. In addition to most of the authors above who focus on clustering, Elenberg et al. [12] approximate 3-profile counts by sampling and use the ER graph as a null model. Unfortunately, the ER random graph model has been described as overly simplistic to model real-world networks (e.g., [1,5,6,16]).

Gao and Lafferty [18] use the ER model to prove a central limit theorem for their estimators of wedges and triangles. However, as pointed out by those same authors [17], the ER model may be an insufficient null hypothesis. Meanwhile, other authors (e.g., [23]) set up a more general null hypothesis of only one single cluster and an alternative hypothesis of more than one.

3 Methods

We develop a test whose successful fulfillment forms a necessary condition for a graph (network) to exhibit a clustered (community) structure. It is founded on statistical hypothesis testing; the reliability of the conclusions it leads to is transparent and rooted in statistical theory.

Unlike many authors' recent work on the topic, our technique does not rely on any assumptions on the structure of the null graph, beyond what is implied by the definition of an unclustered graph structure. In addition, our technique detects anomalies in density, not just specific sub-structures like triangles or other fixed-length structures. It also makes no assumptions on the number of clusters, as in the case of $\kappa - \phi$ models.

3.1 Underlying Assumptions and Densities

Our test only relies on the assumption that graphs composed of more than one cluster are characterized by the presence of subsets of densely connected vertices with sparse connections to the rest of the graph. This assumption is a direct consequence of the fact that a cluster can be described as a subset of vertices that are densely connected to each other but only sparsely connected to vertices outside their cluster. Therefore, a clustered connection pattern translates into heterogeneous local densities across the graph. A clustered graph is characterized by pockets of highly connected nodes leading to high local densities that are significantly higher than the graph's overall density.

The links between density and clustered patterns of connectivity were shown in Miasnikof et al. [26]. Under such a pattern of connectivity, it is expected that the densities of induced subgraphs obtained by sampling vertices within a neighborhood will exhibit, on average, higher densities than the graph's global density. An example of this heterogeneous density is shown in Fig. 2, where the graph's global density is $K = 0.43$, while each of its constituent cluster (subgraph) densities are $k_1 = 0.83, k_2 = 1$.

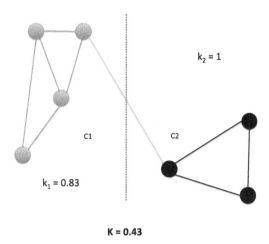

Fig. 2. Heterogeneous densities

On the other hand, if a graph has no meaningful cluster structure, we should not observe a significant number of vertices falling into densely inter-connected subgraphs which are sparsely connected to other vertices on the graph, as seen in Fig. 1b. A non-clustered structure implies that any two vertices have a roughly uniform probability of being connected. This homogeneity can be statistically inferred by the absence of a statistically significant difference between the mean density of a sample of randomly chosen induced subgraphs and the graph's global density.

Unlike much of the current work on this topic (e.g., [4,7,12,32]), our null model does not rely on the assumption of an ER graph and does not make any assumptions on the number of clusters (e.g., [9]). Our proposed technique does not impose any model on the null other than its lack of clustering reflected in a homogeneous density across the graph. Our null graph is simply a general graph with a near constant density, on average. This null hypothesis includes all graphs without clustered structure, not only ER graphs, but any other type of non-clustered graph. The configuration model graph and a graph which contains a small number of clusters while most of the edges are randomly connected are both examples of such general unstructured graphs. Our simulated data sets described in Sect. 3.6 and our empirical tests in Sect. 3.7 include both these examples.

3.2 (Sub)Graph Densities and the δ Statistic

Consider a graph $G = (V, E)$, with $|V| = N$. First, we randomly sample a subset of $S = sN$ vertices, where $s \in (0, 1]$ is a parameter which determines the proportion of total vertices we wish to sample. We call these vertices the root nodes.

Then, for each root node $i \in \{1, \ldots, S\}$, we take all its neighboring vertices and consider the local subgraph induced by them. Let E_i denote the set of edges in the local induced subgraph formed by the neighbors of node i and n_i denote the number of these neighbors. In the case of unweighted graphs, $|E_i|$ (or $|E|$ when dealing with the full graph) represents the cardinality of the set (number of edges). In the case of weighted graphs, $|E_i|$ (or $|E|$) represents the sum of edge weights.

We compute the local induced subgraph density as

$$k_i = \frac{|E_i|}{0.5 \times n_i(n_i - 1)}.$$

Then, the mean local density for S induced local subgraphs is computed as

$$\bar{K}_\ell = \frac{1}{S} \sum_{i=1}^{S} k_i.$$

At the graph level, the global density is computed as

$$K = \frac{|E|}{0.5 \times N(N - 1)}.$$

Finally, we introduce the δ statistic, the normalized mean of the k_i, as a measure of divergence between the global and mean local densities:

$$\delta = \frac{\bar{K}_\ell}{K} - 1.$$

With the help of Fig. 3, we illustrate this sampling procedure and the computation of the δ statistic. We begin by sampling two vertices labelled v_1 and v_2 (we have $S = 2$) and compute the densities of the two local induced subgraphs (k_1, k_2). Our δ statistic is then computed as follows:

$$k_1 = \frac{|E_1|}{0.5 \times n_1(n_1 - 1)} = \frac{2}{0.5 \times 3 \times 2} = 0.67,$$

$$k_2 = \frac{|E_2|}{0.5 \times n_2(n_2 - 1)} = \frac{3}{0.5 \times 3 \times 2} = 1,$$

$$\bar{K}_\ell = \frac{1}{2}(k_1 + k_2) = \frac{1}{2}(0.67 + 1) = 0.84,$$

$$K = \frac{|E|}{0.5 \times N(N - 1)} = \frac{12}{0.5 \times 8 \times 7} = 0.43,$$

$$\delta = \frac{\bar{K}_\ell}{K} - 1 = \frac{0.84}{0.43} - 1 = 0.95.$$

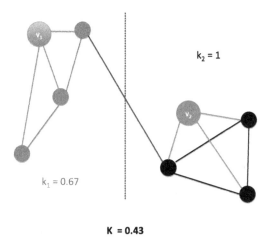

$k_2 = 1$

$k_1 = 0.67$

K = 0.43

Fig. 3. Heterogeneous densities

3.3 Statistical Hypotheses

On the basis of our assumptions on densities, we posit the following relationships between clustering and density:

1. If a graph does not display a clustered structure, then, on average, there should be no statistically significant difference between local densities and the graph's global density (homogeneity);
2. If a graph does display a clustered structure, then, on average, there should be a statistically significant difference between the densities of local induced subgraphs and the graph's global density (heterogeneity);
3. If a graph G_1 has a more strongly clustered structure than a graph G_2, then its mean local density should also be significantly greater, in a statistical sense (relative heterogeneity).

In summary, the existence of meaningful clusters implies that, on average, local connectivity should be significantly stronger than the graph's overall density. Here, we draw the reader's attention to the fact that our technique seeks to statistically determine the presence of locally strong atypical densities. Our approach is more general than the recent work which focuses on triangles or other fixed-length cycles or restrictive $\kappa - \phi$ models (e.g., [9,17,18,23]).

On the basis of the above-mentioned relationships, we formulate the following statistical hypotheses, which can be tested using the appropriate t-tests:

– Under the null hypothesis that a graph does not have a clustered structure, i.e., it has a uniform density, we test for $E(k_i) = K \Rightarrow E(\delta) = 0$ by using a one-tailed Student t-test;
– Alternatively, if the graph does have heterogeneous density, we expect $E(k_i) > K \Rightarrow E(\delta) > 0$, since it is greater than under the null;

- Under the null hypothesis that a graph G_1 has equally strong clustered structure as a graph G_2, we test for $\mathrm{E}(\delta_{G_1}) = \mathrm{E}(\delta_{G_2})$, again by using a one-tailed Student t-test;
- Alternatively, if a graph G_1 has a more heterogeneous density structure than a graph G_2, we expect $\mathrm{E}(\delta_{G_1}) > \mathrm{E}(\delta_{G_2})$.

3.4   Algorithm

As mentioned in the previous section, our δ test statistic can be used to answer two related but distinct questions:

1. Does a graph display heterogeneity in its density?
2. Does a graph G_1 have a more heterogeneous density than a graph G_2?

In the first question, we ask whether a graph meets the necessary condition to have a clustered structure. In the second, we ask if one graph meets this condition more strongly than another. These questions can be answered by following these steps:

- Sample S induced subgraphs, each containing n_i vertices ($i \in \{1, \ldots, S\}$);
- Compute the local densities $k_i = \frac{|E_i|}{0.5 \times n_i(n_i - 1)}$ for each of the S subgraphs;
- Compute graph density $K = \frac{|E|}{0.5 \times N(N-1)}$, for graph $G = (V, E)$ with $|V| = N$;
- Compute the mean of the local densities: $\bar{K}_\ell = \frac{1}{S} \sum_{i=1}^{S} k_i$;
- Normalize the mean and obtain our test statistic: $\delta = \frac{\bar{K}_\ell}{K} - 1$;
- In the case of a single-graph test, perform a one-tailed t-test; the null hypothesis is $\mathrm{E}(\delta) = 0$, the alternative is $\mathrm{E}(\delta) > 0$;
- In the case of a two-graph test, perform a two-sample (unpaired with unequal variance) one-tailed t-test on the δ statistics of graphs G_1, G_2; the null hypothesis is $\mathrm{E}(\delta_{G_1}) = \mathrm{E}(\delta_{G_2})$, the alternative is $\mathrm{E}(\delta_{G_1}) > \mathrm{E}(\delta_{G_2})$

3.5   Methodological Comparison to Other Recently Published Techniques

In Table 1, we compare key features of recently published test techniques that have inspired our own work. In the first column, we display the features associated with the methods proposed by Chiplunkar et al. [9]. In the second, we display the features of Gao and Lafferty's two recently published tests [17,18]. Finally, in the last column we display the features of our own δ test.

In Table 1, we see that our test is more general and versatile than recently published tests. We make no distributional assumptions and impose no generative model on the graph. Our test applies to weighted and unweighted graphs equally and is rooted in statistical testing methodology.

Table 1. A comparison of recently published clustering tests

	Chip et al.	G-L Tests	δ Test
Requires number of clusters as input	Y	N	N
Assumes generative model	Y	N	N
Provides statistical confidence intervals	N	Y	Y
Imposes distribution on null	NA	Y	N
Applies to weighted and unweighted graphs	Y	N	Y

3.6 Empirical Examples

In this section, we begin with the δ for each graph and single graph t-tests for the following ten graphs. We then perform two-sample t-tests (with unequal variance), to compare each pair of graphs (our weighted graph is assessed separately). With the exception of our Stanford University SNAP "real-world" graph [24,35], all graphs were generated using the Networkx library [21]. Details of each graph's characteristics are provided below, in Table 2.

- Scenario 1: generated using the Erdős-Rényi model (ER);
- Scenario 2: generated using the degree-corrected stochastic block model (SBM);
- Scenario 3: (connected) caveman (CC);
- Scenario 4: generated using the Barabási-Albert model with an out-degree of 3 (BA3);
- Scenario 5: generated using the Barabási-Albert model with an out-degree of 5 (BA5);
- Scenario 6: generated using the Watts-Strogatz model (WS);
- Scenario 7: generated using the configuration model with an underlying power law with exponent of 3 (CM);
- Scenario 8: a merger of two connected caveman graphs of 75 vertices each and an ER graph of 850 vertices (MD);
- Scenario 9: a "real-world" graph. Here, we converted the SNAP "Eu-core network" into an undirected graph with no self-loops (EUC);
- Scenario 10: a weighted (connected) caveman (CCw).

3.7 Empirical Results

The tables below contain the p-values for our one-sample and two-sample δ statistics. In Table 3, the diagonal entries are for the one-sample t-tests, while the off diagonal elements are for the two-sample graph vs graph t-tests. These tests were done on graph samples consisting of a set of randomly selected (without replacement) root nodes of size equal to 25% of all graph nodes.

As expected, our one-sample test does not reject the null hypothesis of homogeneous density for the ER and configuration model (CM) graphs, whose edges

Table 2. Test graph characteristics (N denotes total number of vertices, $|c_i|$ number of vertices in cluster i, p edge probability)

Graph	Graph characteristics
ER	$N = 1,000, p = 0.333$
SBM	$N = 1,000, |c_i| = 91, 21, 333, 555$
CC	Num cliques = 10, size of cliques = 100
BA3	$N = 1,000$, out-degree = 3
BA5	$N = 1,000$, out-degree = 5
WS	$N = 1,000, k = 14, p = 0.2$
CM	$N = 1,000$, exp = 3
MD	One ER and two CC, $N = 1,000 = 850 + 2 \times (3 \times 25)$
EUC	$N = 1,005$ in 42 known clusters
CCw	Num cliques = 12, size of cliques = 100, $w \in [0,1]$

Table 3. P-Values for δ (unweighted graphs)

	ER	WS	BA3	BA5	CC	SBM	EUC	CM	MD
ER	0.64	1.00	1.00	1.00	1.00	1.00	1.00	0.75	0.92
WS	0.00	0.00	1.00	1.00	1.00	0.00	1.00	0.25	0.00
BA3	0.00	0.00	0.00	0.15	1.00	0.00	1.00	0.03	0.00
BA5	0.00	0.00	0.85	0.00	1.00	0.00	1.00	0.06	0.00
CC	0.00	0.00	0.00	0.00	0.00	0.00	1.00	0.00	0.00
SBM	0.00	1.00	1.00	1.00	1.00	0.00	1.00	0.69	0.00
EUC	0.00	0.00	0.00	0.00	0.00	0.00	0.00	0.00	0.00
CM	0.25	0.75	0.97	0.94	1.00	0.31	1.00	0.25	0.25
MD	0.08	1.00	1.00	1.00	1.00	1.00	1.00	0.75	0.08

are randomly connected. The null is also not rejected in the case of the "merged graph" in which only a small number of vertices belong to densely connected subgraphs while most of the vertices are randomly connected. On the other hand, the null is correctly rejected in cases where the graph is known to have a strongly clustered structure (SBM, CC, EUC).

Our two-sample test accurately classifies the ER graph as not having a more heterogeneous density than any of the other graphs. It also confirms the EUC and CC graphs are more clustered than all other graphs.

We also perform a one-sample test on a weighted graph, a weighted connected caveman graph. The p-value for that test is $p = 0.00$, which confirms the graph is indeed clustered.

Other Tests. Because of their similar approach, we also perform the EZ and T^2 tests presented by Gao and Lafferty [17,18], as a point of comparison. However, because these tests do not lend themselves to pairwise comparison or weighted graphs, we are only able to compare one-sample tests on unweighted graphs.

Table 4. p-Value comparisons

Graph	EZ test	T^2 test	δ test
ER	1.00	0.82	0.64
WS	1.00	0.00	0.00
BA3	1.00	0.00	0.00
BA5	1.00	0.00	0.00
CC	0.99	0.00	0.00
SBM	0.97	0.00	0.00
EUC	1.00	0.00	0.00
CM	1.00	0.00	0.25
MD	1.00	0.00	0.08
CCw	N.A	N.A	0.00

In Table 4, we observe how our test is more responsive to graph structure than either of the Gao-Lafferty tests. The p-values of the EZ test are completely unaffected by the graph's structure and the null hypothesis of an ER graph is never rejected. The T^2 test performs slightly better, since it correctly rejects the null in all but the ER case. However, unlike our δ test, it is unable to detect the lack of clustering in the configuration model (CM) and "merged" (MD) graphs. Also, neither of these competing tests is designed for graph-to-graph comparisons or defined for weighted graphs.

4   Conclusion

In this article, we present a test for clusterability of a graph that is based on heterogeneity of local densities. We show how variations in densities can be used as indicators of a probable clustered structure. Our tests perform as expected and compare very favorably to recently published competing techniques.

Our future work will focus on sensitivity to sample sizes under various graph structures. We also intend to examine the cases of weighted graphs more closely. Currently, the topics of random and clustered weighted graphs remain subjects of debate.

Acknowledgments. Pierre Miasnikof was supported by a Mitacs-Accelerate PhD award IT05806. He also wishes to thank Lasse Leskelä of Aalto University, for the introduction to the work of Gao and Lafferty. Liudmila Prokhorenkova and Andrei Raigorodskii were supported by The Russian Science Foundation (grant number 16-11-10014).

References

1. Albert, R., Barabási, A.L.: Statistical mechanics of complex networks. Rev. Mod. Phys. **74**, 47–97 (2002)
2. Aleskerov, F., Goldengorin, B., Pardalos, P.: Clusters, Orders, and Trees: Methods and Applications. Springer, Heidelberg (2014). https://doi.org/10.1007/978-1-4939-0742-7. Incorporated
3. Alon, N., Shapira, A.: A characterization of the (natural) graph properties testable with one-sided error. SIAM J. Comput. **37**(6), 1703–1727 (2008)
4. Arias-Castro, E., Verzelen, N.: Community detection in dense random networks. Ann. Statist. **42**(3), 940–969 (2014). https://doi.org/10.1214/14-AOS1208
5. Barabási, A.L., Albert, R.: Emergence of scaling in random networks. Science **286**, 509–512 (1999)
6. Barrat, A., Barthélemy, M., Pastor-Satorras, R., Vespignani, A.: The architecture of complex weighted networks. Proc. Natl. Acad. Sci. **101**, 3747–3752 (2004)
7. Bickel, P.J., Sarkar, P.: Hypothesis testing for automated community detection in networks. J. R. Stat. Soc.: Ser. B (Stat. Methodol.) **78**(1), 253–273 (2016). https://doi.org/10.1111/rssb.12117
8. Butenko, S., Chaovalitwongse, W.A., Pardalos, P.M.: Clustering Challenges in Biological Networks. World Scientific, Singapore (2009). https://doi.org/10.1142/6602
9. Chiplunkar, A., Kapralov, M., Khanna, S., Mousavifar, A., Peres, Y.: Testing graph clusterability: algorithms and lower bounds. ArXiv e-prints, August 2018
10. Czumaj, A., Peng, P., Sohler, C.: Testing cluster structure of graphs. ArXiv e-prints, April 2015
11. Eden, T., Ron, D., Seshadhri, C.: On Approximating the number of k-cliques in sublinear time. ArXiv e-prints, March 2018
12. Elenberg, E.R., Shanmugam, K., Borokhovich, M., Dimakis, A.G.: Beyond triangles: a distributed framework for estimating 3-profiles of large graphs. ArXiv e-prints, June 2015
13. Erdös, P., Rényi, A.: On random graphs I. Publ. Math. Debr. **6**, 290 (1959)
14. Fortunato, S.: Community detection in graphs. Phys. Rep. **486**, 75–174 (2010)
15. Fortunato, S., Hric, D.: Community detection in networks: a user guide. ArXiv e-prints, November 2016
16. Fronczak, A., Hołyst, J.A., Jedynak, M., Sienkiewicz, J.: Higher order clustering coefficients in Barabási-Albert networks. Phys. Stat. Mech. Its Appl. **316**, 688–694 (2002)
17. Gao, C., Lafferty, J.: Testing for global network structure using small subgraph statistics. ArXiv e-prints (Oct 2017)
18. Gao, C., Lafferty, J.: Testing network structure using relations between small subgraph probabilities. ArXiv e-prints, April 2017
19. Gishboliner, L., Shapira, A.: Deterministic vs non-deterministic graph property testing. ArXiv e-prints, April 2013
20. Goldreich, O., Ron, D.: Algorithmic aspects of property testing in the dense graphs model. SIAM J. Comput. **40**(2), 376–445 (2011)
21. Hagberg, A., Schult, D., Swart, P.: Exploring network structure, dynamics, and function using network. In: Varoquaux, G., Vaught, T., Millman, J. (eds.) Proceedings of the 7th Python in Science Conference, Pasadena, CA USA, pp. 11–15 (2008)
22. He, Z., Liang, H., Chen, Z., Zhao, C.: Detecting statistically significant communities. CoRR abs/1806.05602 (2018). http://arxiv.org/abs/1806.05602

23. Jin, J., Ke, Z.T., Luo, S.: Network global testing by counting graphlets. ArXiv e-prints, July 2018
24. Leskovec, J., Kleinberg, J., Faloutsos, C.: Graph evolution: Densification and shrinking diameters. ACM Trans. Knowl. Discov. Data **1**(1), 2 (2007)
25. Lovász, L., Vesztergombi, K.: Nondeterministic graph property testing. Comb. Probab. Comput. **22**, 749–762 (2013)
26. Miasnikof, P., Shestopaloff, A.Y., Bonner, A.J., Lawryshyn, Y.: A statistical performance analysis of graph clustering algorithms. In: Bonato, A., Prałat, P., Raigorodskii, A. (eds.) WAW 2018. LNCS, vol. 10836, pp. 170–184. Springer, Cham (2018). https://doi.org/10.1007/978-3-319-92871-5_11
27. Ostroumova Prokhorenkova, L., Prałat, P., Raigorodskii, A.: Modularity of complex networks models. In: Bonato, A., Graham, F.C., Prałat, P. (eds.) WAW 2016. LNCS, vol. 10088, pp. 115–126. Springer, Cham (2016). https://doi.org/10.1007/978-3-319-49787-7_10
28. Prokhorenkova, L.O., Prałat, P., Raigorodskii, A.: Modularity in several random graph models. Electron. Notes Discret. Math. **61**, 947–953 (2017). http://www.sciencedirect.com/science/article/pii/S1571065317302238. The European Conference on Combinatorics, Graph Theory and Applications (EUROCOMB 2017)
29. Prokhorenkova, L., Tikhonov, A.: Community detection through likelihood optimization: in search of a sound model. In: Proceedings of the 2019 World Wide Web Conference (WWW 2019) (2019)
30. Schaeffer, S.E.: Survey: graph clustering. Comput. Sci. Rev. **1**(1), 27–64 (2007). https://doi.org/10.1016/j.cosrev.2007.05.001
31. Ugander, J., Backstrom, L., Kleinberg, J.: Subgraph frequencies: mapping the empirical and extremal geography of large graph collections. ArXiv e-prints, April 2013
32. Verzelen, N., Arias-Castro, E.: Community detection in sparse random networks. Ann. Appl. Probab. **25**(6), 3465–3510 (2015). https://doi.org/10.1214/14-AAP1080
33. Yang, J., Leskovec, J.: Defining and evaluating network communities based on Ground-truth. CoRR abs/1205.6233 (2012). http://arxiv.org/abs/1205.6233
34. Yin, H., Benson, A.R., Leskovec, J.: Higher-order clustering in networks. ArXiv e-prints (2018)
35. Yin, H., Benson, A., Leskovec, J.: Local higher-order graph clustering. In: Proceedings of the 23rd ACM SIGKDD International Conference on Knowledge Discovery and Data Mining 2017 (2017)

Towards Improving Merging Heuristics for Binary Decision Diagrams

Nikolaus Frohner[✉] and Günther R. Raidl

Institute of Logic and Computation, TU Wien, Vienna, Austria
{nfrohner,raidl}@ac.tuwien.ac.at

Abstract. Over the last years, binary decision diagrams (BDDs) have become a powerful tool in the field of combinatorial optimization. They are directed acyclic multigraphs and represent the solution space of binary optimization problems in a recursive way. During their construction, merging of nodes in this multigraph is applied to keep the size within polynomial bounds resulting in a discrete relaxation of the original problem. The longest path length through this diagram corresponds then to an upper bound of the optimal objective value. The algorithm deciding which nodes to merge is called a *merging heuristic*. A commonly used heuristic for layer-wise construction is *minimum longest path length* (minLP) which sorts the nodes in a layer descending by the currently longest path length to them and subsequently merges the worst ranked nodes to reduce the width of a layer. A shortcoming of this approach is that it neglects the (dis-)similarity between states it merges, which we assume to have negative impact on the quality of the finally obtained bound. By means of a simple tie breaking procedure, we show a way to incorporate the similarity of states into minLP using different distance functions to improve dual bounds for the maximum independent set problem (MISP) and the set cover problem (SCP), providing empirical evidence for our assumption. Furthermore, we extend this procedure by applying similarity-based node merging also to nodes with close but not necessarily identical longest path values. This turns out to be beneficial for weighted problems where ties are substantially less likely to occur. We evaluate the method on the weighted MISP and tune parameters that control as to when to apply similarity-based node merging.

Keywords: Binary decision diagrams · Top-down construction · Merging heuristic · State similarity · Tie breaking

1 Introduction

In the last decade, decision diagrams (DDs) have emerged as a new tool in the field of combinatorial optimization. Originally, they were conceived by Lee [10] in circuit design as a compact representation for binary functions. In the optimization context, they were introduced by Hadzic and Hooker [7] as a tool for post-optimality analysis. Since then, DDs have been used to obtain strong dual

© Springer Nature Switzerland AG 2020
N. F. Matsatsinis et al. (Eds.): LION 13 2019, LNCS 11968, pp. 30–45, 2020.
https://doi.org/10.1007/978-3-030-38629-0_3

bounds by means of a new form of discrete relaxation [6], as constraint stores for advanced constraint propagation in constraint programming [7], for obtaining promising heuristic solutions [4], and for a new branching scheme leading to a general purpose branch-and-bound framework [3]. For a comprehensive book on DDs for optimization, see [2].

A DD for a given problem is a directed acyclic graph $G = (V, A)$ with node set V and arc set A containing dedicated root and target nodes $r, t \in V$. An exact DD represents all feasible solutions of the underlying problem in the sense that there is a one-to-one correspondence between r–t paths and feasible solutions. Therefore exact DDs for hard problems typically have exponential size. In the layer-wise, top-down construction of relaxed DDs, one restricts the size by merging nodes whenever a layer would exceed a specified width. Merging is done in such a way that no feasible solution is lost, but new paths, corresponding to infeasible solutions, may emerge. Assuming maximization, a longest path from the root to the terminal node represents a solution that is usually infeasible for the original problem but yields a dual bound. The tightness of this bound is determined by the maximum width of the layers, the ordering of the decision variables [1] and the merging heuristic, i.e., the selection of the nodes that are merged. Algorithms building on a DD can strongly benefit from a stronger bound or a more compact DD that yields the same bound. The latter holds in particular when the once constructed DD is then traversed many times as in bound strengthening schemes like the value enumeration method [6] that incrementally strengthens integral bounds when there is no path with the current bound that corresponds to a feasible solution.

In this paper, we show how to improve the commonly used merging heuristic *minimum longest path* (minLP) for two benchmark problems, namely the maximum independent set (MISP) and the set cover problem (SCP). Section 2 reviews related work. In Sect. 3, we formally introduce binary decision diagrams (BDDs) based on dynamic programming formulations and provide the concrete modeling of the MISP and SCP. In Sect. 4, we introduce a state similarity-based tie breaking procedure for the minLP merging heuristic with the aim to improve the quality of obtained dual bounds. The approach is specifically instantiated for MISP and SCP. We then generalize the method by applying the similarity-based merging not just in case of ties but already when longest path values of nodes are sufficiently close. This turns out to be particularly meaningful in case of the weighted MISP, since there ties are substantially less likely to occur. More generally, for other problems we also provide suggestions on how to construct meaningful *merging distance functions*. In Sect. 5 we present our computational study, where the effectiveness of our tie breaking approach on compact BDDs with small widths for the MISP, weighted MISP, and SCP is demonstrated. We conclude in Sect. 6.

2   Related Work

Our work builds upon the classic top-down construction method of BDDs as described by Bergman et al. in [5] and [6], whose results we also use as a baseline for the MISP and SCP in our computational study. In limited-width BDDs, nodes are merged to achieve a discrete relaxation of the solution space; the selection of which nodes to merge is called *merging heuristic*, see Sect. 4. The pairwise minLP merging heuristic was introduced in [6], in its bulk form in [5]. The size of a BDD is crucially determined by the order in which the decision variables are processed as elaborated on in [1]. The *minState* variable ordering heuristic selects in each layer dynamically the next decision variable for which the least successor nodes can be derived to aim for keeping the BDD small in a greedy way. Together, the minState variable ordering heuristic and the minLP merging heuristic provide strong bounds for the MISP on random and DIMACS graphs, as presented in [5]. The possible impact of state (dis-)similarity is already addressed and a minimum distance pairwise merging heuristic is suggested in [6], on which we focus in this paper in Sect. 4. In [9], a clustering algorithm is used to partition DD nodes into approximate equivalence classes for solving a multi-dimensional bin packing problem.

3   Binary Decision Diagrams (BDDs)

We consider a combinatorial optimization problem (COP) $\mathcal{C} = \langle S, f \rangle$, where S is the finite search space and $f: S \to \mathbb{R}$ the objective function to be maximized. Every element $x \in S$ is represented by an assignment of values to n binary decision variables $x_i \in \{0, 1\}$, $i = 1, \ldots, n$. Hence, $S \subset \{0, 1\}^n$ and $f: \{0, 1\}^n \to \mathbb{R}$. The goal is to find an optimal solution x^*, i.e., for which the objective value $z^* = f(x^*) \geq f(x') \, \forall x' \in S$:

$$z^* = \max_{x \in S} f(x) \tag{1}$$

We restrict f to be a separable function of the decision variables $f(x) = \sum_{i=1}^{n} f_i(x_i)$ which allows us to state the COP in a recursive formulation. For a well-defined ordering in a recursion, a variable ordering $\pi: \{1, \ldots, n\} \to \{1, \ldots, n\}$, with π being bijective, is assumed. A partial assignment of the decision variables of length k, under the ordering π is then defined by an ordered tuple $(d_{\pi_1}, \ldots, d_{\pi_k}) \in \{0, 1\}^k, k \in \{0, \ldots, n\}$, where $k = 0$ corresponds to an empty assignment.

Definition 1 (State). *A state $s_i \in \mathcal{S}$ is a mapping from an i-partial assignment. It determines the subset $F_{s_i} \subseteq \{0, 1\}^{n-i}$ of feasible decisions for remaining variables $x_{\pi(i+1)}, \ldots, x_{\pi(n)}$, the **feasible completions** of the current partial assignment. If two partial assignments have the same state, they have the same feasible completions.*

The representation of a state needs to be concretely defined for the problem at hand, for example by means of sets or reals. This admits a recursive enumeration of the state space $\mathcal{S}^{n+1} \ni (s_0, \ldots, s_n)$ corresponding to the search space S by defining a state transition function:

$$\tau: \{0, 1\} \times \mathcal{S} \to \mathcal{S} \tag{2}$$

$$(d, s_i) \mapsto \tau(d, s_i) = s_{i+1} \tag{3}$$

We can now formulate our maximization problem recursively over the states via Bellman equations $\forall i \in \{0, \ldots, n-1\}$:

$$z^*(s_i) = \max_{d \in \{0,1\}} \{f_{\pi_i}(d) + z^*(\tau(d, s_i)) \mid d \exists c \in F_{s_i} : c = (d, \ldots)\} \quad (4)$$

$$z^*(s_n) = 0 \quad (5)$$

If for a given s_i there exists a feasible completion $c \in F_{s_i}$ for which we can set the next decision variable $x_{\pi_{i+1}}$ to d, i.e., $\exists c \in F_{s_i} : c = (d, \ldots)$, we say that the state *admits* a d-transition. The root state s_0 corresponds to an empty partial assignment, s_1 to when the first variable x_{π_1} has been assigned, and so forth. Clearly, $F_{s_0} = S$ and $F_{s_n} = \emptyset$.

A binary decision diagram (BDD) in our context is a directed acyclic layered multigraph with layers L, $|L| = n + 1$ and represents this state space enumeration graphically. Layer 0 contains only the root node r representing the root state $s(r) = s_0$ and layer n the terminal node t representing the terminal state $s(t) = s_n$. Each node u in layer l is thus associated with a state $s(u)$. If $s(u) = s(u')$ for nodes u, u' in a given layer, they admit by definition the same feasible completions and can therefore be superimposed to reduce the size of the BDD. Except for the terminal node, each node u has a d-labeled outgoing arc $a = (u, v)$ for each d admissible by $F_{s(u)}$, representing the possible decisions at state $s(u)$. $source(a) = u$ is called the source (node) of the arc and $target(a) = v$ the target (node) respectively. The arcs point downwards, the layer of the target must always be greater than the one of the source.

Every arc receives a label $d(a) \in \{0, 1\}$ to encode a binary decision. If a path starts at the root node and finally leads to some node v, which we denote by p_{rv}, this corresponds to a k-partial assignment $(d((r, u_1)), \ldots, d((u_{k-1}, v)))$, where $d((u, v))$ is the aforementioned label of arc (u, v). Every arc is assigned a weight $f_{\pi_i}(d)$, contributing to the length of paths going trough the decision diagram, for instance $f_{\pi(i)}(d) = c_{\pi_i} d$ when we are given constant objective function contributions c_{π_i} for each decision variables $x_{\pi(i)}$ set to one. In exact BDDs, there is by construction a one-to-one mapping between paths p_{rt} and feasible solutions S. For maximization problems, paths of longest length correspond to its optimal solutions. In general, the exact decision diagrams grow exponential in size in the number of decision variables. The focus in this paper lies on limited-width, *relaxed* DDs, where layers have a maximum number β of nodes to keep the DD size bounded by $\beta|L|$ nodes. The contained paths represent a *superset* of the search space S and, thus, a discrete relaxation of the original problem. This is achieved by also superimposing nodes that have *different* states, which is called *merging*.

Definition 2 (Merging of nodes). *When nodes u, v are merged into a node w, all incoming arcs of u, v are redirected to the new node w and the states $s(u), s(v)$ are merged into $s(w)$ in a way that no feasible paths, i.e., solutions in the search space are lost. Therefore, $F_{s(w)} \supset F_{s(u)} \cup F_{s(v)}$.*

The length of a longest path in a relaxed BDD is an upper bound on the optimal objective value to the original problem. Our first specific problem we

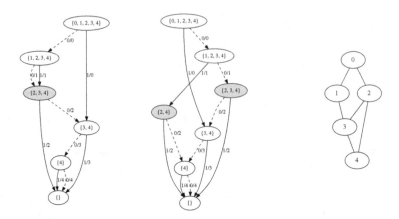

Fig. 1. Two relaxed BDDs for a simple graph instance on the right. Left with maximum width $\beta = 1$, in the center with $\beta = 2$, both having the same longest path length of 2 with optimal solutions of zero-indexed vertices $\{\{0, 3\}, \{0, 4\}, \{1, 2\}, \{1, 4\}\}$.

consider is the maximum independent set problem (MISP). It is defined on an undirected simple graph $G = (V, E)$ as finding a maximum subset of nodes $I \subset V$, s.t. no pair of nodes in I are adjacent. A proper state s_i is the subset of the vertices for which no decision has been yet made and for which no neighbor has been selected so far. The transition function is

$$\tau : \{0, 1\} \times 2^V \rightarrow 2^V \tag{6}$$

$$(0, s_i) \mapsto s_{i+1} = \tau(0, s_i) = s_i - \{\pi_i\} \tag{7}$$

$$(1, s_i) \mapsto s_{i+1} = \tau(1, s_i) = s_i - \{\pi_i\} - N(\pi_i) \tag{8}$$

where $N(\pi_i) \subset V$ is the neighborhood of the i-th considered vertex π_i. The root state s_0 is V, the terminal state $\emptyset$. A natural merging operator $\oplus$ of k states is given by the set union:

$$\oplus (\{u_1, \ldots, u_k\}) \mapsto w : s(w) = \bigcup_{j=1}^{k} s(u_j) \tag{9}$$

Two examplary BDDs for a simple MISP instance of width $\beta = 1$ and $\beta = 2$ are depicted in Fig. 1. For each arc the weight and the corresponding decision variable is shown. The label is indicated by a dotted arc for a 0-transition and a solid arc for a 1-transition. When reducing the maximum width from 2 to 1, we see that a merging is applied in the second layer of states $\{2, 4\}$ and $\{2, 3, 4\}$.

As a second fundamental problem, we consider the classical set cover problem (SCP). Given a universe $\mathcal{U}$ and a set of sets $\mathcal{S}$ with $\mathcal{S} \ni S \subset \mathcal{U}$ and $\bigcup_{S \in \mathcal{S}} S = \mathcal{U}$, we seek to find a $\mathcal{S}^* \subset \mathcal{S}$ with minimum cardinality so that $\bigcup_{S \in \mathcal{S}^*} S = \mathcal{U}$, i.e., a minimum set covering. A proper state s_i is the set of elements that still have to be covered. To ensure that all paths are feasible, a set j has to be selected (i.e.,

```
1  X ← {1, . . . , n}, P ← {r};
2  for l ← 1 to n do
3  |    π_l ← next-decision-variable(l, P, X);
4  |    X ← X − π_l;
5  |    L_l ← P' ⊂ P for which x_{π_l} can bet set to 1;
6  |    while |L_l| > β do
7  |    |    L_l ← merge-nodes(L_l, l);
8  |    end
9  |    foreach u ∈ L_l do
10 |    |    foreach d ∈ {0, 1} do
11 |    |    |    if s(u) admits d-transition then
12 |    |    |    |    create v_d;
13 |    |    |    |    s(v_d) = τ(d, s(u));
14 |    |    |    |    create arc (u, v_d) with label d and weight f_{π(l)}(d);
15 |    |    |    end
16 |    |    end
17 |    end
18 end
```

Algorithm 1. Relaxed limited-width layered binary decision diagram construction algorithm, adapted from [3, p. 12].

its decision variable set to 1) if there exists an element in s_i that can only be covered by selecting j, since all other possible decision variables have been set to 0. A natural merging operator $\oplus$ of k states is given by the set intersection:

$$\oplus(\{u_1, \dots, u_k\}) \mapsto w \colon s(w) = \bigcap_{j=1}^{k} s(u_j) \tag{10}$$

Throughout this paper, we focus on the top-down layer-wise construction algorithm [5] for relaxed binary decision diagrams with maximum width β as described in Algorithm 1.

It facilitates zero-suppressing long arcs, a dynamic variable ordering by the function next-decision-variable and merging of nodes by the function merge-nodes. As concrete variable ordering heuristic, we consider here minState [5] which selects as next decision variable the one that yields the least number of one-transitions from the current nodes for the next layer. A simple, yet effective and commonly used merging heuristic is minLP [5], which sorts the nodes u in a layer by the longest path lengths from the root node to them, denoted by $z^{\mathrm{lp}}(u)$, in decreasing order and merges the tail into one node so that the resulting layer is of maximum width β, see Algorithm 2. In the minLP approaches described in the literature so far, to the best of our knowledge, no tie breaking mechanism for the sorting is explicitly specified which gives rise to the next section.

```
1  Function merge-nodes(L_l, l)
2  |   T ← sorted nodes of L_l in decreasing order of z^lp(u);
3  |   T' ← T without first β − 1 nodes of T;
4  |   L_l = L_l \ T';
5  |   w ← ⊕T;
6  |   if ∃w' ∈ L_l | s(w) = s(w') then
7  |   |   w' ← w' ⊕ w;
8  |   else
9  |   |   L_l = L_l ∪ {w};
10 |   end
```

Algorithm 2. minLP merging heuristic in the bulk variant where all nodes to be merged are merged within one step into one single node.

4 State Similarity

We consider two different merging heuristic patterns: *pairwise* merging and *bulk* merging. Both face a layer l with a set of nodes L_l where $|L_l|$ exceeds the maximum width β. Pairwise merging is a form of *iterative* merging where pairs of nodes are selected and merged until the desired layer width has been reached. In contrast, bulk merging selects and merges the necessary number of nodes in a single iteration. The bulk minLP merging heuristic as introduced in the last section in Algorithm 2 sorts the nodes in a layer according to the longest path length to them and merges the last $|L_l| - \beta + 1$ nodes into one. It generalizes to *rank* based merging, which sorts nodes in a layer according to some criterion and merges the required number of tail nodes. If the criterion can be calculated easily, a clear benefit is the $\mathcal{O}(|L_l| \log |L_l|)$ runtime complexity, whereas pairwise mergings needs in general at least $\mathcal{O}(|L_l|^2)$ time.

The rationale behind minLP is to consider nodes with smaller $z^{lp}(u)$ less promising to be part of an overall longest path in the completed DD and therefore less critical when merged in order to finally obtain a tight upper bound. This strategy is supported by the minState variable ordering heuristic, which keeps the size of the layers before merging as small as possible, therefore reducing the number of nodes that need to be merged.

A shortcoming of this approach is that it neglects information that could be obtained from the states of the nodes themselves, in particular the similarity between states. Intuitively, merging similar states will usually lead to less new paths corresponding to infeasible solutions than merging very different states. If two states are comparable, for instance by the subset relation for sets or the total order for reals, we denote $s(u_1) \succeq s(u_2)$ when $s(u_1)$ is greater than $s(u_2)$. If $s(u_1) \succeq s(u_2)$, then $F_{s(u_1)} \supseteq F_{s(u_2)}$. One way merging of nodes u, v introduces infeasible solutions is by increasing the size of feasible completions $F_{s(w)} \supset F_{s(u)} \cup F_{s(v)}$, which gives rise to a definition for a meaningful distance function:

Definition 3 (Merging distance between two nodes). *A merging distance between two nodes u, v is a non-negative function $d: L_l \times L_l \to \mathbb{R}_0^+$. For any*

```
1  Function merge-nodes(L_l, l)
2  |   T ← pairs of nodes (u, v), u, v ∈ L_l for which (z^lp(u), z^lp(v)) is minimal;
3  |   T' ← pairs of nodes (u, v) ∈ L_l for which d(u, v) is minimal;
4  |   select (u, v) ∈ T' randomly;
5  |   L_l = L_l\{u, v};
6  |   w ← u ⊕ v;
7  |   if ∃w' ∈ L_l|s(w) = s(w') then
8  |   |   w' ← w' ⊕ w;
9  |   else
10 |   |   L_l = L_l ∪ {w};
11 |   end
```

Algorithm 3. Iterative minLP merging function with similarity-based tie breaking.

triple of nodes $u_1, u_2, v \in L_l$, we demand that if $F_{s(u_1 \oplus v)} \supset F_{s(u_2 \oplus v)}$, $d(u_1, v) \geq d(u_2, v)$ should hold.

The goal is to find a distance function *for a specific problem* such that greater distance means a higher probability of introducing new paths and thus new represented solutions in the decision diagram, even if the states of the nodes are uncomparable. To consider a merging distance in the current state-of-the-art merging heuristics, we first look at an iterative minLP variant, where we find the use of the state similarity as a straightforward extension in form of a tie breaking mechanism. This becomes relevant when there are two pairs of nodes $(u_1, v), (u_2, v), s(u_1) \neq s(u_2)$ for which $(z^{lp}(u_1), z^{lp}(v)) = (z^{lp}(u_2), z^{lp}(v))$—then we simply take the pair with minimal distance according to d, see Algorithm 3. In the case of bulk minLP, a tie breaking is necessary when multiple nodes with the same rank go through the *merging boundary*, see Fig. 2, which separates the nodes to be merged from those to be kept as they are. Since the iterative minLP merging always takes pairs of nodes with currently smallest ranks with respect to z^{lp}, an alternative implementation is to first do a bulk merge of the nodes that have rank less than the one causing the need for tie breaking and then switch to merging nodes pairwise:

1. For a given layer l with nodes L_l, sort the nodes according to their current longest path length $z^{lp}(u)$ in decreasing order.
2. If the rank r of the $\beta - 1$-th node equals the rank of the β-th node, then we select all nodes with that rank r into a tie breaking set $T \subset L_l$; otherwise we do a simple minLP bulk merging.
3. Let B be the set of nodes that have a rank $<r$. We merge them yielding a node w with state $s(w)$ that is either still at the end of the ordered list or is absorbed by another node, if there already exists a node w' with $s(w) = s(w')$.
4. Finally, we iteratively merge pairs of nodes out of $T \cup \{w\}$ (or T if w has been absorbed) until the desired width β is reached. In each iteration, we choose the pair u, v that currently has minimal distance $d(u, v)$.

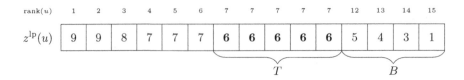

rank(u)	1	2	3	4	5	6	7	7	7	7	7	12	13	14	15
$z^{\mathrm{lp}}(u)$	9	9	8	7	7	7	**6**	**6**	**6**	**6**	**6**	5	4	3	1

Fig. 2. Example layer with $|L_l| = 15$ nodes sorted by longest path length $z^{\mathrm{lp}}(u)$, which is shown in the nodes. Let the maximum width be $\beta = 10$. All nodes with longest path value 6 (bold) are now subject to tie breaking.

When considering weighted problems, ties are in general substantially less likely to occur than in unweighted counterparts—still, we want to take the state similarity into account when differences in the longest path lengths are small. For that purpose, we introduce a parameterized hybrid merging algorithm, which is based on the minLP ordering but artificially introduces a region of nodes of similar longest path value around the merging boundary with which we deal as with the tie breaking region above. This region is determined by parameters δ_l, δ_r. To have meaningful parameters tunable between 0 and 1, regardless of the absolute values of the longest path lengths, we first normalize those according to the following transformation:

$$\tilde{z}^{\mathrm{lp}}(u) = \frac{z^{\mathrm{lp}}(u) - \min_{v \in L_l} z^{\mathrm{lp}}(v)}{\max_{v \in L_l} z^{\mathrm{lp}}(v) - \min_{v \in L_l} z^{\mathrm{lp}}(v)} \tag{11}$$

The reference value is obtained by taking the normalized path value $\tilde{z}^{\mathrm{lp}}_{\mathrm{ref}}$ of the node immediately right to the merging boundary, i.e., the node with the largest value to be merged, if regular minLP would be applied. Now, two regions (contiguous sets of nodes in the ordered view of the layer) are defined:

1. bulk merging region $B := \{u \in L_l \mid \tilde{z}^{\mathrm{lp}}(u) \in (\tilde{z}^{\mathrm{lp}}_{\mathrm{ref}} - \delta_r, 0.]\}$
2. pairwise merging region $T := \{u \in L_l \mid \tilde{z}^{\mathrm{lp}}(u) \in [\tilde{z}^{\mathrm{lp}}_{\mathrm{ref}} + \delta_l, \tilde{z}^{\mathrm{lp}}_{\mathrm{ref}} - \delta_r]\}$

Let $w \leftarrow \oplus B$ be the node resulting from the bulk merging of B, and $L_l - T - B$ are the nodes that are kept as they are. The pairwise merging is now performed iteratively by always selecting a node pair with minimum distance d from $T \cup w$ and replacing the two nodes by the merged node until the desired layer width is reached. Setting $\delta_l = 0.0$, $\delta_r = 0.0$ yields the bulk-iterative hybrid as described before that only considers pairwise merging for real ties, whereas $\delta_l = 1.0$, $\delta_r = 1.0$ would completely ignore the longest path information and only focus on iteratively finding two minimum distance nodes to merge. For the choice of the pairwise merging region T in an example layer, see Fig. 3.

As mentioned before, it is crucial to conceive a meaningful distance function for a concrete problem. Notice that each node u in a layer has a maximum remaining path length $\max_{e \in F_{s_i = s(u)}} f(e)$, where $f(e = (d_{\pi_{i+1}}, \ldots, d_{\pi_n}))$ is the length of the feasible completion (see Definition 1), which is clearly not known to us during the construction of the DD at layer l. Still, a possible construction

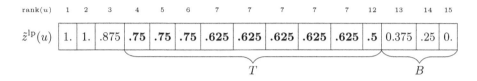

Fig. 3. Example layer with $|L_l| = 15$ nodes sorted by normalized longest path length $\tilde{z}^{\mathrm{lp}}(u)$, which is also shown in the nodes. Let the maximum width be $\beta = 10$, $\delta_l = \delta_r = .125$. All nodes with normalized longest path value $.625 \pm .125$ (bold) are now subject to pairwise merging.

```
 1 Function merge-nodes(L_l, l)
 2    z̃ˡᵖ(u) ← ( zˡᵖ(u) − min_{v∈L_l} zˡᵖ(v) ) / ( max_{v∈L_l} zˡᵖ(v) − min_{v∈L_l} zˡᵖ(v) )  ∀u ∈ L_l;
 3    B ← {u ∈ L_l | z̃ˡᵖ(u) ∈ (z̃ˡᵖ_ref − δ_r, 0.]};
 4    T ← {u ∈ L_l | z̃ˡᵖ(u) ∈ [z̃ˡᵖ_ref + δ_l, z̃ˡᵖ_ref − δ_r]};
 5    L_l = L_l \ B;
 6    w ← ⊕B;
 7    include-node-into-layer(w, L_l, l);
 8    while |L_l| > β do
 9        T' ← pairs of nodes (u, v) ∈ T for which d(u, v) is minimal;
10        select (u, v) ∈ T' for which max{zˡᵖ(u), zˡᵖ(v)} is minimal;
11        L_l = L_l \ {u, v};
12        T = T \ {u, v};
13        w ← u ⊕ v;
14        include-node-into-layer(w, L_l, l);
15    end

16 Function include-node-into-layer(w, L_l, l)
17    if ∃w' ∈ L_l | s(w) = s(w') then
18        w' ← w' ⊕ w;
19    else
20        L_l = L_l ∪ {w};
21    end
```

Algorithm 4. Bulk-iterative minLP-state similarity-based hybrid merging algorithm with parameters δ_l, δ_r.

scheme to formulate a distance between u and v is to consider the maximum increase of the maximum remaining path lengths that u and v experience by being merged to $w = u \oplus v$:

$$d(u,v) = \max\{ \max_{e \in F_{s(w)}} f(e) - \max_{e \in F_{s(u)}} f(e), \max_{e \in F_{s(w)}} f(e) - \max_{e \in F_{s(v)}} f(e) \} \quad (12)$$

This is can be made use of by approximating the maxima by an upper bound function $z^{\mathrm{ub}}(u)$:

$$d_{\mathrm{ub}}(u,v) = \max\{ z^{\mathrm{ub}}(w) - z^{\mathrm{ub}}(u), z^{\mathrm{ub}}(w) - z^{\mathrm{ub}}(v) \} \quad (13)$$

For the MISP a first coarse upper bound to consider is given by the cardinality of the state $|s(u)|$, which is only reasonably tight for sparse graphs, but might still be meaningful since we are only interested in the maximum increase:

$$d_{\text{coarse}}^{\text{MISP}}(u, v) = \max\{|s(u) \cup s(v)| - |s(u)|, |s(u) \cup s(v)| - |s(v)|\} \quad (14)$$

In the weighted MISP (MWISP) case, we sum over the vertex weights of the remaining vertices defined by the state, $z_{\text{MWISP}}^{\text{ub}}(u) = \sum_{j \in s(u)} f_j(x_j = 1)$ to get a coarse upped bound. With SCP, we are facing a minimization problem; there the distance can be defined as maximum *lower bound* change:

$$d_{\text{lb}}(u, v) = \max\{z^{\text{lb}}(u) - z^{\text{lb}}(w), z^{\text{lb}}(v) - z^{\text{lb}}(w)\} \quad (15)$$

In this case, the calculation of a bound on the maximum remaining path length takes a little more work: We go over the remaining elements to be covered and if at the i-th, we increase a counter by one, if none of its covering sets was also a covering set for some $j < i$. The resulting counter value is a lower bound for the number of sets to cover the universe.

Another construction method is to take only the maximum remaining path length after merging $w = u \oplus v$:

$$\tilde{d}(u, v) = \max_{e \in F_{s(w)}} f(e) \quad (16)$$

The rationale is that it should be less likely to merge nodes that have high upper bounds even if they are similar with respect to $d(u, v)$, to balance the resulting upper bounds over the layer:

$$\tilde{d}_{\text{ub}}(u, v) = z^{\text{ub}}(w) \quad (17)$$

As a baseline, we suggest to also include the weighted Hamming distance d_H. It sums the weights of elements that are part of state $s(u)$ but not $s(v)$ or vice versa; for the unweighted case this amounts simply the cardinality of the symmetric set difference.

To summarize, the main idea of these construction methods was to greedily impede the estimated growth in bound by the remaining layers of the decision diagram induced by merging. One subtlety is that a problem specific upper bound does not take future merging operations into account but considers the case when we would continue constructing the BDD *without* merging.

5   Computational Study

We tested the relaxed DD construction applying the minLP merging heuristic with simple tie breaking based on the natural node order as done in [3] (i.e., the classic minLP) and minLP with our new similarity-based tie breaking using different distance functions for the MISP on random graphs from [5] with $n = 200$ and densities from $\{0.1, 0.2 \ldots, 0.9\}$ (20 instances per combination) and on the

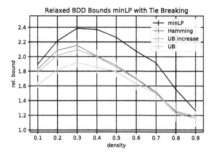 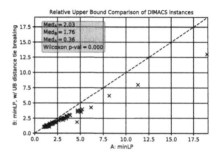

Fig. 4. Comparison of relative bounds of relaxed BDDs with $\beta = 10$ obtained with the classic minLP merging heuristic and with minLP with similarity-based tie breaking using different distance functions. Left: plotted over densities with means and error bars of 1σ; right: scatter plot for classic minLP vs. minLP with $\tilde{d}_{\mathrm{ub}}$ based tie breaking.

DIMACS [8] max clique set instances[1]. For the SCP, we created random instances with $n = 500$ elements that are covered by exactly $k = 20$ sets each following the creation procedure for structured random instances from [6]. The constraint matrices describing which sets cover which elements follow a specific staircase like structure with limited bandwidths from $\{21, \ldots, 27\}$, and there are 20 instances per bandwidth. For the weighted MISP, we used 64 extended DIMACS graphs that we could solve to optimality in which vertex $i \in \{1, \ldots, n\}$ has weight $i \bmod 200 + 1$[2]. All tests were conducted on an Intel Xeon E5-2640 processor with 2.40 GHz in single-threaded mode and a memory limit of 8 GB.

On the left side of Fig. 4 we see the performance of the different tie breaking distance functions from Sect. 4 in comparison to the classic minLP approach in terms of the obtained relative bounds (i.e., obtained bounds divided by known optimal objective values) on the MISP random graph instances when compiling relaxed BDDs of maximum width $\beta = 10$. The tie breaking that seeks for pairs for which merging yields the smallest trivial upper bound (cardinality of state set), gives the strongest results. Differences among the approaches are generally larger for sparser graphs and start to vanish for denser graphs. This is plausible since the trivial upper bound is tighter for sparser graphs. The difference reaches a maximum for density 0.3 of about 40%. On its right side Fig. 4 shows a scatter plot with the relative bounds obtained for the DIMACS graph instances for classic minLP and our minLP with similarity-based tie breaking with the upper bound distance function. The median of the pairwise difference is 36% in favor of our tie breaking. A Wilcoxon signed rank sum test indicated that this difference is significant with an error probability of less than one percent.

Mean values of relative upper bounds and corresponding standard deviations for the different densities and algorithm variants are listed in Table 1 for $\beta = 10$

---

[1] http://www.andrew.cmu.edu/user/vanhoeve/mdd/code/opt_bounds_bdd-instances.tar.gz.

[2] https://github.com/jamestrimble/max-weight-clique-instances/tree/master/DIMACS.

Table 1. Mean relative upper bounds $\bar{u}_{\text{rel}}$ and standard deviations of relaxed BDDs over the 20 random graphs per density p obtained by the different merging heuristics for DD widths $\beta \in \{10, 100\}$.

p	$\beta = 10$								$\beta = 100$							
	minLP		d_H		d_{ub}		$\tilde{d}_{\text{ub}}$		minLP		d_H		d_{ub}		$\tilde{d}_{\text{ub}}$	
	$\bar{u}_{\text{rel}}$	σ_{ub}	$\bar{u}_{\text{rel}}$	σ_{ub}	$\bar{u}_{\text{rel}}$	σ_{ub}	$\bar{u}_{\text{rel}}$	σ_{ub}	$\bar{u}_{\text{rel}}$	σ_{ub}	$\bar{u}_{\text{rel}}$	σ_{ub}	$\bar{u}_{\text{rel}}$	σ_{ub}	$\bar{u}_{\text{rel}}$	σ_{ub}
0.10	1.90	0.03	1.83	0.04	1.79	0.04	**1.61**	0.04	1.63	0.03	1.57	0.04	1.56	0.04	**1.49**	0.03
0.20	2.21	0.07	2.09	0.07	2.02	0.06	**1.82**	0.06	1.80	0.06	1.70	0.04	1.67	0.05	**1.62**	0.06
0.30	2.38	0.10	2.15	0.08	2.09	0.07	**1.92**	0.08	1.80	0.08	**1.63**	0.04	**1.63**	0.06	**1.63**	0.07
0.40	2.37	0.09	2.01	0.08	2.00	0.09	**1.85**	0.09	1.69	0.07	1.51	0.07	1.54	0.06	**1.50**	0.07
0.50	2.26	0.09	1.88	0.07	1.85	0.07	**1.80**	0.08	1.54	0.08	**1.38**	0.06	1.39	0.05	1.40	0.06
0.60	2.07	0.09	1.70	0.06	1.69	0.07	**1.63**	0.08	1.35	0.05	**1.21**	0.05	1.24	0.04	1.22	0.04
0.70	1.91	0.11	1.52	0.08	**1.50**	0.11	**1.50**	0.08	1.22	0.07	1.12	0.05	**1.11**	0.06	**1.11**	0.07
0.80	1.55	0.16	1.24	0.11	1.26	0.12	**1.22**	0.11	**1.00**	0.00	**1.00**	0.00	**1.00**	0.00	**1.00**	0.00
0.90	1.26	0.15	**1.16**	0.12	1.19	0.11	**1.16**	0.12	**1.00**	0.00	**1.00**	0.00	**1.00**	0.00	**1.00**	0.00

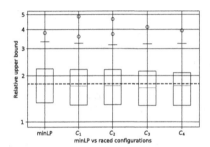 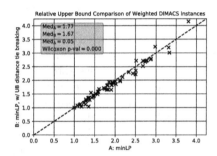

Fig. 5. Comparison of relative bounds of relaxed BDDs with $\beta = 10$ for weighted DIMACS instances for classical minLP vs. minLP with similarity-based merging in configuration C_3 with distance function $\tilde{d}_{\text{ub}}$.

and $\beta = 100$. For selected DIMACS instances relative upper bound values are shown likewise in Table 2. For the weighted DIMACS graph instances, where real ties are virtually non-existent, we tuned the left and right threshold parameters δ_l and δ_r for the region to which similarity-based merging is applied with irace [11] and test on a different set of weighted DIMACS graphs where the vertices have been randomly permuted. Here, we always used the superior upper bound distance function. On the left side of Fig. 5 we see boxplots comparing the raced parameter configurations with the classical minLP approach. The right side of Fig. 5 shows the comparison of the DDs' relative bounds when using the most promising configuration $C_3 = (0.185, 0.043)$. We observe that occasionally worse bounds are obtained but still in the clear majority of the cases the state similarity-based merging yields tighter bounds, which is also confirmed by a Wilcoxon signed rank sum test with an error probability of less than one percent. The median of the pairwise differences is 0.05.

Table 2. Relative upper bounds of relaxed BDDs obtained with different merging heuristics and widths $\beta \in \{10, 100\}$ for selected DIMACS instances.

inst	$\beta = 10$				$\beta = 100$			
	minLP	d_H	d_{ub}	$\tilde{d}_{ub}$	minLP	d_H	d_{ub}	$\tilde{d}_{ub}$
brock200_1	2.29	2.14	2.14	**1.90**	1.81	**1.62**	1.67	1.67
C500.9	3.05	3.00	2.81	**2.47**	2.61	2.46	2.40	**2.28**
gen400_p0.9_55	2.25	2.13	2.04	**1.82**	1.91	1.82	1.80	**1.73**
keller4	1.91	**1.55**	1.64	**1.55**	1.45	**1.18**	**1.18**	**1.18**
MANN_a45	1.34	1.34	**1.21**	1.30	**1.08**	1.32	1.27	1.19
p_hat300-3	2.19	2.11	2.08	**1.86**	1.86	1.75	1.81	**1.69**
p_hat700-2	2.59	2.45	2.32	**2.18**	2.14	1.98	1.95	**1.93**

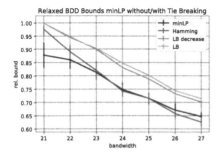 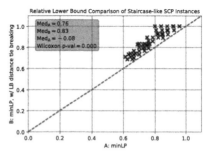

Fig. 6. Comparison of relative bounds of relaxed BDDs with $\beta = 10$ for staircase-like set cover problem instances with $n = 500$ elements to cover with varying bandwidths $b_w \in \{21, \dots, 27\}$ obtained with the classic minLP merging heuristic and with minLP with similarity-based tie breaking using different distance functions. Left: plotted over different bandwidths with error bars for 1σ; right: scatter plot for classic minLP vs. minLP with tie-breaking based on $\tilde{d}_{ub}$.

In Fig. 6, we see the results for analogous comparisons for the set cover problem. As this is a minimization problem, we seek high lower bound values. Again, the lower bound distance turns out to be the most promising and gives statistically significant improvements with a median increase in the lower bound value of 0.08.

6   Conclusion and Future Work

We presented a possibility to improve the minLP merging heuristic in the layer-wise construction of a relaxed BDD. This extension turns in case of ties to a pairwise merging strategy that considers the state similarities for deciding which nodes to merge next. For unweighted problems, ties occur naturally and we obtain significant improvements for MISP random graphs, DIMACS instances, and for the set cover problem with random staircase-like instances.

In the weighted case, due to too few real ties, we generalized the method by considering a range of nodes with close longest path lengths for our similarity-based merging. We see a small but significant improvement for weighted DIMACS instances, after having tuned the corresponding parameters. The computational overhead introduced by our approach depends on the number of ties or the parameters δ_l and δ_r in the generalized variant as well as the applied distance function. However, since minLP still is the dominant criterion for deciding which nodes to merge, the set of nodes to be processed by the pairwise similarity-based merging is typically quite restricted. Our focus was on obtaining relaxed DDs of small width that provide stronger bounds. Such DDs are particularly important when they are used in some further algorithm many times, as frequently is the case in practical applications. Then, an overhead in the DD's construction will quickly pay off. Our ongoing research is concerned with achieving more effect on weighted instances, testing on further problem classes and reducing the time complexity so that state similarity-based approaches become also more effective for larger decision diagram width.

References

1. Bergman, D., Cire, A.A., van Hoeve, W.-J., Hooker, J.N.: Variable ordering for the application of BDDs to the maximum independent set problem. In: Beldiceanu, N., Jussien, N., Pinson, É. (eds.) CPAIOR 2012. LNCS, vol. 7298, pp. 34–49. Springer, Heidelberg (2012). https://doi.org/10.1007/978-3-642-29828-8_3
2. Bergman, D., Cire, A.A., van Hoeve, W.J., Hooker, J.N.: Decision Diagrams for Optimization. Artificial Intelligence: Foundations, Theory, and Algorithms. Springer, Cham (2016). https://doi.org/10.1007/978-3-319-42849-9
3. Bergman, D., Cire, A.A., van Hoeve, W.J., Hooker, J.N.: Discrete optimization with decision diagrams. INFORMS J. Comput. **28**(1), 47–66 (2016)
4. Bergman, D., Cire, A.A., van Hoeve, W.J., Yunes, T.: BDD-based heuristics for binary optimization. J. Heuristics **20**(2), 211–234 (2014)
5. Bergman, D., Cire, A.A., van Hoeve, W.J., Hooker, J.N.: Optimization bounds from binary decision diagrams. INFORMS J. Comput. **26**(2), 253–268 (2013)
6. Bergman, D., van Hoeve, W.-J., Hooker, J.N.: Manipulating MDD relaxations for combinatorial optimization. In: Achterberg, T., Beck, J.C. (eds.) CPAIOR 2011. LNCS, vol. 6697, pp. 20–35. Springer, Heidelberg (2011). https://doi.org/10.1007/978-3-642-21311-3_5
7. Hadzic, T., Hooker, J.: Postoptimality analysis for integer programming using binary decision diagrams. In: GICOLAG Workshop (Global Optimization: Integrating Convexity, Optimization, Logic Programming, and Computational Algebraic Geometry), Vienna. Technical report, Carnegie Mellon University (2006)
8. Johnson, D.S., Trick, M.A.: Cliques, coloring, and satisfiability: second DIMACS implementation challenge, 11–13 October 1993, vol. 26. American Mathematical Soc. (1996)
9. Kell, B., van Hoeve, W.-J.: An MDD approach to multidimensional bin packing. In: Gomes, C., Sellmann, M. (eds.) CPAIOR 2013. LNCS, vol. 7874, pp. 128–143. Springer, Heidelberg (2013). https://doi.org/10.1007/978-3-642-38171-3_9

10. Lee, C.Y.: Representation of switching circuits by binary-decision programs. Bell Syst. Tech. J. **38**(4), 985–999 (1959)
11. López-Ibáñez, M., Dubois-Lacoste, J., Cáceres, L.P., Birattari, M., Stützle, T.: The irace package: Iterated racing for automatic algorithm configuration. Oper. Res. Perspect. **3**, 43–58 (2016)

On Polynomial Solvability of One Quadratic Euclidean Clustering Problem on a Line

Alexander Kel'manov[1,2] and Vladimir Khandeev[1,2(✉)]

[1] Sobolev Institute of Mathematics, 4 Koptyug Ave., 630090 Novosibirsk, Russia
[2] Novosibirsk State University, 2 Pirogova St., 630090 Novosibirsk, Russia
{kelm,khandeev}@math.nsc.ru

Abstract. We consider one problem of partitioning a finite set of points in Euclidean space into clusters so as to minimize the sum over all clusters of the intracluster sums of the squared distances between clusters elements and their centers. The centers of some clusters are given as an input, while the other centers are unknown and defined as centroids (geometrical centers). It is known that the general case of the problem is strongly NP-hard. We show that there exists an exact polynomial algorithm for the one-dimensional case of the problem.

Keywords: Minimum Sum-of-Squares Clustering · Euclidean space · NP-hard problem · One-dimensional case · Polynomial solvability

1 Introduction

The subject of this study is one strongly NP-hard problem of partitioning a finite set of points in Euclidean space into clusters. Our goal is to analyze the computational complexity of the problem in the one-dimensional case. The research is motivated by the openness of the specified mathematical question, as well as by the importance of the problem for some applications, in particular, for Data analysis, Data mining, Pattern recognition, and Data processing.

The paper has the following structure. In Sect. 2, the problem formulation is given. In the same section, a connection is established with a well-known problem that is the closest to we consider one. The next section presents auxiliary statements that reveal the structure of the optimal solution to the problem. These statements allow us to prove the main result. In Sect. 4, our main result of the polynomial solvability of the problem in the 1D case is presented.

2 Problem Formulation, Its Sources and Related Problems

In the well-known clustering K-*Means* problem, an N-element set $\mathcal{Y}$ of points in d-dimension Euclidean space and a positive integer K are given. It is required to

© Springer Nature Switzerland AG 2020
N. F. Matsatsinis et al. (Eds.): LION 13 2019, LNCS 11968, pp. 46–52, 2020.
https://doi.org/10.1007/978-3-030-38629-0_4

find a partition of the input set $\mathcal{Y}$ into non-empty clusters $\mathcal{C}_1, \ldots, \mathcal{C}_K$ minimizing the sum

$$\sum_{k=1}^{K} \sum_{y \in \mathcal{C}_k} \|y - \overline{y}(\mathcal{C}_k)\|^2,$$

where $\overline{y}(\mathcal{C}_k) = \frac{1}{|\mathcal{C}_k|} \sum_{y \in \mathcal{C}_k} y$ is the centroid of the k-th cluster.

Another common name of K-Means problem is MSSC (Minimum Sum-of-Squares Clustering). In statistics, this problem is known from the last century and is associated with Fisher (see, for example, [1,2]). In practice (in a wide variety of applications), this problem arises when there is the following hypothesis on a structure of some given numerical data. Namely, one has assumption that the set $\mathcal{Y}$ of sample (input) data contains K homogeneous clusters (subsets) $\mathcal{C}_1, \ldots, \mathcal{C}_K$, and in all clusters, the points are scattered around the corresponding unknown mean values $\overline{y}(\mathcal{C}_1), \ldots, \overline{y}(\mathcal{C}_K)$. However, the correspondence between points and clusters is unknown. Obviously, in this situation, for the correct application of classical statistical methods (hypothesis testing or parameter estimating) to the processing of sample data, at first it is necessary to divide the data into homogeneous groups (clusters). This situation is typical, in particular, for the above-mentioned (see Sect. 1) applications.

The K-Means strong NP-hardness was proved relatively recently [3]. The polynomial solvability of this problem on a line was proved in [4] in the last century. The cited paper presents an algorithm with $\mathcal{O}(KN^2)$ running time that implements a dynamic programming scheme. This well-known algorithm relies on an exact polynomial algorithm for solving the well-known Nearest neighbor search problem [5]. Note that the polynomial solvability in $\mathcal{O}(KN \log N)$-time of the 1D case of the K-Means problem follows directly from earlier (than [4]) results obtained in [6–9]. In the cited papers, the authors have proved the faster polynomial-time algorithms for some special cases of the Nearest neighbor search problem. Nevertheless, in recent years, for the one-dimensional case of the K-Means problem, some new exact algorithms with $\mathcal{O}(KN \log N)$ running time have been constructed. An overview of these algorithms and their properties can be found in [10,11].

The object of our research is the following problem that is close in its formulation to K-Means and is poorly studied.

Problem 1 (K-Means and Given J-Centers). Given an N-element set $\mathcal{Y}$ of points in d-dimension Euclidean space, a positive integer K, and a tuple $\{c_1, \ldots, c_J\}$ of points. Find a partition of $\mathcal{Y}$ into non-empty clusters $\mathcal{C}_1, \ldots, \mathcal{C}_K$, $\mathcal{D}_1, \ldots, \mathcal{D}_J$ such that

$$F = \sum_{k=1}^{K} \sum_{y \in \mathcal{C}_k} \|y - \overline{y}(\mathcal{C}_k)\|^2 + \sum_{j=1}^{J} \sum_{y \in \mathcal{D}_j} \|y - c_j\|^2 \rightarrow \min,$$

where $\overline{y}(\mathcal{C}_k)$ is the centroid of the k-st cluster.

On the one hand, Problem 1 may be considered as some modification of K-Means. On the other hand, the introduced notation allows us to call Problem 1 as K-*Means and Given J-Centers*.

Unlike K-*Means*, Problem 1 models an applied clustering problem in which for a part of clusters (i.e., for $\mathcal{D}_1, \ldots, \mathcal{D}_J$) the quadratic scatter data centers (i.e., $c_1, \ldots, c_J$) are known in advance, i.e., they are given as input instance. This applied problem is also typical for Data analysis, Data mining, Pattern recognition, and Data processing. In particular, the two-cluster Problem 1, i.e., 1-*Mean and Given* 1-*Center*, is related to the solution of the applied signal processing problem. Namely, this two-clusters problem is related with the problem of joint detecting a quasi-periodically repeated pulse of unknown shape in a pulse train and evaluating this shape under Gaussian noise with given zero value (see [12–14]). In this two-cluster Problem 1, the zero mean corresponds to the cluster with the center specified at the origin. Apparently, the first mention has been made in [12] on this two-cluster Problem 1. It should be noted that simpler optimization problems induced by the applied problems of noise-proof detection and discrimination of impulses of specified shapes are typical, in particular, for radar, electronic reconnaissance, hydroacoustics, geophysics, technical and medical diagnostics, and space monitoring (see, for example, [15–17]).

Problem 1 strong NP-hardness was proved in [18–20]. Note that the K-*Means* problem is not equivalent to Problem 1 and is not a special case of it. Therefore, the solvability of Problem 1 in the 1D case requires independent study. This question until now remained open.

The main result of this paper is the proof of Problem 1 polynomial solvability in the one-dimensional case.

3   Some Auxiliary Statements: Properties of the Problem 1 Optimal Solution in the 1D Case

In what follows, we assume that $d = 1$. Below we will call by Problem 1D the one-dimensional case of Problem 1.

Our proof is based on the few given below auxiliary statements, which reveal the structure of Problem 1D optimal solution. For briefness, we present these statements without proofs, limiting ourselves to the presentation of their ideas.

Denote by $\mathcal{C}_1^*, \ldots, \mathcal{C}_K^*, \mathcal{D}_1^*, \ldots, \mathcal{D}_J^*$ the optimal clusters in Problem 1D.

Lemma 1. *If in Problem 1D $c_m < c_\ell$, where $1 \leq m \leq J$, $1 \leq \ell \leq J$, then for each $x \in \mathcal{D}_m^*$ and $z \in \mathcal{D}_\ell^*$ the inequality $x \leq z$ holds.*

Lemma 2. *If in Problem 1D $\overline{y}(\mathcal{C}_m^*) < \overline{y}(\mathcal{C}_\ell^*)$, where $1 \leq m \leq K$, $1 \leq \ell \leq K$, then for each $x \in \mathcal{C}_m^*$ and $z \in \mathcal{C}_\ell^*$ the inequality $x \leq z$ holds.*

Lemma 3. *For an optimal solution of Problem 1D, the following statements are true:*

(1) *If $\overline{y}(\mathcal{C}_m^*) < c_\ell$, where $1 \leq m \leq K$, $1 \leq \ell \leq J$, then for each $x \in \mathcal{C}_m^*$ and $z \in \mathcal{D}_\ell^*$ the inequality $x \leq z$ holds.*

(2) If $\overline{y}(\mathcal{C}_m^*) > c_\ell$, where $1 \leq m \leq K$, $1 \leq \ell \leq J$, then for each $x \in \mathcal{C}_m^*$ and $z \in \mathcal{D}_\ell^*$ the inequality $x \geq z$ holds.

The proof of Lemmas 1–3 is carried out by the contrary method using the following equality

$$(x - c_m)^2 + (z - c_\ell)^2 = 2(x - z)(c_\ell - c_m) + (z - c_m)^2 + (x - c_\ell)^2.$$

The validity of this equality follows from the well-known formula for the sum of squares of the trapezoid diagonals.

Lemma 4. *In Problem 1D, for each $k \in \{1, \ldots, K\}$ and $j \in \{1, \ldots, J\}$ it is true that $\overline{y}(\mathcal{C}_k^*) \neq c_j$.*

Lemma 5. *In Problem 1D, for each $k, j \in \{1, \ldots, K\}$, $k \neq j$, it is true that $\overline{y}(\mathcal{C}_k^*) \neq \overline{y}(\mathcal{C}_j^*)$.*

The proof of Lemmas 4 and 5 is carried out by the contrary method.

Lemmas 1–5 establish the relative position of the optimal clusters $\mathcal{D}_1^*, \ldots, \mathcal{D}_J^*$ and $\mathcal{C}_1^*, \ldots, \mathcal{C}_K^*$ on a line. These lemmas are the base of the following statement.

Theorem 1. *Let in Problem 1D points $y_1, \ldots, y_N$ of $\mathcal{Y}$, and points $c_1, \ldots, c_J$ be ordered so that*

$$y_1 < \ldots < y_N,$$
$$c_1 < \ldots < c_J.$$

Then optimal partition of $\mathcal{Y}$ into clusters $\mathcal{C}_1^*, \ldots, \mathcal{C}_K^*, \mathcal{D}_1^*, \ldots, \mathcal{D}_J^*$ corresponds to a partition of the positive integer sequence $1, \ldots, N$ into disjoint segments.

4   Polynomial Solvability of the Problem in the 1D Case

The following theorem is the main result of the paper.

Theorem 2. *There exists an algorithm that finds an optimal solution of Problem 1D in polynomial time.*

Our proof of Theorem 1 is constructive. Namely, we justify an algorithm that implements a dynamic programming scheme and allows one to find an exact solution of Problem 1D in $\mathcal{O}(KJN^2)$ time.

The idea of the proof is as follows. Without loss of generality, we assume that the points $y_1, \ldots, y_N$ of $\mathcal{Y}$, as well as the points $c_1, \ldots, c_J$ are ordered as in Theorem 1.

Let $\mathcal{Y}_{s,t} = \{y_s, \ldots, y_t\}$, where $1 \leq s \leq t \leq N$, be a subset of $t - s + 1$ points of $\mathcal{Y}$ with numbers from s to t.

Let

$$f_{s,t}^j = \sum_{i=s}^{t} (y_i - c_j)^2, \quad j = 1, \ldots, J,$$

$$f_{s,t} = \sum_{i=s}^{t}(y_i - \bar{y}(\mathcal{Y}_{s,t}))^2,$$

where $\bar{y}(\mathcal{Y}_{s,t})$ is the centroid of the subset $\mathcal{Y}_{s,t}$.

We prove that the optimal value of the Problem 1 objective function is found by the following formula

$$F^* = F_{K,J}(N),$$

and the values

$$F_{k,j}(n), \quad k = -1, 0, 1, \ldots, K; \quad j = -1, 0, 1, \ldots, J; \quad n = 0, \ldots, N,$$

are calculated by the recurrent formulas. The formula

$$F_{k,j}(n) = \begin{cases} 0, & \text{if } n = k = j = 0; \\ +\infty, & \text{if } n = 0; \ k = 0, \ldots, K; \ j = 0, \ldots, J; \ k+j \neq 0; \\ +\infty, & \text{if } k = -1; \ j = -1, \ldots, J; \ n = 0, \ldots, N; \\ +\infty, & \text{if } j = -1; \ k = -1, \ldots, K; \ n = 0, \ldots, N; \end{cases} \quad (1)$$

sets the initial and boundary conditions for subsequent calculations. Formula (1) follows from the properties of the optimal solution. The basic formula

$$F_{k,j}(n) = \min\left\{ \min_{i=1}^{n}\left\{ F_{k-1,j}(i-1) + f_{i,n} \right\}, \ \min_{i=1}^{n}\left\{ F_{k,j-1}(i-1) + f_{i,n}^j \right\} \right\},$$
$$k = 0, \ldots, K; \ j = 0, \ldots, J; \ n = 1, \ldots, N, \quad (2)$$

defines recursion. In general, the formulas (1), (2) implement the forward algorithm.

Further, we have proved that the optimal clusters $\mathcal{C}_1^*, \ldots, \mathcal{C}_K^*, \mathcal{D}_1^*, \ldots, \mathcal{D}_J^*$ may be found using the following recurrent rule, that implements the backward algorithm.

The step-by-step rule looks as follows:

Step 0. $k := K$, $j := J$, $n := N$.
Step 1. If

$$\min_{i=1}^{n}\left(F_{k-1,j}(i-1) + f_{i,n} \right) \leq \min_{i=1}^{n}\left(F_{k,j-1}(i-1) + f_{i,n}^j \right),$$

then

$$\mathcal{C}_k^* = \{y_{i^*}, y_{i^*+1}, \ldots, y_n\},$$

where

$$i^* = \arg\min_{i=1}^{n}\left(F_{k-1,j}(i-1) + f_{i,n} \right);$$

$k := k - 1$; $n := i^* - 1$.
If, however,

$$\min_{i=1}^{n}\left(F_{k-1,j}(i-1) + f_{i,n} \right) > \min_{i=1}^{n}\left(F_{k,j-1}(i-1) + f_{i,n}^j \right),$$

then
$$\mathcal{D}_j^* = \{y_{i^*}, y_{i^*+1}, \dots, y_n\},$$

where
$$i^* = \arg\min_{i=1}^{n}\left(F_{k,j-1}(i-1) + f_{i,n}^j\right);$$

$j := j - 1;\ n := i^* - 1.$

Step 2. If $k > 0$ or $j > 0$, then go to Step 1; otherwise — the end of calculations.

The validity of this rule we have proved by induction.

Finally, we have proved that the running time of the algorithm is $\mathcal{O}(KJN^2)$, that is, the algorithm is polynomial. The algorithms running time is defined by the complexity of implementation of formula (2). This formula is calculated $\mathcal{O}(KJN)$ times and every calculation of $F_{k,j}(n)$ requires $\mathcal{O}(N)$ operations.

5   Conclusion

In the present paper, we have proved the polynomial solvability of the one-dimensional case of one strongly NP-hard problem of partitioning a finite set of points in Euclidean space. The construction of approximate efficient algorithms with guaranteed accuracy bounds for the general case of Problem 1 and faster polynomial-time exact algorithms for the 1D case of this problem seems to be the directions of future studies.

Acknowledgments. The study presented in Sects. 3 and 4 was supported by the Russian Foundation for Basic Research, projects 19-01-00308 and 18-31-00398. The study presented in the other sections was supported by the Russian Academy of Science (the Program of basic research), project 0314-2019-0015, and by the Russian Ministry of Science and Education under the 5-100 Excellence Programme.

References

1. Fisher, R.A.: Statistical Methods and Scientific Inference. Hafner, New York (1956)
2. MacQueen, J.B.: Some methods for classification and analysis of multivariate observations. In: Proceedings of the 5th Berkeley Symposium on Mathematical Statistics and Probability, vol. 1, pp. 281–297. University of California Press, Berkeley (1967)
3. Aloise, D., Deshpande, A., Hansen, P., Popat, P.: NP-hardness of euclidean sum-of-squares clustering. Mach. Learn. **75**(2), 245–248 (2009)
4. Rao, M.: Cluster analysis and mathematical programming. J. Am. Stat. Assoc. **66**, 622–626 (1971)
5. Bellman, R.: Dynamic Programming. Princeton University Press, Princeton (1957)
6. Glebov, N.I.: On the convex sequences. Discrete Anal. **4**, 10–22 (1965). in Russian
7. Gimadutdinov, E.K.: On the properties of solutions of one location problem of points on a segment. Control. Syst. **2**, 77–91 (1969). in Russian
8. Gimadutdinov, E.K.: On one class of nonlinear programming problems. Control. Syst. **3**, 101–113 (1969). in Russian

9. Gimadi (Gimadutdinov) E.Kh.: The choice of optimal scales in one class of location, unification and standardization problems. Control. Syst. **6**, 57–70 (1970). in Russian

10. Wu, X.: Optimal quantization by matrix searching. J. Algorithms **12**(4), 663–673 (1991)

11. Grønlund, A., Larsen, K.G., Mathiasen, A., Nielsen, J.S., Schneider, S., Song, M.: Fast exact k-means, k-medians and Bregman divergence clustering in 1D. CoRR arXiv:1701.07204 (2017)

12. Kel'manov, A.V., Khamidullin, S.A., Kel'manova, M.A.: Joint finding and evaluation of a repeating fragment in noised number sequence with given number of quasiperiodic repetitions. In: Book of Abstract of the Russian Conference "Discrete Analysis and Operations Research" (DAOR-4), p. 185. Sobolev Institute of Mathematics SB RAN, Novosibirsk (2004)

13. Gimadi, E.K., Kel'manov, A.V., Kel'manova, M.A., Khamidullin, S.A.: A posteriori detection of a quasi periodic fragment in numerical sequences with given number of recurrences. Sib. J. Ind. Math. **9**(1(25)), 55–74 (2006). in Russian

14. Gimadi, E.K., Kel'manov, A.V., Kel'manova, M.A., Khamidullin, S.A.: A posteriori detecting a quasiperiodic fragment in a numerical sequence. Pattern Recogn. Image Anal. **18**(1), 30–42 (2008)

15. Kel'manov, A.V., Khamidullin, S.A.: Posterior detection of a given number of identical subsequences in a guasi-periodic sequence. Comput. Math. Math. Phys. **41**(5), 762–774 (2001)

16. Kel'manov, A.V., Jeon, B.: A posteriori joint detection and discrimination of pulses in a quasiperiodic pulse train. IEEE Trans. Sig. Process. **52**(3), 645–656 (2004)

17. Carter, J.A., Agol, E., et al.: Kepler-36: a pair of planets with neighboring orbits and dissimilar densities. Science **337**(6094), 556–559 (2012)

18. Kel'manov, A.V., Pyatkin, A.V.: On the complexity of a search for a subset of "similar" vectors. Dokl. Math. **78**(1), 574–575 (2008)

19. Kel'manov, A.V., Pyatkin, A.V.: On a version of the problem of choosing a vector subset. J. Appl. Ind. Math. **3**(4), 447–455 (2009)

20. Kel'manov, A.V., Pyatkin, A.V.: Complexity of certain problems of searching for subsets of vectors and cluster analysis. Comput. Math. Math. Phys. **49**(11), 1966–1971 (2009)

How to Use Boltzmann Machines and Neural Networks for Covering Array Generation

Ludwig Kampel[1], Michael Wagner[1], Ilias S. Kotsireas[2],
and Dimitris E. Simos[1(✉)]

[1] SBA Research, 1040 Vienna, Austria
{lkampel,mwagner,dsimos}@sba-research.org
[2] Wilfrid Laurier University, Waterloo, ON, Canada
ikotsire@wlu.ca

Abstract. In the past, combinatorial structures have been used only to tune parameters of neural networks. In this paper, we employ for the first time, neural networks and Boltzmann machines for the construction of covering arrays (CAs). In past works, Boltzmann machines were successfully used to solve set cover instances. For the construction of CAs, we consider the equivalent set cover instances and use Boltzmann machines to solve these instances. We adapt an existing algorithm for solving general set cover instances, which is based on Boltzmann machines and apply it for CA construction. Furthermore, we consider newly designed versions of this algorithm, where we consider structural changes of the underlying Boltzmann machine, as well as a version with an additional feedback loop, modifying the Boltzmann machine. Last, one variant of this algorithm employs learning techniques based on neural networks to adjust the various connections encountered in the graph representation of the considered set cover instances. Supported by an experimental evaluation our findings can act as a beacon for future applications of neural networks in the field of covering array generation and related discrete structures.

Keywords: Neural networks · Boltzmann machines · Covering arrays

1 Introduction

Various approaches have been applied for the construction and optimization of CAs, for a survey see for example [11]. Artificial Neural Networks (ANN) have been applied successfully in various fields of computer science, especially in optimization [10], and recently significant effort has been spent to replicate the decisions of human experts using artificial intelligence [9]. The covering array generation problem is known to be tightly coupled with hard combinatorial optimization problems, see [5] and references therein.

In this paper, to the best of our knowledge, we employ for the very first time neural network models based on Boltzmann machines (BM) to optimization of

© Springer Nature Switzerland AG 2020
N. F. Matsatsinis et al. (Eds.): LION 13 2019, LNCS 11968, pp. 53–68, 2020.
https://doi.org/10.1007/978-3-030-38629-0_5

CA generation, i.e. the construction of CAs with a small number of rows. We would like to note that constructing a BM architecture corresponding to a particular optimization problem is a difficult task, since one has to construct the topology of the architecture to reflect the particular nature of a given problem. This difficulty has been first noted in [1], where the authors tried to combine neural computation with combinatorial optimization for a different class of problems.

This paper is structured as follows. In Sect. 2 we give the preliminaries needed for this paper. Next, in Sect. 3 we describe how Boltzmann machines can be used for the construction of covering arrays. In Sect. 4 this approach is extended by means of allowing for different learning strategies. Finally, in Sect. 5 the results of our experiments are documented while Sect. 6 concludes our work.

2   Preliminaries

In this section we give the definitions needed in this paper. Related notions for covering arrays and set covers can be also found in [4] while the needed definitions regarding graphs can be also found in [12].

Definition 1. *A* covering array $\mathsf{CA}(N; t, k, v)$ *is an* $N \times k$ *array* $(\mathbf{c}_1, \dots, \mathbf{c}_k)$ *with the properties that for all* $j \in \{1, \dots, k\}$ *the values in the j-th column* $\mathbf{c}_j$ *belong to the set* $\{0, \dots, v-1\}$, *and for each selection* $\{\mathbf{c}_{j_1}, \dots, \mathbf{c}_{j_t}\} \subseteq \{\mathbf{c}_1, \dots, \mathbf{c}_k\}$ *of t different columns, the subarray that is comprised by the columns* $\mathbf{c}_{j_1}, \dots, \mathbf{c}_{j_t}$, *has the property that every t-tuple in* $\{0, \dots, v-1\}^t$ *appears at least once as a row. The smallest integer N for which a* $\mathsf{CA}(N; t, k, v)$ *exists is called the* covering array number *for t, k, v and is denoted as* $\mathsf{CAN}(t, k, v)$. *Covering Arrays achieving this bound are called* optimal.

In this work we only consider binary CAs, i.e. CAs over the alphabet $\{0, 1\}$, which we denote as $\mathsf{CA}(N; t, k)$. For given t and k we also say we are given a *CA instance*, when we want to construct a $\mathsf{CA}(N; t, k)$. An example of a $\mathsf{CA}(4; 2, 3, 2)$ is given by the array A in relation (1).

Definition 2. *For positive integers t, k and v, a* t-way interaction *is a set* $\{(p_1, x_1), \dots, (p_t, x_t)\}$ *with the property that* $x_i \in \{0, \dots, v-1\}$, $\forall i \in \{1, \dots, t\}$ *and* $1 \le p_1 < \dots < p_t \le k$.

We represent t-way interactions as vectors of length k with t positions specified and the others unspecified (see Example 1). With this notion CAs can be characterized as arrays which rows cover all t-way interactions for given t and k.

Definition 3. *A* set cover *(SC) of a finite set U is a set $\mathcal{S}$ of non-empty subsets of U whose union is U. In this context, U is called the* universe *and the elements of $\mathcal{S}$ the* blocks.

A typical optimization problem for set covers is the *minimal set cover problem*. That is, for given $(U, \mathcal{S})$, to find a subset $\mathcal{C} \subseteq \mathcal{S}$ of minimal cardinality, such that $\bigcup \mathcal{C} = U$. We call $(U, \mathcal{S})$ also an *SC instance*.

Definition 4. *For a graph $G = (V, E)$ (in this work we only consider undirected graphs) with vertex set V and edges $E \subseteq V \times V$, a* vertex cover *is a subset C of V, such that each edge in E is incident to at least one vertex of C. An* independent set *of G is a set of vertices $I \subseteq V$, such that no two vertices in I are adjacent, i.e. $I \times I \cap E = \emptyset$.*

Finally, we also consider the concept of Boltzmann machines, see also [1].

Definition 5. *A Boltzmann machine (BM) is a (stochastic) neural network, with an underlying undirected graph $G = (V, E)$. The neurons correspond to the vertex set V and can be in two states, either* on *or* off. *The edges of the graph correspond to the connections (synapses) between the neurons, i.e. the edge $\{u_i, u_j\} \in E$ represents the symmetric connection between neuron u_i and u_j. A Boltzmann machine M is now defined as a pair (G, Ω), where $\Omega \subseteq \mathbb{R}^E \times \{0, 1\}^V$ is a set of allowable states. For $\omega = (w_{e_1}, w_{e_2}, \dots, w_{e_{|E|}}, u_{v_1}, \dots, u_{v_{|V|}})$ the vector $w = (w_{e_1}, w_{e_2}, \dots, w_{e_{|E|}})$ describes the weight w_e of each edge $e \in E$, and $\kappa = (u_{v_1}, \dots, u_{v_{|V|}})$ describes for all neurons $v_i \in V$ if it is* on *($u_{v_i} = 1$), or* off *($u_{v_i} = 0$). The* consensus function *of the Boltzmann machine M is the function $F : \Omega \to \mathbb{R}$, defined by $F(\kappa) = \sum_{\{i,j\} \in E} w_{\{i,j\}} u_i u_j$.*

3   A Boltzmann Machine for Covering Arrays

In this section we set up a Boltzmann machine that reflects a given CA instance. To this extent, we first explain how the problem of generating a CA can be interpreted as a set cover problem, following [4]. In a second step we recapitulate the work of [3], where Boltzmann machines where successfully used to compute solutions to set cover problems. Our aim in this work is to combine these works so that we can use Boltzmann machines to compute covering arrays.

3.1   Encoding CA Problems as Set Cover Problems

Next, we explain how to interpret the problem of computing a CA as an SC problem. This connection has been explained in an extensive way, for example, in [4], where the interested reader is referred to for the details. Here we content ourself with repeating the key ideas, guided by means of an example. When we want to construct a CA for given strength t and number of columns k, this can be interpreted as an SC instance $(U, \mathcal{S})$, where the universe U consists of all t-way interactions. Each block in $\mathcal{S}$ corresponds to a row that can appear in a CA and is defined as the set of t-way interactions this row covers. To make this connection more explicit, we review Example 3.3 of [4]:

Example 1. Assume we want to construct a $\mathsf{CA}(N; 2, 3, 2)$ with minimal N. We translate this problem into a minimal set cover problem. Each 2-way interaction needs to be covered, thus $U = \{(0, 0, -), (0, 1, -), (1, 0, -), (1, 1, -), (0, -, 0), (0, -, 1), (1, -, 0), (1, -, 1), (-, 0, 0), (-, 0, 1), (-, 1, 0), (-, 1, 1)\}$. Each vector of $\{0, 1\}^3$ which can appear as a row in a $\mathsf{CA}(N; 2, 3, 2)$ is identified with the set of 2-way interactions it covers, e.g. the row $(0, 0, 1)$ is mapped to the block

$\{(0,0,-),(0,-,1),(-,0,1,)\}$. Thus we get the SC instance $(U,\mathcal{S})$ corresponding to the CA instance with parameters $t = 2$ and $k = 3$, where

$$
\begin{aligned}
\mathcal{S} = \{ & \{(0,0,-),(0,-,0),(-,0,0,)\},\{(0,0,-),(0,-,1),(-,0,1,)\}, \\
& \{(0,1,-),(0,-,0),(-,1,0,)\},\{(0,1,-),(0,-,1),(-,1,1,)\}, \\
& \{(1,0,-),(1,-,0),(-,0,0,)\},\{(1,0,-),(1,-,1),(-,0,1,)\}, \\
& \{(1,1,-),(1,-,0),(-,1,0,)\},\{(1,1,-),(1,-,1),(-,1,1,)\}\}.
\end{aligned}
$$

Provided this correspondence, it is therefore possible to map the minimal
set cover $\mathcal{C} = \{\{(0,0,-),(0,-,0),(-,0,0,)\},\{(0,1,-),(0,-,1),(-,1,1,)\},\{(1,0,-),(1,-,1),(-,0,1,)\},\{(1,1,-),(1,-,0),(-,1,0,)\}\}$ of $(U,\mathcal{S})$ to the optimal CA$(4;2,3,2)$

$$
A = \begin{pmatrix} 0 & 0 & 0 \\ 0 & 1 & 1 \\ 1 & 0 & 1 \\ 1 & 1 & 0 \end{pmatrix}. \tag{1}
$$

3.2 Boltzmann Machines for Set Cover Problems

We now give an overview of how SC instances can
be solved with Boltzmann machines, following the
work presented in [3], which serves us as a point
of reference. A high level view of the procedure we
follow can be found in Fig. 1.

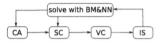

Fig. 1. An overview.

In the following paragraphs we make explicit the connections between set covers, vertex covers and independent sets. Consider a given SC instance $(U,\mathcal{S})$, we
can construct an edge labelled graph $G_S = (V, E, \ell)$ representing this instance,
as follows. The node set V is defined to be the set of blocks $\mathcal{S}$, such that each
block is represented by a node of the graph. For the set of (undirected) edges
E, we define $E := \{\{S_i, S_j\} | S_i \cap S_j \neq \emptyset\}$ and the labelling function of the edges
$\ell : E \to \mathcal{P}(U) : \{S_i, S_j\} \mapsto S_i \cap S_j$, i.e. we label each edge with the set of
elements of U its adjacent vertices cover in common. We call these labels also
label sets. At this point we would like to remark, that we can assume without
loss of generality, that each element of the universe U appears in at least two
blocks and hence in at least one label set.

Assume now we are given a vertex cover $\mathcal{V} = \{S_1, \ldots, S_r\}$ of G_S, then $\mathcal{V}$ represents already a set cover of U. This holds, since the vertices (i.e. sets) $S_1, \ldots, S_r$
cover all edges of G_S, the labels of which are subsets of the S_i and already cover
the whole universe U. Further in [3] reduced graphs G'_S are considered, where
for each element $u \in U$ *exactly one* edge of $E(G_S)$ is allowed to keep u in its
label set. Hence, a vertex cover of the reduced graph still constitutes a set cover
of $(U,\mathcal{S})$, see Proposition 1 of [3]. Generalizing this approach, in our work we
consider reduced graphs $G'_S = (V, E(G'_S))$, where for each $u \in U$ *at least* one
edge of $E(G'_S)$ has u in its label set. We thus maintain the property that a vertex
cover of a reduced graph G'_S constitutes a set cover of U. We give an example

of these two different types of reduced graphs in Example 3 and Fig. 3. Considering that a vertex cover C of a graph $G = (V, E)$ is exactly the complement of an independent set $V \setminus C$ of G (see also Remark 1 of [3]), the analogue of Proposition 1 of [3] also holds for reduced graphs as considered in this work:

Proposition 1. *The complement of a maximal independent set of a reduced graph G'_S (where each element of U appears in the label of at least one edge) is a set cover for (U, S).*

Sketch of Proof. The complement of an independent set is a vertex cover. A vertex cover contains vertices such that each edge is incident to at least one vertex. The label set of an edge is a subset of the sets corresponding to its adjacent vertices. Since each element of U appears in at least one label of an edge of the reduced graph G'_S, the sets corresponding to the nodes of the vertex cover are a set cover of U. □

Before we describe how Boltzmann machines can be used to find independent sets of graphs which yield set covers, respectively CAs for our purpose, we fix some notations and consider some examples. We use the notation $G_{t,k}$ for the graph that corresponds to the set cover instance (U, S), which corresponds again to a CA instance for given t and k, and call it the *underlying graph* of the CA instance.

Example 2. Continuing Example 1 the graph $G_{2,3}$ corresponding to the set cover (U, S) is depicted in Fig. 2. Although self-contained, graph $G_{2,3}$ is not very representative for the general CA instances we have to deal with, since there are exactly two rows that share a 2-way interaction, there is a one-to-one correspondence between edges and 2-way interactions. This is also the reason why we omitted the set notation for the labels, as label sets are singletons in this case.

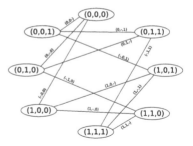

Fig. 2. $G_{2,3}$ underlying the CA instance $t = 2$, $k = 3$.

Example 3. To give a better impression on the instances we have to deal with, we also give an example of a partial graph in Fig. 3, where a subgraph of the graph $G_{2,5}$ is given. Labels occur right next to the edges they belong to. The green coloured 2-way interactions are selected to reside in the label set of the edges. In the middle of Fig. 3 we depict the reduced graph $G'_{2,5}$ as it is described in [3]. Rightmost we show the reduced graph $G'_{2,5}$ as we consider them in this work.

3.3   Computing CAs with Boltzmann Machines

Following [3], the neural network is constructed isomorphic to the graph G'_S, where neurons correspond to vertices and synapses correspond to edges. Each neuron S_i is in state u_i, which can be either *on* or *off*, represented by $u_i \in \{0, 1\}$. Synapses, i.e. edges of G'_S, are assigned weights according to Eq. (2). At any time the configuration of the network is defined by the vector of all states $(u_i)_{i \in V}$ of the neurons. For any configuration $\kappa = (u_i)_{i \in V}$ the consensus $F(\kappa)$ is defined as the following quadratic function of the states of the neurons, and the weights of the synapses: $F(\kappa) = \sum_{i,j} \omega(e_{i,j}) u_i u_j$. The defined weights (loops $e_{i,i}$ have positive, other edges have negative weights) and the consensus function F reflect the target of finding a maximal set of vertices, that are not adjacent.[1] Respectively, we want to activate as many non-connected neurons as possible. In fact local maxima of F correspond to maximal (not necessary *maximum*) independent sets of the graph G'_S, which in turn yield set covers, considering the complement on G_S. We refer the interested reader to [3] for more details.

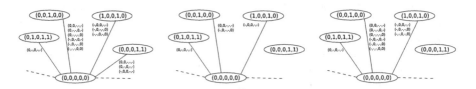

Fig. 3. From left to right: subgraph of the underlying graph of the CA instance $t = 2$, $k = 5$; a subgraph of a reduced graph as it can occur due to the method described in [3]; a subgraph of the reduced graph as consider in our work.

Remark 1. To better illustrate the connections between the different structures, we give an overview of the introduced concepts and notions as follows:

- Rows of CAs correspond to blocks of SCs, which are further mapped to vertices of BMs. These serve as neurons for the devised neural network.
- Analogue t-way interactions correspond to elements of the universe in terms of set covers. These serve as labels of edges that define the weight of the synapses of the devised neural network.

In Algorithm 2 we give a high level description of the algorithm developed in [3], modified in this work in order to be applied to CA instances. The initial weights of the edges (line 3) are set according to Eq. (2). Thereafter a simulated annealing (SA) algorithm is run on the Boltzmann Machine to find a local maximum of the consensus F, where initially all synapses are in state *off* (i.e. the state vector κ equals the all zero vector). A pseudo code of such a simulated annealing algorithm is given in Algorithm 1, taking as input a graph G with weights for the edges, denoted by $\omega(G)$. Further, a starting temperature T_0, a final temperature T_f and a factor α for the cooling schedule is required. In each step a random neuron is selected to change its state. In case the change in the consensus function $\Delta F(\kappa) = (1 - 2u_i)(\omega(e_{ii}) + \sum_j \omega(e_{ij}) u_j)$ is positive the change in state u_i is accepted, otherwise it is refuted with probability $(1 - 1/(1 + \exp(-\Delta F/T))$.

---

[1] Note that we do not consider vertices as being adjacent to themselves by their loops. In this work we rather use loops to represent the weight of vertices.

The cooling schedule of the SA algorithm is based on the schedule developed by Lundy and Mees [6]. In this cooling schedule it is required that only one iteration is executed at each temperature. In particular, we implemented the cooling schedule (line 11) according to the recursive function $T_{n+1} = T_n/(1 + \alpha T_n)$ where n is the iteration number and α, a small positive value close to zero depending on the instance, that allows for fast convergence.

We describe our algorithmic design using a variety of building blocks. In this way, a compact presentation of the devised algorithms is ensured and also flexibility in terms of their evaluation which is presented in Sect. 5.

The first building block introduced is that of INITIALGRAPH, which is a procedure that transforms the underlying graph $G_{t,k}$ to a subgraph $G'_{t,k}$, to which the Boltzmann machine is reduced and on which the simulated annealing algorithm runs. In all presented versions we instantiate this building block with a randomized procedure, which selects for each t-way interaction a random edge of $E(G_{t,k})$, such that the t-way resides in the label set of the edge. Edges selected this way reside in $G'_{t,k}$ and keep all their labels, where Edges that get not selected are deleted. See also Example 3.

The second building block used to devise our algorithms is that of INITIAL-WEIGHT, which is a procedure that assigns a weight to each edge of G'_S. One way considered to instantiate this building block is via BMWEIGHT, assigning the weights as described in [3]

$$\omega(e_{ij}) = \begin{cases} -(\max\{1/|S_i|, 1/|S_j|\} + \epsilon), & i \neq j \\ 1/|S_i|, & i = j \end{cases} \tag{2}$$

Note, that this weighting comes down to a uniform weighting, as $|S_i| = |S_j|$ for all i, j when considering CA instances.

Using these algorithmic building blocks, we can describe the algorithm of [3] applied to CA instances as an instance of Algorithm 2, Instantiating INITIALGRAPH with RANDOMGRAPH and INITIALWEIGHTS according to (2). The simulated annealing algorithm is run once, to find a maximal independent set I on $G'_{t,k}$, the complement of which is returned and constitutes a CA.

Combining the reductions of CAs to SCs and SCs to independent sets on reduced graphs G'_S (Fig. 1), it is possible to state the following corollary of the works presented in [4] and [3] (Theorem 1), which proves the correctness of the previously described algorithm.

Corollary 1. *Maxima of F induce configurations of the BM-network corresponding to Covering Arrays.*

Algorithm 2 serves as a base line for the development of our own learning algorithms, which we describe in the next section.

4   Finding Covering Arrays with Neural Networks

In this section, we describe the algorithms we have devised to find CAs with neural networks. We start with the algorithm described in [3], which serves as a starting point for the development of the algorithms presented in this work, where we incrementally built upon this baseline algorithm and extend it with various features. In detail, as a first extension we consider a weight update for the underlying BM and a second extension introduces a notion of graph update for the underlying graph (which is part of the BM).

Algorithm 1. SA

1: INPUT: $G, \omega(G)$
Require: T_0, T_f, α
2: $T \leftarrow T_0$, $\kappa \leftarrow \mathbf{0}$ ▷ κ is the vector of states
3: **while** $T > T_f$ **do**
4:     randomly choose neuron S_i
5:     change state: $u_i \leftarrow u_i + 1(mod\ 2)$
6:     **if** $\Delta F(\kappa) > 0$ **then**
7:         keep κ
8:     **else**
9:         with probability $1/(1 + \exp(-\Delta F/T))$ keep κ
10:     **end if**
11:     $T \leftarrow T/(1 + \alpha T)$
12: **end while**
13: **return** κ

Algorithm 2. BMFORCA

1: INPUT: t, k
Require: ϵ
2: $G'_{t,k} \leftarrow$ INITIALGRAPH$(G_{t,k})$
3: $\omega(G'_{t,k}) \leftarrow$ INITIALWEIGHT$(G'_{t,k}, \epsilon)$ ▷ Assign weights
4: $I \leftarrow$ SA$(G'_{t,k}, \omega(G'_{t,k}))$
5: **return** CA$(|V| - |I|; t, k, 2) = V \setminus I$

Before we describe the algorithmic extensions we made to Algorithm 2, we would like to mention that our initial experiments with Algorithm 2 were not satisfactory. Due to the good experimental results in [3], reporting to find smaller SCs than other heuristics, we expected that Algorithm 2 would produce CAs with a small number of rows. Yet, our objective is still not achieved, i.e. having a learning algorithm capable of further reducing the number of rows in a CA. We believe that this is the case, since the approach of finding small set covers as the complements of large independent sets of vertices on corresponding graphs is badly suited for graphs that have a relative high *density*, i.e. on average, vertices are highly connected. It seems that the condition of finding an independent set of nodes on the reduced graph G'_S, is too strong of a sufficient condition to actually find small SCs for such instances. To illustrate this we give the following example of a general SC instance, which is highly connected and the algorithm described in [3] is badly suited.

Example 4. Consider the following set cover instance $(U, \mathcal{S})$, $U = \{a, b, c, \ldots, k, l, m\}$ and $\mathcal{S} = \{S_1, \ldots, S_{13}\}$ where $a \in S_i$ for all $i = 1, \ldots,$ 13 and $S_1 = \{a, m\}, S_2 = \{a, m, b, c\}, S_3 = \{a, b, c, d\}, S_4 = \{a, c, d, e\}, S_5 = \{a, d, e, f\}, S_6 = \{a, e, f, g\}, S_7 = \{a, f, g, h\}, S_8 = \{a, g, h, i\}, S_9 = \{a, h, i, j\}, S_{10} = \{a, i, j, k\}, S_{11} = \{a, j, k, l\}, S_{12} = \{a, k, l, m\}, S_{13} = \{a, l, m, b\}$. The

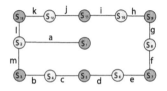

graph representing the set cover instance $(U, \mathcal{S})$ is the complete graph with 13 vertices. In the figure above we give an example of a reduced graph of this set cover instance, in the sense of [3], i.e. each element of the universe appears as a label of exactly one edge. A maximal (even maximum) independent set of nodes can be identified as $I = \{S_1, S_3, S_5, S_7, S_9, S_{11}, S_{13}\}$. Then the complement $\mathcal{C} = \{S_2, S_4, S_6, S_8, S_{10}, S_{12}\}$ constitutes a minimal vertex cover of this graph, and hence $\mathcal{C}$ is a cover of the universe U. Though, $\mathcal{C}$ is not a minimal set cover, since $\mathcal{C}' = \{S_2, S_5, S_8, S_{11}\}$ also constitutes a cover of U of smaller size. In fact it is not hard to see that $\mathcal{C}'$ is a minimal set cover of $(U, \mathcal{S})$.

We take this example as a further motivation to modify the approach of [3], towards relaxing the target of finding an independent set on the reduced graph G_S'. Finding independent sets is encoded in the consensus function F (as introduced in Subsect. 3.3), that characterizes independent sets through local maxima, as long as weights of vertices are positive, and the weights of the edges are smaller than the negative vertex weights. Using the same consensus function, our approach is to increase the edge weights, such that local maxima of F can originate also from vertex sets containing adjacent vertices. From this we gain, that when maximizing consensus F more neurons are in state *on*, hence the complement, the neurons in state *off* will be less than in the original approach of [3]. On the downside, we might lose the property that these neurons in state *off* translate to a set cover, respectively a CA in our case. We address this issue by evaluating the returned solution, and updating the weights of the edges. Then we maximize the consensus F for the updated instance. The key idea behind this approach is, that the neural network decreases the weights of those edges that carry elements as labels that where not covered in the previous iteration. This modifies the network, such that in the next iteration it is less likely that all neurons connected by such edges are turned *on* and hence some will remain turned *off*, which means they will be part of the suggested solution of the set cover. We detail our edge updates and additional learning features in the next section. The experimental results provided in Sect. 5 fully justify this approach.

4.1   Weight Updates: A First Step Towards Learning

New Initial Weights. The first change we made to the algorithm as it is presented in [3] is that we changed the computation of the edge weights. This is done by assigning the weights as a function of $|S_i \cap S_j|$ instead of $\max(|S_i|, |S_j|)$. The number of t-way interactions two rows S_i and S_j cover in common depends on the number of positions in which these rows are equal, we hence can compute

$|S_i \cap S_j| = \binom{k-d_{ij}}{t}$, where d_{ij} denotes the *hamming distance*[2] of the two rows S_i and S_j. We thus considered two additional instantiations of the building block INITIALWEIGHTS:

- HDWEIGHT$_1$: $\omega(e_{ij}) = -\binom{k-d_{ij}}{t} \cdot 1/\binom{k}{t}$
- HDWEIGHT$_1$: $\omega(e_{ij}) = -\dfrac{\binom{k}{d_{ij}}-1}{\binom{k}{\lfloor k/s \rfloor}}$

for $i \neq j$ in both cases and $\omega(e_{ii}) = 1$ for the loops. In Sect. 5 we will also compare the results when the initial edge weighting HDWEIGHT$_1$ and HDWEIGHT$_2$ are used in Algorithm 2.

Weight Updates: Learning in Epochs. Another improvement to the algorithm presented in [3] was achieved by extending it by means of epochs in which the weights of the edges connecting neurons get updated. This algorithmic extension was implemented for two reasons: First and foremost we wanted the neural network to be able to adapt to given problem instances. Second, since we gave the neural network more freedom by weakening the consensus function F by assigning larger weights to edges using our newly introduced versions of INITIALWEIGHTS, we are not guaranteed anymore that the output of the SA algorithm constitutes an independent set and hence its complement must not constitute a CA.

In short, we lose the guarantee of a feasible solution as it was guaranteed by Corollary 1. Therefore we enabled the neural network with the capability to increase or decrease the weight of edges, depending on whether the elements in their label sets were covered in the solution returned in the previous epoch or not. This new algorithmic building block WEIGHTUPDATE can be described as procedure that modifies the weight of the edges of the underlying graph $G_{t,k}$, in the following way. Whenever a t-way interaction is covered more than twice in the solution of the previous epoch, all edges that have this interaction in its label set get an increment of $1/cov$ in weight (recall that edge weights are initialized negative), where cov is the total number of covered t-way interactions. Opposite, every edge carrying an interaction that was not covered by the solution returned in the previous epoch gets a proportional decrement in weight. The weights of some edges get smaller and in the next epoch it is less likely that both vertices adjacent to such an edge will be in the independent set to be constructed. This in turn means, that at least one of the vertices will be in the complement, i.e. the return of the next epoch.

We present Algorithm 3 in terms of a pseudocode. First a reduced graph $G'_{t,k}$ is constructed and initial weights are assigned. Further a global best solution is recorded in I_{max}, which is initially set empty. Then a number e of epochs is run, where in each epoch x runs of SA are executed, where we keep the solution I maximizing the consensus F over these x runs. The weight update COVERWEIGHT is based on this solution I. If I is larger than I_{max} we store it accordingly, before entering the next epoch. Finally if $V \setminus I$ covers all t-way interactions, a CA is found and returned.

---

[2] The hamming distance of two vectors is defined as the number of positions in which these two disagree.

4.2 Graph Updates: An Additional Layer for Learning

In our experiments we recognized that the quality of the solution produced by Algorithm 3 highly depends on the random graph that is chosen in the initialization step.

Algorithm 3. BMFORCALEARNING

1: INPUT: t, k
Require: e, x
2: $G'_{t,k} \leftarrow$ INITIALGRAPH$(G_{t,k})$, $\omega(G'_{t,k}) \leftarrow$ INITIALWEIGHTS$(G'_{t,k})$ ▷ Initialization
3: $I_{max} \leftarrow \emptyset$
4: **while** epoch count $\leq$ e **do**
5:      run SA on $G'_{t,k}$ x times, store I maximizing consensus
6:      $\omega(G'_{t,k}) \leftarrow$ WEIGHTUPDATE$(G'_{t,k}, I)$
7:      **if** $|I_{max}| < |I|$ **then**
8:           $I_{max} \leftarrow I$
9:      **end if**
10: **end while**
11: **if** $V \setminus I$ covers all t-way interactions **then return** CA$(|V| - |I_{max}|; t, k, 2) = V \setminus I_{max}$
12: **else return** $V \setminus I_{max}$ with additional coverage information
13: **end if**

Thus we strived to enhance the learning rate of the neural network with a functionality capable to update the reduced graph that the Boltzmann machine runs on. We describe this additional layer of learning next. A pseudo code description can be seen in Algorithm 4. The graph updates essentially happen in an additional layer of learning phases, built upon Algorithm 3. Therefore, the initialization as well as lines 6–12 of Algorithm 4 are the same as the initialization and the lines 4–10 of Algorithm 3 where variable I_{max} gets renamed to I_{learn}). Around these epochs n learning phases are run, where at the beginning of each learning phase the I_{learn} parameter is reset to the empty set. At the end of each learning phase a graph update based on I_{max} occurs and a bias update based to the best solution I_{learn} found during this learning phase. Both procedures act on the underlying graph $G_{t,k}$ and are explained more detailed as follows.

For the key procedure GRAPHUPDATE we introduce the following instances:

- RANDOMGRAPH: This procedure selects a new random graph, just as in the random initialization.
- BESTEDGES: In each learning phase a subset L of the nodes (respectively rows) of $V \setminus I_{max}$ is randomly selected. For each row in L we flip a random position to create a second row, to which we draw an edge in the graph. By only flipping one position we generate a row that shares the maximal number of t-way interactions with the original row. The edge thus constructed has a large label set. Thereafter for each t-way interaction that is not covered by any of the rows in L, we generate a random edge having this interaction as a label, just as in INITIALGRAPH. With this strategy the neural network can reduce the number of edges in the new reduced graph.

To guide the neural network and enable it to learn from solutions previously found, we added the additional functionality of BIASUPDATE. The bias update acts on the neurons, rather than on the synapses of the neural network. In our encoding it can be realized as a weight update, acting exclusively on the loops, by adding a certain, relatively small, δ to the weight of the loops. The bias update is a way to reward vertices that were part of previous solutions, so that the Boltzmann network has a larger tendency to include them in future solutions. This is due to the structure of the consensus function F (see Subsect. 3.3) which value increases whenever a vertex with an increased weight $\omega(e_{ii})+\delta$ is activated, instead of a vertex with edge weight $\omega(e_{ii})$. Vertices being part of I_{learn} in several learning phases are incrementally rewarded through this bias update.

Remark 2. Note that due to bias updates, and also updates of edge weights the cumulative weight in the whole network is not constant over several learning phases. Adopting our weight updates such that the total weight of the network is constant over time is considered as part of future work.

Algorithm 4. BMFORCALEARNINGGRAPH

1: INPUT: t, k
Require: e, x, n, δ
2: $G'_{t,k} \leftarrow$ INITIALGRAPH$(G_{t,k})$, $\omega(G'_{t,k}) \leftarrow$ INITIALWEIGHTS$(G'_{t,k})$ ▷ Initialization
3: $I_{max} \leftarrow \emptyset$
4: **while** learning phases $\leq n$ **do**
5:     $I_{learn} \leftarrow \emptyset$
6:     **while** epoch count $\leq e$ **do**
7:         run SA on $G'_{t,k}$ x times, store I maximizing consensus
8:         $\omega(G'_{t,k}) \leftarrow$ WEIGHTUPDATE$(G'_{t,k}, I)$
9:         **if** $|I_{learn}| < |I|$ **then**
10:            $I_{learn} \leftarrow I$
11:        **end if**
12:    **end while**
13:    **if** $|I_{max}| < |I_{learn}|$ **then**
14:        $I_{max} \leftarrow I_{learn}$
15:    **end if**
16:    $G'_{t,k} \leftarrow$ GRAPHUPDATE$(G_{t,k}, I_{max})$
17:    $G_{t,k} \leftarrow$ BIASUPDATE(I_{learn}, δ)
18: **end while**
19: **return** $V \setminus I_{max}$ with coverage information

---

5 Experimental Evaluation

In this section we report experimental results for different configurations of our algorithms which serves as a proof of concept for their validity and efficiency. Tuning the parameters of neural networks for search problems has been subject to a number of related works (e.g. with *genetic algorithms* [2] or combinatorial approaches [7,8]) but an evaluation in that direction is beyond the scope of this paper, and is left for future work. Once, again here we want to demonstrate the premise of our approach especially when compared to past works related with BMs and SCs. We implemented the algorithms in C# and performed the experiments in a workstation equipped with $8\,GB$ of RAM and an i5-Core. In the experiments conducted we used the following settings regarding the simulated annealing algorithm. Temperatures and the factor α were set as $T_0 = 1$, $T_f = 0.001$, $\alpha = 0.005$. The number of inner SA cycles x for configurations of Algorithm 3 was set to 5. One SA cycle takes around 20 ms of execution time. Finally we would like to remark that although the numbers seem very small, bear in mind that for a CA instance (t, k), the underlying graphs grow exponentially in k, e.g. for the CA instance $(2, 10)$ the underlying graph $G_{2,10}$ has 1024 vertices, each having 1012 edges to other vertices.

5.1   Tested Configurations of Algorithm 2

With our first experiments we compare different Configurations of Algorithm 2, using different instantiations of INITIALWEIGHTS. Configuration 2.1 uses BMWEIGHT, Configuration 2.2 uses HDWEIGHT$_1$ and Configuration 2.3 uses HDWEIGHT$_2$. Table 1 documents the results of our experiments. In the first column we specify the CA instance, the second column headed by CAN lists the number of rows of the respective optimal CAs. In the remaining columns, headed by *min, avg, max* we document the smallest, average and maximal size of the generated *covering arrays* for each configuration. Additionally in the column headed by *avg % cov* we document the average percentage of t-way interactions covered over all generated arrays. Since we deal with randomized algorithms (recall that the procedure INITIALGRAPH is randomized), we executed 100 individual runs of each configuration. Note that Configuration 2.1 always returns a CA, due to Corollary 1. As discussed in Sect. 4 we abandoned the conceptions of [3] with our weight initialization, which is the reason why the other two configurations also return arrays with less than 100% coverage of t-way interactions. Nevertheless considering the sizes of the generated CAs, we can see that the algorithms with the initial weights computed with HDWEIGHT$_1$ and HDWEIGHT$_2$ generate much smaller covering arrays. Especially Configuration 2.2, producing the smallest covering arrays of these three versions, seems to prioritize smaller size over coverage of t-way interactions, having also the smallest percentage in t-way interactions covered over all returned CAs. In our evaluation of Configuration 2.1 we could not produce amazing results as documented in [3] achieved for general set cover instances. We believe this is mostly due to the graphs underlying the CA instances, being very dense compared to underlying graphs of general SC instances.

Summarizing these experiments we can see that the initial weighting of edges in the graph, respectively of synapses in the neural network, is crucial for the quality of the output of the tested algorithms.

Table 1. Results of the experiments with configurations of Algorithm 2.

(t, k)	CAN	Configuration 2.1				Configuration 2.2				Configuration 2.3			
		min	*avg*	*max*	*avg %* *cov*	*min*	*avg*	*max*	*avg %* *cov*	*min*	*avg*	*max*	*avg %* *cov*
(2,3)	4	4	4	4	100	4	4	4	51.5	4	4	4	100
(2.4)	4	7	8.3	10	100	5	6.14	7	85.04	5	6.15	7	93.71
(2.5)	5	13	15.16	18	100	6	8.22	10	94.75	8	10.25	12	98.83
(2.6)	6	23	26.9	31	100	9	11.90	15	98.42	12	15.22	19	99.63
(2.7)	6	41	45.62	51	100	11	13.63	17	98.33	18	23.39	29	99.95
(2.8)	6	65	72.37	81	100	11	14.60	19	97.26	23	28.69	34	99.98
(2.9)	6	97	106.04	116	100	12	13.77	15	94.74	24	31.75	39	99.99
(2.10)	6	138	146.36	158	100	12	12.67	14	86.23	25	33.22	42	99.99

5.2   Tested Configurations of Algorithm 3

In this subsection we document the results of our experiments with different configurations of Algorithm 3, to evaluate the efficiency of the introduced weight update, in combination with the different weight initializations. Thus we compared Configuration 3.1 using BMWEIGHT, Configuration 3.2 using HDWEIGHT$_1$ and Configuration 3.3 using HDWEIGHT$_2$, each using the WEIGHTUPDATE. The results can be found in Table 2, which have been generated over 10 individual runs for each configuration. Summarizing, first and foremost it is remarkable, that the deployed learning in form of weight updates nullified the severe difference in the number of rows of the generated CAs, as it is witnessed in the versions of Algorithm 2. Further it is notable that due to the weight update the number of rows of the smallest CAs generated decreases, even when comparing to Configuration 2.2 which scaled the best in the previous subsection. Additionally all Configurations always returned CAs (attained 100% coverage of t-way interactions). Hence, we omit the column with the average coverage information.

Table 2. Results of experiments with configurations of Algorithm 3.

(t, k)	CAN	Configuration 3.1			Configuration 3.2			Configuration 3.3		
		min	*avg*	*max*	*min*	*avg*	*max*	*min*	*avg*	*max*
(2,3)	4	4	4	4	4	4	4	4	4	4
(2.4)	4	5	6.2	7	5	5.9	6	5	6	7
(2.5)	5	6	6.8	8	6	6.2	7	6	6.5	8
(2.6)	6	6	7.9	9	6	7.8	9	6	7.6	9
(2.7)	6	8	9.6	11	6	9.2	12	7	9.9	12
(2.8)	6	10	11.2	13	10	11.1	12	9	10.8	12
(2.9)	6	9	11.6	13	11	13	15	11	12.6	14
(2.10)	6	11	13.1	15	11	12.9	14	11	13.1	14

5.3   Tested Configurations of Algorithm 4

We evaluated the two configurations of Algorithm 4 as given by Table 3, Configuration 4.1 uses the graph update BESTEDGES and Configuration 4.2 uses the graph update RANDOMGRAPH. We ran them for the CA instances $(t = 2, k = 6)$ and $(t = 3, k = 5)$, where we limited them to 60 and 70 learning phases respectively. Each learning phase contains 20 epochs, and a bias update at the end, where a the weight of the vertices (i.e. rows) in I_{learn} is increased by 0.01. For the BESTEDGES graph update 50% of the rows in $V \setminus I_{max}$ were used to construct the new graph. The graphs in Figs. 4a and b depict the evaluation of the best found solution after each learning phase which is normalized to the ideal solution. These experiments show that the Configuration using BESTEDGE as graph update converges faster towards the ideal solution.

Table 3. Benchmark configurations for Algorithm 4.

Building block	Configuration	
	Config. 4.1	Config. 4.2
INITIALWEIGHT	HDWEIGHT$_1$	HDWEIGHT$_1$
WEIGHTUPDATE	On	On
GRAPHUPDATE	BESTEDGES	RANDOMGRAPH
BIASUPDATE	On	On

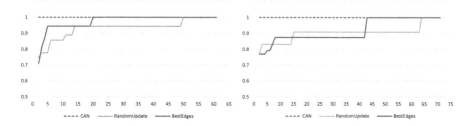

Fig. 4. Configuration 4.1 (green) and 4.2 (red) for the CA instances $(t = 2, k = 6)$ (left) and $(t = 3, k = 5)$ (right). (Color figure online)

6   Conclusion

The cornerstone of this paper is the use of artificial neural networks for CA problems, where we presented for the first time neural models for the construction of covering arrays. Combining the works of [4] and [3], we were able to devise Boltzmann machines for the construction of covering arrays and enhance them with learning capabilities in form of weight updates, graph updates and bias updates. The first experiment results confirm that the application of neural networks to the CA generation problem can lead to optimal solutions and pave the way for future applications.

Acknowledgements. This research was carried out as part of the Austrian COMET K1 program (FFG).

References

1. Aarts, E., Korst, J.: Simulated Annealing and Boltzmann Machines. Wiley, Hoboken (1988)
2. Bashiri, M., Geranmayeh, A.F.: Tuning the parameters of an artificial neural network using central composite design and genetic algorithm. Scientia Iranica **18**(6), 1600–1608 (2011)
3. Hifi, M., Paschos, V.T., Zissimopoulos, V.: A neural network for the minimum set covering problem. Chaos, Solitons Fractals **11**(13), 2079–2089 (2000)
4. Kampel, L., Garn, B., Simos, D.E.: Covering arrays via set covers. Electron. Notes Discrete Math. **65**, 11–16 (2018)
5. Kampel, L., Simos, D.E.: A survey on the state of the art of complexity problems for covering arrays. Theoret. Comput. Sci. **800**, 107–124 (2019)
6. Lundy, M., Mees, A.: Convergence of an annealing algorithm. Math. Program. **34**(1), 111–124 (1986)
7. Ma, L., et al.: Combinatorial testing for deep learning systems. arXiv preprint arXiv:1806.07723 (2018)
8. Pérez-Espinosa, H., Avila-George, H., Rodriguez-Jacobo, J., Cruz-Mendoza, H.A., Martínez-Miranda, J., Espinosa-Curiel, I.: Tuning the parameters of a convolutional artificial neural network by using covering arrays. Res. Comput. Sci. **121**, 69–81 (2016)
9. Silver, D., et al.: Mastering the game of Go without human knowledge. Nature **550**(7676), 354 (2017)
10. Smith, K.A.: Neural networks for combinatorial optimization: a review of more than a decade of research. INFORMS J. Comput. **11**(1), 15–34 (1999)
11. Torres-Jimenez, J., Izquierdo-Marquez, I.: Survey of covering arrays. In: 2013 15th International Symposium on Symbolic and Numeric Algorithms for Scientific Computing, pp. 20–27, September 2013
12. Vazirani, V.V.: Approximation Algorithms. Springer, Heidelberg (2013). https://doi.org/10.1007/978-3-662-04565-7

Adaptive Sequence-Based Heuristic
for the Three-Dimensional Bin Packing Problem

Óscar Oliveira$^{(\boxtimes)}$ ⓘ, Telmo Matos ⓘ, and Dorabela Gamboa ⓘ

CIICESI – Center for Research and Innovation in Business Sciences and Information Systems,
School of Technology and Management, Polytechnic of Porto, Porto, Portugal
{oao,tsm,dgamboa}@estg.ipp.pt

Abstract. We consider the three-dimensional Bin Packing Problem in which a set of boxes must be packed into the minimum number of identical bins. We present a heuristic that iteratively creates new sequences of boxes that defines the packing order used to generate a new solution. The sequences are generated retaining, adaptively, characteristics of previous sequences for search intensification and diversification. Computational experiments of the effectiveness of this approach are presented and discussed.

Keywords: Three-dimensional · Non-guillotine · Bin Packing · Heuristics

1 Introduction

In this paper, we address the Bin Packing Problem (BPP) in which it is intended to pack, orthogonally and without overlapping, a given set of m small boxes into the minimum number of identical bins. We consider the three-dimensional case where boxes and bins are characterized by their length, height and depth. The boxes cannot be rotated, and no additional constraints are considered, e.g. cargo stability (see Bortfeldt and Wäscher [1] for a recent overview on constraints encountered in practice).

The three-dimensional BPP is strongly NP-hard as it generalizes the strongly NP-hard one-dimensional BPP (see Martello et al. [2]). Following the typology of Wäscher et al. [3], the problem addressed in this paper is classified as Single Bin Size Bin Packing Problem (SBSBPP).

Several solution methods to solve this problem were proposed and are next briefly described in chronological order. Martello and Vigo [2] demonstrated that the Continuous Lower Bound $\left(CLB = \lceil \frac{\sum_{i=1}^{m} l_i h_i d_i}{LHD} \rceil \right)$ for the SBSBPP has a $\frac{1}{8}$ worst-case performance ratio and presented new lower bounds that are used in a Branch-and-Bound algorithm. Fekete and Schepers [4] presented lower bounds that dominate those presented in [2].

Lodi *et al.* [5] presented a Tabu Search (see Glover [6]) framework based on a new constructive heuristic for the neighbourhood evaluation.

© Springer Nature Switzerland AG 2020
N. F. Matsatsinis et al. (Eds.): LION 13 2019, LNCS 11968, pp. 69–76, 2020.
https://doi.org/10.1007/978-3-030-38629-0_6

Faroe *et al.* [7] presented a Guided Local Search (GLS, see Voudouris and Tsang [8]) heuristic that starts with a number of bins obtained through a greedy heuristic, and iteratively decreases this number searching for a feasible packing.

Boschetti [9] proposed new lower bounds that dominate the previous ones.

Fekete *et al.* [10] presented a two-level tree search algorithm for multi-dimensional packing problems using the data structure for characterizing feasible packings proposed by Fekete and Schepers [11].

Crainic *et al.* [12] proposed a two-level heuristic based on Tabu Search using the graph-theoretical characterization of feasible packings proposed by Fekete and Schepers [11]. The first level of this heuristic tries to reduce the number of bins, while the second optimizes the packing of the bins.

Parreño *et al.* [13] presented a Greedy Randomized Adaptive Search Procedure (GRASP, see Feo and Resende [14]) with Variable Neighbourhood Descent (VND, see Hansen *et al.* [15]) for the two- and three-dimensional SBSBPP.

Crainic *et al.* [16] proposed a greedy heuristic, called Greedy Adaptive Search Procedure (GASP), that allies greedy algorithms with learning mechanisms.

Gonçalves and Resende [17] presented a Biased Random-key Genetic Algorithm (BRKGA, see Gonçalves and Resende [18]) for the two- and three-dimensional SBSBPP.

Hifi *et al.* [19] presented two heuristics using a simple strategy based on integer linear programming heuristics without using any improvement based on metaheuristics.

Zudio *et al.* [20] presented a BRKGA/VND based on the work of Gonçalves and Resende [17].

Zhao *et al.* [21] presented a recent review on the solution methods found in the literature for three-dimensional cutting and packing problems and an extensive comparison of the performance of these solutions methods on standard benchmark instances.

This paper presents a heuristic for solving the three-dimensional SBSBPP and is organized as follows. Section 2 presents a description of our heuristic. Computational experiments on standard benchmark instances are presented in Sect. 3 and in Sect. 4 conclusions and future work directions are provided.

2   Adaptive Sequence-Based Heuristic

We propose a heuristic, denoted hereafter as Adaptive Sequence-based Heuristic (ASH), that iteratively creates a new sequence of boxes that defines the packing order. An ASH based heuristic was presented by Oliveira and Gamboa [23] to solve the two-dimensional SBSBPP.

The main concept behind the proposed heuristic is that if a good solution was obtained packing the boxes following some ordering (S_{base}), it is possible that a better solution exists changing the order of few boxes, creating a new packing sequence $S_{current}$. If no improvement is obtained with this new ordering, it may be the case that an ordering with more changes can lead to a better solution. So, we incrementally allow more changes to the base ordering. When an ordering generates a new best solution, this will replace the base ordering to be used in the next iterations. The main steps of the proposed heuristic are shown in Fig. 1.

S_{base} ← Boxes ordered by non-increasing volume
α ← α_{min}
Generate a new solution with sequence S_{base}
While stopping criteria are not met **do**
   $S_{current}$ ← Generate a new sequence of items based on S_{base} and α
   Generate a new solution with sequence $S_{current}$
   If a new best solution is found **then**
     S_{base} ← $S_{current}$
     α ← α_{min}
   Otherwise
     α ← $\min\{\alpha + \alpha_{inc}, \alpha_{max}\}$

Fig. 1. ASH main steps.

Initializing the base sequence of boxes (S_{base}) ordered by non-increasing volume, iteratively, a new sequence ($S_{current}$) is generated that is used to create a new solution.

We generate new sequences with a method based on the BubbleSearch proposed by Lesh and Mitzenmacher [22] that creates new orderings based on the Kendall-tau distance between permutations. In [22], BubbleSearch-based heuristics compared favourably against similar GRASP-based variations for the two-dimensional rectangular Strip Packing Problem and the Job Shop Scheduling Problem. The Randomized BubbleSearch can be implemented as a simple stochastic process (see Belov *et al.* [23] and Zubaran and Ritt [24]), depicted in Fig. 2 and next briefly described.

Input: sequence *In*, probability α
Out ← $\emptyset$
n ← $|In|$
for i ← $1, ..., n$ **do**
  j ← 1
  Out_i ← $\emptyset$
  while $Out_i = \emptyset$ **do**
    if $\alpha \leq$ generated random value **then**
      Out_i ← In_j
      In ← $In \setminus \{In_j\}$
    j ← $(j \bmod |In|) + 1$
return Out

Fig. 2. Sequence generator.

While there are items of the base sequence (S_{base}) that are not in the new sequence ($S_{current}$), a random number is generated iteratively for each item of S_{base} that is not already in $S_{current}$. If the generated number is greater than the permutation probability (α), the item is copied to $S_{current}$ at the next empty position. Otherwise, the item will be copied in a following iteration.

If the solution generated with $S_{current}$ is the best one so far, this ordering replaces S_{base} and α is set to its minimum value α_{min}. Otherwise, S_{base} is inaltered and α is updated to min $\{\alpha + \alpha_{inc}, \alpha_{max}\}$. When a new best solution is found, we seek to intensify the search near to this solution, which is accomplished by setting α to its minimum value, aiming to generate sequences that are very similar to the base. While incrementally higher α allows to diversify the search space producing sequences that differ more from the base sequence.

The boxes are packed, following the ordering given by the current box sequence, one at a time, into the empty rectangular space (ERS), i.e., maximal empty space interval, that have space to pack it inside and that has the highest number of common edges (highest volume as tie-breaker) among all the available ERSs present in the current opened bin. The boxes are placed at the bottom-left-back position of the ERSs. A new bin is opened when no more boxes can be packed inside the currently opened bin, if any. The Difference Process, presented in Lai and Chan [25], is used to keep track of the ERSs that are generated during the packing. This process, first, places the box inside the given ERS, then generates the new ERSs that result from the intersection of the box with the existing ERS and removes intersected ERSs. The last phase removes the ERSs that are infinitely thin or are totally inscribed by other ERSs.

ASH iterates until a maximum number of iterations has been performed or the optimality is guaranteed, i.e., if the solution value is equal to the Continuous Lower Bound $(CLB = \lceil \frac{\sum_{i=1}^{m} l_i h_i d_i}{LHD} \rceil$, see Martello and Vigo [2]).

3 Computational Results

The proposed heuristic was implemented in C and the tests were run on a computer with an Intel Core i7-4800MQ at 2.70 GHz with 8 Gb RAM and operating system Linux Ubuntu 18.04.

We tested our algorithm on the dataset generated by Martello *et al.* [2] which generator is available at http://hjemmesider.diku.dk/~pisinger/codes.html. This dataset contains 320 instances divided into 8 classes. The classes are composed of 4 groups of 10 instances each with an identical number of boxes to pack ($m \in [50, 100, 150, 200]$).

Each instance was run only once generating at most 2500 solutions with $\alpha_{min}, \alpha_{max}$, and α_{inc} set to 0.1, 0.65, and 0.002, respectively.

Table 1 presents the computational results obtained by ASH solving all the instances of the dataset considered. Each row gives the number of boxes (third column), the cumulative number of bins (fourth column) and the running time in seconds (fifth column) solving the 10 instances of each class. The first and second columns denote the class and dimensions of the bins for each class, respectively.

Table 2 shows the number of bins obtained solving the instances of the dataset above described by other state-of-the-art algorithms (identified in the first column). The second column presents the results solving classes 1, 4, 5, 6, 7 and 8, while the last column of this table presents the results obtained solving all the classes. The last line of this table presents the results obtained by ASH. The symbol - denotes that the respective algorithm does not present results for the complete dataset.

Comparing with the most similar heuristic to ASH, namely the Randomized Constructive presented by Parreño et al. [13], the results are effectively very promising as

Table 1. ASH results.

Class	Bin size	m	Bins	Time (s)
1	$100 \times 100 \times 100$	50	134	0.9
		100	270	2.3
		150	372	4.0
		200	516	6.1
2	$100 \times 100 \times 100$	50	141	0.8
		100	258	2.4
		150	374	4.1
		200	502	6.2
3	$100 \times 100 \times 100$	50	133	0.9
		100	265	2.3
		150	380	4.1
		200	506	6.3
4	$100 \times 100 \times 100$	50	294	0.7
		100	590	2.0
		150	870	3.7
		200	1192	5.7
5	$100 \times 100 \times 100$	50	84	1.6
		100	151	3.6
		150	203	6.0
		200	273	8.4
6	$10 \times 10 \times 10$	50	99	0.7
		100	192	1.6
		150	293	2.6
		200	377	3.9
7	$40 \times 40 \times 40$	50	74	1.6
		100	124	3.6
		150	159	5.9
		200	237	7.9
8	$100 \times 100 \times 100$	50	92	1.3
		100	189	3.0
		150	242	5.2
		200	299	7.6

Table 2. Literature results.

Algorithm	Classes 1, 4-8	Classes 1-8
L2002-TS [5]	7320	9863
F2003 [7]	7302	–
C2009 [12]	7298	–
Constructive [13]	7514	10148
Randomized Constructive [13]	7330	9877
GRASP/VND [13]	7286	9813
GASP [16]	7364	–
BRKGA [17]	7258	9777
EHGH2 [19]	7565	–
BRKGA/VND [20]	7253	9771
ASH	7326	9885

the latter were obtained with (1) a pre-processing stage that tries to create an equivalent problem easiest to solve, (2) imposing a limit of 50000 iterations and (3) choosing randomly one of the corner selection strategies at each iteration.

ASH, with a maximum of 2500 iterations, obtained results equivalent to those obtained by the Randomized Constructive and setting the maximum number of iterations with the same value, i.e., 50000, ASH obtained 7309 and 9858 bins for classes 1, 4–8 and 1–8, respectively. The results obtained by ASH with 50000 iterations were slightly better than the ones obtained by the Tabu Search proposed by Lodi *et al.* [5].

ASH obtained very good results, comparing favourably against more complex approaches. Considering most of the other approaches, ASH is simpler to implement and to parametrize (i.e., maximum number of iterations, and minimum, maximum and increment of the probability α). GRASP needs a Restricted Candidate List and improvement methods, Tabu Search requires neighbourhood structures and memory strategies, and Genetic Algorithms needs chromosomes and an evolutionary process. The results show that this heuristic, although simple, can generate high-quality solutions using small computing times. ASH can be considered a competitive approach for solving the SBS-BPP as the literature algorithms obtain their results through a higher number of iteration or population size, use of improvement methods, and/or requiring a high execution time (most of the referred algorithms have set a computational time limit).

Although not directly comparable between approaches because in addition to the different technologies used, e.g., hardware, programming language and operating systems, most of the approaches used for comparison impose a time limit for their execution, we consider that the execution time of ASH is very low, thereby giving a large margin for improvements.

ASH is fast, simple to implement and can be an effective approach to solve the 3D SBSBPP. We consider that the results presented instigate more research for a solution method based on the ASH. Extensions of this work will be done trying to improve the results obtained. Combinations between ASH and search methods (in particular with other metaheuristics) will be part of our future research. ASH can be further enhanced for even better results with the inclusion of search strategies, such as local search after the cutting plan generation, or with more sophisticated ones such as a Path Relinking (see Glover et al. [26]) as a final phase of this heuristic.

Another research direction that will be interesting to follow is to evaluate alternative placement methods, as the selection of the ERS to use and the selection of the position inside the ERS to place the current box can originate very distinct packing layouts.

4   Conclusion

We present a heuristic for the three-dimensional Single Bin Size Bin Packing Problem. This heuristic creates, at each iteration, a new sequence of boxes that will be used to create a new solution. The generated sequences retain, in an adaptive manner, characteristics of previous sequences providing intensification and diversification on the explored solution space. To assess the performance of the proposed heuristic computational experiments have been performed on standard benchmark instances, validating its effectiveness. The computational results show that our approach is competitive with other proposed solution methods for the problems considered. Noteworthy the implementation simplicity and the little parameterization required by this approach. Combinations between ASH and search strategies, and new placement rules can be studied to improve the results. Also, as the proposed approach seems to be very promising, we intend to apply it to other combinatorial optimization problems.

Acknowledgement. This project is funded by Portuguese funds through FCT/MCTES (PIDDAC) under the project CIICESI_2017-03.

References

1. Bortfeldt, A., Wäscher, G.: Constraints in container loading – a state-of-the-art review. Eur. J. Oper. Res. **229**, 1–20 (2013)
2. Martello, S., Pisinger, D., Vigo, D.: The Three-dimensional bin packing problem. Oper. Res. **48**, 256–267 (2000)
3. Wäscher, G., Haußner, H., Schumann, H.: An improved typology of cutting and packing problems. Eur. J. Oper. Res. **183**, 1109–1130 (2007)
4. Fekete, S.P., Schepers, J.: New classes of fast lower bounds for bin packing problems. Math. Program. **91**, 11–31 (2001)
5. Lodi, A., Martello, S., Vigo, D.: Heuristic algorithms for the three-dimensional bin packing problem. Eur. J. Oper. Res. **141**, 410–420 (2002)
6. Glover, F.: Tabu search—part I. ORSA J. Comput. **1**, 190–206 (1989)
7. Faroe, O., Pisinger, D., Zachariasen, M.: Guided local search for the three-dimensional bin-packing problem. INFORMS J. Comput. **15**, 267–283 (2003)
8. Voudouris, C., Tsang, E.P.K.: Guided Local Search. In: Glover, F., Kochenberger, G.A. (eds.) Handbook of Metaheuristics. ISOR, vol. 57, pp. 185–218. Springer, Boston (2003). https://doi.org/10.1007/0-306-48056-5_7

9. Boschetti, M.A.: New lower bounds for the three-dimensional finite bin packing problem. Discrete Appl. Math. **140**, 241–258 (2004)
10. Fekete, S.P., Schepers, J., van der Veen, J.C.: An exact algorithm for higher-dimensional orthogonal packing. Oper. Res. **55**, 569–587 (2007)
11. Fekete, S.P., Schepers, J.: A combinatorial characterization of higher-dimensional orthogonal packing. Math. Oper. Res. **29**, 353–368 (2004)
12. Crainic, T.G., Perboli, G., Tadei, R.: TS2PACK: a two-level tabu search for the three-dimensional bin packing problem. Eur. J. Oper. Res. **195**, 744–760 (2009)
13. Parreño, F., Alvarez-Valdés, R., Oliveira, J.F., Tamarit, J.M.: A hybrid GRASP/VND algorithm for two- and three-dimensional bin packing. Ann. Oper. Res. **179**, 203–220 (2010)
14. Feo, T., Resende, M.G.C.: Greedy randomized adaptive search procedures. J. Glob. Optim. **6**, 109–133 (1995)
15. Hansen, P., Mladenović, N., Brimberg, J., Pérez, J.A.M.: Variable neighborhood search. Comput. Oper. Res. 57–97 (2019)
16. Crainic, T.G., Perboli, G., Tadei, R.: A greedy adaptive search procedure for multi-dimensional multi-container packing problems (2012)
17. Gonçalves, J.F., Resende, M.G.C.: A biased random key genetic algorithm for 2D and 3D bin packing problems. Int. J. Prod. Econ. **145**, 500–510 (2013)
18. Gonçalves, J.F., Resende, M.G.C.: Biased random-key genetic algorithms for combinatorial optimization. J. Heuristics **17**, 487–525 (2011)
19. Hifi, M., Negre, S., Wu, L.: Hybrid greedy heuristics based on linear programming for the three-dimensional single bin-size bin packing problem. Int. Trans. Oper. Res. **21**, 59–79 (2014)
20. Zudio, A., da Silva Costa, D.H., Masquio, B.P., Coelho, I.M., Pinto, P.E.D.: BRKGA/VND hybrid algorithm for the classic three-dimensional bin packing problem. Electron. Notes Discret. Math. **66**, 175–182 (2018)
21. Zhao, X., Bennell, J.A., Bektaş, T., Dowsland, K.: A comparative review of 3D container loading algorithms. Int. Trans. Oper. Res. **23**, 287–320 (2016)
22. Lesh, N., Mitzenmacher, M.: BubbleSearch: a simple heuristic for improving priority-based greedy algorithms. Inf. Process. Lett. **97**, 161–169 (2006)
23. Belov, G., Scheithauer, G., Mukhacheva, E.A.: One-dimensional heuristics adapted for two-dimensional rectangular strip packing. J. Oper. Res. Soc. **59**, 823–832 (2008)
24. Zubaran, T., Ritt, M.: A simple, adaptive bubble search for improving heuristic solutions of the permutation flowshop scheduling problem. In: XLV SBPO - Simpósio Brasileiro de Pesquisa Operacional, pp. 1847–1856 (2013)
25. Lai, K.K., Chan, J.W.M.: Developing a simulated annealing algorithm for the cutting stock problem. Comput. Ind. Eng. **32**, 115–127 (1997)
26. Glover, F., Laguna, M., Martí, R.: Fundamentals of scatter search and path relinking. Control Cybern. **39**, 653–684 (2000)

Optimizing Partially Defined Black-Box Functions Under Unknown Constraints via Sequential Model Based Optimization: An Application to Pump Scheduling Optimization in Water Distribution Networks

Antonio Candelieri$^{(\boxtimes)}$ (ID), Bruno Galuzzi (ID), Ilaria Giordani (ID), Riccardo Perego (ID), and Francesco Archetti (ID)

University of Milano-Bicocca, Viale Sarca 336, 20126 Milan, Italy
{antonio.candelieri,bruno.galuzzi,ilaria.giordani,
riccardo.perego,francesco.archetti}@unimib.it

Abstract. This paper proposes a Sequential Model Based Optimization framework for solving optimization problems characterized by a black-box, multi-extremal, expensive and partially defined objective function, under unknown constraints. This is a typical setting for simulation-optimization problems, where the objective function cannot be computed for some configurations of the decision/control variables due to the violation of some (unknown) constraint. The framework is organized in two consecutive phases, the first uses a Support Vector Machine classifier to approximate the boundary of the unknown feasible region within the search space, the second uses Bayesian Optimization to find a globally optimal (feasible) solution. A relevant difference with traditional Bayesian Optimization is that the optimization process is performed on the estimated feasibility region, only, instead of the entire search space. Some results on three 2D test functions and a real case study for the Pump Scheduling Optimization in Water Distribution Networks are reported. The proposed framework proved to be more effective and efficient than Bayesian Optimization approaches using a penalty for function evaluations outside the feasible region.

Keywords: Sequential Model Based Optimization · Bayesian Optimization · Constrained Global Optimization · Pump Scheduling Optimization · Water Distribution Networks

1 Introduction

Sequential Model Based Optimization (SMBO), and more precisely Bayesian Optimization (BO) [1], has been proven to be effective and sample efficient in optimizing black-box expensive functions [2–11]. Moreover, it is currently the standard approach, in the Machine Learning community, for solving automatic algorithm configuration and hyper-parameter tuning [12–14]. Its building blocks are a *probabilistic surrogate model*

© Springer Nature Switzerland AG 2020
N. F. Matsatsinis et al. (Eds.): LION 13 2019, LNCS 11968, pp. 77–93, 2020.
https://doi.org/10.1007/978-3-030-38629-0_7

– usually a Gaussian Process (GP) – to approximate the black-box function and an *acquisition function* to choose the next promising point to evaluate, balancing between exploitation and exploration.

Although SMBO has been usually adopted for solving black-box optimization within bounded-box search spaces [1–3, 7, 8, 15], more interesting and realistic cases are related to Constrained Global Optimization (CGO) [16–18], where the search space – aka feasible region – is defined by a set of more complex constraints than simple thresholds on the value of the variables. A specific setting is related to *partially defined objective functions*, meaning that the function to be optimized cannot be computed (i.e., it is undefined) outside the feasible region [16, 19–21]. Furthermore, the most challenging setting arises when constraints are unknown. Relevant prior results on BO with unknown constraints are presented in [22–26], with some of them proposing new acquisition functions, such as Integrated Expected Conditioned Improvement (IECI) [27] and a modification of the Predictive Entropy Search (PES) [28]. Most of these results use independent GPs to model the objective function and the constraints, requiring two strong assumptions: a priori knowledge about the number of constraints and the independence among objective function and all the constraints. The assumption of independence permits to compute a probability of feasibility simply as the product of individual probability with respect to every constraint. The result is multiplied for the acquisition function, usually Expected Improvement (EI), in order to optimize the objective function while satisfying, with high probability, the constraints.

The main contribution of this paper is the development of a method which does not require any of the two previous assumptions. The basic idea proposed is to use Support Vector Machine (SVM) [29, 30] to sequentially estimate and model the unknown feasible region (Ω), aka "feasibility determination", without any assumption on the number of constraints as well as their independence. Indeed, only one SVM classifier is learned to model the boundary of the feasible region instead of one GP for every constraint. The adoption of SVM has been previously suggested in [31] where a Probabilistic SVM (PSVM) is used to calculate the overall probability of feasibility while the optimization schema alternates between (*i*) a global search for the optimal solution, depending on both this probability and the estimated value of the objective function – modelled through a GP – and (*ii*) a local refinement of the PSVM through an adaptive local sampling schema.

Analogously, our approach, namely SVM-CBO (SVM constrained BO), is organized in two phases, but they are consecutive and not alternating: the first is aimed to provide a first estimate of Ω (feasibility determination) and the second is "vanilla" BO [1] performed on such an estimate, only. The motivation is that we are interested in obtaining a good approximation of the overall feasible region, and not only close to the optimal solution (i.e., feasibility determination is a goal per-se, in our approach). Another relevant difference is that SVM-CBO uses more efficiently the available "budget" (i.e., maximum number of function evaluations): at every iteration, of both phase 1 and 2, we perform just one function evaluation, while in the boundary refinement of [31] a given number n_p of function evaluations (namely, auxiliary samples) is performed in the neighborhood of the next point to evaluate (namely, update region), with the aim to locally refine the boundary estimate.

The importance of feasibility determination has been specifically highlighted in real-life problems, such as Pump Scheduling Optimization (PSO) in a Water Distribution Networks (WDN) [32, 33]. While [32] presents an adaptive random search approach to approximate feasible region while optimizing the objective function, [33] proposes a BO framework where infeasibility is treated by assigning a fixed penalty as value of the objective function. As reported in [34], penalty can be also assigned directly to the acquisition function, with the aim to quickly move away from infeasible regions (we refer as "BO with penalty").

The SVM-CBO approach proposed in this paper was tested on a set of test functions and a PSO case study. As main result, the proposed SVM-CBO approach proved to be more effective and efficient when compared to the BO with penalty.

2 Support Vector Machine Based Constrained Bayesian Optimization

2.1 Problem Formulation

We start with the definition of the problem, that is:

$$\min_{x \in \Omega \subset X \subset \mathbb{R}^d} f(x)$$

where $f(x)$ has the following properties: it is black-box, multi-extremal, expensive and partially defined. The last feature means that $f(x)$ is undefined outside that feasible region Ω [16, 19–21], which is a subset of overall bounded-box search space $X \subset \mathbb{R}^d$. Moreover, we consider the case that constraints defining the feasible region are also black-box. This can be a typical setting in optimization problems where the objective function is computed by a, usually time consuming, simulation process/software for each $x \in X$ [35].

We introduce some notation that will be used in the following:

- $D_n^{\Omega} = \{(x_i, y_i)\}_{i=1,...,n}$ is the **feasibility determination** dataset;
- $D_l^f = \{(x_i, f(x_i))\}_{i=1,...,l}$ is the **function evaluations** dataset, with $l \le n$ (because $f(x)$ is partially defined on X) and where l is the number of feasible points out of the n evaluated so far;

and x_i is the i-th evaluation point and $y_i = \{+1, -1\}$ defines if x_i is feasible or infeasible, respectively. Thus, it is easy to notice that the feasibility determination dataset, D_n^{Ω}, is exactly in the form of any generic dataset which SVM classification can be applied on.

2.2 Phase 1 – Feasibility Determination

The first phase of the approach aims to find an estimate $\tilde{\Omega}$ of the actual feasible region Ω, in M function evaluations ($\tilde{\Omega}_M = \tilde{\Omega}$). The sequence of function evaluations is determined according to an SMBO process whose surrogate model approximates the boundary of the feasible region, instead of the objective function. The current estimate of the feasible region is denoted by $\tilde{\Omega}_n$. As surrogate model we use the (non-linear) separation hyperplane of an SVM classifier, trained on the set D_n^Ω. The SVM classifier uses an RBF kernel to model feasible regions with non-linear boundaries. Let denote with $h_n(x)$ the argument of the SVM-based classification function [29, 30]:

$$h_n(x) = \sum_{i=1}^{n_{SV}} \alpha_i y_i k(x_i, x) + b$$

where α_i and y_i are the Lagrangian coefficient and the "feasibility label" of the i-th support vector, respectively, $k(.,.)$ is the kernel function (i.e., an RBF kernel, in this study), b is the offset and n_{SV} is the number of support vectors. The boundaries of the estimated feasible region $\tilde{\Omega}_n$ are given by $h_n(x) = 0$ (i.e. non-linear separation hyperplane). The SVM-based classification function provides the estimated feasibility for any $x \in X$:

$$\tilde{y} = \text{sign}(h_n(x)) = \begin{cases} +1 \ if \ x \in \tilde{\Omega}_n \\ -1 \ if \ x \notin \tilde{\Omega}_n \end{cases}$$

With respect to the aim of the first phase, we propose an acquisition function aimed at identifying the next promising point depending on two different goals:

- Improving the estimate of the feasible region;
- Discovering possible disconnected feasible regions

To deal with the first goal, we use the distance from the boundaries of the currently estimated feasible region $\tilde{\Omega}_n$, using the following formula from the SVM classification theory:

$$d_n(h_n(x), x) = |h_n(x)| = \left| \sum_{i=1}^{n_{SV}} \alpha_i y_i k(x_i, x) + b \right|$$

To deal with the second goal, we introduce the concept of "coverage of the search space", defined by:

$$c_n(x) = \sum_{i=1}^{n} e^{-\frac{\|x_i - x\|^2}{2\sigma_c^2}}$$

So, $c_n(x)$ is a sum of n RBF functions centred on the points evaluated so far, with σ_c a parameter to set the width of the corresponding bell-shaped curve.

Finally, the acquisition function for phase 1 is given by the sum of $d_n(h_n(x), x)$ and $c_n(x)$, and the next promising points is identified by solving the following optimization problem:

$$x_{n+1} = \underset{x \in X}{\text{argmin}} \{d_n(h_n(x), x) + c_n(x)\}$$

Thus, we want to select the point associated to minimal coverage (i.e., max uncertainty) and minimal distance from the boundary of the current estimated feasible region.

This allows to balance between improving the estimate of the feasible region and discovering possible disconnected feasible regions (in less explored areas of the search space). It is important to highlight that, in phase 1, the optimization is performed on the overall bounded-box search space X. After the function evaluation of the new point x_{n+1}, the following information is available:

$$y_{n+1} = \begin{cases} +1 \ if \ x_{n+1} \in \Omega; \quad f(x_{n+1}) \ is \ defined \\ -1 \ if \ x_{n+1} \notin \Omega : f(x_{n+1}) \ is \ not \ defined \end{cases}$$

and the following updates are performed:

- Feasibility determination dataset and estimated feasible region $\tilde{\Omega}_{n+1}$

$$D^{\Omega}_{n+1} = D^{\Omega}_n \cup \{(x_{n+1}, y_{n+1})\}$$
$$h_{n+1}(x)|D^{\Omega}_{n+1}$$
$$n \leftarrow n + 1$$

- Only if $x \in \Omega$, function evaluations dataset

$$D^{f}_{l+1} = D^{f}_l \cup \{(x_{l+1}, f(x_{l+1}))\}$$
$$l \leftarrow l + 1$$

The SMBO process for phase 1 is repeated until $n = M$.

2.3 Phase 2 – Bayesian Optimization in the Estimated Feasible Region

In this phase a traditional BO process is performed, but with the following relevant differences:

- the search space is not a bounded-box but the estimated feasible region $\tilde{\Omega}_n$ identified in phase 1
- the probabilistic surrogate model of the objective function – a GP, in this study – is fitted only using the feasible solutions observed so far, D^{f}_l
- the acquisition function – Lower Confidence Bound (LCB), in this study – is defined on $\tilde{\Omega}_n$, only

Thus, the next point to evaluate is given by:

$$x_{n+1} = \operatorname*{argmin}_{x \in \tilde{\Omega}_n}\{LCB_n(x) = \mu_n(x) - \beta_n \sigma_n(x)\}$$

where $\mu_n(x)$ and $\sigma_n(x)$ are the mean and the standard deviation of the current probabilistic surrogate model and β_n is the inflate parameter to deal with the trade-off between exploration and exploitation for this phase. Theoretically motivated guidelines for setting and scheduling β_n to achieve optimal regret are presented in [36]. It is important to highlight that, contrary to phase 1, the acquisition function is here minimized on $\tilde{\Omega}_n$, only, instead of the entire bounded-box search domain X. The point x_{n+1} is just expected to be feasible, according to $\tilde{\Omega}_n$, but the information on its actual feasibility is known only after having checked whether $f(x_{n+1})$ can or cannot be computed (i.e. it is defined or not in x_{n+1}). Subsequently, the feasibility determination dataset is updated as follows:

$$D_{n+1}^{\Omega} = D_n^{\Omega} \cup \{(x_{n+1}, y_{n+1})\}$$

and according to the two alternative cases:

- x_{n+1} is actually feasible: $x_{n+1} \in \Omega$, $y_{n+1} = +1$;
  the function evaluations dataset is updated as follows: $D_{l+1}^f = D_l^f \cup \{(x_{l+1}, f(x_{l+1}))\}$, with $l \leq n$ is the number of the feasible solutions with respect to all the points observed so far. The current estimated feasible region $\tilde{\Omega}_n$ can be considered accurate and retraining of the SVM classifier can be avoided: $\tilde{\Omega}_{n+1} = \tilde{\Omega}_n$
- x_{n+1} is actually infeasible: $x_{n+1} \notin \Omega$, $y_{n+1} = -1$;
  the estimated feasible region must be updated to reduce the risk for further infeasible evaluations

$$h_{n+1}(x)|D_{l+1}^f \Rightarrow \tilde{\Omega}_{n+1}$$

The phase 2 continues until the overall available budget $n = N$ is reached.

3   Experimental Setting

3.1   Test Functions

The SVM-CBO approach has been validated on three well-known 2D test functions for CGO, that are: Rosenbrock constrained to a disk [37], Rosenbrock constrained to a line and a cubic [37, 38], and Mishra's Bird constrained [39]. Table 1 summarizes the three test problems considered, more precisely summarizes the analytical expressions of both objective functions and relative constraints (i.e. feasible region). This information is considered unknown for the experimental tests: both objective functions and constraints are all black-box. Table 2 summarizes other information about the test functions, more precisely the initial bounded box search space, the global minimizer, the associated value of objective function and a graphical representation of the partially defined test functions.

Table 1. Test functions: analytical expression of the test objective functions and relative constraints

Test function	$f(x)$	(Unknown) constraints
Rosenbrock constrained to a disk	$f(x_1, x_2) =$ $(1 - x_1)^2 + 100\left(x_1 + x_2^2\right)^2$	$x_1^2 + x_2^2 \leq 2$
Rosenbrock constrained to a cubic and a line	$f(x_1, x_2) =$ $(1 - x_1)^2 + 100\left(x_1 + x_2^2\right)^2$	$(x_1 - 1)^3 - x_2 + 1 < 0$ and $x_1 + x_2 - 2 < 0$
Mishra's bird constrained	$\sin(x_2)e^{(1-\cos(x_1))^2} +$ $\cos(x_1)e^{(1-\sin(x_2))^2} +$ $(x_1 - x_2)^2$	$(x_1 + 5)^2 + (x_2 + 5)^2 < 25$

3.2 Pump Scheduling Optimization Case Study

The goal of PSO is to identify the pump schedule (i.e., the status of each pump over time) associated to the lowest energy cost while satisfying hydraulic feasibility (i.e., water demand satisfaction, pressure level within a given operational range, etc.). Status of a pump is its activation – in the case of an ON/OFF pump – or its speed – in the case of a Variable Speed Pump (VSP), leading to discrete or continuous decision variables, respectively. Thus, a general formulation of the PSO, involving both the two types of decision variables (i.e. a WDN with both ON/OFF pumps and VSP), is reported as follows:

OBJECTIVE FUNCTION (minimizing the energy costs)

$$\min\left\{C = \Delta t \sum_{t=1}^{T} c_E^t \left(\sum_{k_b=1}^{N_b} \gamma x_{k_b}^t Q_{k_b}^t \frac{H_{k_b}^t}{\eta_{k_b}} + \sum_{k_v=1}^{N_v} \gamma x_{k_v}^t Q_{k_v}^t \frac{H_{k_v}^t}{\eta_{k_v}}\right)\right\}$$

where:

- C – total energy cost over time steps
- Δt – time step width
- C_E^t – energy cost at time t
- γ – specific weight of water
- $Q_{k_j}^t$ – water flow provided through the pump k_j at time t
- $H_{k_j}^t$ – head loss on pump k_j at time t
- η_{k_j} – efficiency of the pump k_j

and the decision variables are:

- $x_{k_b}^t \in \{0, 1\}$ with $t = 1, \ldots, T$ and $k_b = 1, \ldots, N_b$ – status of pump k_b at time t (i.e. ON = 1, OFF = 0)
- $x_{k_v}^t \in [0, 1]$ with $t = 1, \ldots, T$ and $k_v = 1, \ldots, N_v$ – speed of pump k_v at time t

where N_b is the number of ON/OFF pumps and N_v is the number of VSPs.

Table 2. Test functions: search space, minimizer, function value at the global optimum and graphical representation

Test function	Search Space X	x^*	$f(x^*)$	Visualization
Rosenbrock constrained to a disk	$-1.5 \leq x_1 \leq 1.5$ $-1.5 \leq x_2 \leq 1.5$	$x_1 = x_2 = 1.0$	0.0	
Rosenbrock constrained to a cubic and a line	$-1.5 \leq x_1 \leq 1.5$ $-0.5 \leq x_2 \leq 2.5$	$x_1 = x_2 = 1.0$	0.0	
Mishra's bird constrained	$-10 \leq x_1 \leq 0$ $-6.5 \leq x_2 \leq 0$	$x_1 = -3.1302468$ $x_2 = -1.5821422$	-106.7645367	

The time horizon usually adopted in PSO is one day with hourly time steps, leading to $T = 24$ time steps overall. This choice is basically motivated by the energy tariffs which are typically characterized by a different energy price depending on the hour of the day.

Although the PSO objective function has an apparently closed expression, its exact resolution through mathematical programming requires to approximate the flow equations (linearization or convexification) which regulates the hydraulic behaviour of a pressurized WDN. More recent approaches address the optimization of the objective function by considering it as "black-box", computing the value associated to a given schedule through a hydraulic simulation software which solves all the flow equations. Naturally, hydraulic simulation is performed at the selected time resolution – usually hourly basis for 24 h – which is adopted for the input data (i.e. demand at consumption points) as well as the outputs obtained at the end of a simulation run (i.e. pressure at nodes, flow/velocity on the pipes). Depending on the size and complexity of the WDN, time required for a complete simulation can significantly change. This means that in the case of very large and complex WDN evaluating a schedule may also become significantly expensive in terms of time.

Although not expressed in the formulation above, there exists a set of hydraulic constraints which are checked by the hydraulic simulation software. Even in papers reporting the analytical formulation of the constraints, such as in [40], this check is usually verified by running a simulation. In the case a given pump schedule does not satisfy at least one of the hydraulic constraints the value of the objective function cannot be computed, therefore it is partially defined. The hydraulic simulation software used in this study is EPANET 2.0 and a detailed list of errors/warnings provided by EPANET, with respect to hydraulic unfeasibility of a pump schedule, is reported in [41].

PSO is computationally hard to solve through an exhaustive search, due to the large number of possible pump schedules even for a very tiny WDN. A simple WDN, consisting of just 1 ON/OFF pump, and considering an hourly resolution and a 24-hours horizon, leads to an optimization problem with 24 discrete decision variables and, consequently, 2^{24} (i.e., more than 16 million) possible pump schedules.

In this paper we have addressed PSO on a real-life WDN in Milan, Italy. The water distribution service in the urban area of Milan is managed by Metropolitana Milanese (MM) through a cyber-physical system consisting of: a transmission network (550 wells and pipelines bringing water to 29 storage and treatment facilities located inside the city) and a distribution network (26 different pump stations spread over the city - 95 pumps overall - which take water from the storage and treatment facilities and supply it to the final customers, about 1.350.000 habitants). Financial burden for MM is approximately 16.000.000 € for energy costs, with 45% of them due to pumping operations in the distribution network [42]. During the project ICeWater - co-founded by the European Union - a Pressure Management Zone (PMZ), namely "Abbiategrasso", was identified in the South of the city and isolated from the rest of the network. The EPANET model of the Abbiategrasso PMZ, reported in the following Fig. 1, was presented in [42]: it consists of 7855 nodes, 6073 pipes, 1961 valves, 1 reservoir and 4 pumps. Two of them, the oldest ones, are ON/OFF, while the other two, the newest ones installed in 2014, are VSPs. As previously reported in [42], preliminary results showed that two pumps are enough to match supply and demand within the PMZ. This consideration allows for limiting the number of decision variables (only two pumps are, at most, active simultaneously) and quantify which is the economic return of MM in using VSPs instead of ON/OFF pumps. Already in the ON/OFF pumps case (i.e., binary decision variables) the overall number of possible pump schedules is huge, more precisely $2^{2*24} = 2^{48}$ possible pump schedules, that is about $2.81 * 10^{14}$, in any case too large to perform an exhaustive search.

Fig. 1. EPANET model of the Pressure Management Zone (PMZ) "Abbiategrasso", in Milan, Italy

The Time-Of-Use (TOU) tariff in place for the real-life case study is:

- (low-price) 0.0626 €/kWh between 00:00–7:00 and between 23:00–24:00
- (mid-price) 0.0786 €/kWh between 08:00–19:00
- (high-price) 0.0806 €/kWh between 07:00–08:00 and between 19:00–23:00

All the available data for the study, including the reported energy tariff, are related to the year 2014.

3.3 SVM-CBO Settings

Experimental Setting for Test Functions
The proposed constrained SMBO framework was validated by considering an overall budget of 100 function evaluations, divided as follows:

- 10 for initialization through Latin Hypercube Sampling (LHS);
- 60 for feasibility estimation (phase 1)
- and 30 for SMBO constrained to estimated feasible region (phase 2)

This budget is clearly insufficient to solve the CGO test functions, that is SMBO cannot guarantee an optimal solution differing less than a given ε from the optimum value function [5]. On the other hand, it is important to highlight that the main benefit of SMBO is to obtain a good solution under very limited budget. Thus, the goal of the experiments was to validate whether the proposed approach is more effective, and efficient, than BO approaches which use penalties for function evaluations outside the feasible region.

In the case of BO with penalty, the overall budget has been divided as follows:

- 10 evaluations for initialization (LHS)
- and 90 for BO (that is the sum of budget used for phase 1 and phase 2 in the proposed approach).

It is important to highlight that, for each independent run, the initial set of solutions identified through LHS is the same for SVM-CBO and BO with penalty. This allows to avoid differences in the values of the gap metric due to different initialization.

To compare the approaches we have considered the so-called Gap metric [43, 44] that measures the improvement obtained along the SMBO process with respect to global optimum $f(x^*)$ and the initial best solution $f(x_0)$ obtained from the initialization step:

$$G_n = \frac{|f(x^+) - f(x_0)|}{|f(x^*) - f(x_0)|}$$

where $f(x^+)$ is the "best seen" up to iteration n. Gap metrics varies in the range $[0, 1]$.

For statistical significance, the gap metrics has been computed on 30 different runs, performed for every test function and for both SVM-CBO and BO with penalty.

Experimental Setting for Pump Scheduling Optimization
The number of overall objective function evaluations (i.e. EPANET simulation runs) has been fixed at 1200, organized as follows, for the SVM-CBO:

- 400 for the initial design, via LHS;
- 400 for phase 1 (feasibility determination)
- 400 for phase 2 (Constrained Bayesian Optimization on the estimate of the feasible region)

As far as the BO with penalty is concerned, the overall budget is dived in:

- 400 for initial design, via LHS
- 800 for the BO process

For a robust comparison, SVM-CBO and BO with penalty share the same initial design. They also share all the other choices, more precisely the probabilistic surrogate model (GP) and the acquisition function (LCB, with $\beta_n = 1$), so that the difference in the result can be only motivated by the usage, or not, of the estimate of the feasible region. Since the value $f(x^*)$ is not known for this case study, the Gap metrics cannot be used as metric. The minimum value of the objective function observed at the end of the approaches, that is the energy cost associated to the best seen pump schedule, has been directly considered as metric.

3.4 Computational Setting

All the experiments have been performed on the following system: intel i7 (3.4 GHz) processor with 4 cores and 16 Gb of RAM. The proposed approach has been developed starting from the R package named "mlrMBO", a flexible and comprehensive toolbox for model-based optimization (MBO). This package does not address CGO: we have extended it in order to deal with unknown constraints, feasibility determination and partially defined acquisition function.

4 Results

4.1 Results on Test Functions

The following set of figures show the Gap metric computed for every test function with respect to the number of function evaluations, excluding the initialization through LHS.

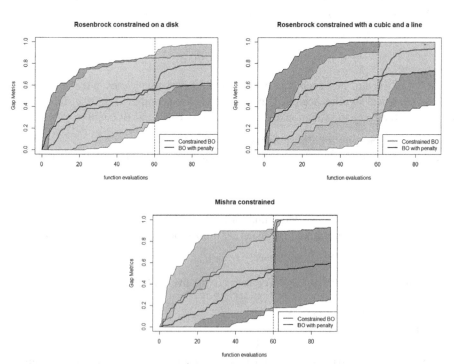

Fig. 2. Comparison of gap metrics for the proposed SVM-CBO approach vs BO with penalty, for each one of the five test functions, respectively

The value of Gap metric at iteration 0 is the best seen at the end of the initialization step (i.e., $f(x_0)$ in the Gap metric formula). Each graph compares the Gap metric provided by SVM-CBO and BO with penalty, respectively. Both the average and the standard deviation of the Gap metric, computed on 30 independent runs for each approach, are depicted. The higher effectiveness of the proposed approach is clear, even considering the variance in the performances. End of phase 1 is represented by a vertical dotted line in the charts; a significant improvement of the SVM-CBO's Gap metric is observed after this phase, in every test case. It is important to remind that phase 1 of the SVM-CBO is aimed at approximating the unknown feasible region, while the optimization process only starts with phase 2. Thus, the relevant shift in the Gap metric is motivated by the explicit model of the feasible region learned in phase 1.

Results on the two Rosenbrock test cases are particularly interesting: although the SVM-CBO's Gap metric is initially lower, on average, than that offered by BO with penalty, this situation abruptly changes when phase 2 starts (Fig. 2).

4.2 Results on the PSO Case Study

For this experiment the Gap metrics has no sense, because the BO with penalty was not able to provide any improvement with respect to the energy cost identified on the initial design, that is 172.89€. On the contrary, SVM-CBO was able to further reduce the energy to 168.60€, by identifying a new and more efficient pump schedule. A significantly relevant result is that the improvement has been registered exactly at the starting of phase 2, just like in the test functions. This result further proves that solving feasibility determination and constraining the following sequential optimization process to an accurate estimate of the feasible region can significantly improve the effectiveness and efficiency of the SMBO process.

As already reported by the authors in their previous work [33], assigning a penalty to function evaluations outside the feasible region has the aim to move towards feasible points/regions. Although this is usual workaround adopted in the PSO literature, it is quite naïve because the penalty does not depend on the entity of violation. This leads to almost "flat" objective functions, especially when the unknown feasible region results extremely narrow with respect to the overall search space. As already reported, the real-life PSO case study seems to be characterized by this kind of situation: in the following figure it is possible to see the how much narrow is feasible region with respect first (x) and the second (y) VSP, respectively. To have a 2D representation, the first hour of the day is considered, only, while the value of the z-value is the energy cost at the end of the day (only for feasible pump schedules) (Fig. 3).

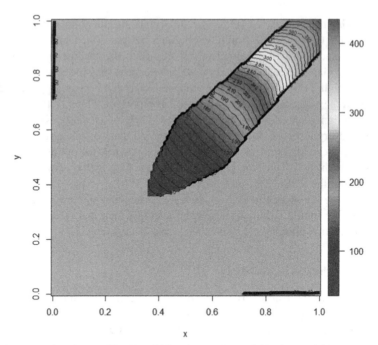

Fig. 3. Objective function and feasible region for a subset of decision variables of the real life PSO case study: *x* and *y* are the speed of the first and second pump, respectively, for the first hour of the day only.

Since BO gives, along the sequential optimization process, some chance to explore unseen regions of the search space, there is some opportunity to get to the feasible region. This chance increases when the optimization process is performed only within the current estimate of the feasible region (phase 2 of SVM-CBO), as proven by the results obtained.

5   Conclusions

The proposed SVM-CBO proved to be an effective and efficient approach for the constrained global optimization of black-box, multi-extremal, expensive and partially defined objective functions under unknown constraints. The capability to accurately estimate the boundary of the feasible region allows to both solve the feasibility estimation problem and to significantly improve with respect to Bayesian Optimization with penalty. The main limitation of SVM-CBO is that it does not allow any sensitivity analysis, since the SVM classifier models the overall boundary of the feasible region instead of each individual constraint.

Acknowledgements. This study has been partially supported by the Italian project "PerFORM WATER 2030" – programme POR (Programma Operativo Regionale) FESR (Fondo Europeo di Sviluppo Regionale) 2014–2020, innovation call "Accordi per la Ricerca e l'Innovazione" ("Agreements for Research and Innovation") of Regione Lombardia, (DGR N. 5245/2016 - AZIONE I.1.B.1.3 – ASSE I POR FESR 2014–2020) – CUP E46D17000120009.

References

1. Frazier, P.I.: Bayesian Optimization. In: Recent Advances in Optimization and Modeling of Contemporary Problems - INFORMS, pp. 255–278 (2018)
2. Jones, D.R., Schonlau, M., Welch, W.J.: Efficient global optimization of expensive black-box functions. J. Global Optim. **13**(4), 455–492 (1998)
3. Shahriari, B., Swersky, K., Wang, Z., Adams, R.P., De Freitas, N.: Taking the human out of the loop: A review of Bayesian Optimization. Proc. IEEE **104**(1), 148–175 (2016)
4. Žilinskas, A., Žilinskas, J.: Global optimization based on a statistical model and simplicial partitioning. Comput. Math Appl. **44**(7), 957–967 (2002)
5. Sergeyev, Y.D., Kvasov, D.E., Mukhametzhanov, M.S.: On the efficiency of nature-inspired metaheuristics in expensive global optimization with limited budget. Sci. Rep.-UK **8**(1), 453 (2018)
6. Sergeyev, Y.D., Strongin, R.G., Lera, D.: Introduction to Global Optimization Exploiting Space-Filling Curves. Springer, New York (2013). https://doi.org/10.1007/978-1-4614-8042-6
7. Sergeyev, Y.D., Kvasov, D.E.: Deterministic Global Optimization: An Introduction to the Diagonal Approach. Springer, Berlin (2017). https://doi.org/10.1007/978-1-4939-7199-2
8. Zhigljavsky, A., Zilinskas, A.: Stochastic Global Optimization, vol. 9. Springer, Berlin (2008). https://doi.org/10.1007/978-0-387-74740-8
9. Archetti, F., Betrò, B.: A priori analysis of deterministic strategies. In: Towards Global Optimization 2 (1978)
10. Archetti, F., Betrò, B.: Stochastic models and optimization. Bollettino dell'Unione Matematica Italiana **5**(17-A), 295–301 (1980)
11. Archetti, F., Betrò, B.: A probabilistic algorithm for global optimization. Calcolo **16**, 335–343 (1979)
12. Thornton, C., Hutter, F., Hoos, H.H., Leyton-Brown, K.: Auto-WEKA: combined selection and hyperparameter optimization of classification algorithms. In: Proceedings of ACM-SIGKDD, pp. 847–855 (2013)
13. Feurer, M., Klein, A., Eggensperger, K., Springenberg, J., Blum, M., Hutter, F.: Efficient and robust automated machine learning. In: Advances in Neural Information, pp. 2962–2970 (2015)
14. Candelieri, A., Archetti, F.: Global optimization in machine learning: the design of a predictive analytics application. Soft Comput. 1–9 (2018)
15. Galuzzi, B.G., Perego, R., Candelieri, A., Archetti, F.: Bayesian optimization for full waveform inversion. In: Daniele, P., Scrimali, L. (eds.) New Trends in Emerging Complex Real Life Problems. ASS, vol. 1, pp. 257–264. Springer, Cham (2018). https://doi.org/10.1007/978-3-030-00473-6_28
16. Sergeyev, Y.D., Pugliese, P., Famularo, D.: Index information algorithm with local tuning for solving multidimensional global optimization problems with multiextremal constraints. Math. Program. **96**(3), 489–512 (2003)
17. Paulavičius, R., Žilinskas, J.: Advantages of simplicial partitioning for Lipschitz optimization problems with linear constraints. Optim. Lett. **10**(2), 237–246 (2016)

18. Strongin, R.G., Sergeyev, Y.D.: Global Optimization with Non-Convex Constraints: Sequential and Parallel Algorithms. Springer, Boston (2013). https://doi.org/10.1007/978-1-4615-4677-1

19. Donskoi, V.I.: Partially defined optimization problems: an approach to a solution that is based on pattern recognition theory. J. Sov. Mathematics **65**(3), 1664–1668 (1993)

20. Rudenko, L.I.: Objective functional approximation in a partially defined optimization problem. J. Math. Sci. **72**(5), 3359–3363 (1994)

21. Sergeyev, Y.D., Kvasov, D.E., Khalaf, F.M.: A one-dimensional local tuning algorithm for solving GO problems with partially defined constraints. Optim. Lett. **1**(1), 85–99 (2007)

22. Hernández-Lobato, J.M., Gelbart, M.A., Adams, R.P., Hoffman, M.W., Ghahramani, Z.: A general framework for constrained Bayesian optimization using information-based search. J. Mach. Learn. Res. **17**(1), 5549–5601 (2016)

23. Gorji Daronkolaei, A., Hajian, A., Custis, T.: Constrained bayesian optimization for problems with piece-wise smooth constraints. In: Bagheri, E., Cheung, J.C.K. (eds.) Canadian AI 2018. LNCS (LNAI), vol. 10832, pp. 218–223. Springer, Cham (2018). https://doi.org/10.1007/978-3-319-89656-4_18

24. Picheny, V., Gramacy, R.B., Wild, S., Le Digabel, S.: Bayesian optimization under mixed constraints with a slack-variable augmented Lagrangian. In Advances in Neural Information Processing Systems, pp. 1435–1443 (2016)

25. Feliot, P., Bect, J., Vazquez, E.: A bayesian approach to constrained single-and multi-objective optimization. J. Global Optim. **67**(1–2), 97–133 (2017)

26. Gramacy, R.B., Lee, H.K.M., Holmes, C., Osborne, M.: Optimization Under Unknown Constraints. In: Bayesian Statistics 9 (2012)

27. Bernardo, J., et al.: Optimization under unknown constraints. Bayesian Stat. **9**(9), 229 (2011)

28. Hernández-Lobato, J.M., Gelbart, M.A., Hoffman, M.W., Adams, R.P., Ghahramani, Z.: Predictive entropy search for bayesian optimization with unknown constraints. In: Proceedings of the 32nd International Conference on Machine Learning, vol. 37 (2015)

29. Scholkopf, B., Smola, A.J.: Learning with Kernels: Support Vector Machines, Regularization, Optimization, and Beyond. MIT Press, Cambridge (2001)

30. Steinwart, I., Christmann, A.: Support Vector Machines. Springer, New York (2008). https://doi.org/10.1007/978-0-387-77242-4

31. Basudhar, A., Dribusch, C., Lacaze, S., Missoum, S.: Constrained efficient global optimization with support vector machines. Struct. Multidiscip. Optim. **46**(2), 201–221 (2012)

32. Tsai, Y.A., et al.: Stochastic optimization for feasibility determination: an application to water pump operation in water distribution network. In: Winter Simulation Conference 2018 (WSC 2018) December 9–12, Gothenburg, Sweden (2018)

33. Candelieri, A., Perego, R., Archetti, F.: Bayesian optimization of pump operations in water distribution systems. J. Global Optim. **71**(1), 213–235 (2018)

34. Letham, B., Karrer, B., Ottoni, G., Bakshy, E.: Constrained Bayesian optimization with noisy experiments. Bayesian Anal. **14**, 495–519 (2018)

35. Hartfiel, D.J., Curry, G.L.: On optimizing certain nonlinear convex functions which are partially defined by a simulation process. Math. Program. **13**(1), 88–93 (1977)

36. Srinivas, N., Krause, A., Kakade, S.M., Seeger, M.: Gaussian process optimization in the bandit setting: No regret and experimental design. In: Proceedings of International Conference on Machine Learning, pp. 1015–1022 (2010)

37. Neve, A.G., Kakandikar, G.M., Kulkarni, O.: Application of grasshopper optimization algorithm for constrained and unconstrained test functions. Int. J. Swarm Intel. EvolComput. **6**(165), 2 (2017)

38. Simionescu, P.A., Beale, D.G.: New concepts in graphic visualization of objective functions. In ASME 2002 International Design Engineering Technical Conferences and Computers and Information in Engineering Conference, pp. 891–897 (2002)

39. Mishra, S.K.: Some new test functions for global optimization and performance of repulsive particle swarm method. MPRA Paper No. 2718 (2008)
40. Castro Gama, M.E., Pan, Q., Salman, M.A.: Jonoski, Multivariate optimization to decrease total energy consumption in the water supply system of Abbiategrasso (Milan, Italy). Environ. Eng. Manag. J. **14**(9), 2019–2029 (2015)
41. Rossman, L.A.: EPANET2 User's Manual. U.S. Environmental Protection Agency, Washington, DC (2000)
42. Castro-Gama, M., Pan, Q., Lanfranchi, E.A., Jomoski, A., Solomatine, D.P.: Pump scheduling for a large water distribution network. Milan, Italy. Procedia Eng. **186**, 436–443 (2017)
43. Huang, D., Allen, T.T., Notz, W.I., Zheng, N.: Global optimization of stochastic black-box systems via sequential Kriging meta-models. J. Global Optim. **3**(34), 441–466 (2006)
44. Hoffman, M.D., Brochu, E., De Freitas, N.: Portfolio Allocation for Bayesian Optimization. In: UAI, pp. 327–336 (2011)

A Hessian Free Neural Networks Training Algorithm with Curvature Scaled Adaptive Momentum

Flora Sakketou$^{(\boxtimes)}$ and Nicholas Ampazis

Department of Financial and Management Engineering, University of the Aegean,
Chios, Greece
{fsakketou,n.ampazis}@fme.aegean.gr

Abstract. In this paper we propose an algorithm for training neural network architectures, called Hessian Free algorithm with Curvature Scaled Adaptive Momentum (HF-CSAM). The algorithm's weight update rule is similar to SGD with momentum but with two main differences arising from the formulation of the training task as a constrained optimization problem: (i) the momentum term is scaled with curvature information (in the form of the Hessian); (ii) the coefficients for the learning rate and the scaled momentum term are adaptively determined. The implementation of the algorithm requires minimal additional computations compared to a classical SGD with momentum iteration since no actual computation of the Hessian is needed, due to the algorithm's requirement for computing only a Hessian-vector product. This product can be computed exactly and very efficiently within any modern computational graph framework such as, for example, Tensorflow. We report experiments with different neural network architectures trained on standard neural network benchmarks which demonstrate the efficiency of the proposed method.

Keywords: Training algorithms · Deep neural networks · Optimization

1 Introduction

The most commonly used method for training deep neural networks is, undoubtedly, Stochastic Gradient Descent (SGD) [15]. At any time step t, SGD computes the gradient of the network's loss/cost function $\mathcal{L}(\boldsymbol{w}_t)$ over subsets of the training set (known as *mini-batches*) with respect to the network's synaptic weights (which we can group into a single column vector $\boldsymbol{w}$). The iterations of SGD can then be described as:

$$\boldsymbol{w}_{t+1} = \boldsymbol{w}_t + d\boldsymbol{w}_t = \boldsymbol{w}_t - \eta \boldsymbol{G}_t \tag{1}$$

where $\boldsymbol{G}_t$ is the gradient at the current time step t and η the leaning rate which is heuristically determined.

© Springer Nature Switzerland AG 2020
N. F. Matsatsinis et al. (Eds.): LION 13 2019, LNCS 11968, pp. 94–105, 2020.
https://doi.org/10.1007/978-3-030-38629-0_8

SGD can be extended with the incorporation of a momentum term into its update rule in order to accelerate learning [17]. The momentum term is simply vector of weight updates at the previous time step, and thus its iterations can be described as:

$$\boldsymbol{w}_{t+1} = \boldsymbol{w}_t + d\boldsymbol{w}_t \tag{2}$$

$$d\boldsymbol{w}_t = -\eta \boldsymbol{G}_t + \alpha d\boldsymbol{w}_{t-1} \tag{3}$$

where α is known as the momentum coefficient. The momentum term actually mimics the inclusion of second order information in the training process and provides iterations whose form is similar to the Conjugate Gradient (CG) method [4]. The major difference with the CG method, however, is that the coefficients regulating the weighting between the gradient and the momentum term are heuristically selected, whereas in the CG algorithm these coefficients are adaptively determined.

One disadvantage of SGD is that the gradient is scaled (by the learning rate) uniformly in all directions, which is known to hinder learning for most real world problems. Thus, due to these problematic weights updates, the weights of the network themselves eventually become poorly scaled and increasingly harder to optimize [9]. It is interesting to note that the incorporation of the momentum term in the SGD weight update rule does not alleviate this problem since it simply adds an extra term which is also uniformly scaled in all directions (at the previous time step). In order to address this problem, a number of adaptive methods have been proposed which attempt to diagonally scale the gradient via estimates of the cost function's curvature. The most well know examples of such methods are Adam [6] and RMSprop [18]. In essence these methods compute an adaptive learning rate for scaling the gradient of each individual weight of the network.

Ideally, the gradient should be rescaled by taking into account the curvature information, which is exactly the update rule given by Newton's method [1]. Unfortunately the computational cost for computing the Hessian matrix and its inverse renders the method prohibitive for most practical problems. However regardless of our inability to practically utilize Newton's method, we could still exploit the implicit scaling provided by the curvature information (in the form of the Hessian).

In this paper we illustrate that such a methodology can be achieved by formulating the training task as a constrained optimization problem whose solution effectively offers the necessary framework for successfully incorporating a curvature scaled momentum term into an SGD with momentum iteration. In our derived learning rule, the momentum term is scaled by the local curvature information given by the Hessian which effectively shifts the whole weight update vector towards eigendirections of the curvature matrix. This has the beneficial effect to shift updates away from directions of low-curvature which, by definition, persistently provide slower changes and are amplified by classical SGD with momentum [17]. The proposed algorithm derived from our framework, called Hessian Free with Curvature Scaled Adaptive Momentum (HF-CSAM) requires no actual computation of the Hessian due to its requirement for computing only a Hessian-vector product. This product can be computed exactly and very efficiently within any modern computational graph framework such as, for example, within the Tensorflow framework [14]. An additional benefit of the proposed algorithm is that unlike SGD with momentum, where the values of the hyperparameters for the learning rate and the momentum coefficient are constant, its corresponding hyperparameters are adaptively determined.

2   The HF-CSAM Algorithm

The main idea in the formulation of the algorithm is that while trying to minimize the network's loss/cost function $\mathcal{L}(\boldsymbol{w}_t)$ at time step t, then a one-dimensional minimization in the direction $\boldsymbol{dw}_{t-1}$ at the previous time step $t-1$ followed by a second minimization in the direction $\boldsymbol{dw}_t$ at the current time step t, does not guarantee that the cost function has been minimized on the subspace spanned by both of these directions. This can be achieved by the selection of *conjugate* directions which form the basis of the CG method.

Two vectors $\boldsymbol{dw}_t$ and $\boldsymbol{dw}_{t-1}$ are non-interfering or mutually conjugate with respect to $\boldsymbol{H}_t = \nabla^2 \mathcal{L}(\boldsymbol{w}_t)$ when

$$\boldsymbol{dw}_t{}^T \boldsymbol{H}_t \boldsymbol{dw}_{t-1} = 0 \tag{4}$$

Therefore, the objective is to reach a minimum of the cost function $\mathcal{L}(\boldsymbol{w}_t)$ (from now on abbreviated simply as $\mathcal{L}_t$) with respect to the synaptic weights, and simultaneously to maximize *incrementally* at each epoch the following quantity, without compromising the need for a decrease in the cost function:

$$\Phi_t = \boldsymbol{dw}_t{}^T \boldsymbol{H}_t \boldsymbol{dw}_{t-1} \tag{5}$$

The strategy which we adopt for the solution of this constrained optimization problem follows the methodology for incorporating additional knowledge in the form of constraints in neural network training originally proposed in [13]. To this end, we will adopt an epoch-by-epoch (i.e. *mini-batch*) optimization framework with the following objectives:

1. At each epoch t of the learning process, the vector $\boldsymbol{w}_t$ will be incremented by dw_t, so that the search for an optimum new point in the space of $\boldsymbol{w}_t$ is restricted to a hypersphere of known radius δP centered at the point defined by the current $\boldsymbol{w}_t$

$$\boldsymbol{dw}_t^T \boldsymbol{dw}_t = (\delta P)^2 \tag{6}$$

2. At each epoch t, the objective function $\mathcal{L}_t$ must be decremented by a quantity δQ_t, so that, at the end of learning, $\mathcal{L}$ is rendered as small as possible. To first order, we can substitute the change in $\mathcal{L}_t$ by its first differential and demand that

$$d\mathcal{L}_t = \delta Q_t \tag{7}$$

Consequently the learning rule can be derived by solving the following constrained optimization problem:

$$\text{Maximize equation (5)} \ (\Phi_t = max) \ \text{w.r.t} \ d\boldsymbol{w}_t$$

$$\text{subject to} \quad d\boldsymbol{w}_t^T d\boldsymbol{w}_t = (\delta P)^2$$

$$d\mathcal{L}_t = \delta Q_t$$

This constrained optimization problem can be solved analytically by a method similar to the constrained gradient ascent technique introduced in optimal control in [2], and leads to a generic update rule for $\boldsymbol{w}$ as follows:

First, we introduce suitable *Lagrange* multipliers λ_1 and λ_2 to take into account Eqs. (7) and (6) respectively. If δP is small enough, the changes to Φ_t induced by changes in $\boldsymbol{w}_t$ can be approximated by the first differential $d\Phi_t$. Thus, secondly, we introduce the function ϕ_t

$$\phi_t = \Phi_t + \lambda_1(\delta Q_t - d\mathcal{L}_t) + \lambda_2[(\delta P)^2 - d\boldsymbol{w}_t^T d\boldsymbol{w}_t] \tag{8}$$

On evaluating the differentials involved in the right hand side and substituting Φ_t, we readily obtain

$$\phi_t = d\boldsymbol{w}_t^T \cdot \boldsymbol{F}_t + \lambda_1(\delta Q_t - \boldsymbol{G}_t \cdot d\boldsymbol{w}_t) + \lambda_2[(\delta P)^2 - d\boldsymbol{w}_t^T d\boldsymbol{w}_t] \tag{9}$$

where $\boldsymbol{F}_t$ and $\boldsymbol{G}_t$ are given by

$$\boldsymbol{F}_t = \partial \Phi_t / \partial \boldsymbol{w}_t = \boldsymbol{H}_t d\boldsymbol{w}_{t-1} \ , \quad \boldsymbol{G}_t = \partial \mathcal{L}_t / \partial \boldsymbol{w}_t \tag{10}$$

To maximize ϕ_t at each epoch, we demand that

$$d\phi_t = (\boldsymbol{F}_t - \lambda_1 \boldsymbol{G}_t - 2\lambda_2 d\boldsymbol{w}_t) \cdot d^2 \boldsymbol{w}_t = 0 \tag{11}$$

and

$$d^2 \phi_t = -2\lambda_2 d^2 \boldsymbol{w}_t^T d^2 \boldsymbol{w}_t < 0 \tag{12}$$

Hence, the factor multiplying $d^2 \boldsymbol{w}_t$ in Eq. (11) should vanish, and therefore we obtain

$$d\boldsymbol{w}_t = -\frac{\lambda_1}{2\lambda_2}\boldsymbol{G}_t + \frac{1}{2\lambda_2}\boldsymbol{F}_t$$

$$= -\frac{\lambda_1}{2\lambda_2}\boldsymbol{G}_t + \frac{1}{2\lambda_2}\boldsymbol{H}_t \cdot d\boldsymbol{w}_{t-1} \tag{13}$$

Equation (13) constitutes the weight update rule, provided that λ_1 and λ_2 can be evaluated in terms of known quantities. This can be done as follows:

From Eqs. (6), (10) and (11) we obtain

$$\lambda_1 = \frac{I_{GF} - 2\lambda_2 \delta Q_t}{I_{GG}} \tag{14}$$

with I_{GG} and I_{GF} given by

$$I_{GG} = ||\boldsymbol{G}_t||^2, \quad I_{GF} = \boldsymbol{G}_t^T \boldsymbol{F}_t \tag{15}$$

It remains to evaluate λ_2. To this end, we substitute (13) into (7) to obtain

$$4\lambda_2^2(\delta P)^2 = I_{FF} + \lambda_1^2 I_{GG} - 2\lambda_1 I_{GF} \tag{16}$$

where I_{FF} is given by

$$I_{FF} = \boldsymbol{F}_t^T \boldsymbol{F}_t \tag{17}$$

Finally, we substitute (14) into (16) and solve for λ_2 to obtain

$$\lambda_2 = \frac{1}{2}\left[\frac{I_{GG}(\delta P)^2 - (\delta Q_t)^2}{I_{FF}I_{GG} - I_{GF}^2}\right]^{-1/2} \tag{18}$$

Note that the positive square root value has been chosen for λ_2 in order to satisfy Eq. (12).

In effect, the weight updates as given by Eq. (13) are formed at each epoch as a linear combination of the cost function's gradient $\boldsymbol{G}_t$ with respect to the network's weights and of the product between the Hessian at the current epoch $\boldsymbol{H}_t$ and the vector of weight updates $d\boldsymbol{w}_{t-1}$ at the immediately preceding epoch. This weight update rule is similar to that of SGD with momentum but with two major differences.

Firstly, the most obvious difference is that the momentum term $d\boldsymbol{w}_{t-1}$ is weighted by the current curvature of the cost function (in the form of the Hessian). This has the advantage of taking into account the underlying geometry of the space defined by the synaptic weights. The weighting of the momentum term by a matrix which has the same shape with the local quadratic approximation of the cost function, reflects the scaling of the problem and allows for the correct weighting of its coordinates among all possible directions. This has the effect that directions for which the model may differ most from the true function are restricted more than those directions for which the curvature is small [5]. An important observation here is that we do not actually need to explicitly compute the Hessian matrix, but we rather need its product with the vector $d\boldsymbol{w}_{t-1}$. For functions that can be computed using a computational graph there are automatic methods available for computing Hessian-vector products exactly, e.g. by using the $\mathcal{R}_v\{.\}$ operator [11].

The second important difference of the proposed algorithm, is that unlike SGD with momentum, where the values of the hyperparameters for the learning rate and the momentum coefficient are constant, the corresponding hyperparameters in the weight update rule are chosen adaptively as shown by Eqs. (14) and (18).

Let us now discuss our choice for δQ. This choice is dictated by the demand that the quantity under the square root in Eq. (18) should be positive. It is easy to show that the term $I_{FF}I_{GG} - I_{GF}^2$ is always positive by the Cauchy-Schwarz inequality [12]. Now, since $I_{GG} = ||\boldsymbol{G}_t||^2 \geq 0$, it follows that care must be taken to ensure that $I_{GG}(\delta P)^2 > (\delta Q)^2$. The simplest way to achieve this is to select δQ adaptively by setting

$$\delta Q = -\xi \delta P \sqrt{I_{GG}} \tag{19}$$

with $0 < \xi < 1$. Consequently, the proposed generic weight update algorithm has two free parameters, namely δP and ξ.

An important observation is that the algorithm's objective is to reach a minimum of the cost function with respect to the weight vector w while simultaneously trying to maintain the conjugacy between successive minimization directions, through the maximization of the quantity Φ_t given by Eq. (5). Since this conjugacy can be achieved only when $\Phi_t = 0$, this means that we have already made the assumption that Φ_t is bounded above by zero, that is:

$$\Phi_t \leq 0 \tag{20}$$

In order to test the validity of this assumption, we can substitute (5), (13), (15) and (17) in the above relation, so that we can directly obtain

$$-\frac{\lambda_1}{2\lambda_2}G_t^T dw_{t-1} + \frac{1}{2\lambda_2}dw_{t-1}^T F_t \leq 0$$
$$-\lambda_1 I_{GF} + I_{FF} \leq 0 \tag{21}$$

Lagrange multiplier λ_1 is given by (14) in which the expression for the second Lagrange multiplier λ_2 is involved. Therefore, by substituting (19) into (18) we can obtain

$$\lambda_2 = \frac{1}{2}\left[\frac{A}{I_{GG}(\delta P)^2(1-\xi^2)}\right]^{1/2} \tag{22}$$

where

$$A = I_{FF}I_{GG} - I_{GF}^2 \tag{23}$$

Based on the relations (19), (22) and (23), Eq. (14) can, therefore, be written as

$$\lambda_1 = \frac{I_{GF} + \left[\frac{A}{(1-\xi^2)}\right]^{1/2}\xi}{I_{GG}} \tag{24}$$

Substituting the above expression into relation (21) and taking into account (23) we can obtain the result

$$I_{GF}\left[\frac{A}{(1-\xi^2)}\right]^{1/2}\xi \geq A \tag{25}$$

Due to the fact that A and ξ are positive, the above inequality can hold only when $I_{GF} > 0$. The quantity I_{GF} is the inner product between the current gradient vector G_t and the vector computed by the product of the Hessian with the momentum term dw_{t-1}. If, at every epoch, the size of the weight changes was determined by an exact line minimization technique then this inner product would be equal to zero. In our case where the size of the step is limited within a hypersphere, due to (6), the sign of I_{GF} is determined by the value of the parameter δP. If this value is large then the movement along the direction dw_{t-1} overshoots the minimum resulting in the sign of I_{GF} being negative. On the other hand if the size of δP is conservatively selected, then I_{GF} is positive and hence Eq. (25) holds. Therefore extra care should be taken at every epoch for monitoring the sign of I_{GF}. In case that the sign is positive then the weight update rule is given by Eq. (13) due to the demand for maximization of Φ_t. On the contrary, if due to the size of δP the sign of I_{GF} is negative, then in this case we demand the minimization of the quantity Φ_t. The effect that this demand has on the learning rule is actually minimal since the only expression that changes in the optimization formalism is the sign on the right-hand side of (18), which is the expression for calculating Lagrange multiplier λ_2 (which in this case should be negative). Therefore, in this way, we not only ensure that the quantity Φ_t is maximized (or minimized) appropriately, but also that the minimum (or maximum) that we seek is equal to zero.

3   Experimental Results

The performance of the proposed HF-CSAM algorithm was evaluated on the training of multilayer networks (MLP) and Convolutional Neural Networks (CNN) on two standard benchmark datasets, namely the MNIST [8] and CIFAR10 [7] datasets. Details on the network architectures for each of these benchmarks are mentioned on the corresponding paragraphs of this section discussing the performance of the algorithms for each of these problems.

Due to its nature, the performance of HF-CSAM was evaluated against the following neural network training algorithms: SGD with momentum, Adam and RMSprop. All experiments were carried within the Tensorflow framework using the Keras high-level API [3]. The algorithm's source code as well as all the helper scripts for running the experiments are available at Github[1]. For each dataset and for each network architecture we performed 5 training trials starting from different random initializations of the networks' weights. In all trials Adam and RMSProp were employed with their default learning rate values as defined in their Keras API implementation, that is 0.001. For SGD we utilized the default learning rate value defined in the Keras API implementation (0.01), and we chose a sensible fixed value for the momentum coefficient (0.7) for all trials. For HF-CSAM the value of ξ was set to 0.99 for all trials. For the MNIST dataset we set a value of $\delta P = 0.1$ for the MLP network and $\delta P = 0.3$ for the CNN network. For the CIFAR10 dataset the value of δP was set to 0.07 for both networks. However similar performances were recorded with $0.8 < \xi < 0.99$ and $0.05 < \delta P < 0.5$, indicating that the results are not very sensitive to the exact values of these parameters.

In order to obtain fair comparisons, the weights initializations of the networks trained with the aforementioned algorithms were initialized from the same random seed for the 5 trials.

---

[1] https://github.com/flo3003/HF-CSAM.

3.1   Results on the MNIST Dataset

Table 1. Neural networks architectures for the MNIST dataset: (a) MLP and (b) CNN

```
model = Sequential()
model.add(Dense(512, bias_initializer='zeros', activation='relu', input_shape
                =(784,)))
model.add(Dropout(0.2))
model.add(Dense(512, bias_initializer='zeros', activation='relu'))
model.add(Dropout(0.2))
model.add(Dense(num_classes, bias_initializer='zeros', activation='softmax'))
```

(a)

```
model = Sequential()
model.add(Conv2D(32, bias_initializer='zeros', kernel_size=(3, 3),
                 activation='relu',
                 input_shape=input_shape))
model.add(Conv2D(64, (3, 3), bias_initializer='zeros', activation='relu'))
model.add(MaxPooling2D(pool_size=(2, 2)))
model.add(Dropout(0.25))
model.add(Flatten())
model.add(Dense(128, bias_initializer='zeros', activation='relu'))
model.add(Dropout(0.5))
model.add(Dense(num_classes, bias_initializer='zeros', activation='softmax'))
```

(b)

Table 1 shows the Keras API definitions of the neural network architectures utilized for the MNIST dataset. As shown in Table 1(a) the MLP network consists of two dense hidden layers, each with 512 units with RELU activations [10] and a dropout rate [16] of 0.2, and a dense output softmax layer for the 10 classes of the problem. Table 1(b) shows the CNN architecture for the same problem which consists of two sequential 2D convolution layers with RELU activation functions, of size 32 and 64 respectively, followed by a max pooling 2D layer with dropout rate of 0.25. This feeds into a dense hidden layer of 128 units with RELU activations and a dropout rate of 0.5. Finally, as is the case with MLP, the CNN network has a dense output softmax layer for the 10 classes of the problem.

Figure 1 shows the experimental results obtained when training the neural network architectures with the different optimization algorithms. For each algorithm we report the results obtained for the average validation loss and accuracy over the 5 training trials.

From Fig. 1(a) we can see that, for the MLP case, HF-CSAM achieves the lowest validation loss, followed (in ascending order) by Adam, SGD, and RMSprop. HF-CSAM achieves around 37% lower validation loss compared to the worst performer (RMSprop), and around 12% lower validation loss compared to Adam, which is usually considered to be the state of the art neural networks training algorithm. For the CNN architecture HF-CSAM achieves 24% lower validation loss than that of the worst performer (in this case SGD) and comparable performance (to the order of 3 decimal places) to that of Adam and RMSProp.

From Fig. 1(b) we can see that for the MLP architecture, and at this level of validation loss, HF-CSAM outperforms SGD in terms of validation accuracy,

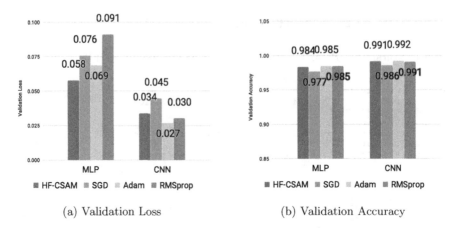

(a) Validation Loss   (b) Validation Accuracy

Fig. 1. Experimental results of the MNIST dataset: (a) Validation Loss, (b) Validation Accuracy

and is tied with both Adam and RMSprop. The exact same behavior also holds for the CNN architecture where HF-CSAM, Adam, and RMSprop practically achieve the same level of validation accuracy.

3.2   Results on the CIFAR10 Dataset

Table 2 shows the Keras API definitions of the neural network architectures utilized for the CIFAR10 dataset. As shown in Table 2(a) the MLP network consists of three dense hidden layers, the first with 1024 units and the second and third with 512 units with RELU activations and a dropout rate of 0.2, and a dense output softmax layer for the 10 classes of the problem. Table 2(b) shows the CNN architecture for the same problem which consists of four sequential 2D convolution layers with RELU activation functions, the first two of size 32 and next two of size 64, the second and fourth layers are followed by a max pooling 2D layer with dropout rate of 0.25. This feeds into a dense hidden layer of 512 units with RELU activations and a dropout rate of 0.5. Finally, as is the case with MLP, the CNN network has a dense output softmax layer for the 10 classes of the problem.

Figure 2 shows the experimental results obtained when training the neural network architectures with the different optimization algorithms. Again, for each algorithm we report the results obtained for the average validation loss and accuracy over the 5 training trials.

From Fig. 2(a) we can see that, for the MLP case, HF-CSAM achieves the lowest validation loss, followed (in ascending order) by SGD, Adam, and RMSprop. HF-CSAM achieves around 15% lower validation loss compared to the worst performer (RMSprop), and around 6% lower validation loss compared to Adam. For the CNN architecture all algorithms practically achieve the same level of validation loss.

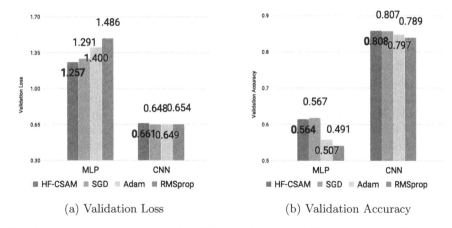

(a) Validation Loss   (b) Validation Accuracy

Fig. 2. Experimental results of the CIFAR10 dataset: (a) Validation Loss, (b) Validation Accuracy

Table 2. Neural networks architectures for the CIFAR10 dataset: (a) MLP and (b) CNN

```
model = Sequential()
model.add(Flatten(input_shape=(32, 32, 3)))
model.add(Dense(1024, bias_initializer='zeros'))
model.add(Activation('relu'))
model.add(Dropout(0.2))
model.add(Dense(512, bias_initializer='zeros'))
model.add(Activation('relu'))
model.add(Dropout(0.2))
model.add(Dense(512, bias_initializer='zeros'))
model.add(Activation('relu'))
model.add(Dropout(0.2))
model.add(Dense(10, bias_initializer='zeros'))
model.add(Activation('softmax'))
```

(a)

```
model = Sequential()
model.add(Conv2D(32, (3, 3), padding='same', input_shape=x_train.shape[1:],
                    bias_initializer='zeros'))
model.add(Activation('relu'))
model.add(Conv2D(32, (3, 3), bias_initializer='zeros'))
model.add(Activation('relu'))
model.add(MaxPooling2D(pool_size=(2, 2)))
model.add(Dropout(0.25))
model.add(Conv2D(64, (3, 3), padding='same', bias_initializer='zeros'))
model.add(Activation('relu'))
model.add(Conv2D(64, (3, 3), bias_initializer='zeros'))
model.add(Activation('relu'))
model.add(MaxPooling2D(pool_size=(2, 2)))
model.add(Dropout(0.25))
model.add(Flatten())
model.add(Dense(512, bias_initializer='zeros'))
model.add(Activation('relu'))
model.add(Dropout(0.5))
model.add(Dense(num_classes, bias_initializer='zeros'))
model.add(Activation('softmax'))
```

(b)

From Fig. 2(b) we can see that for the MLP architecture, and at this level of validation loss, HF-CSAM achieves similar validation accuracy level with that of SGD which is around 12% higher than the accuracy achieved by Adam and around 16% higher than that of RMSprop. For the CNN architecture, HF-CSAM achieves the best validation accuracy, followed (in descending order) by SGD, Adam, and RMSprop. It is very interesting to note that for this problem, and for both the MLP and CNN architectures, the validation accuracy performances of Adam and RMSprop are inferior to the performance of SGD. This is in agreement with the findings of [19] which highlights the inability of adaptive methods such as Adam and RMSprop to perform on par with SGD in terms of generalization ability. The authors of [19] suggest that the lack of generalization performance of adaptive methods stems from the non-uniform scaling of the gradient. However as we observe from the figure, HF-CSAM generalizes in a fashion comparable to that of SGD despite its non-uniform scaling in its weight update rule. This is quite intriguing, and we believe that it can be attributed to the uniform scaling of the gradient of HF-CSAM's weight updates. HF-CSAM does not scale the gradient directly in a non-uniform fashion, but utilizes the curvature information in the momentum term. Thus, its sequence of updates may allow HF-CSAM to shift away from paths followed by adaptive methods such as Adam and RMSprop, and to converge to points with similar generalization performance with that of SGD, while maintaining its adaptive benefits. We obviously plan to investigate this interesting behavior in more detail in the near future.

4   Conclusions

In this paper we have proposed an efficient algorithm (HF-CSAM) for the training of deep neural networks. The algorithm has been derived from the formulation of the training task as a constrained optimization problem attempting to introduce conjugate directions of motion within a framework similar to that of the CG algorithm and SGD with momentum. The algorithm produces iterations similar to SGD with an additional adaptive momentum term that is scaled by the Hessian. However no actual computation of the Hessian is needed, due to the requirement for computing only a Hessian-vector product, which can be computed exactly and very efficiently within any modern computational graph framework. In addition the algorithm's hyperparameters for the gradient scaling and the momentum coefficient are adaptively determined, even though its formalism involves two free parameters which should be adjusted to achieve optimum performance. Experiments with different neural network architectures on standard neural network benchmarks revealed that its results are not very sensitive to the exact values of the free parameters, and that its performance is competitive to that of other adaptive training algorithms. The same results also revealed that its generalization ability is similar to that of SGD which in recent studies has been shown to be superior to that of more advanced adaptive methods. All these findings point to the conclusion that the proposed algorithm stands as a promising new tool for the efficient training of deep neural networks whenever the employment of adaptive methods is required.

Funding. Mrs. Sakketou is supported by a Ph.D. Scholarship by the State Scholarships Foundation (IKY), Greece.

References

1. Battiti, R.: First and second-order methods for learning: between steepest descent and Newton's method. Neural Comput. **4**, 141–166 (1992)
2. Bryson, A.E., Denham, W.F.: A steepest-ascent method for solving optimum programming problems. J. Appl. Mech. **29**, 247–257 (1962)
3. Chollet, F., et al.: Keras (2015). https://keras.io
4. Gilbert, J.C., Nocedal, J.: Global convergence properties of conjugate gradient methods for optimization. SIAM J. Optim. **2**(1), 21–42 (1992)
5. Gould, N.I.M., Nocedal, J.: The modified absolute-value factorization norm for trust-region minimization. In: De Leone, R., Murli, A., Pardalos, P.M., Toraldo, G. (eds.) High Performance Algorithms and Software in Nonlinear Optimization. Applied Optimization, vol. 24, pp. 225–241. Springer, Boston (1998). https://doi.org/10.1007/978-1-4613-3279-4_15
6. Kingma, D.P., Ba, J.: Adam: a method for stochastic optimization. CoRR, abs/1412.6980 (2014)
7. Krizhevsky, A.: Learning multiple layers of features from tiny images. Technical report, Canadian Institute for Advanced Research (2009)
8. LeCun, Y., Cortes, C.: MNIST handwritten digit database (2010)
9. Martens, J.: Deep learning via Hessian-free optimization. In: Proceedings of the 27th International Conference on Machine Learning (ICML-2010), Haifa, Israel, 21–24 June 2010, pp. 735–742 (2010)
10. Nair, V., Hinton, G.E.: Rectified linear units improve restricted Boltzmann machines. In: Proceedings of the 27th International Conference on International Conference on Machine Learning, ICML 2010, USA, pp. 807–814. Omnipress (2010)
11. Pearlmutter, B.A.: Fast exact multiplication by the Hessian. Neural Comput. **6**(1), 147–160 (1994)
12. Perantonis, S.J., Ampazis, N., Virvilis, V.: A learning framework for neural networks using constrained optimization methods. Ann. Oper. Res. **99**, 385–401 (2000)
13. Perantonis, S.J., Karras, D.A.: An efficient constrained learning algorithm with momentum acceleration. Neural Networks **8**(2), 237–239 (1995)
14. Ramsundar, B., Zadeh, R.B.: TensorFlow for Deep Learning: From Linear Regression to Reinforcement Learning, 1st edn. O'Reilly Media Inc., Sebastopol (2018)
15. Robbins, H., Monro, S.: A stochastic approximation method. Ann. Math. Stat. **22**(3), 400–407 (1951)
16. Srivastava, N., Hinton, G., Krizhevsky, A., Sutskever, I., Salakhutdinov, R.: Dropout: a simple way to prevent neural networks from overfitting. J. Mach. Learn. Res. **15**, 1929–1958 (2014)
17. Sutskever, I., Martens, J., Dahl, G., Hinton, G.: On the importance of initialization and momentum in deep learning. In: Proceedings of the 30th International Conference on International Conference on Machine Learning, ICML 2013, vol. 28, pp. III–1139–III–1147 (2013). www.JMLR.org
18. Tieleman, T., Hinton, G.: Lecture 6.5—RMSProp: Divide the gradient by a running average of its recent magnitude. COURSERA: Neural Networks for Machine Learning (2012)
19. Wilson, A.C., Roelofs, R., Stern, M., Srebro, N., Recht, B.: The marginal value of adaptive gradient methods in machine learning. In: Guyon, I., et al. (eds.) Advances in Neural Information Processing Systems, vol. 30, pp. 4148–4158. Curran Associates Inc. (2017)

Irreducible Bin Packing: Complexity, Solvability and Application to the Routing Open Shop

Ilya Chernykh[1,2,3]($\boxtimes$) and Artem Pyatkin[1,2]

[1] Sobolev Institute of Mathematics, Koptyug Avenue 4, Novosibirsk 630090, Russia
{idchern,artem}@math.nsc.ru
[2] Novosibirsk State University, Pirogova Street 2, Novosibirsk 630090, Russia
[3] Novosibirsk State Technical University,
Marksa Avenue 20, Novosibirsk 630073, Russia

Abstract. We introduce the following version of an "inefficient" bin packing problem: maximize the number of bins under the restriction that the total content of any two bins is larger than the bin capacity. There is a trivial upper bound on the optimum in terms of the total size of the items. We refer to the decision version of this problem with the number of bins equal to the trivial upper bound as Irreducible Bin Packing. We prove that this problem is NP-complete in an ordinary sense and derive a sufficient condition for its polynomial solvability. The problem has a certain connection to a routing open shop problem which is a generalization of the metric TSP and open shop, known to be NP-hard even for two machines on a 2-node network. So-called job aggregation at some node of a transportation network can be seen as an instance of a bin packing problem. We show that for a two-machine case a positive answer to the Irreducible Bin Packing problem question at some node leads to a linear algorithm of constructing an optimal schedule subject to some restrictions on the location of that node.

Keywords: Bin packing to the maximum · Irreducible Bin Packing · Routing open shop · Job aggregation · Superoverloaded node · Polynomially solvable subcase · Efficient normality

1 Introduction

The classical Bin Packing Problem (BPP) is probably one of the most studied and discussed combinatorial optimization problems. It can be formulated as follows. Given a *bin capacity* B and a set of n integer numbers from $(0, B]$ (representing *sizes* of items) one have to pack items into some number of bins

This research was supported by the program of fundamental scientific researches of the SB RAS No. I.5.1., project No. 0314-2019-0014, by the Russian Foundation for Basic Research, projects 17-01-00170 and 18-01-00747, and by the Russian Ministry of Science and Education under the 5–100 Excellence Programme.

N. F. Matsatsinis et al. (Eds.): LION 13 2019, LNCS 11968, pp. 106–120, 2020.
https://doi.org/10.1007/978-3-030-38629-0_9

not exceeding each bin's capacity. The classical objective is to minimize the number of bins used. In other words, the set of items should be partitioned into the minimum number of subsets so that the total size of the items in each subset would be at most B. The BPP is one of the first known NP-hard problems [17,20]. A survey of various approximation algorithms for it can be found in [13].

Many variants and generalizations of BPP have been studied over the last decades, including but not limited to the following list.

- Geometric BPP including 2-dimensional [25], 3-dimensional [26] and multidimensional [11] BPP, various forms of bins and items [4–6].
- Online version of BPP [13,14,28].
- Additional restrictions on elements and bins (conflicts, precedence constraints, etc) [16,24,27].
- Inverse BPP—maximizing the number or the total size of packed items into a fixed number of bins [12].
- Bin Covering problem (BCP) [1,15,19]—the capacity of bins is unbounded and the goal is to maximize the number of bins containing items of the total size of at least given threshold W.
- Maximum resource BPP [7]—to maximize the number of bins provided there exists such an ordering of bins that no item in a current bin can fit into any previous bin.

The latter two problems can be seen as examples of bin packing with "inefficient" objective to maximize the number of bins, which stands in contrast to the classical BPP goal. Some motivation for that is suggested in [7]: consider the situation from the point of view of some transportation company which receives payment proportional to the number of trucks used to ship a set of items for the client. The objective to maximize the number of trucks serves in that company's best interest. However some restrictions on the feasibility of packing are necessary for at least the two following reasons: so that the client does not consider himself deceived, and so that the optimization problem wouldn't be too trivial to investigate. The Bin Covering problem and the Maximum resource BPP offer different restrictions for that matter.

The problem considered in this paper has the same objective to maximize the number of bins with the following restriction: the total size of the contents of any two bins should be strictly greater than the bin capacity. Note that such a formulation was suggested at [7], but to the best of our knowledge has not been studied yet. We call our problem the *Irreducible Bin Packing* (IBP) to stress out that the number of bins in any solution cannot be reduced by combining of any two bins. The motivation from [7] fits our problem as well. However we have totally different reason to study such a problem, as it turned out to be very helpful for the research of the *routing open shop* problem—a generalization of the metric traveling salesman problem (TSP) and the open shop scheduling problem. The metric TSP hardly needs a special introduction. The open shop problem ([18]) can be described as follows. Given a set of machines $\mathcal{M} = \{M_1, \ldots, M_m\}$

and a set of jobs $\mathcal{J} = \{J_1, \ldots, J_n\}$ one has to perform operations of each job by each machine. A non-negative processing time p_{ji} is given for each *operation* (which is essentially a pair (J_j, M_i)). Operations of the same job or of the same machine cannot be processed simultaneously. The goal is to construct a feasible schedule of processing all the operations minimizing the *makespan*, denoted as $C_{\max}$—the maximum completion time of an operation. Following the standard three-field notation for scheduling problems (see for instance [23]) the open shop problem with m machines is denoted as $Om||C_{\max}$. The notation $O||C_{\max}$ is used in the case when the number of machines is not bounded by any constant. The $O2||C_{\max}$ problem is solvable in linear time by a well-known Gonzalez-Sahni algorithm [18], and the optimal makespan always coincides with the *standard lower bound* $\bar{C} = \max\left\{ \max_i \sum_{j=1}^{n} p_{ji}, \max_j \sum_{i=1}^{m} p_{ji} \right\}$. Following [21], this property of an instance is referred to as the *normality* (see Sect. 2 for details). In case $m \geqslant 3$ the normality is not guaranteed: for $O3||C_{\max}$ the optimal makespan can reach as much as $\frac{4}{3}\bar{C}$ [30]. The $Om||C_{\max}$ problem is NP-hard for the case of three and more machines [18]. A polynomial time approximation scheme for that problem can be found in [29]. On the other hand, the $O||C_{\max}$ problem is NP-hard in a strong sense. Moreover, for any $\rho < \frac{5}{4}$ no ρ-approximation algorithm for the $O||C_{\max}$ problem exists unless $P = NP$ [31].

The input of the routing open shop problem combines inputs of the TSP and $Om||C_{\max}$ in the following way. The machines from set $\mathcal{M}$ are mobile, and have to perform operations of jobs from $\mathcal{J}$ in the open shop environment. Jobs are located at the nodes of transportation network described by an edge-weighted graph $G = \langle V, E \rangle$. Machines are initially located at a specific node referred to as *the depot* and have to come back after completing all the operations. The goal is to minimize the makespan $R_{\max}$ which is in this case defined as the completion time of last machine's activity—either returning of a machine to the depot or processing of some job from the depot. We denote the routing open shop problem with m machines as $ROm||R_{\max}$ or $ROm|G = X|R_{\max}$ if we want to specify the structure of the transportation network. This model was introduced in [2]. It is obviously NP-hard even for a single machine as it contains a metric TSP as a special case. Moreover it was proved in [3] that even a simplest version of the problem with two machines and two nodes (i.e. $RO2|G = K_2|R_{\max}$ there K_p is the complete graph with p nodes) is NP-hard. An FPTAS for $RO2|G = K_2|R_{\max}$ was described in [22]. A series of approximation algorithms and detailed review for $ROm||R_{\max}$ can be found in [8] and references therein.

The *standard lower bound* $\bar{R}$ on the optimal makespan of $ROm||R_{\max}$ was introduced in [2]. It generalizes the known bound $\bar{C}$, taking the necessary travel times into account (see Sect. 2). Following [9] we extend the definition of normality on the routing open shop problem: we call the instance *normal* if its optimal makespan equals $\bar{R}$. In contrast with the $O2||C_{\max}$ problem, the normality is not guaranteed even for the $RO2|G = K_2|R_{\max}$. It is shown in [2] that for that simplest case the optimal makespan can reach $\frac{6}{5}\bar{R}$ bound. However, there is a bunch of known normal special cases of the problem $RO2||R_{\max}$. Two sufficient normality conditions of an instance of $RO2|G = K_2|R_{\max}$ problem are described in [22] in terms of the properties of the Gonzalez-Sahni schedule for the underlying open shop instance. On top of that, a wide special class of instances of $RO2|G = tree|R_{\max}$ problem with guaranteed normality can be found in [10].

We investigate the routing open shop problem in order to find more normal classes of instances. We discovered that there are such classes that can be described with the help of the IBP problem. The main goal of this paper is to study the relations between these two problems.

The paper is organized as follows. The formulation of the routing open shop problem together with necessary notation and definitions can be found in Sect. 2. Section 3 presents the formulation and investigation of the IBP problem, including complexity and sufficient condition of polynomial solvability. The main result—the connection between these two problems—is presented in Sect. 4, followed by some conclusions and open questions in Sect. 5.

2   Routing Open Shop: Formal Description and Notation

We use the following notation to describe an instance of the $ROm||R_{\max}$ problem. The transportation network is represented by an undirected graph $G = \langle V; E \rangle$, $V = \{v_0, \ldots, v_{p-1}\}$, node v_0 is the depot. Jobs from set $\mathcal{J} = \{J_1, \ldots, J_n\}$ are distributed among the nodes and the set of jobs located at $v \in V$ is denoted by $\mathcal{J}(v)$. We assume that $\mathcal{J}(v)$ is not empty for each node with possible exception of the depot. A nonnegative weight function τ is defined on the set E, $\tau(e)$ represents travel time of each machine over an edge $e \in E$. Machines from set $\mathcal{M} = \{M_1, \ldots, M_m\}$ are initially located at the depot v_0 and have to travel between the nodes to execute jobs. Operation of a machine M_i on a job J_j takes p_{ji} time units, no preemption is allowed. Different operations of the same machine or of the same job cannot be processed simultaneously. A machine has to be at a node v in order to process operations of any job from $\mathcal{J}(v)$. Each machine has to return to the depot after processing of all the operations.

Machines are allowed to visit each node multiple times, therefore we assume each machine takes the shortest path while traveling between the nodes. The length of the shortest path between nodes v and u is denoted by $dist(v, u)$.

A schedule S can be described by specifying *starting times* of each operation:

$$S = \{s_{ji}|i = 1, \ldots, m; j = 1, \ldots, n\}.$$

A *completion time* of each operation is denoted by $c_{ji} = s_{ji} + p_{ji}$.

Let the job $J_j \in \mathcal{J}(v)$ be the last job processed by the machine M_i in a schedule S. Then the *return moment* of machine M_i is $R_i(S) = c_{ji} + dist(v, v_0)$. The *makespan* of the schedule S is the maximal return moment $R_{\max}(S) = \max R_i(S)$.

Definition 1. *A schedule S is referred to as* feasible *if each of the following conditions holds:*

1. *for each $i_1 \neq i_2 \in \{1, \ldots, m\}$, $j_1 \neq j_2 \in \{1, \ldots, n\}$*

$$(s_{j_1 i_1}, c_{j_1 i_1}) \cap (s_{j_1 i_2}, c_{j_1 i_2}) = (s_{j_1 i_1}, c_{j_1 i_1}) \cap (s_{j_2 i_1}, c_{j_2 i_1}) = \emptyset;$$

2. *if the machine M_i processes the job $J_{j_1} \in \mathcal{J}(v)$ before the job $J_{j_2} \in \mathcal{J}(u)$ then*

$$s_{j_2 i} \geqslant c_{j_1 i} + \mathrm{dist}(v, u);$$

3. *if the job $J_j \in \mathcal{J}(v)$ is the first job to be processed by the machine M_i then*

$$s_{ji} \geqslant \mathrm{dist}(v_0, v).$$

The goal is to construct a feasible schedule with minimum makespan. We use the following notation.

$\ell_i = \sum\limits_{j=1}^{n} p_{ji}$—load of machine M_i;

$\ell_{\max} = \max \ell_i$—maximum machine load;

$d_j = \sum\limits_{i=1}^{m} p_{ji}$—length of job J_j;

$d_{\max}(v) = \max\limits_{j \in \mathcal{J}(v)} d_j$—maximum length of job from v;

$\Delta(v) = \sum\limits_{j \in \mathcal{J}(v)} d_j$—load of node v;

$\Delta = \sum\limits_{v \in V} \Delta(v) = \sum\limits_{j \in \mathcal{J}} d_j$—total load of the instance;

T^*—length of the shortest route over the graph G (TSP optimum).

The following *standard lower bound* was introduced in [3]:

$$\bar{R} = \max\left\{ \ell_{\max} + T^*, \max_{v \in V}\left(d_{\max}(v) + 2\mathrm{dist}(v_0, v) \right) \right\} \qquad (1)$$

Note that (1) implies

$$\Delta = \sum_{i=1}^{m} \ell_i \leqslant m\ell_{\max} \leqslant m(\bar{R} - T^*). \qquad (2)$$

We use the following definition inherited from [21].

Definition 2. *A feasible schedule S for a problem instance I is referred to as normal if $R_{\max}(S) = \bar{R}(I)$. Instance I is normal if it admits constructing of a normal schedule.*

A class $\mathcal{K}$ of instances of the routing open shop problem is normal if each instance from $\mathcal{K}$ is normal. A normal class $\mathcal{K}$ is referred to as efficiently normal if there exists a polynomial-time algorithm constructing a normal (and hence optimal) schedule for each instance from $\mathcal{K}$.

The main goal of the paper is to describe new normal and efficiently normal classes of instances of the routing open shop problem based on the properties of the Irreducible Bin Packing problem introduced in the next section.

3   Irreducible Bin Packing

In this section we derive some properties of the IBP problem that are helpful for solving $RO2||R_{max}$. An input of a bin packing problem includes the bin capacity B and a set of the items sizes $\mathcal{D} = \{\delta_1, \ldots, \delta_n\}$ where $\delta_i \leqslant B$ for all i. We use notation $\langle B; \delta_1, \ldots, \delta_n \rangle$ or $\langle B; \mathcal{D} \rangle$ for such an instance.

Definition 3. *Let* $I = \langle B; \mathcal{D} \rangle$ *be an instance of bin packing problem and* $\mathcal{B} = (\mathcal{B}_1, \ldots, \mathcal{B}_N)$ *is a partition of* $\mathcal{D}$. *Partition* $\mathcal{B}$ *is referred to as a* feasible *solution for instance* I *if* $B_i \doteq \sum_{\delta \in \mathcal{B}_i} \delta \leqslant B$ *for all* $i \in \{1, \ldots, N\}$.

The classical BPP can be formulated as looking for a feasible solution minimizing number of bins N. In our problem we search for a feasible solution with maximum N under a certain additional restriction.

Definition 4. *A feasible solution* $\mathcal{B}$ *for instance* I *is called* irreducible *if* $B_i + B_j > B$ *for all* $i \neq j \in \{1, \ldots, N\}$.

The **Irreducible Bin Packing (IBP)** problem can be described in the following way: given an input $I = \langle B; \mathcal{D} \rangle$, find an irreducible solution $\mathcal{B}$ for I maximizing the number of bins N.

We use the following auxiliary notation for any problem instance I:

$\Delta = \sum_{i=1}^{n} \delta_i$;

$\delta_{max} = \max\{\delta_1, \ldots, \delta_n\}$;

$M = \lceil \frac{\Delta}{B} \rceil$.

Note that the notation Δ is also used as a total load of an instance of the routing open shop problem (see Sect. 2). This is not accidental and will be explained in Sect. 4.

The IBP problem has some similarity to a Bin Covering Problem with a threshold $W = \left\lceil \frac{B+1}{2} \right\rceil > \frac{B}{2}$. Obviously any solution feasible to the BCP is irreducible, however the opposite is not true. To be precise, in any irreducible solution $B_i \geqslant W$ for each i with a possible exception of one bin, therefore the BCP optimum is a lower bound on the IBP optimum and those optima differ by at most 1. Nevertheless the IBP problem is strongly NP-hard (same as BCP; can be shown by reduction from **3-Partition** problem [17]). On the other hand, that reduction doesn't work if M is bounded by some constant, which is exactly the case interesting for us. Hereafter we assume M is an integer constant greater than 1.

There is a trivial upper bound on a number of bins in any irreducible solution.

Proposition 1. *For any irreducible solution* $\mathcal{B} = (\mathcal{B}_1, \ldots, \mathcal{B}_N)$

$$M \leqslant N \leqslant 2M - 1. \tag{3}$$

Proof. The first inequality is straightforward from Definition 3. Lets prove the second one. Note that

$$\Delta \leqslant MB. \tag{4}$$

Suppose $N \geqslant 2M$. Then we have at least M disjunctive pairs of bins. By Definition 4, for each pair (i, j) we have $B_i + B_j > B$ and therefore $\Delta > MB$, a contradiction with (4). □

In fact, we are mostly interested in studying a decision version of the IBP problem. The formulation depends on an integer parameter $M \geqslant 2$, and we use notation IBP(M) for that problem.

IBP(M).
Input: $I = \langle B; \mathcal{D} \rangle$, $\Delta \leqslant MB$.
Question: Does there exist an irreducible solution for I with $N = 2M - 1$?

Definition 5. *And instance $I = \langle B; \mathcal{D} \rangle$ with $\Delta \leqslant MB$ is referred to as M-irreducible if the answer to IBP(M) for instance I is positive.*

A class $\mathcal{K}$ of M-irreducible instances is called efficiently *M-irreducible if there exists a polynomial time algorithm for construction of an irreducible solution with $N = 2M - 1$ for any instance from $\mathcal{K}$.*

Theorem 1. *For any $M \geqslant 2$ the IBP(M) problem is NP-complete.*

Proof. We use reduction from a well-known NP-complete **Partition** problem [17]. Let $\mathcal{E} = \{e_1, \ldots, e_k\}$ be an input for the **Partition**, $\sum_{i=1}^{k} e_i = 2E$. Without loss of generality assume

$$E \geqslant 2M - 1, \tag{5}$$

since we can scale up the values from $\mathcal{E}$ to accommodate this inequality. Consider the following instance I for an IBP(M) problem:

$$I = \langle 2E - 2; \delta_0, \ldots, \delta_{k+2M-4} \rangle,$$

there $\delta_0 = E - 1$ is a *special* item and $\delta_i = e_i$ for $i = 1, \ldots, k$ are *small* items. In case $M > 2$ the instance also contains *large* items $\delta_{k+1} = \cdots = \delta_{k+2M-4} = E$.

Note that due to (5) for this instance $\Delta = (2M-1)E - 1 = 2ME - (E+1) \leqslant 2ME - 2M = MB$, so the instance is valid for the IBP(M) problem.

Suppose a partition of $\mathcal{E}$ into two equally sized subsets $\mathcal{E}_1$ and $\mathcal{E}_2$ exists. Then the following solution is M-irreducible for I:

$$\mathcal{B}_1 = \{\delta_0\}, \mathcal{B}_2 = \{\delta_{k+1}\}, \ldots, \mathcal{B}_{2M-3} = \{\delta_{k+2M-4}\},$$

$\mathcal{B}_{2M-2}$ and $\mathcal{B}_{2M-1}$ correspond to $\mathcal{E}_1$ and $\mathcal{E}_2$.

Now consider a M-irreducible solution $\langle \mathcal{B}_1, \ldots, \mathcal{B}_{2M-1} \rangle$ for I. Let us prove that in this case a partition of $\mathcal{E}$ exists.

Without loss of generality assume $\delta_0 \in \mathcal{B}_1$ and large items belong to sets $\mathcal{B}_2, \ldots, \mathcal{B}_{2M-3}$. By Definition 4 we have $B_{2M-2} + B_{2M-1} > B = 2E - 2$. On the other hand the last two bins only contain small items, hence

$$2E - 1 \leqslant B_{2M-2} + B_{2M-1} \leqslant \sum e_i = 2E. \tag{6}$$

Consider two cases.

Case 1. $\mathcal{B}_1 = \{\delta_0\}$.

In this case $B_1 = E - 1$ and hence by the irreducibility of the solution $\min\{B_{2M-2}, B_{2M-1}\} > E - 1$. Due to (6) we have $B_{2M-2} = B_{2M-1} = E$ implying the partition of $\mathcal{E}$ into two equally sized sets.

Case 2. $\mathcal{B}_1$ contains more than one element.

The only possibility in this case is that there is one small item of size exactly 1 in the first bin (otherwise $B_{2M-2} + B_{2M-1} \leqslant B$). Therefore $B_1 = E$ and by irreducibility $\min\{B_{2M-2}, B_{2M-1}\} > E - 2$ implying $\{B_{2M-2}, B_{2M-1}\} = \{E - 1, E\}$. This means that the set $\mathcal{E}$ is partitioned into three subsets of sizes $1, E - 1$ and E, hence the equal partition exists.   □

It is easy to observe that the following condition is necessary for the M-irreducibility of the instance I:

$$\Delta(I) > \left(M - \frac{1}{2}\right) B. \tag{7}$$

A sufficient condition, together with a polynomial-time procedure to obtain an irreducible solution with $N = 2M - 1$, is given in the next.

Theorem 2. *Let I be an instance of the IBP(M) problem and*

$$\Delta(I) > \left(M - \frac{1}{2}\right) B + (2M - 3)\delta_{\max}, \tag{8}$$

Then I is M-irreducible and such solution for I can be found in linear time.

Proof. Suppose $\mathcal{D} = \{\delta_1, \ldots, \delta_n\}$. Construct a solution for instance I using the following procedure.

0. **Set** $t = 0$.
1. **For each** $i = 1, \ldots, 2M - 3$
   (a) Find $j > t$ such that $\sum\limits_{k=t+1}^{j-1} \delta_j \leqslant \dfrac{B}{2} < \sum\limits_{k=t+1}^{j} \delta_j$.
   (b) **Set** $\mathcal{B}_i = \{\delta_{t+1}, \ldots, \delta_j\}$, $B_i = \sum\limits_{k=t+1}^{j} \delta_j$, $x_i = B_i - \frac{B}{2}$, $t = j$.
2. **Set** $x^* = \min\{x_1, \ldots, x_{2M-3}\}$.
3. Find $j > t$ such that $\sum\limits_{k=t+1}^{j-1} \delta_j \leqslant \dfrac{B}{2} - x^* < \sum\limits_{k=t+1}^{j} \delta_j$.
4. **Set** $\mathcal{B}_{2M-2} = \{\delta_{t+1}, \ldots, \delta_j\}$, $B_{2M-2} = \sum\limits_{k=t+1}^{j} \delta_j$, $y = B_{2M-2} - \frac{B}{2} + x^*$, $t = j$.

5. **Set** $\mathcal{B}_{2M-1} = \{\delta_{t+1}, \ldots, \delta_n\}$, $B_{2M-1} = \sum_{k=t+1}^{j} \delta_n$.

It is sufficient to prove that the solution $\mathcal{B} = (\mathcal{B}_1, \ldots, \mathcal{B}_{2M-1})$ is irreducible. Note that (4) and (8) imply $\delta_{\max} < \frac{B}{2}$, therefore for each $i = 1, \ldots, 2M - 2$ we have $B_i \leqslant B$. Let

$$\Delta' = \sum_{i=1}^{2M-2} B_i = (M-1)B + \sum x_i - x^* + y > (M-1)B, \qquad (9)$$

thus $B_{2M-1} = \Delta - \Delta' < B$ and the solution $\mathcal{B}$ is feasible. On the other hand $\Delta' \leqslant (M-1)B + (2M-3)\delta_{\max}$, and by (8) $B_{2M-1} > \frac{B}{2}$. By the construction $B_i > \frac{B}{2}$ and $B_i + B_{2M-2} > B$ for each $i = 1, \ldots, 2M - 3$. By (9) and (8)

$$B_{2M-2} + B_{2M-1} > B - \sum_{i=1}^{2M-3} x_i + (2M-3)\delta_{\max} \geqslant B,$$ therefore the solution $\mathcal{B}$

is irreducible.   □

Note that the condition (8) is presented in its best form: the equality $\Delta = \left(M - \frac{1}{2}\right) B + (2M-3)\delta_{\max}$ is not sufficient for the M-irreducibility of an instance. Consider the following counterexample. Let $I = \langle 4M - 6; \underbrace{1, \ldots, 1}_{4M^2-6M} \rangle$, in this case

$\Delta = 4M^2 - 6M = \left(M - \frac{1}{2}\right) B + (2M-3)\delta_{\max}$. Assume there is an irreducible solution $(\mathcal{B}_1, \ldots, \mathcal{B}_{2M-1})$ and without loss of generality for each $i = 1, \ldots, 2M-2$ we have $B_i > \frac{B}{2}$. Let $B_{\min} = \min\{B_1, \ldots, B_{2M-2}\} = \frac{B}{2} + 1 + x = 2M - 2 + x$, $x \geqslant 0$. Then

$$B_{2M-1} \leqslant \Delta - (2M-2)B_{\min} = 2M - 4 - (2M-2)x.$$

Therefore we have

$$B_{\min} + B_{2M-1} \leqslant 2M - 2 + x + 2M - 4 - (2M-2)x = B - (2M-3)x \leqslant B$$

contradicting the irreducibility of $\mathcal{B}$.

So, the answer to the IBP(M) question is negative in case $\Delta \leqslant \left(M - \frac{1}{2}\right) B$ and positive in case $\Delta > \left(M - \frac{1}{2}\right) B + (2M-3)\delta_{\max}$. For $\left(M - \frac{1}{2}\right) B < \Delta \leqslant \min\{MB, \left(M - \frac{1}{2}\right) B + (2M-3)\delta_{\max}\}$ the IBP(M) problem is NP-complete.

4   The Main Result: The Connection Between IBP(2) and $RO2||R_{\max}$

In this section we use the following simplified notation for the two-machine routing open shop problem. The processing times of operations of the job J_j are denoted as a_j and b_j instead of p_{j1} and p_{j2}. We also use the same notation (a_j, b_j) to denote the operations themselves. Starting and completion times of operations are denoted as $s(a_j)$, $s(b_j)$, $c(a_j)$ and $c(b_j)$.

The main result is based on the instance reduction procedure by means of the following jobs aggregation operation.

Definition 6. *Let I be an instance of $ROm||R_{\max}$ problem, $v \in V$ and $\mathcal{K} \subseteq \mathcal{J}(v)$. By the operation of jobs aggregation of set $\mathcal{K}$ we understand the following instance transformation:*

$$\mathcal{J}(v) \rightarrow \mathcal{J}(v) \setminus \mathcal{K} \cup \{J_{j_\mathcal{K}}\}, \ p_{j_\mathcal{K} i} = \sum_{J_j \in \mathcal{K}} p_{ji}.$$

Basically that means that the set of jobs $\mathcal{K}$ is substituted with a new job $J_{j_\mathcal{K}}$ with processing times equal to the total processing times if the respective operations of jobs from $\mathcal{K}$.

This operation and its properties in application to the $RO2||R_{\max}$ problem are described in detail in [9]. One obvious property of the jobs aggregation operation is its *reversibility*: any feasible schedule for the transformed instance can be easily treated as a feasible schedule for the initial instance with the same makespan.

Machine loads are preserved by any jobs aggregation operation, however the standard lower bound might increase in case then the length of a new job is greater than $d_{\max}(v)$. Moreover, let $\tilde{I}$ is an instance obtained from I by aggregation of set $\mathcal{K} \subseteq \mathcal{J}(v)$. Then

$$\bar{R}(\tilde{I}) = \bar{R}(I) \iff \sum_{J_j \in \mathcal{K}} d_j + 2\text{dist}(v_0, v) \leqslant \bar{R}(I). \tag{10}$$

Definition 7. *A jobs aggregation of set $\mathcal{K}$ is referred to as* legal *if it preserves the standard lower bound. An instance is called* irreducible *if no legal jobs aggregation is possible.*

Suppose an instance I' is obtained from I by legal aggregation operations. In this case, if I' is normal then I is normal as well. In this section we study irreducible instances and sufficient conditions of their normality.

Definition 8. *A node v is referred to as* overloaded *if the aggregation of set $\mathcal{J}(v)$ is not legal. Otherwise the node v is* underloaded.

It was shown in [9] that any instance of $RO2||R_{\max}$ contains at most one overloaded node.

Due to (10) the procedure of legal aggregations at node $v \in V$ can be seen as an instance of a bin packing problem with $\delta_j = d_j$ and $B = \bar{R} - 2\text{dist}(v_0, v)$. Moreover, as soon as $2\text{dist}(v_0, v) \leqslant T^*$, (2) implies $\sum \delta_j \leqslant 2B$ and we actually have an instance of the IBP(2) problem. The node is called *superoverloaded* if the correspondent instance of IBP(2) problem is 2-irreducible. Note that due to (7) any superoverloaded node is overloaded as well.

By Proposition 1 there always exists a legal jobs aggregation into at most three jobs at an overloaded node. (It was shown in [9] that such an aggregation can be done in linear time.) Any irreducible instance contains single job at each underloaded node and two or three jobs at overloaded one (if any). In the latter case that node is superoverloaded.

Theorem 3. *Let I be an instance of $RO2||R_{\max}$ with a superoverloaded node v and one of the following conditions holds:*

1. *$v = v_0$;*
2. *there exists an optimal Hamiltonian walk σ on the graph G such that v is adjacent to v_0 in σ.*

Then I is normal.

Proof. Consider the following enumeration of the nodes of G. In case $v = v_0$ enumerate nodes according to any optimal Hamiltonian walk $(v_0, v_1, \ldots, v_{p-1}, v_0)$. Otherwise we use enumeration consistent with $\sigma = (v_0 = v_1, v_2, \ldots, v_{p-1}, v, v_0)$.

Denote $\tau = \text{dist}(v_0, v)$ and $T = T^* - \tau$. Note that $T \geqslant \tau$ due to metric properties of distances.

Perform legal jobs aggregation at each node in the following way: for each underloaded node v_k, $k = 1, \ldots, p - 1$ jobs from $\mathcal{J}(v_k)$ are aggregated into a single job J_k, and jobs from $\mathcal{J}(v)$ are aggregated into three jobs J_α, J_β and J_γ according to an irreducible solution of the underlying IBP(2) problem. Denote the obtained irreducible instance by $\tilde{I}$. Let us prove that $\tilde{I}$ is normal.

Without loss of generality assume

$$a_\alpha \leqslant \min\{b_\alpha, a_\gamma, b_\gamma\}. \tag{11}$$

(This can be achieved by renaming of machines and/or jobs J_α, J_γ.)

As soon as $\tilde{I}$ is irreducible, the following inequality holds:

$$d_\alpha + d_\beta + 2\tau > \bar{R}, \tag{12}$$

which due to (2) implies

$$\sum_{j=1}^{p-1} d_j + d_\gamma + 2T < \bar{R}. \tag{13}$$

Further we describe schedules by specifying a partial order of operations, represented graphically by so-called *schemes*. On such schemes nodes represent operations and linear orders of operations of every job and every machine are given. Two auxiliary nodes S and F represent starting and finishing event of the schedule and have null weight. Weights of other nodes are respective processing times. A weight of arc connecting two operations of the same machine represent the distance between the respective nodes. The resulting schedule is an *early* one: each operation starts as early as possible without violation of feasibility (see Definition 1) and of the partial order given by the scheme. It is a well-known fact that the makespan of such an early schedule coincides with the weight of the *critical path* in the scheme, i.e. a path of maximum weight.

Construct early schedules S_1 and S_2 according to the schemes from Figs. 1 and 2 respectively.

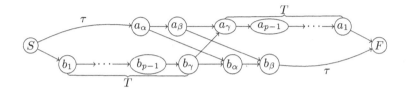

Fig. 1. A scheme of the schedule S_1.

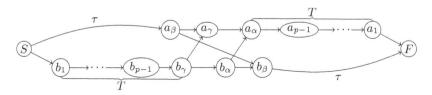

Fig. 2. A scheme of the schedule S_2.

Note that due to (11) in schedule S_1 we have $c(a_\alpha) \leqslant c(b_\gamma)$ and hence $2\tau + a_\alpha + b_\alpha + b_\beta \leqslant \ell_2 + T^*$, therefore

$$R_{\max}(S_1) = \max\left\{ \ell_1 + T^*, \ell_2 + T^*, \sum_{j=1}^{p-1} d_j + d_\gamma + 2T, 2\tau + a_\alpha + a_\beta + b_\beta \right\}.$$

Due to (1) and (13) we have

$$R_{\max}(S_1) \leqslant \max\{\bar{R}, 2\tau + a_\alpha + a_\beta + b_\beta\}. \tag{14}$$

Accordingly,

$$R_{\max}(S_2) \leqslant \max\left\{ \bar{R}, \sum_{j=1}^{p-1} d_j + b_\gamma + a_\alpha + \max\{a_\gamma, b_\alpha\} + 2T \right\}. \tag{15}$$

It sufficient to prove that at least one of the schedules S_1 and S_2 is normal. Assume otherwise, then by (14) and (15) we have

$$R_{\max}(S_1) = 2\tau + a_\alpha + a_\beta + b_\beta,$$

due to (11)

$$R_{\max}(S_2) = \sum_{j=1}^{p-1} d_j + b_\gamma + a_\alpha + \max\{a_\gamma, b_\alpha\} + 2T \leqslant \sum_{j=1}^{p-1} d_j + b_\gamma + a_\gamma + b_\alpha + 2T,$$

and due to (2)

$$R_{\max}(S_1) + R_{\max}(S_2) \leqslant 2\tau + a_\alpha + d_\beta + \sum_{j=1}^{p-1} d_j + d_\gamma + b_\alpha + 2T = \Delta + 2T* \leqslant 2\bar{R}.$$

Theorem is proved by contradiction.   □

Conditions of the Theorem 3 describes a normal class of instances of the $RO2\|R_{\max}$ problem. However this class is not efficiently normal: in order to construct a normal schedule we need an optimal solution of an underlying TSP, which might be hard to obtain, and an irreducible solution for an IBP problem, which is also hard to get. In the next theorem we consider a special case of the routing open shop problem denoted as $RO2|easy - TSP|R_{\max}$ (see [8]). The $easy - TSP$ option refers to a case when the underlying TSP problem on the graph G is solvable in polynomial time (due to specific graph structure or to the properties of the distance function), or the time complexity of the TSP is not taken into account.

Theorem 4. *Let I be an instance of $RO2\|R_{\max}$ with an overloaded node v with*

$$\Delta(v) > \frac{3}{2}\left(\bar{R} - 2\mathrm{dist}(v_0, v)\right) + d_{\max}(v) \qquad (16)$$

and one of the following condition holds:

1. *$v = v_0$;*
2. *there exists an optimal Hamiltonian walk σ on the graph G such that v is adjacent to v_0 in σ.*

Then a normal schedule for I can be constructed in $O(n + t_{\mathrm{TSP}})$ there t_{TSP} is a running time of solving the underlying TSP.

Proof. Straightforward from Theorems 2 and 3. □

In particular we have the following

Corollary 1. *The existence of a node v satisfying (16) is sufficient for efficient normality of classes of instances of $RO2|G = K_2|R_{\max}$, $RO2|G = K_3|R_{\max}$ and $RO2|G = tree|R_{\max}$ problems.*

5   Concluding Remarks

We introduced the IBP(M) problem and described new normal (Theorem 3) and efficiently normal (Theorem 4) classes of instances of $RO2\|R_{\max}$ based on the properties of the IBP(2).

We plan the following directions for future research.

1. Find new sufficient conditions of 2-irreducibility and efficient 2-irreducibility of an instance of the IBP(2) problem to extend normal and efficiently normal classes from Theorems 3 and 4.
2. Investigate the connection between IBP(m) and $ROm\|R_{\max}$ for $m > 2$.

References

1. Assmann, S.F., Johnson, D.S., Kleitman, D.J., Leung, J.Y.T.: On the dual of the one-dimensional bin packing problem. J. Algorithms **5**(4), 502–525 (1984). https://doi.org/10.1016/0196-6774(84)90004-X
2. Averbakh, I., Berman, O., Chernykh, I.: A 6/5-approximation algorithm for the two-machine routing open-shop problem on a two-node network. Eur. J. Oper. Res. **166**(1), 3–24 (2005). https://doi.org/10.1016/j.ejor.2003.06.050
3. Averbakh, I., Berman, O., Chernykh, I.: The routing open-shop problem on a network: complexity and approximation. Eur. J. Oper. Res. **173**(2), 531–539 (2006). https://doi.org/10.1016/j.ejor.2005.01.034
4. Bennell, J.A., Oliveira, J.: A tutorial in irregular shape packing problems. J. Oper. Res. Soc. **60**, S93–S105 (2009). https://doi.org/10.1057/jors.2008.169
5. Bennell, J.A., Song, X.: A beam search implementation for the irregular shape packing problem. J. Heuristics **16**(2), 167–188 (2010). https://doi.org/10.1007/s10732-008-9095-x
6. Birgin, E.G., Martnez, J.M., Ronconi, D.P.: Optimizing the packing of cylinders into a rectangular container: a nonlinear approach. Eur. J. Oper. Res. **160**, 19–33 (2005). https://doi.org/10.1016/j.ejor.2003.06.018
7. Boyar, J., et al.: The maximum resource bin packing problem. Theor. Comput. Sci. **362**(1), 127–139 (2006). https://doi.org/10.1016/j.tcs.2006.06.001
8. Chernykh, I., Kononov, A.V., Sevastyanov, S.: Efficient approximation algorithms for the routing open shop problem. Comput. Oper. Res. **40**(3), 841–847 (2013). https://doi.org/10.1016/j.cor.2012.01.006
9. Chernykh, I., Lgotina, E.: The 2-machine routing open shop on a triangular transportation network. In: Kochetov, Y., Khachay, M., Beresnev, V., Nurminski, E., Pardalos, P. (eds.) DOOR 2016. LNCS, vol. 9869, pp. 284–297. Springer, Cham (2016). https://doi.org/10.1007/978-3-319-44914-2_23
10. Chernykh, I., Lgotina, E.: Two-machine routing open shop on a tree: instance reduction and efficiently solvable subclass (2019). Submitted to Optimization Methods and Software
11. Christensen, H.I., Khan, A., Pokutta, S., Tetali, P.: Multidimensional bin packing and other related problems: a survey (2016). https://people.math.gatech.edu/~tetali/PUBLIS/CKPT.pdf
12. Coffman Jr., E.G., Leung, J.Y., Ting, D.W.: Bin packing: maximizing the number of pieces packed. Acta Inf. **9**(3), 263–271 (1978). https://doi.org/10.1007/BF00288885
13. Coffman Jr., E.G., Csirik, J., Galambos, G., Martello, S., Vigo, D.: Bin packing approximation algorithms: survey and classification. In: Pardalos, P.M., Du, D.-Z., Graham, R.L. (eds.) Handbook of Combinatorial Optimization, pp. 455–531. Springer, New York (2013). https://doi.org/10.1007/978-1-4419-7997-1_35
14. Epstein, L., Favrholdt, L.M., Kohrt, J.S.: Comparing online algorithms for bin packing problems. J. Sched. **15**(1), 13–21 (2012). https://doi.org/10.1007/s10951-009-0129-5
15. Epstein, L., Imreh, C., Levin, A.: Bin covering with cardinality constraints. Discrete Appl. Math. **161**, 1975–1987 (2013). https://doi.org/10.1016/j.dam.2013.03.020
16. Epstein, L., Levin, A.: On bin packing with conflicts. SIAM J. Optim. **19**, 1270–1298 (2008). https://doi.org/10.1137/060666329
17. Garey, M.R., Johnson, D.S.: Computers and Intractability: A Guide to the Theory of NP-Completeness. W. H. Freeman & Co., New York (1979)

18. Gonzalez, T.F., Sahni, S.: Open shop scheduling to minimize finish time. J. ACM **23**(4), 665–679 (1976). https://doi.org/10.1145/321978.321985

19. Jansen, K., Solis-Oba, R.: An asymptotic fully polynomial time approximation scheme for bin covering. Theoret. Comput. Sci. **306**, 543–551 (2003). https://doi.org/10.1016/s0304-3975(03)00363-3

20. Karp, R.M.: Reducibility among combinatorial problems. In: Miller, R.E., Thatcher, J.W., Bohlinger, J.D. (eds.) Complexity of Computer Computations. The IBM Research Symposia Series, pp. 85–103. Springer, Boston (1972). https://doi.org/10.1007/978-1-4684-2001-2_9

21. Kononov, A., Sevastianov, S., Tchernykh, I.: When difference in machine loads leads to efficient scheduling in open shops. Ann. Oper. Res. **92**, 211–239 (1999). https://doi.org/10.1023/a:1018986731638

22. Kononov, A.: On the routing open shop problem with two machines on a two-vertex network. J. Appl. Ind. Math. **6**(3), 318–331 (2012). https://doi.org/10.1134/s1990478912030064

23. Lawler, E.L., Lenstra, J.K., Rinnooy Kan, A.H.G., Shmoys, D.B.: Chapter 9. Sequencing and scheduling: algorithms and complexity. In: Logistics of Production and Inventory, Handbooks in Operations Research and Management Science, vol. 4, pp. 445–522. Elsevier (1993). https://doi.org/10.1016/S0927-0507(05)80189-6

24. Levin, M.S.: Bin packing problems (promising models and examples). J. Commun. Technol. Electron. **63**, 655–666 (2018). https://doi.org/10.1134/s1064226918060177

25. Lodi, A., Martello, S., Monaci, M.: Two-dimensional packing problems: a survey. Eur. J. Oper. Res. **141**(2), 241–252 (2002). https://doi.org/10.1016/S0377-2217(02)00123-6

26. Martello, S., Pisinger, D., Vigo, D.: The three-dimensional bin packing problem. Oper. Res. **48**, 256–267 (2000). https://doi.org/10.1287/opre.48.2.256.12386

27. Muritiba, A.E.F., Iori, M., Malaguti, E., Toth, P.: Algorithms for the bin packing problem with conflicts. INFORMS J. Comput. **22**, 401–415 (2010). https://doi.org/10.1287/ijoc.1090.0355

28. Seiden, S.S., van Stee, R., Epstein, L.: New bounds for variable sized online bin packing. SIAM J. Comput. **32**, 455–469 (2003). https://doi.org/10.1137/s0097539702412908

29. Sevastianov, S.V., Woeginger, G.J.: Makespan minimization in open shops: a polynomial time approximation scheme. Math. Program. **82**(1-2, Ser. B), 191–198 (1998). https://doi.org/10.1007/BF01585871

30. Sevastianov, S.V., Tchernykh, I.D.: Computer-aided way to prove theorems in scheduling. In: Bilardi, G., Italiano, G.F., Pietracaprina, A., Pucci, G. (eds.) ESA 1998. LNCS, vol. 1461, pp. 502–513. Springer, Heidelberg (1998). https://doi.org/10.1007/3-540-68530-8_42

31. Williamson, D.P., et al.: Short shop schedules. Oper. Res. **45**(2), 288–294 (1997). https://doi.org/10.1287/opre.45.2.288

Learning Probabilistic Constraints for Surgery Scheduling Using a Support Vector Machine

Thomas Philip Runarsson[✉]

School of Engineering and Natural Sciences, University of Iceland, Reykjavik, Iceland
tpr@hi.is

Abstract. The problem of generating surgery schedules is formulated as a mathematical model with probabilistic constraints. The approach presented is a new method for tackling probabilistic constraints using machine learning. The technique is inspired by models that use slacks in capacity planning. Essentially support vector classification is used to learn a linear constraint that will replace the probabilistic constraint. The data used to learn this constraint is labeled using Monte Carlo simulations. This data is iteratively discovered, during the optimization procedure, and augmented to the training set. The linear support vector classifier is then updated during search, until a feasible solution is discovered. The stochastic surgery model presented is inspired by real challenges faced by many hospitals today and tested on real-life data.

Keywords: Support vector classification · Surgery scheduling · Stochastic mixed integer programming · Monte Carlo simulations

1 Introduction

Hospitals worldwide are facing the pressure of an aging society and increasing costs while simultaneously dealing with nursing shortages. Patient flow from the operating rooms (ORs) highly influences the workload on downstream resources such as wards, post anesthesia care unit (PACU) and intensive care unit (ICU). It is important to balance the patient flow to the downstream resources, not only to level the workload, but also to hedge against last minute cancellations. Scheduling surgeries is undeniable a challenging task where several stakeholders [1,5,10,16], with competing objectives, are involved. The focal point of a scheduler is to keep the operating room well utilized. In general the goal is to maximize the overall throughput and so minimize the overall waiting time for patients. Scheduling too many surgeries may cause an unbalanced flow which may result in resource blocking.

In practice, due to their complexity, nurses create schedules by hand. When scheduling they typically use the expected surgery time and the expected length of stay. This information is available using historical data and the nurse's experience. However, as pointed out by [8] using the expected surgery times will leave

© Springer Nature Switzerland AG 2020
N. F. Matsatsinis et al. (Eds.): LION 13 2019, LNCS 11968, pp. 121–134, 2020.
https://doi.org/10.1007/978-3-030-38629-0_10

the ORs underutilized. Studies have revealed the importance of incorporating stochastic elements to the scheduling process and especially to surgery times [3,9,11–13]. The models and techniques that have been proposed vary significantly. Each hospital is unique and typically requires specialized models. The flexibility of a mixed integer programming (MIP) model makes it possible to tailor the solution to the different requirements made by the hospitals. Nevertheless, parts of the problem, remain identical [15].

Assuming that the surgery time distributions are available, for any given surgery, a number of different optimization approaches can be applied. Essentially filling the operating room can be considered a stochastic knapsack problem. In this case the capacity of the knapsack is analogous to the time available in the OR. In [12] a stochastic MIP model for the assignment of surgeries to operating days and rooms is proposed. This model is complex and considers not only the uncertainty in surgery duration but also emergency arrivals and operator's capacity. The main goal is to minimize the under- and the overtime costs of the operating rooms and the costs of exceeding the defined capacity of the system. A two-stage stochastic programming model and a robust model are presented in [3]. The results indicate that a robust model can provide high quality solutions in a shorter amount of computational time. In stochastic programming it is desirable to provide the probability distribution functions of the underlying stochastic parameters. Alternatively, realizations from these distributions, or historical data, can be used to create scenarios. In robust optimization the uncertainty is described by an upper and lower limit on stochastic parameters. The technique uses then a min-max approach guaranteeing the feasibility and optimality of the solution against all instances of the parameters within an uncertainty set. Both techniques will require various parameter settings and will not necessarily guarantee that a given OR will go over the time limit set for the OR. An alternative approach to tackle the OR time capacity constraint is by introducing a planned slack to each surgery day and room. The slack is used to hedge against uncertainty in the total surgery duration [7]. By planning a slack to each surgery day and room, the risk of overtime is minimized.

Similarly to the methods cited above a MIP formulation will be used, however, all generated operating room day schedules (ORDS) will be validated using Monte-Carlo simulations. The technique has many similarities with the technique described in [14] where a column generation technique is used to generate ORDS. A column generation approach is needed since generating all possible ORDS is intractable in general. In order to apply the column generation approach the uncertainty is tackled using a planned slack [14], similar to that in [7]. Our ORDS will be verified by a Monte-Carlo simulation during search and so a completely different approach is needed. The idea is then to learn to classify which ORDS are feasible and non-feasible using a linear support vector (SV) classifier. Then this classifier is added to the MIP model as a linear constraint.

Our analysis is based on real life data. The data is obtained from the National University Hospital of Iceland, which contains information on all surgical activities spanning the last ten years. We will use this data to generate different instances, which will be used for our experiments. This work is the result of a close collaboration with the hospital and is inspired by the practical challenges faced by the hospital.

The paper is organized as follows. The following section provides a discussion on the sources of uncertainty faced in surgery scheduling and a complete description of the MIP model for the surgery scheduling problem. In Sect. 3 the SV ORDS classifier is introduced and the procedure for building this model while iteratively solving the MIP model. The potential and properties of this technique are investigated in Sect. 4. The paper concludes discussion and conclusion.

2   Stochastic Surgery Scheduling

In practice, when schedules are created by hand, the expected time for the surgery is commonly used. The mean is usually larger than the median, as the time typically follows the log-normal distribution [17]. As a result the operating room is underutilized and effectively a slack is introduced to the schedule. Nevertheless, the variance on the time of completion of the last surgery of the day is quite high. Therefore, the likelihood of going over the set time limit may be high. For example, the Erasmus MC tolerates a 30% probability of going into overtime [7]. For the frequently performed surgeries accurate simulations are possible using historical data. Less frequent operations will have less reliable data and may require insights from the team to estimate the time required. When estimating the mean time of a surgery the operator is commonly taken into consideration. Nevertheless, as pointed out by [2], there may also exist other causes for variations in surgery time that depend on the patient's characteristics. As for the time between surgery, it may not necessarily be due to cleaning alone but the different setup required by the surgery to follow. Separating the data by these different characteristics will reduce variability.

Some ORs may be kept open longer to accommodate for acute surgeries and so going into overtime may be acceptable. However, when too many operating rooms are observed to be going over then surgeries will be canceled. Keeping the probability of going over the set time limit is, therefore, of paramount importance. Let $z_{o,r,d,s} \in \{0,1\}$ indicate when surgery s is performed by operator o in an operating room r on day d with capacity $d_{r,d}$. Then the probability α of going into overtime may be bounded by the probabilistic constraint

$$\Pr\left[f(z_{r,d}) \geq d_{r,d}\right] \leq \alpha \tag{1}$$

where $f(z_{r,d})$ denotes the distribution function for the stochastic sum of all surgical procedures, including the time between surgeries, for room r on day d.

Assuming the time distributions are available, for any given surgery, a number of different optimization approaches can be applied. However, the only exact way of solving this problem is by trying all operating room day schedules (ORDS). A technique described in [14] attempts to do precisely this. In this case a column generation approach is needed since generating all possible ORDS is intractable in general. When generating the columns a sub-problem is solved and the stochastic capacity constraint (1) is approximated using a planned slack. As a result this technique must validate the solution found using a Monte-Carlo simulation. The planned slack technique essentially assumes that the variance of the total surgery time and mean are known. The amount of slack depends on the accepted probability of overtime. When using a planned slack $\delta_{o,r,d}$ the capacity constraint is reduced to

$$\sum_{s \in \mathcal{S}^o} z_{o,r,d,s} \mu_s + \delta_{o,r,d} \le d_{r,d} \quad \forall o \in \mathcal{O} \tag{2}$$

where μ_s is the mean time needed for surgery s. Determining the planned slack can be done in a straight forward manner when the distributions are known. For example in [7], the surgery times are assumed to be normally distributed. Then

$$\delta_{o,r,d} = \beta \sqrt{\sum_{s \in \mathcal{S}^o} z_{o,r,d,s} \sigma_s^2} \tag{3}$$

where σ_s^2 is the variance of surgery s. The parameter β is adjusted to achieve a suitable probability of going into overtime. However, in general the central limit theorem will have limited application here, since the number of surgeries performed will commonly range from one to four. Furthermore, the individual surgery times are far from normal. Any reliable estimate on the planned slack, in practice, will require Monte Carlo simulations based on historical data.

The flexibility of using the MIP formulation for the stochastic surgery problem will now be illustrated. In this model we will consider the elective surgery for general surgery at the National University Hospital of Iceland. Each patient is assigned to an operator (surgeon) and so there are as many patient lists as there are operators. Each operator must then be assigned to a room and day. Furthermore, patients are selected from the patient list and assigned to one of the days assigned to their operator. In general the operator room day schedule belongs to a single operator, however, is is possible to create shared ORDS in the case where two surgeons share a room during the day.

Any given surgery can only be performed once, so

$$\sum_{d \in \mathcal{D}, r \in \mathcal{R}, o \in \mathcal{O}: s \in \mathcal{S}^o} z_{o,r,d,s} \le 1, \quad \forall, s \in \mathcal{S} \tag{4}$$

where $\mathcal{S}^o \subseteq \mathcal{S}$ are the surgeries on operator o's patient list. A surgeon can also not be operating in more than one room on any given day. This can be forced by the condition

$$\sum_{r \in \mathcal{R}} y_{o,r,d} \le 1, \quad \forall d \in \mathcal{D}, o \in \mathcal{O} \tag{5}$$

where the variable $y_{o,r,d} \in \{0,1\}$ indicates that the operator is using room r on day d and is determined by the constraint

$$z_{o,r,d,s} \le y_{o,r,d} \quad \forall (o,r,d,s) \in \mathcal{O} \times \mathcal{R} \times \mathcal{D} \times \mathcal{S}^o \tag{6}$$

Furthermore, only one operator can be in any room at any given day,

$$\sum_{o \in \mathcal{O}} y_{o,r,d} \leq 1, \quad \forall d \in \mathcal{D}, r \in \mathcal{R} \tag{7}$$

Additional conditions could be added here to force operators or patients to specific days or rooms. The flexibility of the MIP formulation is such that it will easily accommodate for any additional hospital specific conditions. One such condition, we will force, is that an operator should perform at least one operation per week, or for a given week v with working days $\mathcal{D}^v$,

$$\sum_{s \in \mathcal{S}^o, d \in \mathcal{D}^v, r \in \mathcal{R}} z_{o,r,d,s} \geq 1, \quad \forall o \in \mathcal{O} \tag{8}$$

This condition can also be easily extended to allow for the operator's planned roster and days off by using operator specific days $\mathcal{D}_o^v$.

Now we define a continuous variable w_d denoting the expected number in the ward at any given day d. For every surgery it is assumed that it is known whether a surgery will require the patient to enter the ward. Historical data can be used to estimate the probability of being in the ward on day j for surgery s, denoted by $W_{j,s}$. Then w_d may be computed as follows,

$$w_d = \sum_{\substack{r \in \mathcal{R}, s \in \mathcal{S}^o, j \in \{0,\ldots,m\}: \\ d-j \in \mathcal{D}}} W_{j,s} z_{o,r,d-j,s}, \quad \forall d \in \mathcal{D} \tag{9}$$

where m is some maximal number of expected ward days, for example one week. Typically the ward is a limited resource and so an upper bound may be forced on $w_d \leq \overline{w}$ and typically also a lower bound $w_d \geq \underline{w}$, as one does not want the limited resource to be underutilized. A similar condition may be used for the PACU and ICU and other costly resources in use by the hospital.

Unlike the master surgery scheduling (MSS) problem presented in [14], where the goal is to solve for the allocated time per specialty within the hospital, the formulation here focuses on master operator schedule. The problem is in essence the same in nature, however, here we will be forcing the ward capacities and limiting the number of operating rooms. Both are limited hospital resources. These conditions will have a significant effect on the schedules created since the operators have different case mixes.

The object will now simply be maximizing the throughput, this may be defined by the objective

$$\max_z \sum_{(o,r,d,s) \in \mathcal{O} \times \mathcal{R} \times \mathcal{D} \times \mathcal{S}^o} z_{o,r,d,s} \tag{10}$$

which will be the maximum number of surgeries performed.

Let us now turn to the problem of determining the planned slack. Determining the planned slack can be almost impossible to achieve in practice and so an alternative technique must be devised. Here we will propose using a support vector (SV) machine that predicts which ORDS are feasible using a linear classifier. The classifier is then added to our MIP model as follows:

$$\sum_{s \in S^o} \omega_s z_{o,r,d,s} + b \geq 0 \quad \forall (o,r,d) \in \mathcal{O} \times \mathcal{R} \times \mathcal{D} \tag{11}$$

and an alternative version will be illustrated using one SV ORDS classifier for each operator o, or

$$\sum_{s \in S^o} \omega_{s,o} z_{o,r,d,s} + b_o \geq 0 \quad \forall (o,r,d) \in \mathcal{O} \times \mathcal{R} \times \mathcal{D} \tag{12}$$

The training of the weights ω_o and bias b_o is presented in the following section.

3   SV-ORDS Classifier

The number of feasible and infeasible ORDS can be generated using random sampling and labelling using Monte-Carlo simulation. The initial training data generated must be balanced for each operator, nevertheless, the number of infeasible solutions will typically be far greater than the feasible ones. However, finding feasible solutions will in general be possible when assigning fewer surgeries to the ORDS. Once the initial data set has been constructed the first classifier may be learned. The SVM classifier applied here solves the 2-norm soft margin optimization problem:

$$\min_{\omega,b} \langle \omega \cdot \omega \rangle + C \sum_{i=1}^{\ell} \xi_i^2$$

subject to

$$u_i \left(\langle \omega \cdot x_i \rangle + b \right) \geq 1 - \xi_i$$

where $u_i \in \{1, -1\}$ labels the ORDS x_i ($x \in \{0, 1\}$) as feasible or not and $\xi \geq 0$, $i = 1, \ldots, \ell$. When all the x_i ORDS are for a specific operator o only, then we denote the classifier by the pair (ω_o, b_o) using the subscript o.

The entire SVORDS procedure may now be described as follows:

1. Generate the initial training data (X, u) of size ℓ using randomly generated ORDS and labelling using Monte-Carlo simulations, based on historical data. Set counter k to zero.
2. Generate the classifier (ω, b) using (X, u).
3. Solve the MIP problem described in the previous section.
4. Extract the ORDS from the MIP solution and use a Monte Carlo simulation to verify that they will not result in overtime with α probability. Denote this new data by $(X, u)^k$ (the size of this training set will be the number of ORs times planned days).
5. If the new ORDS are not all feasible augment them to the training data $(X, u) = (X, u) \cup (X, u)^k$, let $k = k + 1$ and return to step 2.

When there is a need to re-train the classifier it may be necessary to put more weight on the new ORDS data created. One would like them to be classified correctly, so that the infeasible ORDS do not re-appear in the next iteration. A simple approach would to append them more than once to the data set. For example, if you would like these new ORDS to weight ten times more than any other single ORDS in the current data set, then just add them ten times.

4   Experimental Study

For the experiments we will create a master operating schedules for one surgical speciality known as general surgery. The experiments will be performed on a 32 GB Intel core i7-7700 3.60 GHz with 4 cores. The MIP model is programmed in Python 3.7 using Gurobi version 8.1.0 [6] and the SVM classifier used is LIBLINEAR version 2.21 [4].

4.1   Instance Generation

In this study a large data set was obtained from the National University Hospital of Iceland. The data consists of information on all surgical activities spanning 10 years. Even though one might have years of historical data at hand, only the most recent data should be used, as surgery times and length of stay have been shown to either decrease or increase with time and so too do the surgeries performed.

To create a good master operator schedule, one must identify the surgeries that are most likely to occur within each planning cycle. As each patient is assigned to an operator o, each will have their own waiting list S^o of patients. Typically, no two surgeons will perform exactly the same set of procedures. Hence, differentiating between the surgeons is required in practice when finding the frequencies of surgeries. However, for simplification, it is assumed that each waiting list is equal in size for all of the surgeons. Typically the operator's list is in the 100s and so one might consider the shortened lists used in this study as patients with higher and equal priority. The length of the list will be limited to twice the maximum number required to fit the available OR surgery time. In Table 1 a summary of the main characteristics of the most frequently performed surgeries, by the operators, is presented.

For the computational experiments we will examine how different parameter settings will influence the results of the models with respect to total number of patients scheduled. Based on the scheduling practices of Landspitalinn, it is assumed that each $d_{r,d}$ has a capacity of 450 min and that ORs are closed during the weekend. The number of operating rooms available to general surgery is two. The maximum number of ORDS created is, therefore, 10 for a 5-day working week. As discussed previously, it is not necessarily undesirable to run an OR into overtime. In the real life settings, two out of the seven ORs are kept open longer for acute surgeries that continue, when possible, after the elective program. It can be assumed here that $\alpha \approx 0.25$. The upper limit on the ward capacity will

Table 1. Summary of main characteristics of most frequently performed surgeries by the nine operators

Operator	Number of surgery types	Mean surgery time	Mean ward probability	Mean ward length of stay
A	9	184 (min.)	0.4	2.4 (days)
B	6	137	0.5	2.9
C	9	188	0.5	2.6
D	10	167	0.7	3.2
E	11	162	0.6	3.4
F	5	214	0.4	3.7
G	7	90	0.1	2.0
H	12	143	0.3	2.5
I	13	84	0.1	2.3

be forced at $\overline{w} = 8$, as is the case at the hospital in question. A lower bound $\underline{w} = 2$ will ensure that the ward is utilized.

4.2   Results

The first result presented is the classification accuracy of the linear support vector machine when all feasible ORDS are known. This model will then be used as the starting model for our SVORDS procedure where the MIP is solved using this classifier. The final experiment investigates the performance of SVORDS when the initial model is weak, that is, initially created from a smaller training set.

Classification Accuracy. When the number of patients per operator is 20 it is possible to generate all ORDS for our test function. Each ORDS is validated using a Monte Carlo simulation using 1000 realizations. In this case they are 544.131 and the number of feasible ORDS are 31.843. When sampling the infeasible ORDS uniformly and randomly one can create a balanced data set of 63.686 samples. The resulting classifier gives the confusion matrix presented in Table 2 for the training data. The accuracy of the classifier is very high or greater than 99%. We would expect using single classifiers for each operator to be of equal accuracy or even greater. Out of the 63.686 samples, 439 examples that were actually infeasible were predicted to be feasible. This case is perhaps not as serious as in the case where a solution is indeed feasible and is predicted to be infeasible. In the former case these ORDS can be trained to be excluded and the MIP solved again. In the latter case, one would be unaware of these missing ORDS. If they were part of the optimal solution, the optimal solution would no longer be reachable. To get a better understanding of the performance of this approach, to understand how critical these misclassified ORDS are, we need to solve some problems.

Table 2. Confusion matrix using all feasible ORDS for training a single linear classifier.

$\ell = 63.686$	Predicted: feasible	Predicted: infeasible	
Actual: feasible	31.839	4	31.843
Actual: infeasible	439	31404	31.843
	32.278	31408	

Model Using All Feasible ORDS. In this first experiment we use the model described in the previous section. The size of the balanced training data set is 63.686 but the size of the entire ORDS space is 544.131. The number of patients is 20 per operator, in total 180 patients. The optimal solution to this problem may be found by searching all possible ORDS and has the value of 42 patients. The five step procedure described Sect. 3 was executed until a feasible ORDS solution is found. The procedure is repeated from step 2. as long as an optimal solution was not found within 200 s or the ORDS in the solution found were verified as being infeasible using the Monte Carlo simulation. In the case when the ORDS in the solution are feasible and the optimization was stopped early, it might be desirable to continue the search. Otherwise, early stopping may speed up the generation of new infeasible ORDS data. The infeasible ORDS were given a weight 10 times that of any other ORDS in the training data set. The result of this run is given in Table 3. The table shows how the number of infeasible ORDS is reduced as the SVORDS procedure advances. At each iterations the table also gives the number of infeasible solution ORDS remaining after re-training. When not zero there is a chance that they will re-appear in the next iteration. At the 26th iteration a feasible solution is found and so no new information is added to our classification model. At each SVORDS iteration the MIP has returned an optimal solution. The table gives the time required for each complete SVORDS iteration. The geometric distance from origin, $\|\omega\|/b$ increases slowly and converges. This defines the distance from the origin to the hyperplane separating all the infeasible ORDS from the feasible ones.

When training multiple classifiers (ω_o, b_o), one for each operator, a similar result is observed, indeed after 18 iterations a feasible solution is found with an objective value of 41. No statistical difference was observed between these experiments.

Models Using a Subset of Feasible ORDS. In this experiment the initial training data does not contain the entire set of all feasible ORDS, but rather a small subset of 2.000 training samples, balanced for each operator. The total size is 18.000 ORDS validated and labelled using Monte Carlo simulations. When using a smaller subset of infeasible ORDS the hyperplane separating all infeasible ORDS is at a smaller distance from the origin than when using the entire set of feasible ORDS and an equivalent amount of infeasible ORDS. The number of iterations needed to find a feasible set of optimal ORDS is now greater. The geometric distance expands and converges to a distance of 4.359, which is

Table 3. The iterations of the SVORDS procedure, at iterations 26 a feasible solution is found with objective value 41, when the number of infeasible ORDS are 0. The computation time in seconds is for a complete iteration of SVORDS.

IP iteration	Infeasible ORDS	Infeasible ORDS after re-train	Best objective	$\|\omega\|/b$	Computation time (sec)
1	3	0	42	4.398	37
2	3	0	42	4.416	50
3	3	0	41	4.411	96
4	5	1	41	4.403	90
5	5	0	42	4.404	154
6	5	1	41	4.407	83
7	3	0	42	4.416	144
8	2	1	41	4.415	102
9	3	1	42	4.412	102
10	2	0	41	4.421	112
11	3	0	41	4.425	111
12	3	0	41	4.434	103
13	1	0	42	4.437	130
14	3	0	41	4.429	117
15	3	1	41	4.412	74
16	4	0	41	4.422	78
17	2	0	42	4.434	69
18	1	0	41	4.434	117
19	1	0	41	4.439	70
20	3	0	41	4.444	88
21	1	0	41	4.433	98
22	1	0	41	4.433	115
23	1	0	41	4.441	72
24	1	0	41	4.445	131
25	2	0	41	4.452	95
26	**0**	0	**41**	**4.461**	116

close but smaller than the previous classifier, at 4.461. Each time an infeasible ORDS is added to the training this distance expands. This is clearly shown in Table 4, starting at 3.629 and ending in 4.359. Again when using different classifiers for each operator there is no clear advantage over using a single classifier and repeated runs show no statistical difference in the quality of solutions found (40 being the typical result).

Table 4. The iterations of the SVORDS procedure, at iterations 73 a feasible solution is found with objective value 40.

IP iteration	Infeasible ORDS	Infeasible ORDS after re-train	Objective	Best bound	MIP gap	$\|\omega\|/b$
1	9	0	54		0	3.629
2	10	0	50		0	3.527
3	10	0	48	50	416.67	3.533
4	10	0	49		0	3.52
5	6	0	45		0	3.508
6	8	0	43		0	3.515
7	8	0	48		0	3.569
8	8	0	44		0	3.567
9	8	0	45		0	3.583
10	6	0	44		0	3.629
11	5	0	42		0	3.639
12	5	0	43		0	3.668
13	8	0	44		0	3.723
14	7	0	43		0	3.743
15	7	0	44	55	2500	3.767
16	7	0	45		0	3.793
17	7	0	44	52	1818.18	3.804
18	6	0	40	49	2250	3.824
19	5	0	41		0	3.866
⋮	⋮	⋮	⋮	⋮	⋮	
50	2	0	40		0	4.238
51	2	0	41		0	4.252
52	3	0	41		0	4.257
53	1	0	40		0	4.262
54	1	0	40		0	4.269
55	2	0	40		0	4.29
56	4	0	41		0	4.289
57	4	0	41	50	2195.12	4.305
58	3	0	40		0	4.322
59	1	0	41	50	2195.12	4.345
60	5	0	40		0	4.346
61	2	0	40		0	4.33
62	2	0	40		0	4.333
63	2	0	40		0	4.349
64	2	0	40		0	4.345
65	1	0	40		0	4.357
66	2	0	40		0	4.356
67	1	0	40		0	4.358
68	1	0	40		0	4.359
69	4	0	40		0	4.358
70	1	0	40		0	4.361
71	2	0	40		0	4.363
72	2	0	40		0	4.357
73	**0**	0	**40**		0	**4.359**

Looking closer at all solutions we notice that in all cases the classifiers found have $b > 0$ and $\omega_s < 0$, this implies that constraint (11) takes the form

$$\sum_{s \in \mathcal{S}^o} \bar{\omega}_s z_{o,r,d,s} \leq 1 \quad \forall \, (o,r,d) \in \mathcal{O} \times \mathcal{R} \times \mathcal{D} \tag{13}$$

where $(-\omega_s/b) = \bar{\omega}_s \geq 0$ which can be interpreted as a normalized time with a fixed room capacity of one.

5 Discussion and Conclusion

The experimental results presented are based on one particular realization from our instance generator, however, similar results were obtained using different instances from our generator. In order to provide a clearer picture of the nature of the SVORDS procedure a typical instance was chosen and the result illustrated. When using a weaker classification model the optimal solution found was of a poorer quality (median 40) to that of the classifications model using all feasible ORDS (median 41) in the training data. The known optimal solution to this problem is 42 and so the technique has clearly eliminated the optimal solution. Nevertheless, the quality of the solutions found is high. Typically the general surgery will schedule no more than 38–39 surgeries per week, which is regarded as a high number.

It was assumed that the stochastic capacity constraints for the ORs could be replaced with a linear classifier. This is not necessarily the case, as the confidence bounds for the sum of all surgery times planned, for any given room, may not be linear. Ideally one would need to know the ORDS from which one could estimate the mean time for the surgeries and their variance. From this the confidence bounds could be accurately determined or estimated using Monte Carlo simulation. Nevertheless, knowing all ORDS is intractable in general and so the problem of determining the slack for the capacity constraint remains unknown. It has been shown that the linear SV classifier is able to classify feasible ORDS with high accuracy. Nevertheless, the ORDS that are close to the separating hyperplane are the most interesting ones. The SVORDS procedure was able to identify most of these quality ORDS and arrange them in such a way as to satisfy the ward constraints. Thus creating high quality surgery schedules.

The number of infeasible ORDS clearly outnumber the feasible ones. Great care was taken to balance the number of feasible and infeasible ORDS in the training data. Once too many infeasible ORDS are used, the classifier may be tempted to classify all ORDS as infeasible and so the entire search space. This is not the desired effect we seek, but clearly we need to add more infeasible ORDS to avoid the infeasible space. SVORDS efficiently selects the most relevant infeasible ORDS for the training set.

The result of our linear support vector classifier is an estimate of a pseudo slack or bias, and classification weights for each surgery. This has some nice practical consequences. For instance, if we assume that a surgery is postponed it is possible to find another surgery with a similar or larger classification weight.

Given that care is taken not to cause any ward congestion down-stream a replacement may be found without having to re-optimize. This is important, since a drastic change in the plan will not be possible once the patients have been alerted about their upcoming surgery appointment. Finding a replacement for surgery s is then reduced to looking at candidates s', not currently planned, with $\bar{\omega}'_s \leq \bar{\omega}_s$ and where $\bar{\omega}'_s$ takes a maximal value.

Acknowledgement. The author would like to acknowledge the National University Hospital of Iceland for providing data, insights and support for this project.

References

1. Cappanera, P., Visintin, F., Banditori, C.: Addressing conflicting stakeholders' priorities in surgical scheduling by goal programming. Flex. Serv. Manuf. J. **30**(1), 252–271 (2018). https://doi.org/10.1007/s10696-016-9255-5
2. Cardoen, B., Demeulemeester, E., Beliën, J.: Operating room planning and scheduling: a literature review. Eur. J. Oper. Res. **201**(3), 921–932 (2010). https://doi.org/10.1016/j.ejor.2009.04.011. http://www.sciencedirect.com/science/article/pii/S0377221709002616
3. Denton, B., Miller, A.J., Balasubramanian, H.J., Huschka, T.R.: Optimalallocation of surgery blocks to operating rooms under uncertainty. Oper. Res. **58**(4–part–1), 802–816 (2010). https://doi.org/10.1287/opre.1090.0791
4. Fan, R.E., Chang, K.W., Hsieh, C.J., Wang, X.R., Lin, C.J.: LIBLINEAR: a library for large linear classification. J. Mach. Learn. Res. **9**(Aug), 1871–1874 (2008)
5. Guido, R., Conforti, D.: A hybrid genetic approach for solving an integrated multi-objective operating room planning and scheduling problem. Comput. Oper. Res. **87**(Supplement C), 270–282 (2017). https://doi.org/10.1016/j.cor.2016.11.009. http://www.sciencedirect.com/science/article/pii/S0305054816302684
6. Gurobi Optimization, Inc.: Gurobi optimizer reference manual (2018). http://www.gurobi.com
7. Hans, E., Wullink, G., van Houdenhoven, M., Kazemier, G.: Robust surgery loading. Eur. J. Oper. Res. **185**(3), 1038–1050 (2008). https://doi.org/10.1016/j.ejor.2006.08.022. <Go to ISI>://WOS:000251070500011
8. Kroer, L.R., Foverskov, K., Vilhelmsen, C., Hansen, A.S., Larsen, J.: Planning and scheduling operating rooms for elective and emergency surgeries with uncertain duration. Oper. Res. Health Care (2018). https://doi.org/10.1016/j.orhc.2018.03.006
9. Lamiri, M., Xie, X., Dolgui, A., Grimaud, F.: A stochastic model for operating room planning with elective and emergency demand for surgery. Eur. J. Oper. Res. **185**(3), 1026–1037 (2008). https://doi.org/10.1016/j.ejor.2006.02.057. http://www.sciencedirect.com/science/article/pii/S0377221706005832
10. Marques, I., Captivo, M.E.: Different stakeholders' perspectives for a surgical case assignment problem: deterministic and robust approaches. Eur. J. Oper. Res. **261**(1), 260–278 (2017). https://doi.org/10.1016/j.ejor.2017.01.036. http://www.sciencedirect.com/science/article/pii/S0377221717300711
11. Min, D., Yih, Y.: Scheduling elective surgery under uncertainty and downstream capacity constraints. Eur. J. Oper. Res. **206**(3), 642–652 (2010)

12. Molina-Pariente, J.M., Hans, E.W., Framinan, J.M.: A stochastic approach for solving the operating room scheduling problem. Flex. Serv. Manuf. J. (2016). https://doi.org/10.1007/s10696-016-9250-x
13. Neyshabouri, S., Berg, B.P.: Two-stage robust optimization approach to elective surgery and downstream capacity planning. Eur. J. Oper. Res. **260**(1), 21–40 (2017). https://doi.org/10.1016/j.ejor.2016.11.043. http://www.sciencedirect.com/science/article/pii/S0377221716309687
14. van Oostrum, J.M., Van Houdenhoven, M., Hurink, J.L., Hans, E.W., Wullink, G., Kazemier, G.: A master surgical scheduling approach for cyclic scheduling in operating room departments. OR Spectr. **30**(2), 355–374 (2008). https://doi.org/10.1007/s00291-006-0068-x
15. Riise, A., Mannino, C., Burke, E.K.: Modelling and solving generalised operational surgery scheduling problems. Comput. Oper. Res. **66**, 1–11 (2016)
16. Samudra, M., Van Riet, C., Demeulemeester, E., Cardoen, B., Vansteenkiste, N., Rademakers, F.E.: Scheduling operating rooms: achievements, challenges and pitfalls. J. Sched. **19**(5), 493–525 (2016). https://doi.org/10.1007/s10951-016-0489-6
17. Spangler, W.E., Strum, D.P., Vargas, L.G., May, J.H.: Estimating procedure times for surgeries by determining location parameters for the lognormal model. Health Care Manag. Sci. **7**(2), 97–104 (2004)

Exact Algorithm for One Cardinality-Weighted 2-Partitioning Problem of a Sequence

Alexander Kel'manov[1,2]($\boxtimes$) (iD), Sergey Khamidullin[1]($\boxtimes$) (iD),
and Anna Panasenko[1,2]($\boxtimes$) (iD)

[1] Sobolev Institute of Mathematics, 4 Koptyug Avenue, 630090 Novosibirsk, Russia
{kelm,kham,a.v.panasenko}@math.nsc.ru
[2] Novosibirsk State University, 2 Pirogova Street, 630090 Novosibirsk, Russia

Abstract. We consider a problem of 2-partitioning a finite sequence of points in Euclidean space into two clusters of the given sizes with some additional constraints. The solution criterion is the minimum of the sum (over both clusters) of weighted intracluster sums of squared distances between the elements of each cluster and its center. The weights of the intracluster sums are equal to the cardinalities of the desired clusters. The center of one cluster is given as input, while the center of the other one is unknown and is determined as a geometric center, i.e. as a point of space equal to the mean of the cluster elements. The following constraints hold: the difference between the indices of two subsequent points included in the first cluster is bounded from above and below by given some constants. It is shown that the considered problem is the strongly NP-hard one. An exact algorithm is proposed for the case of integer-valued input of the problem. This algorithm has a pseudopolynomial running time if the space dimension is fixed.

Keywords: Euclidean space · Sequence of points · Weighted 2-partition · Quadratic variation · NP-hard problem · Integer coordinates · Exact algorithm · Fixed space dimension · Pseudopolynomial time

1 Introduction

The subject of the study is one quadratic cardinality-weighted problem of partitioning a sequence of points in Euclidean space into two subsequences of the given sizes with some additional constraints. The goal of our study is to analyze the computational complexity of the problem and to substantiate an efficient

The study presented in Sects. 3 and 4 was supported by the Russian Foundation for Basic Research, projects 19-07-00397, 19-01-00308 and 18-31-00398. The study presented in the other sections was supported by the Russian Academy of Science (the Program of basic research), project 0314-2019-0015, and by the Russian Ministry of Science and Education under the 5–100 Excellence Programme.

N. F. Matsatsinis et al. (Eds.): LION 13 2019, LNCS 11968, pp. 135–145, 2020.
https://doi.org/10.1007/978-3-030-38629-0_11

algorithm for the solution to this problem. Our study is motivated by the importance of considered problem for some applications, in particular, for data mining and data clustering, when the data having in the hands is a time series.

The paper is organized as follows. In Sect. 2 we present the problem formulation and its interpretation. Also, here one closely related problem is presented. In addition, in the same section, we analyze the problem complexity. In Sect. 3 we formulate an auxiliary problem and prove some statements which underlie quality estimates for the proposed algorithm. The next Sect. 4 contains the algorithm for the solution to the problem considered. The analysis of the algorithm properties is also in this section. In Sect. 5 some results of numerical experiments are proposed.

2   Problem Formulation, Related Problem, and Complexity

Everywhere below $\mathbb{R}$ denotes the set of real numbers, $\|\cdot\|$ denotes the Euclidean norm, and $\langle \cdot, \cdot \rangle$ denotes the scalar product.

We consider the following

Problem 1. Given a sequence $\mathcal{Y} = (y_1, \ldots, y_N)$ of points in $\mathbb{R}^d$ and some positive integers $T_{\min}$, $T_{\max}$, and $M > 1$. *Find* a subset $\mathcal{M} = \{n_1, n_2, \ldots\} \subset \mathcal{N} = \{1, \ldots, N\}$ of indices in $\mathcal{Y}$ such that

$$F(\mathcal{M}) = |\mathcal{M}| \sum_{j \in \mathcal{M}} \|y_j - \overline{y}(\{y_n | n \in \mathcal{M}\})\|^2 + |\mathcal{N} \setminus \mathcal{M}| \sum_{i \in \mathcal{N} \setminus \mathcal{M}} \|y_i\|^2 \longrightarrow \min, \quad (1)$$

where $\overline{y}(\{y_n | n \in \mathcal{M}\}) = \frac{1}{|\mathcal{M}|} \sum_{i \in \mathcal{M}} y_i$ is the centroid of $\{y_n | n \in \mathcal{M}\}$ with the following constraints

$$1 \leq T_{\min} \leq n_m - n_{m-1} \leq T_{\max} \leq N, \quad m = 2, \ldots, |\mathcal{M}|, \quad (2)$$

and $|\mathcal{M}| = M$.

Problem 1 has the following applied interpretation. We have a sequence $\mathcal{Y}$ of N time-ordered measurement results (i.e., time series or discrete signal) for d characteristics of some object in two different states (active and passive, for example). Each measurement has an error and nobody knows the correspondence between the elements of the input sequence and the states. But it is known that the time interval between every two consequent active states is bounded from below and above by some constants $T_{\min}$ and $T_{\max}$. In addition, it is known that exactly M times the object was in the active state (or the probability of the active state is $\frac{M}{N}$). It requires to find 2-partition of the input sequence and evaluate the object characteristics (i.e., $\overline{y}(\{y_n | n \in \mathcal{M}\})$) in accordance with (1)).

What can we say on this application problem is it is very typical for processing time-series or discrete signals. One can see this problem, for example, in technical and medical diagnostics, in distant object monitoring and in geophysics, etc. (see, for example, [1–5]).

In Fig. 1 one can see an example of the input 2-dimensional sequence (200 points) for Problem 1. Each point of the sequence corresponds to a vertical strip in the tape. The given values, in this case, are $T_{\min} = 2$, $T_{\max} = 20$, $M = 17$. The points of the same input sequence are presented on a plane in Fig. 2.

Fig. 1. Example of the input sequence for $d = 2$, $N = 200$, $T_{\min} = 2$, $T_{\max} = 20$, $M = 17$.

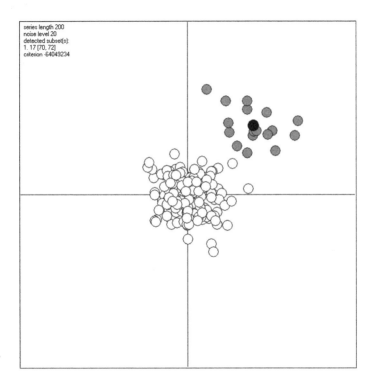

Fig. 2. $N = 200$, $T_{\min} = 2$, $T_{\max} = 20$, $M = 17$ on a plane.

In the mathematical sense, the considered problem is closely related to the following problem.

Problem 2. (Cardinality-weighted variance-based 2-clustering with given center).
Given an N-element set $\mathcal{Y}$ of points in $\mathbb{R}^d$, and positive integer number M. *Find* a partition of $\mathcal{Y}$ into two non-empty clusters $\mathcal{C}$ and $\mathcal{Y} \setminus \mathcal{C}$ such that

$$|\mathcal{C}| \sum_{y \in \mathcal{C}} \|y - \overline{y}(\mathcal{C})\|^2 + |\mathcal{Y} \setminus \mathcal{C}| \sum_{y \in \mathcal{Y} \setminus \mathcal{C}} \|y\|^2 \to \min ,$$

where $\overline{y}(\mathcal{C}) = \frac{1}{|\mathcal{C}|} \sum\limits_{y \in \mathcal{C}} y$ is the centroid of $\mathcal{C}$, subject to constrain $|\mathcal{C}| = M$.

The strong NP-hardness of Problem 2 was established in [6]. The strong NP-hardness of Problem 1 follows from this result, as Problem 2 is the special case of Problem 1 when $T_{\min} = 1$ and $T_{\max} = N$.

Problem 2 has been studied in algorithmic sense in [7–11].

In [7], an exact pseudo-polynomial algorithm was constructed for the case of integer components of the input points and fixed dimension d of the space. The running time of this algorithm is $\mathcal{O}(N(MD)^d)$, where D is the maximum absolute value of coordinates of the input points.

In [8], an approximation scheme that allows one to find $(1 + \varepsilon)$-approximate solution in $\mathcal{O}\left(dN^2\left(\sqrt{\frac{2d}{\varepsilon}} + 2\right)^d\right)$ time was proposed. It implements an FPTAS in the case of the fixed space dimension.

Moreover, in [9], the modification of this algorithm with improved time complexity: $\mathcal{O}\left(\sqrt{d}N^2\left(\frac{\pi e}{2}\right)^{d/2}\left(\sqrt{\frac{2}{\varepsilon}} + 2\right)^d\right)$, was proposed. The algorithm implements an FPTAS in the case of fixed space dimension and remains polynomial for instances of dimension $\mathcal{O}(\log n)$. In this case, it implements a PTAS with $\mathcal{O}\left(N^{C\,(1.05 + \log(2 + \sqrt{\frac{2}{\varepsilon}}))}\right)$ time, where C is a positive constant.

In [10], an approximation algorithm that allows one to find a 2-approximate solution to the problem in $\mathcal{O}\left(dN^2\right)$ time was constructed.

In [11], a randomized algorithm was constructed. It allows one to find $(1+\varepsilon)$-approximate solution with probability not less than $1 - \gamma$ in $\mathcal{O}(dN)$ time for an established parameter value, a given relative error ε and fixed γ. The conditions are found under which the algorithm is asymptotically exact and runs in $\mathcal{O}(dN^2)$ time.

In this paper, we present the first result for strongly NP-hard Problem 1. Namely, we present an exact pseudo-polynomial algorithm for the case of integer instances and fixed d of space dimension. This algorithm is based on the approaches and results presented in [7,12,13]. The running time of this algorithm is $\mathcal{O}(N(M(T_{\max} - T_{\min} + 1) + d)(2MD + 1)^d)$, where D is the maximum absolute value of coordinates of the input points.

3   Foundations of the Algorithm

In this section, we formulate some statements about indices of $\mathcal{M}$ and formulate one more auxiliary problem which can be solved in polynomial time. All these statements and auxiliary problem are necessary for substantiation of our algorithms.

Two following lemmas were proved in [12,13].

Lemma 1. *If the elements of $\mathcal{M} = \{n_1, \ldots, n_M\}$ belong to $\mathcal{N} = \{1, \ldots, N\}$ and satisfy the system of constraints (2), then for every fixed $M \in \{2, \ldots, N\}$ we have:*

(1) the parameters of this system are related by inequality

$$(M - 1)T_{\min} \leq N - 1 , \qquad (3)$$

(2) the element n_m in $\{n_1, \ldots, n_m, \ldots, n_M\}$ belongs to the set

$$\omega_m = \{n | 1 + (m - 1)T_{\min} \leq n \leq N - (M - m)T_{\min}\}, \; m = 1, \ldots, M , \quad (4)$$

(3) the feasibility domain of components n_{m-1} from this set under condition $n_m = n$ is defined by formula

$$\gamma_{m-1}(n) = \{j | \max\{1 + (m - 2)T_{\min}, n - T_{\max}\} \leq j \leq n - T_{\min}\} , \qquad (5)$$

where $n \in \omega_m$, $m = 2, \ldots, M$.

Lemma 2. *For every $M \in \{2, \ldots, N\}$ the system of constraints (2) is feasible if and only if inequality (3) holds.*

Consider the following function:

$$S(\mathcal{M}, b) = M \sum_{n \in \mathcal{M}} \|y_n - b\|^2 + (N - M) \sum_{n \in \mathcal{N} \setminus \mathcal{M}} \|y_n\|^2, \; b \in \mathbb{R}^d, \; \mathcal{M} \subset \mathcal{N} .$$

It is similar to the objective function of Problem 1, since $|\mathcal{M}| = M$ and $|\mathcal{N} \setminus \mathcal{M}| = N - M$. The only difference is the point b instead of the centroid $\bar{y}(\{y_n | n \in \mathcal{M}\})$. This function can be rewritten as follows:

$$S(\mathcal{M}, b) = (N - M) \sum_{n \in \mathcal{N}} \|y_n\|^2$$

$$- \left(2M \sum_{n \in \mathcal{M}} \langle y_n, b \rangle - (2M - N) \sum_{n \in \mathcal{M}} \|y_n\|^2 - M^2 \|b\|^2 \right)$$

$$= (N - M) \sum_{n \in \mathcal{N}} \|y_n\|^2 - \sum_{n \in \mathcal{M}} \left(2M \langle y_n, b \rangle - (2M - N) \|y_n\|^2 - M \|b\|^2 \right) .$$

Note that the first summand is a constant, hence the minimum of $S(\mathcal{M}, b)$ is reached on the subsequence that maximizes the second summand. This expression motivates us to formulate auxiliary:

Problem 3. Given a sequence $\mathcal{Y} = (y_1, \ldots, y_N)$ of points in $\mathbb{R}^d$, a point $b \in \mathbb{R}^d$, and some positive integers $T_{\min}$, $T_{\max}$ and M. Find a subset $\mathcal{M} = \{n_1, \ldots, n_M\} \subset \mathcal{N} = \{1, \ldots, N\}$ of indices in the sequence $\mathcal{Y}$ such that

$$G^b(\mathcal{M}) = \sum_{n \in \mathcal{M}} g^b(n) \longrightarrow \max , \qquad (6)$$

where

$$g^b(n) = 2M \langle y_n, b \rangle - (2M - N) \|y_n\|^2 - M \|b\|^2, \; n \in \mathcal{N} , \qquad (7)$$

with additional constraints (2) on the elements of $\mathcal{M}$, if $M \neq 1$.

Let us define the set Ψ_M of subsets of admissible index tuples in the auxiliary problem:

$$
\Psi_M = \begin{cases}
\{(n_1)\,|\,n_1 \in \mathcal{N}\}, & \text{if } M = 1; \\
\{(n_1,\dots,n_M)\,|\,n_i \in \mathcal{N},\ i = 1,\dots,M; \\
\quad 1 \le T_{\min} \le n_m - n_{m-1} \le T_{\max} \le N, \\
\quad\quad m = 2,\dots,M\}, & \text{if } 1 < M \le N.
\end{cases} \tag{8}
$$

For $M = 1$ the set Ψ_M is not empty for any parameters $T_{\min}$ and $T_{\max}$ by definition (8). For other feasible values of M we have ([12,13]):

Lemma 3. *If $M \ge 2$, then the set Ψ_M is not empty if and only if an inequality (3) holds.*

Proofs of the following lemma and its corollary in [12,13] do not use (7) and so they hold for our case too.

Lemma 4. *Let $\Psi_M \ne \emptyset$ for some $M \ge 1$. Then for this M, the optimal value $G_{\max}^b = \max\limits_{\mathcal{M}} G^b(\mathcal{M})$ of objective function (6) can be found by formula*

$$
G_{\max}^b = \max_{n \in \omega_M} G_M^b(n) . \tag{9}
$$

The values $G_M^b(n)$, $n \in \omega_M$, can be calculated by the following recurrent formulae:

$$
G_m^b(n) = \begin{cases}
g^b(n), & \text{if } n \in \omega_1, m = 1, \\
g^b(n) + \max\limits_{j \in \gamma_{m-1}^-(n)} G_{m-1}^b(j), & \text{if } n \in \omega_m, m = 2,\dots,M,
\end{cases} \tag{10}
$$

where sets ω_m and $\gamma_{m-1}^-(n)$ are defined by formulae (4) and (5).

Corollary 1. *The elements $n_1^b,\dots,n_M^b$ of the optimal set $\mathcal{M}^b = \arg\max\limits_{\mathcal{M}} G^b$ $(\mathcal{M})$ can be found by the following recurrent formulae:*

$$
n_M^b = \arg \max_{n \in \omega_M} G_M^b(n) , \tag{11}
$$

$$
n_{m-1}^b = \arg \max_{n \in \gamma_m^-(n_m^b)} G_m^b(n),\ m = M, M - 1,\dots,2 . \tag{12}
$$

The following algorithm finds an optimal solution for auxiliary Problem 3. The step-by-step description of the algorithm looks like as follows.

Algorithm $\mathcal{A}_1$.

Input: a sequence $\mathcal{Y}$, a point b, some positive integer $T_{\min}$, $T_{\max}$, M.

Step 1. Compute $g^b(n)$, $n \in \mathcal{N}$, using formula (7).

Step 2. Using recurrent formulae (10), compute $G_m^b(n)$ for each $n \in w_n$ and $m = 1,\dots,M$.

Step 3. Find the maximal value $G_{\max}^b$ of the objective function G^b using formula (9) and the optimal set $\mathcal{M}^b = \{n_1^b,\dots,n_M^b\}$ by (11) and (12) from Corollary 1; exit.

Output: the value of $G_{\max}^b$, the set $\mathcal{M}^b$.

The following theorem has been established in [13].

Theorem 1. *Algorithm $\mathcal{A}_1$ finds an optimal solution of Problem 3 in $\mathcal{O}(N (M(T_{\max} - T_{\min} + 1) + d))$ time.*

4 Exact Algorithm

In our algorithm, we construct a multidimensional grid in the area defined by the maximal absolute value of the coordinates of the input set. We choose the node spacing of the grid such that the geometric center of one of the desired clusters coincides with one of the nodes. For each node of the constructed grid, we solve Problem 3 with the help of Algorithm $\mathcal{A}_1$ presented in the previous section and then choose the best solution.

Assume now that all the coordinates of the points of $\mathcal{Y}$ are integers. Put

$$D = \max_{y \in \mathcal{Y}} \max_{j \in \{1,\dots,d\}} |(y)^j|, \tag{13}$$

where $(y)^j$ is the j-th coordinate of y. Define the set

$$\mathcal{D} = \left\{ x \in \mathbb{R}^d \mid (x)^j = \frac{1}{M}(v)^j, \ (v)^j \in \mathbb{Z}, \ |(v)^j| \le MD, \ j = 1,\dots,d \right\} \tag{14}$$

as a multidimensional grid with a uniform rational node spacing equal to $1/M$ in each coordinate. Note that $|\mathcal{D}| = (2MD + 1)^d$.

The following statement is obvious.

Lemma 5. *Assume that the elements of $\mathcal{Y}$ have the integer values in the interval $[-D, D]$. Then the centroid of every subset $\mathcal{C} \subset \mathcal{Y}$ of the size M lies in $\mathcal{D}$.*

Algorithm $\mathcal{A}$.

Input: a sequence $\mathcal{Y}$, some positive integers $T_{\min}$, $T_{\max}$, M.

Step 1. Find D by (13) and construct the grid $\mathcal{D}$ by (14).

Step 2. For each point $x \in \mathcal{D}$, find the optimal solution $\mathcal{M}^x$ and the maximal value $G^x_{\max}$ of objective function (6) using algorithm $\mathcal{A}_1$.

Step 3. Find the point $x_{\mathcal{A}} = \arg\max_{x \in \mathcal{D}} G^x_{\max}$ and the corresponding subset $\mathcal{M}_{\mathcal{A}} = \mathcal{M}^{x_{\mathcal{A}}}$, centroid $\overline{y}(\{y_n | n \in \mathcal{M}_{\mathcal{A}}\}) = \frac{1}{M} \sum_{n \in \mathcal{M}_{\mathcal{A}}} y_n$, and the value of the objective function $F(\mathcal{M}_{\mathcal{A}})$ by (1). If there are several solutions, choose any; exit.

Output: the set $\mathcal{M}_{\mathcal{A}}$, the point $x_{\mathcal{A}}$.

Theorem 2. *Assume that the conditions of Lemma 5 hold. Then Algorithm $\mathcal{A}$ finds an optimal solution of Problem 1 in*

$$\mathcal{O}(N(M(T_{\max} - T_{\min} + 1) + d)(2MD + 1)^d)$$

time.

Proof. Let $\mathcal{M}^*$ be an optimal solution of Problem 1, $y^* = \bar{y}(\{y_n | n \in \mathcal{M}^*\})$ be the centroid of this optimal solution, and $\mathcal{M}_{\mathcal{A}}$ be an output of Algorithm $\mathcal{A}$. Let us show that $F(\mathcal{M}^*) = F(\mathcal{M}_{\mathcal{A}})$.

Lemma 5 implies that a centroid y^* is in $\mathcal{D}$.

$$F(\mathcal{M}_{\mathcal{A}}) = S(\mathcal{M}_{\mathcal{A}}, \bar{y}(\{y_n | n \in \mathcal{M}_{\mathcal{A}}\})) \leq S(\mathcal{M}_{\mathcal{A}}, x_{\mathcal{A}})$$
$$= (N - M) \sum_{n \in \mathcal{N}} \|y_n\|^2 - G^{x_{\mathcal{A}}}(\mathcal{M}_{\mathcal{A}}) \leq (N - M) \sum_{n \in \mathcal{N}} \|y_n\|^2 - G^{y^*}(\mathcal{M}^{y^*})$$

Indeed, one can check by the differentiation that the minimum of $S(\mathcal{M}_{\mathcal{A}}, \cdot)$ (for the fixed $\mathcal{M}_{\mathcal{A}}$) is attained in $\bar{y}(\{y_n | n \in \mathcal{M}_{\mathcal{A}}\})$, so the first inequality holds. The last inequality holds by the definition of Step 3.

Theorem 1 implies that

$$G^{y^*}(\mathcal{M}^{y^*}) \geq G^{y^*}(\mathcal{M}^*) .$$

Hence,

$$F(\mathcal{M}_{\mathcal{A}}) \leq (N - M) \sum_{n \in \mathcal{N}} \|y_n\|^2 - G^{y^*}(\mathcal{M}^*) = S(\mathcal{M}^*, y^*) = F(\mathcal{M}^*) .$$

On the other hand, $F(\mathcal{M}^*) \leq F(\mathcal{M}_{\mathcal{A}})$ since $\mathcal{M}_{\mathcal{A}}$ is a feasible solution to Problem 1. So $F(\mathcal{M}^*) = F(\mathcal{M}_{\mathcal{A}})$ and the algorithm finds an optimal solution.

Let us estimate the time complexity. At Step 1 we need $\mathcal{O}(dN)$ operations to find the value D and $\mathcal{O}(d|\mathcal{D}|)$ operations to construct the grid $\mathcal{D}$. Step 2 is executed for $|\mathcal{D}|$ times. In each iteration, we use Algorithm $\mathcal{A}_1$ that requires $\mathcal{O}(N(M(T_{\max} - T_{\min} + 1) + d))$ time. At Step 3 we need $\mathcal{O}(|\mathcal{D}|)$ operations. Thus, the total time complexity of the algorithm is $\mathcal{O}(N(M(T_{\max} - T_{\min} + 1) + d)(2MD + 1)^d)$.   □

Remark 1. Note that $(2MD + 1)^d \leq (3MD)^d = 3^d(MD)^d$. This inequality implies that if the dimension d of the space is fixed, then Algorithm $\mathcal{A}$ is pseudopolynomial.

5   Examples of Numerical Simulation

The figures presented below show the robustness of the algorithm for Problem 1 of searching for a subsequence.

The first example is the following. In Fig. 3 (upper tape) one can see the input 2-dimensional sequence of integer-valued points (out of 100 points). Each point of the sequence corresponds to a vertical strip in the tape. The subsequence of points found by the algorithm for $T_{\min} = 1$, $T_{\max} = 5$, $M = 33$ is presented in the same Fig. 3 on the lower tape.

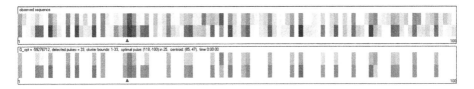

Fig. 3. The input of Algorithm $\mathcal{A}$ (upper tape) and the found subsequence (lower tape): $N = 100$, $T_{\min} = 1$, $T_{\max} = 5$, $M = 33$.

The points of the same input sequence are presented on a plane in Fig. 4 at the left-hand side. At the right-hand side, it is shown a subset (light/dark points) that corresponds to the subsequence presented in Fig. 3 on the lower tape.

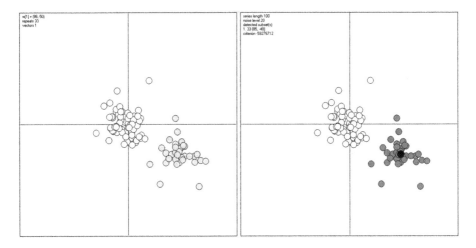

Fig. 4. The input of Algorithm $\mathcal{A}$ (left part) and the found subsequence (right part, light/dark points) shown on a plane: $N = 100$, $T_{\min} = 1$, $T_{\max} = 5$, $M = 33$.

Fig. 5. The input of Algorithm $\mathcal{A}$ (upper tape) and the found subsequence (lower tape): $N = 1000$, $T_{\min} = 2$, $T_{\max} = 20$, $M = 92$.

The second example has the same structure and looks like as follows in Figs. 5 and 6.

Here we have the input 2-dimensional sequence of integer-valued points (out of 1000 points). The subsequence of points was found by the algorithm for $T_{\min} = 2$, $T_{\max} = 20$, $M = 92$.

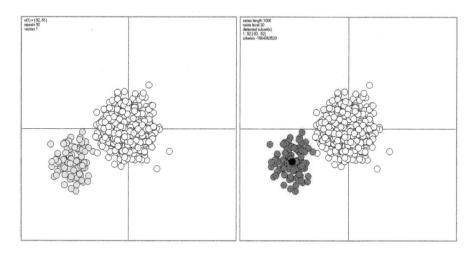

Fig. 6. The input of Algorithm $\mathcal{A}$ (left part) and the found subsequence (right part, light/dark points) shown on a plane: $N = 1000$, $T_{\min} = 2$, $T_{\max} = 20$, $M = 92$.

6   Conclusion

In this paper, we have shown the strong NP-hardness of one cardinality-weighted quadratic partitioning problem of a sequence of points in Euclidean space into two clusters when the center of one of the desired clusters is fixed. We also presented the first algorithmic result for the problem considered. This result is the exact algorithm for the case of integer components of the input points. Our algorithm on the one hand based on an adaptive-grid-approach and on the other hand based on the ideas of the dynamic programming. The running time of this algorithm is $\mathcal{O}(N(M(T_{\max} - T_{\min} + 1) + d)(2MD + 1)^d)$, where D is the maximum absolute value of coordinates of the input points. When the dimension d of the space is fixed, the algorithm is the pseudopolynomial one. Thus, we have established the conditions under which the problem is solvable in pseudopolynomial time.

It is clear that the algorithm can be used to solve practical problems having integer instances of small dimensions only. It seems important to continue studying the questions on algorithmic approximability of the problem since the considered problem is poorly studied in the algorithmic sense.

References

1. Fu, T.: A review on time series data mining. Eng. Appl. Artif. Intell. **24**(1), 164–181 (2011)
2. Kuenzer, C., Dech, S., Wagner, W. (eds.): Remote Sensing Time Series. RSDIP, vol. 22. Springer, Cham (2015). https://doi.org/10.1007/978-3-319-15967-6
3. Liao, T.W.: Clustering of time series data – a survey. Pattern Recogn. **38**(11), 1857–1874 (2005)
4. Kel'manov, A.V., Jeon, B.: A posteriori joint detection and discrimination of pulses in a quasiperiodic pulse train. IEEE Trans. Signal Process. **52**(3), 645–656 (2004)
5. Carter, J.A., Agol, E., et al.: Kepler-36: a pair of planets with neighboring orbits and dissimilar densities. Science **337**(6094), 556–559 (2012)
6. Kel'manov, A.V., Pyatkin, A.V.: NP-hardness of some quadratic euclidean 2-clustering Problems. Doklady Math. **92**(2), 634–637 (2015)
7. Kel'manov, A.V., Motkova, A.V.: Exact Pseudopolynomial Algorithms for a Balanced 2-clustering problem. J. Appl. Ind. Math. **10**(3), 349–355 (2016)
8. Kel'manov, A., Motkova, A.: A fully polynomial-time approximation scheme for a special case of a balanced 2-clustering problem. In: Kochetov, Y., Khachay, M., Beresnev, V., Nurminski, E., Pardalos, P. (eds.) DOOR 2016. LNCS, vol. 9869, pp. 182–192. Springer, Cham (2016). https://doi.org/10.1007/978-3-319-44914-2_15
9. Kel'manov, A., Motkova, A., Shenmaier, V.: An approximation scheme for a weighted two-cluster partition problem. In: van der Aalst, W., et al. (eds.) AIST 2017. LNCS, vol. 10716, pp. 323–333. Springer, Cham (2018). https://doi.org/10.1007/978-3-319-73013-4_30
10. Kel'manov, A.V., Motkova, A.V.: Polynomial-time approximation algorithm for the problem of cardinality-weighted variance-based 2-clustering with a given center. Comp. Math. Math. Phys. **58**(1), 130–136 (2018)
11. Kel'manov, A., Khandeev, V., Panasenko, A.: Randomized algorithms for some clustering problems. In: Eremeev, A., Khachay, M., Kochetov, Y., Pardalos, P. (eds.) OPTA 2018. CCIS, vol. 871, pp. 109–119. Springer, Cham (2018). https://doi.org/10.1007/978-3-319-93800-4_9
12. Kel'manov, A.V., Khamidullin, S.A.: Posterion detection of a given number of identical subsequences in a quasi-periodic sequence. Comp. Math. Math. Phys. **41**(5), 762–774 (2001)
13. Kel'manov, A.V., Khamidullin, S.A.: An approximation polynomial algorithm for a sequence partitioning problem. J. Appl. Ind. Math. **8**(2), 236–244 (2014)

A Hybrid Variation Harmony Search Algorithm for the Team Orienteering Problem with Capacity Limitations

Eleftherios Tsakirakis$^{(\boxtimes)}$, Magdalene Marinaki, and Yannis Marinakis

School of Production Engineering and Management, Decision Support Systems Laboratory, Technical University of Crete, University Campus, 73100 Chania, Greece
etsakirakis@isc.tuc.gr, magda@dssl.tuc.gr, marinakis@ergasya.tuc.gr

Abstract. This paper addresses a new optimization method for a variant of the category of orienteering problems (OP), which is well-known as the capacitated team orienteering problem (CTOP). The main objective of CTOP focuses on the maximization of the total collected profit from a set of candidate nodes or customers by taking into account the limitations of vehicle capacity and time upper boundary of a constructed route. To solve CTOP, we present a new optimization algorithm called the Similarity Hybrid Harmony Search. This methodology includes an innovative "similarity process" technique, which takes advantage the most profitable nodes/customers during the algorithmic procedure aiming to extend the diversification in the solution area. The experimental tests were conducted in the most popular set of instances and the obtained results are compared with most competitive algorithms in the literature.

Keywords: Capacitated Team Orienteering Problem · Harmony Search algorithm · Vehicle routing problems with profits

1 Introduction

Throughout years, a big part of the scientific community have developed innovative and robust methods for solving VRP and its variations. However, there is a growing interest for two specific categories of vehicle routing. Precisely, we refer to the Capacitated Vehicle Routing Problem (CVRP) and the VRPs with profits. Both categories contain a series of NP-Hard problems, which are enough popular not only to scientists, due to the their close relation with real life applications. In this paper we deal with the Capacitated Team Orienteering Problem (CTOP) and present our approach the Similarity Hybrid Harmony Search algorithm (SHHS) for solving it. The CTOP is a recent routing problem in contrast with the vast majority of VRP, which introduced officially by [1]. Furthermore, the proposed SHHS method based on the classic Harmony Search (HS) algorithm, which developed by [5].

The Capacitated Team Orienteering problem (CTOP) constitutes an extension of the Team Orienteering Problem (TOP), which both are included to the

© Springer Nature Switzerland AG 2020
N. F. Matsatsinis et al. (Eds.): LION 13 2019, LNCS 11968, pp. 146–156, 2020.
https://doi.org/10.1007/978-3-030-38629-0_12

category of Orienteering Problems (OP). Main objective of these type of problems is the maximum collection of rewards (profits) from a subset customers. The OP was initially introduced by [2]. The goal was the collection of checkpoints as much as possible, which were associated with points within the already given time period. With a few differences, the basic concept of the OP was similar to the Traveling Salesman Problem (TSP). The OP stands as a single route with an initial and final point, targeting the maximization of the profits collected from a number of visited customers in a specific time limit. The TOP is considered to be the expansion of the OP, while it includes the same objectives and limitations. The significant difference lies into the number of routes (vehicles), which is more than single one. Considering TOP a huge variety of optimization algorithms has been developed in order to provide sufficient solutions.

Combinatorial optimization contains a series of strong NP-Hard problems. Due to their complexity, finding the optimum solutions is always a difficult task. To solve CTOP we propose the Similarity Hybrid Harmony Search (SHHS) algorithm. Our new solution method is a variation of the meta-heuristic Harmony search (HS) algorithm. [4] introduced a new meta-heuristic algorithm, the Harmony Search (HS) algorithm. The philosophy of the HS was based on the procedure of music improvisation. In the process of music inspiration, the musicians play different music tones in order to find the appropriate acoustic frequencies. While the musicians match the frequencies, a new harmony is composed. The SHHS iherits the philosophy of HS and contains specific tools, such as memory, parameter tuning and randomization procedures, which offer further exploration and intense exploitation in the solution areas. Additionally, we add a new technique called "similarity process", which is applied for solution diversification and enhancement. Our goal through the SHHS algorithm is to provide an alternative solution approach for the CTOP and generally for discreet optimization problems. The experimental results show that our proposed methodology can withstand the best algorithms in the literature.

The rest details of the paper are organized as follows. Further references of the existing methods solving the CTOP and its mathematical model are presented in Sect. 2. Section 3 is devoted to the proposed Similarity Hybrid Harmony Search (SHHS) algorithm. The analysis of the computational results and the comparison with other solution approaches are provided in Sect. 4. Our future directions and plans are concluded in Sect. 5.

2   Capacitated Team Orienteering Problem (CTOP)

2.1   A Literature Review

The Capacitated Team Orienteering Problem is a new variant in comparison with the rest of OP or CVRP problems. It was introduced as CTOP by the [1] back to 2009. As a combinatorial optimization problem with NP-Hard complexity, it has been proposed a rich variety of metaheuristics approaches to reach the best or close to the best solutions. For this kind of problem we proposed our solution method, the Similarity Hybrid Harmony Search algorithm (SHHS). We focus on the algorithms of [1,7] and [6], which have been tested in the same benchmark instances proposed by [1].

To solve CTOP, [1] proposed three methodologies. Two are based on the Tabu Search (TS) optimization method. The first searches in feasible solution areas and denoted as TSF, while the second explores in non-feasible solution areas and denoted as TSA. The third one is a classic Variable Neighborhood Search (VNS) algorithm. All three methods create feasible solutions by adding new customers in each one of the created vehicle routes. An alternative and effective approach is the Bi-level Filter and Fan methodology of [7]. They introduced a bi-level search framework, which is divided into two parts, the master (upper) and the subordinate (lower). The upper expands the search in the solution areas in order to maximize the total collected profit, while the latter enhances the solutions by adding unvisited customers. The algorithm begins with a greedy parallel insertion-based construction heuristic method for generating the initial solutions. Referring to the solution methods of the CTOP, [6] designed an adaptive ejection pool with toggle-rule diversification (ADEPT-RD) approach achieving excellent performance according to the obtained results. The algorithm is based on the concept of the ejection pool framework including specific mechanisms for diversification and optimization of the solutions.

2.2   Mathematical Model

CTOP can be modeled as an undirected graph $G = (V, A)$, in which $V = \{1, \cdots, N\}$ is the set of nodes and $A = \{(i,j)|i,j \in V\}$ represents the set of arcs. The first node in V ($1 \in V$) is designated as the depot. The rest of the nodes are the candidate customers and are included in the set $V_{wd} = \{2, \cdots, N\}$ as subset of V ($V_{wd} \subset V$). The connected arcs between every pair of i, j customers describe the travel cost, which is denoted t_{ij}. For each customer i in V_{wd} is associated with a profit p_i and a demand d_i. The fleet for servicing the customers consists of a predefined number of M available vehicles. Each vehicle $m \in M$ has a specific deposit space for adding d_i without exceeding its maximum store capacity Q_{max}. Lastly, a vehicle route m is considered to be feasible, when its travel time is below or equal with the total time bound T_{max}. The aim of CTOP is the maximization of the total profit, which will be collected from the visited customers.

Of course there are some limitations, which must be declared in order to be solved the CTOP properly. To begin with, the depot node is not assigned with a profit or a demand. The travel cost t_{ij} is calculated from Euclidean distance mathematical type and naturally is symmetric. Additionally, we impose, that the triangle inequality for every distance is satisfied. Each candidate customer in V_{wd} can be served only once and by one vehicle $m \in M$. All vehicles from the M fleet begin and finish their route at the depot without exceeding the limits of maximum capacity Q_{max} and travel time T_{max}. For instance, a perfect vehicle route contains as many as possible candidate customers with high profits, while the corresponding demand and travel cost are equal or almost equal with their respectively upper limit values.

The CTOP can be denoted as an extended version with capacity limitations of TOP. It can be modeled as an integral programming problem. The constrains of the decision variables for the problem are the following:

- $x_{ijm} = 1$ if vehicle route $m \in M$ includes edges $i, j \in V$,
  $x_{ijm} = 0$ otherwise.
- $y_{im} = 1$ if node $i \in V$ belongs to vehicle route $m \in M$,
  $y_{id} = 0$ otherwise.

The objective function and the constrains of the CTOP are presented below:

$$z = \max \sum_{i \in V_{wd}} \sum_{m \in M} p_i y_{im} \tag{1}$$

$$s.t.$$

$$\sum_{m \in M} \sum_{j \in V_{wd}} x_{0jm} = \sum_{m \in M} \sum_{i \in V_{wd}} x_{i0m} = M \tag{2}$$

$$\sum_{m \in M} y_{im} \leq 1, \ \forall i \in V_{wd} \tag{3}$$

$$\sum_{j \in V} x_{ijm} = \sum_{j \in V} x_{jim} = y_{im}, \ \forall i \in V_{wd}, \forall m \in M \tag{4}$$

$$\sum_{i \in V} \sum_{j \in V} t_{ij} x_{ijm} \leq T_{max}, \ \forall m \in M \tag{5}$$

$$\sum_{i \in V_{wd}} d_i y_{im} \leq Q_{max}, \ \forall m \in M \tag{6}$$

$$\sum_{(i,j) \in S} x_{ijm} \leq |S| - 1, \ \forall S \subseteq V_{wd}, \ 2 \leq |S|, \ \forall m \in M \tag{7}$$

The objective function (1) aims to the maximization of the total collected profit. Constraint (2) ensures that the vehicle routes of the M vehicles will begin and end to the depot node. Constraint (3) eliminates the possibility to visit twice each of the i customers. Constraint (4) secures the connection of the vehicle route. The time and the capacity limitations of the vehicle routes are secured by the constraints (5), (6). And the constraint (7) forbids any sub-routes.

3   Proposed Optimization Algorithm

3.1   Stages of the Methodology

In this subsection the Similarity Hybrid Harmony Search algorithm is presented. The proposed method combines the mechanisms of the classic HS with the new added technique "similarity process". The functionality of SHHS algorithm is detailed analyzed for each of the following stages. This subsection is organized with the illustration of the algorithmic parameters and the main phases of the methodology as follows:

- Stage 1: Parameter definition and initialization
- Stage 2: Initial harmonies and Harmony Memory construction
- Stage 3: The Improvisation stage
- Stage 4: New technique called "**similarity process**"
- Stage 5: Update Harmony Memory with best harmonies

3.2   Initialization of Parameters

The SHHS includes a series of different parameters, which some of them are inherited from the classic HS algorithm and others. We take into the account not only the already existed parameters, but also the new added ones, which are essential for main algorithmic procedures. Before the algorithm enters in the main loop to execute the basic functions, the harmony memory list (HML) is created. It is a important parameter, which is vital for the following operations of the algorithm. As it called, it is a list, which contains all the routes from the stored solutions in the HM. As the problem requires, the feasible solutions consist of M routes. For this reason, the size of the HML will depend on M, which obtains different values according to the current example.

3.3   Initial Harmonies and Harmony Memory Construction

The SHHS algorithm starts with the creation of the initial solutions. The construction of a feasible solution depends on the limitations of travel time and vehicle capacity. Each of the c customers are checked in order to enter in a vehicle of the current solution. The procedure for customer insertion is simple. We follow the concept to add as many as possible candidate customers to the solutions. To achieve this, we rank the customers according to their amount of profit. Then, the algorithm inserts them alternately, until all the M routes are created. When this procedure is complete, the solutions are stored in the harmony memory HM. This strategy of creating solutions with big amount of customers is proved that, it is very promising for the core functions of the algorithm.

3.4   Improvisation Stage

The Improvisation phase contains one of the core functions in the main loop of the Similarity Hybrid Harmony Search algorithm. The proposed methodology includes a set of parameters, which are existed in the classic Harmony Search algorithm. In addition, we add an essential parameter the harmony memory list (HML). During this phase, new solutions are produced with or without the usage of the HML. Whether or not to be taken into account the usage of the HML depends on the $HMCR$ and PAR probabilities. Through this function feasible routes are constructed via three options.

The Improvisation phase obtain the HML list and uses the $HMCR$ and PAR probabilities. If the $HMCR$ probability is occurred, the algorithm will consider the HML to pick a route. It follows the PAR possibility. If the PAR is accepted, the algorithm performs the 1-1 Exchange local search option to improve the profit of the m picked route. Otherwise, none operation is applied to the selected route and the solution is formed. On the other hand, when the $HMCR$ is not occurred, the HML is not taken into account and the algorithm produces a route with randomly inserted customers. At the end of the process, the created solutions are stored in a temporarily memory.

3.5   "Similarity Process"

The "similarity process" is a new technique, which focuses on enhancement of the solution's quality by increasing the total collected profit. It is an alternative matching selection process, which allows the algorithm to pick different routes in order to create more profitable solutions. This function is divided into two parts. In first the algorithm selects the suitable routes by using the similarity parameter (SP) and the similarity matrix (SM). In the second part, the nearest insertion (NI) heuristic is used to increase the total amount of the collected profit.

In this algorithmic phase are applied the two new added parameters, the similarity parameter (SP) and the similarity matrix (SM). Before the begin of the process, the algorithm constructs the SM. The rows and columns of the SM represent the corresponding routes from the harmony memory list (HML). Additionally, each row or column contains the numbers of the similar nodes for all the candidate routes. SP is an integer number, which takes various values between the minimum and the maximum value of the SM. The selected SP value indicates the number of similar nodes, which the candidate routes must contain. The algorithm seeks all the routes with the same SP and picks the most profitable one. The selected route becomes a part of the solution and the similar customers are removed. Moreover, it increases the available time travel and load capacity of the routes. For this reason, NI is applied to each one of the solutions. The NI inserts customers on the basis of the profit amount and the available time travel.

4   Analysis of the Experimental Results

In this part of the paper, the results of our SHHS algorithm are presented. We test the efficiency of our solution method in the benchmark instances proposed by [1], which have been modified from the original data set of problems for the CVRP proposed by [3]. CTOP includes ten sets of data instances (p03-p16). The characteristics of each set of instances are the fleet of the vehicles, maximum capacity Q_{max}, total distance time T_{max}, location of the depot (x,y), amount of customers. These features are varied for all of the sets.

In the following three tables are presented the obtained experimental results of the Similarity Hybrid Harmony Search (SHHS) algorithm alongside with the most efficient solutions methods in the literature for the Capacitated Team Orienteering Problem. Tables 1, 2 and 3 are organized in nine columns. In the first column the name of the current instance (Problem) is defined. Each of the instances is referred with the characteristics of the set, the amount of vehicles, the Q_{max} and the T_{max} limitations. For example, p03-2-50-50 denotes an instance from set p03 with $M = 2$ vehicles and values of $Q_{max} = 50$ and $T_{max} = 50$. In the second column the optimal solution value (B*) in the literature is presented. Through third and ninth columns are shown the results from the compared optimization methodologies. These are the following as they appear in Tables 1, 2 and 3:

- Variable neighborhood search (**VNS**), [1]
- Tabu search with only feasible solutions (**TSF**), [1]
- Tabu search considering infeasible solutions (**TSA**), [1]
- Adaptive ejection pool with toggle rule diversification (**ADEPT-RD**), [6]
- Bi-level Filter-and-Fan fast configuration (**BiF&F-f**), [7]
- Bi-level Filter-and-Fan slow configuration (**BiF&F-s**), [7]
- Similarity Hybrid Harmony Search (**SHHS**)

The Similarity Hybrid Harmony Search algorithm was executed 90 times as the total number of the instances. Our proposed methodology reaches the optimal solutions 63 out of 90 achieving 70% in efficiency. Each one of Tables 1, 2 and 3 is divided according to the number of the M available vehicles. In Table 1 our algorithm gains almost all the best known solutions reaching 27 out of 30. In the three examples (p09/10/13-2-100-100) SHHS looses the optimal value by the gap of just two points. Table 2 presents the results of the instances with $M = 3$ of vehicles routes. The proposed method reaches the optimal known values in 20 instances out of 30. It can be observed, that in the examples with $Q_{max}, T_{max} = 50$ SHHS obtains all the optimal solutions, while with increased Q_{max} and T_{max} values it struggles to reach them. In Table 3 are presented the results in the most tough complexity. The obtained values are not so satisfactorily as in the previous two tables. Our algorithm finds the optimal solutions in slightly above the half of the total examples (16/30 instances) with 53,3% efficiency. However, the experimental results of SHHS are quite close not only to the best values, but in some cases outperforms the some of the other optimization methods. For example, in instance p16-4-100-100 our method reaches 555 points and surpasses the VNS and the two TS approaches. In general, the performance of our proposed methodology can be consider almost antagonistic and competitive with the rest of the compared optimization methods.

Table 1. Regular Set with $M = 2$ vehicles

Problem	B*	VNS	TSF	TSA	ADEPT-RD	BiF & F-f	BiF & F-s	SHHS
p03-2-50-50	133	133	133	133	133	133	133	133
p06-2-50-50	121	121	121	121	121	121	121	121
p07-2-50-50	126	126	126	126	126	126	126	126
p08-2-50-50	133	133	133	133	133	133	133	133
p09-2-50-50	137	137	137	137	137	137	137	137
p10-2-50-50	134	134	134	134	134	134	134	134
p13-2-50-50	134	134	134	134	134	134	134	133
p14-2-50-50	124	124	124	124	124	124	124	124
p15-2-50-50	134	134	134	134	134	134	134	134
p16-2-50-50	137	137	137	137	137	137	137	137
p03-2-75-75	208	208	208	208	208	208	208	208
p06-2-75-75	183	183	183	183	183	183	183	183
p07-2-75-75	193	193	193	193	193	193	193	193
p08-2-75-75	208	208	208	208	208	208	208	208
p09-2-75-75	210	210	210	210	210	210	210	210
p10-2-75-75	208	208	208	208	208	208	208	208
p13-2-75-75	193	193	193	193	193	193	193	193
p14-2-75-75	190	190	190	190	190	190	190	190
p15-2-75-75	211	210	211	211	211	211	211	211
p16-2-75-75	212	212	212	212	212	212	212	212
p03-100-100	277	277	277	277	277	277	277	277
p06-2-100-100	252	252	252	251	252	252	252	252
p07-2-100-100	266	266	266	266	266	266	266	266
p08-2-100-100	277	277	277	276	277	277	277	277
p09-2-100-100	279	279	278	279	279	279	279	277
p10-2-100-100	282	282	280	279	282	282	282	280
p13-2-100-100	253	253	253	253	253	253	253	251
p14-2-100-100	271	271	271	271	271	271	271	271
p15-2-100-100	282	282	280	279	282	282	282	282
p16-2-100-100	285	284	285	284	285	284	285	285

Table 2. Regular Set with $M = 3$ vehicles

Problem	B*	VNS	TSF	TSA	ADEPT-RD	BiF & F-f	BiF & F-s	SHHS
p03-3-50-50	198	198	198	198	198	198	198	198
p06-3-50-50	177	177	177	177	177	177	177	177
p07-3-50-50	187	187	187	187	187	187	187	187
p08-3-50-50	198	198	198	198	198	198	198	198
p09-3-50-50	201	201	201	201	201	201	201	201
p10-3-50-50	200	200	200	200	200	200	200	200
p13-3-50-50	193	193	193	193	193	193	193	193

(continued)

Table 2. (*continued*)

Problem	B*	VNS	TSF	TSA	ADEPT-RD	BiF & F-f	BiF & F-s	SHHS
p14-3-50-50	184	184	184	184	184	184	184	184
p15-3-50-50	200	200	200	199	200	200	200	200
p16-3-50-50	203	203	203	203	203	203	203	203
p03-3-75-75	307	307	307	307	307	307	307	307
p06-3-75-75	269	269	269	269	269	269	269	269
p07-3-75-75	287	287	287	287	287	287	287	287
p08-3-75-75	307	307	307	307	307	307	307	305
p09-3-75-75	312	310	310	310	312	312	312	311
p10-3-75-75	311	310	310	310	311	311	311	311
p13-3-75-75	265	265	265	265	265	265	265	263
p14-3-75-75	279	279	279	279	279	279	279	279
p15-3-75-75	315	315	315	315	315	315	315	315
p16-3-75-75	317	317	317	317	317	317	317	317
p03-3-100-100	408	407	408	407	408	407	407	408
p06-3-100-100	369	369	369	369	369	369	369	369
p07-3-100-100	397	397	397	391	397	397	397	393
p08-3-100-100	408	407	408	407	408	407	407	407
p09-3-100-100	415	413	414	412	415	413	415	410
p10-3-100-100	418	416	417	412	418	418	418	415
p13-3-100-100	344	344	344	343	344	344	344	339
p14-3-100-100	399	399	399	399	399	399	399	399
p15-3-100-100	418	417	416	416	418	418	418	416
p16-3-100-100	423	420	421	421	423	422	423	421

Table 3. Regular Set with $M = 4$ vehicles

Problem	B*	VNS	TSF	TSA	ADEPT-RD	BiF & F-f	BiF & F-s	SHHS
p03-4-50-50	260	260	260	260	260	260	260	260
p06-4-50-50	222	222	222	222	222	222	222	222
p07-4-50-50	240	240	240	240	240	240	240	240
p08-4-50-50	260	260	260	260	260	260	260	260
p09-4-50-50	262	262	262	262	262	262	262	262
p10-4-50-50	265	265	265	265	265	265	265	265
p13-4-50-50	243	243	243	242	243	243	243	239
p14-4-50-50	241	241	241	241	241	241	241	238
p15-4-50-50	266	266	266	266	266	266	266	266
p16-4-50-50	269	269	269	269	269	269	269	269
p03-4-75-75	403	401	403	402	403	403	403	403
p06-4-75-75	349	349	348	348	403	349	349	349
p07-4-75-75	349	349	348	348	403	349	349	349

(*continued*)

Table 3. (*continued*)

Problem	B*	VNS	TSF	TSA	ADEPT-RD	BiF & F-f	BiF & F-s	SHHS
p08-4-75-75	403	401	403	402	403	403	403	403
p09-4-75-75	408	407	407	407	408	407	408	406
p10-4-75-75	411	410	410	407	411	410	411	409
p13-4-75-75	323	323	323	323	323	323	323	320
p14-4-75-75	366	366	366	366	366	366	366	366
p15-4-75-75	415	414	414	413	415	415	415	415
p16-4-75-75	420	419	420	419	420	420	420	420
p03-4-100-100	532	529	531	529	532	531	532	526
p06-4-100-100	482	481	482	481	482	482	482	482
p07-4-100-100	521	521	521	514	521	518	521	517
p08-4-100-100	532	529	531	529	532	531	532	526
p09-4-100-100	546	545	539	536	546	545	546	536
p10-4-100-100	553	548	549	550	553	553	553	549
p13-4-100-100	419	419	419	416	419	419	419	419
p14-4-100-100	525	525	523	525	525	525	525	522
p15-4-100-100	549	548	549	545	549	548	549	545
p16-4-100-100	558	554	554	553	558	556	558	555

5   Conclusion

The present work for this paper is based on a new optimization method for the Capacitated Team Orienteering Problem. Our proposed Similarity Hybrid Harmony Search (SHHS) algorithm embraces the mechanisms of the Improvisation stage for expanding the variety of the produced solutions. Adding the "similarity process" to the whole algorithmic procedure in order to intense the exploration in the solution areas and to increase the total collected profit. It is a construction strategy, which exploits the similar nodes of each available solution. Taking into account that solutions consist of vehicle routes, through this strategy are combined different routes in order to synthesize new feasible solutions.

The performance of our proposed methodology in CTOP shows, that further modifications on the algorithmic procedure can be applied in order to enhance even more the dynamic and the impact of the algorithm. Furthermore, an extended evaluation of the already existed and the new added parameters considering the way of their calculation during the algorithmic procedures should be examined in order to improve the efficiency of the this method. Further improvements of the SHHS algorithm will be tested in similar discreet combinatorial problems. Our future scientific work will be directed to the development of innovative optimization methodologies for providing feasible solutions and prototype mechanisms likewise "similarity process".

References

1. Archetti, C., Feillet, D., Hertz, A., Speranza, M.G.: The capacitated team orienteering and profitable tour problems. J. Oper. Res. Soc. **60**, 831–842 (2009)
2. Chao, I.M., Golden, B., Wasil, E.A.: The team orienteering problem. Eur. J. Oper. Res. **88**, 464–474 (1996)
3. Christofides, N., Mingozzi, A., Toth, P.: The vehicle routing problem. In: Combinatorial Optimization, pp. 315–338 (1979)
4. Geem, Z.W., Kim, J.H., Loganathan, G.V.: A new heuristic optimization algorithm: harmony search. Simulation **76**(2), 60–68 (2001)
5. Lee, K.S., Geem, Z.W.: A new meta-heuristic algorithm for continuous engineering optimization: harmony search theory and practice. Comput. Methods Appl. Mech. Eng. **194**, 3902–3933 (2004)
6. Luo, Z., Cheang, B., Lim, A., Zhu, W.: An adaptive ejection pool with toggle-rule diversification approach for the capacitated team orienteering problem. Eur. J. Oper. Res. **229**(3), 673–682 (2013)
7. Tarantilis, C.D., Stavropoulou, F., Repoussis, P.P.: The capacitated team orienteering problem: a bi-level filter-and-fan method. Eur. J. Oper. Res. **224**(1), 65–78 (2013)

Adaptive GVNS Heuristics for Solving the Pollution Location Inventory Routing Problem

Panagiotis Karakostas[1] , Angelo Sifaleras[2] ,
and Michael C. Georgiadis[1(✉)]

[1] Department of Chemical Engineering, Aristotle University of Thessaloniki,
University Campus, 54124 Thessaloniki, Greece
pkarakost@cheng.auth.gr, mgeorg@auth.gr
[2] Department of Applied Informatics, School of Information Sciences,
University of Macedonia, 156 Egnatia Str., 54636 Thessaloniki, Greece
sifalera@uom.gr

Abstract. This work proposes Adaptive General Variable Neighborhood Search metaheuristic algorithms for the efficient solution of Pollution Location Inventory Routing Problems (PLIRPs). A comparative computational study, between the proposed methods and their corresponding classic General Variable Neighborhood Search versions, illustrates the effectiveness of the intelligent mechanism used for automating the re-ordering of the local search operators in the improvement step of each optimization method. Results on 20 PLIRP benchmark instances show the efficiency of the proposed metaheuristics.

Keywords: Adaptive General Variable Neighborhood Search ·
Intelligent optimization methods · Pollution Location Inventory
Routing Problem · Green logistics

1 Introduction

The Pollution Location Inventory Routing Problem (PLIRP) is an NP-hard combinatorial optimization problem, which involves both economic and environmental decisions [7]. It simultaneously addresses strategic decisions, such as the location of candidate depots and the allocation of customers to the opened depots, tactical decisions, as the inventory levels and the replenishment rates and finally operational decisions, such as routing schedules. The objective of this problem is the minimization of the total cost, which consists of facilities opening costs, inventory control costs, general routing costs and fuel consumption costs.

It should be mentioned that, there are several factors affecting the fuel consumption. The main factors are the speed, the acceleration of the vehicle, the traveled distance and the total weight of the vehicle, consisting of the curb and freight weight [2].

© Springer Nature Switzerland AG 2020
N. F. Matsatsinis et al. (Eds.): LION 13 2019, LNCS 11968, pp. 157–170, 2020.
https://doi.org/10.1007/978-3-030-38629-0_13

In this work, several Adaptive General Variable Neighborhood Search (AGVNS) heuristic algorithms have been developed for the efficient solution of recently proposed PLIRP instances [7]. The proposed AGVNS schemes are compared both with their corresponding GVNS methods as well as with the only available heuristic algorithm in the literature for this problem variant. The remainder of this paper is organized as follows: Sect. 2 provides the mathematical formulation of the problem. Section 3 describes the developed solution approaches and their algorithmic details, while Sect. 4 provides extensive numerical analyses for testing the efficiency of the proposed methods on 20 PILRP instances. Finally, Sect. 5 draws up main concluding remarks and highlights direction for future work.

2   Problem Statement

For the readers clarity sake, the mathematical formulation of the problem is presented in this section. Model notations are summarized in Tables 1, 2 and 3.

Table 1. Model sets.

Indices & Explanation
V is the set of nodes
J is the set of candidate depots
I is the set of customers
H is the set of discrete and finite planning horizon
K is the set of vehicles
R is the set of speed levels

Table 2. Model decision variables.

Notation	Explanation
y_j	1 if j is opened; 0 otherwise
z_{ij}	1 if customer i is assigned to depot j; 0 otherwise
x_{ijkt}	1 if node j is visited after i in period t by vehicle k
q_{ikt}	Product quantity delivered to customer i in period t by vehicle k
w_{itp}	Quantity delivered to customer i in period p to satisfy its demand in period t
a_{vikt}	Load weight by travelling from node v to the customer i with vehicle k in period t
$zz_{v_1 v_2 ktr}$	1 if vehicle k travels from node $v1$ to $v2$ in period t with speed level r

The following utilization of formulas simplifies the fuel consumption components of the objective function: $\lambda = \frac{HVDF}{\psi}$, $\gamma_k = \frac{1}{1000VDTE\eta}$, $\alpha = \tau + gCR\sin\theta + gCR\cos\theta$ and $\beta_k = 0.5CAD\rho FSA_k$.

Table 3. Model parameters.

Notation	Explanation	Value
f_j	Fixed opening cost of depot j	Instance-depended
C_j	Storage capacity of depot j	Instance-depended
h_i	Unit inventory holding cost of customer i	Instance-depended
Q_k	Loading capacity of vehicle k	Instance-depended
d_{it}	Period variable demand of customer i	Instance-depended
c_{ij}	Travelling cost of locations pair (i,j)	Instance-depended
s_r	The value of the speed level r	Instance-depended
ϵ	Fuel-to-air mass ratio	1
g	Gravitational constant (m/s^2)	9.81
ρ	Air density (kg/m^3)	1.2041
CR	Coefficient of rolling resistance	0.01
η	Efficiency parameter for diesel engines	0.45
f_c	Unit fuel cost (e/L)	0.7382
f_e	Unit CO_2 emission cost (e/kg)	0.2793
σ	CO_2 emitted by unit fuel consumption (kg/L)	2.669
$HVDF$	Heating value of a typical diesel fuel (kj/g)	44
ψ	Conversion factor (g/s to L/s)	737
θ	Road angle	0
τ	Acceleration (m/s^2)	0
CW_k	Curb weight (kg)	3500
EFF_k	Engine friction factor (kj/rev/L)	0.25
ES_k	Engine speed (rev/s)	39
ED_k	Engine displacement (L)	2.77
CAD_k	Coefficient of aerodynamics drag	0.6
FSA_k	Frontal surface area (m^2)	9
$VDTE_k$	Vehicle drive train efficiency	0.4

$$\min \sum_{j\in J} f_j y_j + \sum_{i\in I} h_i \sum_{t\in H} \left(\tfrac{1}{2} d_{it} + \sum_{p\in H, p<t} w_{itp}(t-p) + \sum_{p\in H, p>t} w_{itp}(t-p+|H|) \right)$$
$$+ \sum_{i\in V}\sum_{j\in V}\sum_{t\in H}\sum_{k\in K} c_{ij} x_{ijkt} + \sum_{i\in V}\sum_{j\in V}\sum_{k\in K}\sum_{t\in H} \left\{ \lambda (f_c + (f_e\sigma)) \left(\sum_{r\in R} \frac{(zz_{ijktr} EFF_k ES_k ED_k c_{ij})}{s_r} \right) \right.$$
$$\left. + \left(\alpha\gamma_k (CW_k x_{ijkt} + a_{ijkt}) c_{ij} \right) + \left(\beta_k\, \gamma_k \sum_{r\in R} (s_r\, zz_{ijktr})^2 \right) \right) \right\}$$

Subject to $\hfill (1)$

$$\sum_{r\in R} zz_{ijktr} = 0 \quad \forall i,j \in V, \forall k \in K, \forall t \in H \hfill (2)$$

$$\sum_{i \in V} a_{ijkt} - \sum_{i \in V} a_{jikt} = q_{jkt} PW \quad \forall j \in I, \forall k \in K, \forall t \in H \tag{3}$$

$$\sum_{j \in V} x_{ijkt} - \sum_{j \in V} x_{jikt} = 0 \qquad \forall i \in V, \forall k \in K, \forall t \in H \tag{4}$$

$$\sum_{j \in V} \sum_{k \in K} x_{ijkt} \leq 1 \quad \forall t \in H, \forall i \in I \tag{5}$$

$$\sum_{j \in V} \sum_{k \in K} x_{jikt} \leq 1 \quad \forall t \in H, \forall i \in I \tag{6}$$

$$\sum_{i \in I} \sum_{j \in J} x_{ijkt} \leq 1 \quad \forall k \in K, \forall t \in H \tag{7}$$

$$x_{ijkt} = 0 \quad \forall i, j \in J, \forall k \in K, \forall t \in H, i \neq j \tag{8}$$

$$\sum_{i \in I} q_{ikt} \leq Q_k \quad \forall k \in K, \forall t \in H \tag{9}$$

$$\sum_{j \in J} z_{ij} = 1 \quad \forall i \in I \tag{10}$$

$$z_{ij} \leq y_j \quad \forall i \in I, \forall j \in J \tag{11}$$

$$\sum_{i \in I} \left(z_{ij} \sum_{t \in H} d_{it} \right) \leq C_j \quad \forall j \in J \tag{12}$$

$$\sum_{u \in I} x_{ujkt} + \sum_{u \in V \setminus \{i\}} x_{iukt} \leq 1 + z_{ij} \quad \forall i \in I, \forall j \in J, \forall k \in K, \forall t \in H \tag{13}$$

$$\sum_{i \in I} \sum_{k \in K} \sum_{t \in H} x_{jikt} \geq y_j \quad \forall j \in J \tag{14}$$

$$\sum_{i \in I} x_{jikt} \leq y_j \quad \forall j \in J, \forall k \in K, \forall t \in H \tag{15}$$

$$\sum_{p \in H} w_{itp} = d_{it} \quad \forall i \in I, \forall t \in H \tag{16}$$

$$\sum_{t \in H} w_{itp} = \sum_{k \in K} q_{ikp} \quad \forall i \in I, \forall p \in H \tag{17}$$

$$q_{ikt} \leq M \sum_{j \in V} x_{ijkt} \quad \forall i \in I, \forall t \in H, \forall k \in K \tag{18}$$

$$\sum_{j \in V} x_{ijkt} \leq M q_{ikt} \quad \forall i \in I, \forall t \in H, \forall k \in K \tag{19}$$

$$x_{ijkt} \in \{0, 1\} \quad \forall i \in I, \forall j \in J, \forall t \in H, \forall k \in K \tag{20}$$

$$y_j \in \{0, 1\} \quad \forall j \in J \tag{21}$$

$$z_{ij} \in \{0, 1\} \quad \forall i \in I, \forall j \in J \tag{22}$$

$$q_{ikt} \leq \min \left\{ Q_k, \sum_{p \in H} d_{ip} \right\} \quad \forall i \in I, \ \forall j \in J, \forall k \in K \qquad (23)$$

$$w_{itp} \leq d_{ip} \quad \forall i \in I, \ \forall t, p \in H \qquad (24)$$

The objective goal of the problem is the minimization of the total cost which consists of facilities opening costs, inventory holding costs, general routing costs and fuel consumption and CO_2 emissions costs. The constraints of the model can be grouped into routing-, inventory- and location-related constraints. For example, the routing-related set of Constraints 8 prevent vehicles from traveling between two depots in each time period, while the location-related set of Constraints 10 guarantee that a customer will be served by exactly one depot. Also, an example of inventory-related constraints are the set of Constraints 9 which they ensure that the delivered amount of product quantities will not exceed the capacity of the used vehicle in a specific time period.

3   Solution Approach

3.1   Initialization

A three-phase construction heuristic has been developed in order to build quick initial feasible solutions for the PLIRP. Location and allocation decisions are determined in the first phase and inventory-routing decisions are made in the second one. Finally, the speed levels for travelling through the nodes of a network are selected.

More specifically, the opening of the required depots is based on a ratio-based selection technique. For each candidate depot, the ratio $\frac{fixed\_opening\_cost}{Capacity}$ is calculated, where the "fixed_opening_cost" represents the cost for opening each depot, and the "capacity" is the maximum amount of product units that the selected depot can offer. The depot with the minimum ratio is selected to be opened. The number of the opened depots depends on their capacities and the total demand of customers. The allocation of customers to the opened depots is made in a serial way. Serial allocation means that for a selected depot, the set of customers is being ran and the first unallocated customer, whose total demand does not exceed the remaining capacity of the depot, is assigned to that depot. The allocation process is completed when all customers have been assigned to the opened depots.

In order to determine the inventory-routing decisions, a Random Insertion method [3] is applied for building the vehicles routes and each customer receives the demanded quantity of product in each time period. The selection of speed levels is performed randomly.

3.2   Neighborhood Structures

Three neighborhood structures are proposed for the efficient exploration of the solution space, as follows:

Inter-route Relocate: In this local search operator two customers (i and b) from different routes are selected. The Inter-route Relocated removes i from his route R_i and moves him in the route servicing customer b, R_b in the next position of b, in each time period. Both customers can be serviced by the same depot or by different depots over the time horizon. A replenishment shifting may be performed in order to avoid vehicle capacity violation in route R_b.

Exchange Opened-Closed Depots: In this neighborhood a closed depot i is being exchanged with an opened one j. The capacity of the depot i must be at least equal to the capacity of j for having a valid move. Also, a reordering of the routes allocated on depot j may occur according to the minimum insertion cost criterion of depot i.

2-2 Replenishment Exchange: Two time periods t_1 and t_2 are randomly selected in this operator, and then, the two most distant customers i and b are found. Both of those customers must be serviced in t_1 and t_2. The cost changes of removing i and b from their routes in periods t_1 and t_2 respectively and shifting their receiving deliveries from t_1 to t_2 for customer i and from t_2 to t_1 for b are calculated. The move is applied only if no violations over the vehicles capacities occurred.

3.3 Shaking Scheme

The main scope of a shaking procedure is to help the algorithm escaping from local optimum solutions [5]. In each shaking iteration, a new random solution S' is obtained by a randomly selected neighborhood from a predefined set of neighborhoods and according to a given solution S. In this work, an intensified shaking method has been developed with two neighborhood structures, the Exchange Opened-Closed Depots and the Intra-route Relocate.

The Exchange Opened-Closed Depots operator is applied as described in Subsect. 3.2. The Intra-route Relocate operator removes a randomly selected customer from its current position in its route and moves him in a different position in the same route. The pseudo-code of this diversification method is summarized in Algorithm 1. The Shake function receives an incumbent solution S and the number of iterations k (where $1 < k < k_{max}$ and $k_{max} = 12$), which indicates the times that, one randomly selected neighborhood operator (of the two in total) will be applied for generating a new solution S'.

3.4 Adaptive General Variable Neighborhood Search

Variable Neighborhood Descent (VND). The VND is the deterministic variant of the well-known metaheuristic framework Variable Neighborhood Search (VNS). In a VND method the local search operators are ordered in a specific sequence and applied successively until no more improvements can be noticed. According to the neighborhood change step, there are different VND schemes. Two of the most well-known VND schemes are the cyclic VND (cVND) and the pipe VND (pVND). In the first one, the search continuous in the next

Algorithm 1. Shaking Procedure

```
1: procedure SHAKE(S, k)
2:     l = random_integer(1, 2)
3:     for i ← 1, k do
4:         select case(l)
5:         case(1)
6:             S' ← Intra − route_Relocate(S)
7:         case(2)
8:             S' ← Exchange_OpenedClosed_Depots(S)
9:         end select
10:     end for
11:     Return S'
12: end procedure
```

neighborhood in the set independently of the improvement criterion, while in the pVND the exploration continuous in the same neighborhood while an improvement is occurred [5]. In this work both cVND and pVND are used. Moreover, it should be mentioned that the parameter l_{max} in both VND pseudo-codes denotes the number of the used neighborhood structures. The pseudo-codes of the proposed VND schemes are given in Algorithms 2 and 3.

Algorithm 2. cyclic VND

```
1: procedure cVND(S, l_max)
2:     l = 1
3:     while l ≤ l_max do
4:         select case(l)
5:         case(1)
6:             S' ← Inter_Relocate(S)
7:         case(2)
8:             S' ← Exchange_OpenedClosed_Depots(S)
9:         case(3)
10:            S' ← 2 − 2ReplenishmentExchange(S)
11:        end select
12:        if f(S') < f(S) then
13:            S ← S'
14:            l = l + 1
15:        else
16:            l = l + 1
17:        end if
18:     end while
19:     Return S
20: end procedure
```

The local search operators are applied with an adaptive search strategy, which combines the first and best improvement search strategies [8]. More specifically, if the number of customers in a problem instance is more than 90, the fist improvement search strategy is applied, otherwise the algorithm explores the neighborhoods with the best improvement strategy.

General Variable Neighborhood Search (GVNS). The GVNS is a variant of the VNS, which combines deterministic and stochastic components during the search. More specifically, it adopts one of the VND schemes as its main improvement step [5,10]. Based on the two proposed VND methods, two GVNS schemes are shaped and provided in the following pseudo-codes.

Algorithm 3. pipe VND

```
1: procedure PVND(S, l_max)
2:     l = 1
3:     while l ≤ l_max do
4:         select case(l)
5:         case(1)
6:             S' ← Inter_Relocate(S)
7:         case(2)
8:             S' ← Exchange_OpenedClosed_Depots(S)
9:         case(3)
10:            S' ← 2 − 2ReplenishmentExchange(S)
11:        end select
12:        if f(S') < f(S) then
13:            S ← S'
14:        else
15:            l = l + 1
16:        end if
17:    end while
18:    Return S
19: end procedure
```

Algorithm 4. $GVNS_{cVND}$

```
1: procedure GVNS_cVND(S, k_max, l_max, max_time)
2:     while time ≤ max_time do
3:         S* = Shake(S, k)
4:         S' = cVND(S*, l_max)
5:         if f(S') < f(S) then
6:             S ← S'
7:         end if
8:     end while
9:     return S
10: end procedure
```

Algorithm 5. $GVNS_{pVND}$

```
1: procedure GVNS_pVND(S, k_max, l_max, max_time)
2:     while time ≤ max_time do
3:         S* = Shake(S, k)
4:         S' = pVND(S*, l_max)
5:         if f(S') < f(S) then
6:             S ← S'
7:         end if
8:     end while
9:     return S
10: end procedure
```

Adaptive Mechanism. The order of the neighborhood structures is a crucial component for the successful performance of a VNS-based algorithm [4,6]. Consequently, it is important to employ an intelligent mechanism for the re-ordering of the neighborhood structures. According to the literature, some adaptive variants of the VNS have been proposed for that case. Todosijevic et al. [12] proposed an Adaptive GVNS in which a re-ordering of the neighborhoods is applied in each iteration based on their success in previous solution process. Li and Tian [9] also, proposed an adaptive version of VNS in which a probabilistic selection mechanism is used for deciding the sequence of the neighborhoods. In this work, an adaptive neighborhoods re-ordering mechanism is proposed. In each iteration, the sequence of the neighborhoods is re-formed based on the number of achieved improvements in the previous iteration. The parameter "Improvements_Counter" is an array and its positions keep the improvements achieved by each neighborhood structure. The initial order is based on the complexity of each local search operator. Thus, the initial order is the following:

1. Inter-route Relocate.
2. Exchange Opened-Closed Depots.
3. 2-2 Replenishment Exchange.

The same order is adopted any time all neighborhoods are unable to provide any improved solution. The adaptive mechanism is summarized in Algorithm 6.

Algorithm 6. Adaptive_Order

1: **procedure** ADAPTIVE_ORDER(N_Order, $Improvements\_Counter$)
2:   **if** no improvement is found in any neighborhood **then**
3:       *Keep the same order*
4:   **end if**
5:   **if** an improvement is found **then**
6:       $New\_N\_Order \leftarrow Descending\_Order(N\_Order, Improvements\_Counter)$
7:   **end if**
8:   $N\_Order \leftarrow New\_N\_Order$
9:   **return** N_Order
10: **end procedure**

The pseudo-codes of the Adaptive GVNS schemes, proposed in this work, are provided in Algorithms 7 and 8.

Algorithm 7. $AGVNS_{cVND}$

1: **procedure** $AGVNS_{cVND}(S, k_{max}, l_{max}, max\_time, N\_Order, Improvements\_Counter)$
2:   **while** $time \leq max\_time$ **do**
3:       $S^* = Shake(S, k)$
4:       $N\_Order \leftarrow Adaptive\_Order(N\_Order, Improvements\_Counter)$
5:       $S' = cVND(S^*, l_{max})$
6:       **if** $f(S') < f(S)$ **then**
7:           $S \leftarrow S'$
8:       **end if**
9:   **end while**
10:   **return** S
11: **end procedure**

Furthermore, it is examined if the initial order of the neighborhoods affects the performance of the AGVNS schemes. Consequently, an alternative of the previous mentioned adaptive mechanism is applied, which it uses random re-ordering either in the first iteration or each time no improvements achieved through the VND methods. The random re-ordering is achieved by applying a shuffle method over the neighborhoods order set.

Algorithm 8. $AGVNS_{pVND}$

1: **procedure** $AGVNS_{pVND}(S, k_{max}, l_{max}, max\_time, N\_Order, Improvements\_Counter)$
2:    **while** $time \leq max\_time$ **do**
3:        $S^* = Shake(S, k)$
4:        $N\_Order \leftarrow Adaptive\_Order(N\_Order, Improvements\_Counter)$
5:        $Improvements\_Counter \leftarrow 0$
6:        $S' = pVND(S^*, l_{max})$
7:        **if** $f(S') < f(S)$ **then**
8:            $S \leftarrow S'$
9:        **end if**
10:    **end while**
11:    **return** S
12: **end procedure**

4   Computational Results

4.1   Computer Environment and Benchmark Instances

The proposed algorithms were implemented in Fortran. They ran using Intel Fortran compiler 18.0 with optimization option /O3 on a desktop PC running Windows 7 Professional 64-bit with an Intel Core i7-4771 CPU at 3.5 GHz and 16 GB RAM. The parameter k_{max} was set at 12 and the maximum execution time for each VNS-based algorithm is 60 s.

The benchmark instances used for testing the efficiency of the proposed algorithms was initially proposed in [7] and can be found at: http://pse.cheng. auth.gr/index.php/publications/benchmarks. The form of each instance name is $X - Y - Z$, where X is the number of candidate depots, Y the number of customers and Z the number of time periods.

4.2   Numerical Analysis

In Table 4, the AGVNS1 represents an adaptive GVNS scheme with the complexity-based initial neighborhoods order, while the AGVNS2 represents the scheme with the random initial order.

The results illustrate that all the Adaptive GVNS perform better than the classic GVNS schemes. However, the $AGVNS1_{pVND}$ is the method which provides the best solutions in average. The $AGVNS2_{pVND}$ is ranked as the second method and the methods $AGVNS1_{cVND}$ and $AGVNS2_{cVND}$ hold the third and the fourth place respectively. The $GVNS_{pVND}$ takes the fifth place and the last one is the $GVNS_{cVND}$.

Karakostas et al. [7], recently proposed a Basic Variable Neighborhood Search (BVNS) heuristic algorithm for solving the tested PLIRP instances. Their algorithm used two local search operators, the Inter-route Exchange and the Exchange Opened-Closed Depots.

Figure 1 illustrates the performance of the $AGVNS1_{pVND}$ and the BVNS on the 20 PLIRP instances. It is clear that the $AGVNS1_{pVND}$ outperforms the BVNS algorithm, especially on larger PLIRP instances.

Table 4. Computational results of the proposed methods

Instance	$GVNS_{CVND}$	$AGVNS1_{CVND}$	$AGVNS2_{CVND}$	$GVNS_{PVND}$	$AGVNS1_{PVND}$	$AGVNS2_{PVND}$
4-8-3	23069.96	22944.88	22936.8	22937.11 '	22935.84	**22935.53**
4-8-5	19829.12	19693.68	19731.48	19489.22	**19373.37**	19475.87
4-10-3	17415.26	17452	17697.38	17612.43	**17604.76**	17630.7
4-10-5	23954.56	**23932.88**	23968.77	23944.69	23937.55	23946.91
4-15-5	22115.26	22090.42	22270.71	**22049.63**	22068.88	22174.31
5-9-3	18496.52	18488.41	18570.3	18496.69	**18431.97**	18456.86
5-12-3	24752.46	24747.32	24748.96	24738.99	**24735.39**	24741.45
5-15-3	17502.49	**17469.69**	17495.32	17487.35	17481.19	17479.19
5-18-5	19485.86	19358.93	19312.34	19082.68	**19048.77**	19079.43
5-20-3	17238.63	**17082.93**	17202.05	17248.24	17159.07	17179.59
6-22-7	20042.48	19998.92	**19982.33**	20008.2	19998.96	20023.52
6-25-5	22031.12	21877.13	21864.83	21877.2	21738.51	**21706.41**
7-25-5	29958.96	29798.39	29965.02	29226.52	29183.52	**29173.52**
7-25-7	22559.16	**22239.8**	22366.23	22273.87	22253.14	22331.7
8-25-5	19103.74	19029.96	19178.87	18982.77	**18925.32**	19099.61
8-30-7	20714.09	22706.62	20653.14	20594.12	**20454.58**	20544.96
8-50-5	23106.05	23054.47	23097.66	22922.63	**22365.3**	22533.46
8-65-7	26419.61	**23980.33**	24985.91	27288.72	25496.35	26857.53
9-40-7	21456.76	20908.32	**20860.35**	21243.43	20996.73	20929.3
9-55-5	23254.2	**22563.17**	22734.83	23296.72	22754.5	22703.21
Average	21625.57	21470.91	21481.16	21540.06	**21347.19**	21450.45

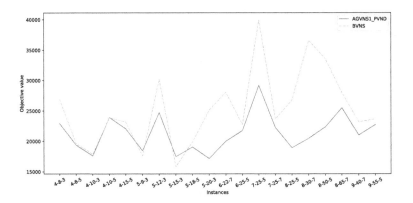

Fig. 1. $AGVNS1_{pVND}$ vs BVNS on 20 PLIRP instances

Table 5 depicts the best values found by all the proposed methods. Most of them were produced using the $AGVNS1_{pVND}$ algorithm.

The second column of Table 6 presents the current best known values of the 20 PLIRP instances. In the third column the overall best values achieved by the proposed methods of this work are provided. As it can be seen, new best solutions have been reported in 13 out of 20 PLIRP instances.

Table 5. Best found values of the proposed methods

Instance	$GVNS_{CVND}$	$AGVNS1_{CVND}$	$AGVNS2_{CVND}$	$GVNS_{PVND}$	$AGVNS1_{PVND}$	$AGVNS2_{PVND}$
4-8-3	23033.9	22935.29	**22934.45**	**22934.45**	**22934.45**	**22934.45**
4-8-5	19600.14	19622.18	19717.81	19436.24	**19368.56**	19379.93
4-10-3	**17327.69**	17411.45	17504.14	17607	17587.21	17607.18
4-10-5	23935.3	23922.35	23942.93	23926.63	**23921.99**	23931.95
4-15-5	22091.05	22067.85	22219.5	**21907.34**	22048.38	22118.42
5-9-3	18480.68	18480.68	18515.44	18494.14	**18425.61**	18439.89
5-12-3	24748.28	24746.24	24746.24	24732.45	**24730.78**	24735.6
5-15-3	17470.64	17468.54	17470.64	**17465.72**	17475.51	**17465.72**
5-18-5	19370.63	19342.65	19300.07	19037.44	19009.86	**18960.25**
5-20-3	17121.62	**17063.64**	17178.92	17221.55	17130.9	17129.98
6-22-7	**19967.57**	**19967.57**	19967.58	20008.2	19980.53	20008.21
6-25-5	21859.47	21830.73	21787.44	21779.89	21701.27	**21623.56**
7-25-5	29854.82	29754.79	29936.99	29158.77	29154.19	**29119.54**
7-25-7	22288.16	**22124.58**	22229.73	22199.59	22239.25	22297.28
8-25-5	**18729.09**	**18729.09**	19021.24	18827.73	18838.97	18775.41
8-30-7	20629.6	20606.69	20553.74	20420.14	**20414.56**	20515.41
8-50-5	23090.5	22980.6	23066.54	**22284.9**	22348.25	22306.72
8-65-7	25744.94	**24424.65**	**24424.65**	26908.92	25176.64	25176.64
9-40-7	21181.06	20700.37	20773.95	21041.51	**20655.25**	20846.73
9-55-5	22698.68	22372.37	22437.05	22649.32	22503.26	**22358.33**
Average	21461.19	21327.62	21386.45	21402.1	**21282.27**	21286.56

Table 6. BKS vs Best found values of the proposed methods

Instance	BKS	Overall_Best
4-8-3	22647.63	22934.45
4-8-5	18282.71	19368.56
4-10-3	16929.96	17327.69
4-10-5	23895.99	23921.99
4-15-5	22013.99	**21907.34**
5-9-3	16700.29	18425.61
5-12-3	24152.36	24730.78
5-15-3	15842.7	17465.72
5-18-5	19891.27	**18960.25**
5-20-3	24605.64	**17063.64**
6-22-7	28074.69	**19967.57**
6-25-5	22747.42	**21623.56**
7-25-5	39914.72	**29119.54**
7-25-7	23675.7	**22124.58**
8-25-5	26773.1	**18729.09**
8-30-7	36582.34	**20414.56**
8-50-5	33536.73	**22284.9**
8-65-7	27986.69	**24424.65**
9-40-7	23176.14	**20655.25**
9-55-5	23688.55	**22358.33**

5   Conclusions

This work presents several Adaptive GVNS-based algorithms for solving PLIRP instances. Two variants of the adaptive mechanism were developed based on the initial order of the neighborhoods. The proposed algorithms were compared both with their corresponding classic GVNS methods and other heuristic methods [7] for this specific problem. The computational results reveal the superiority of the Adaptive GVNS, which uses the pipe-VND as its main improvement step and the complexity-based initial neighborhoods order in the adaptive re-ordering mechanism. Furthermore, new best values have been reported for 13 out of 20 PILRP instances.

Future work can focus on the determination of lower bounds in order to evaluate the efficiency of the proposed methods. An other future research direction can examine the use of more sophisticated adaptive mechanisms for the re-ordering of the neighborhoods structure. Also, the adaptability may be applied both on the improvement and the shaking step, in order to generate neighborhood patterns that they will produce high quality solutions. Furthermore, this work can be generalized in order to address other environmental implications, such as the emission of pollutants during production [11]. Finally, such computational difficult problems that combine decisions for location, inventory and routing, can be benefited a lot by using parallel computing techniques [1].

Acknowledgement. The second author has been funded by the University of Macedonia Research Committee as part of the "Principal Research 2019" funding scheme (ID 81307).

References

1. Antoniadis, N., Sifaleras, A.: A hybrid CPU-GPU parallelization scheme of variable neighborhood search for inventory optimization problems. Electron. Notes Discrete Math. **58**, 47–54 (2017)
2. Cheng, C., Yang, P., Qi, M., Rousseau, L.M.: Modeling a green inventory routing problem with a heterogeneous fleet. Transp. Res. Part E **97**, 97–112 (2017)
3. Glover, F., Gutin, G., Yeo, A., Zverovich, A.: Construction heuristics for the asymmetric TSP. Eur. J. Oper. Res. **129**, 555–568 (2001)
4. Hansen, P., Mladenovic, N.: Variable neighborhood search. In: Burke, E., Kendall, G. (eds.) Search Methodologies: Introductory Tutorials in Optimization and Decision Support Techniques, chap. 12, pp. 313–337. Springer, New York (2014). https://doi.org/10.1007/978-1-4614-6940-7
5. Hansen, P., Mladenovic, N., Todosijevic, R., Hanafi, S.: Variable neighborhood search: basics and variants. EURO J. Comput. Optim. **5**, 423–454 (2017)
6. Huber, S., Geiger, M.: Order matters - a variable neighborhood search for the swap-body vehicle routing problem. Eur. J. Oper. Res. **263**, 419–445 (2017)
7. Karakostas, P., Sifaleras, A., Georgiadis, M.C.: Basic VNS algorithms for solving the pollution location inventory routing problem. In: Sifaleras, A., Salhi, S., Brimberg, J. (eds.) ICVNS 2018. LNCS, vol. 11328, pp. 64–76. Springer, Cham (2019). https://doi.org/10.1007/978-3-030-15843-9_6

8. Karakostas, P., Sifaleras, A., Georgiadis, C.: A general variable neighborhood search-based solution approach for the location-inventory-routing problem with distribution outsourcing. Comput. Chem. Eng. **126**, 263–279 (2019)
9. Li, K., Tian, H.: A two-level self-adaptive variable neighborhood search algorithm for the prize-collecting vehicle routing problem. Appl. Soft Comput. **43**, 469–479 (2016)
10. Sifaleras, A., Konstantaras, I.: General variable neighborhood search for the multi-product dynamic lot sizing problem in closed-loop supply chain. Electron. Notes Discrete Math. **47**, 69–76 (2015)
11. Skouri, K., Sifaleras, A., Konstantaras, I.: Open problems in green supply chain modeling and optimization with carbon emission targets. In: Pardalos, P.M., Migdalas, A. (eds.) Open Problems in Optimization and Data Analysis. SOIA, vol. 141, pp. 83–90. Springer, Cham (2018). https://doi.org/10.1007/978-3-319-99142-9_6
12. Todosijevic, R., Mladenovic, M., Hanafi, S., Mladenovic, N., Crevits, I.: Adaptive general variable neighborhood search heuristics for solving the unit commitment problem. Electr. Power Energy Syst. **78**, 873–883 (2016)

A RAMP Algorithm for Large-Scale Single Source Capacitated Facility Location Problems

Óscar Oliveira$^{(\boxtimes)}$ (ID), Telmo Matos (ID), and Dorabela Gamboa (ID)

CIICESI – Center for Research and Innovation in Business Sciences and Information Systems,
School of Technology and Management, Polytechnic of Porto, Porto, Portugal
{oao,tsm,dgamboa}@estg.ipp.pt

Abstract. We propose a Relaxation Adaptive Memory Programming (RAMP) algorithm for the solution of the Single Source Capacitated Facility Location Problem (SSCFLP). This problem considers a set of possible locations for opening facilities and a set of clients whose demand must be satisfied. The objective is to minimize the cost of assigning the clients to the facilities, ensuring that all clients are served by only one facility without exceeding the capacity of the facilities. The RAMP framework efficiently explores the relation between the primal and the dual sides of combinatorial optimization problems. In our approach, the dual problem, obtained through a lagrangean relaxation, is solved by subgradient optimization. Computational experiments of the effectiveness of this approach are presented and discussed.

Keywords: Single Source Capacitated Facility Location Problems · SSCFLP · Relaxation Adaptive Memory Programming · Dual RAMP

1 Introduction

The Single Source Capacitated Facility Location Problem (SSCFLP) considers a set of possible locations for opening facilities and a set of clients whose demand must be satisfied. The objective is to minimize the cost of assigning the clients to the facilities, ensuring that all clients are served by only one facility without exceeding the capacity of the facilities. The SSCFLP has several practical applications, such as the design of delivery systems, computer networks, among many others. This problem is a special case of Capacitated Facility Location Problem (CFLP) which belongs to the class of NP-hard problems [1]. Solving the SSCFLP is, generally, more difficult than the CFLP, since it considers that all decision variables are binary, whereas the CFLP considers continuous decision variables for the client's assignments.

To solve the SSCFLP, we propose a Relaxation Adaptive Memory Programming (RAMP) algorithm. RAMP, proposed by Rego [2], efficiently exploit primal and dual relationships of combinatorial optimization problems. To guide the search, principles of adaptive memory are used, taking advantage of information generated by both primal and dual side of the problem. RAMP implementations proceed as follows. At each iteration, a new solution for the relaxed problem is obtained. When a dual solution is

© Springer Nature Switzerland AG 2020
N. F. Matsatsinis et al. (Eds.): LION 13 2019, LNCS 11968, pp. 171–183, 2020.
https://doi.org/10.1007/978-3-030-38629-0_14

infeasible in the primal side, a projection method is employed to create a feasible primal solution, using information gathered from the current dual solution. Then, the primal solution is subjected to an improvement method. Rego [2] presents different strategies for the primal and dual components allowing different levels of sophistication. RAMP algorithms can be implemented incrementally, starting with the simplest version and successively move to more complex approaches to strengthen the primal-dual relationship that, usually, leads to better results. The simplest versions perform a more thorough exploration on the dual side and are usually referred to as Dual-RAMP (or just RAMP). The more sophisticated versions explore both solution spaces intensely and are designated as Primal-Dual RAMP (PD-RAMP). The RAMP metaheuristic demonstrated to be very effective for a variety of combinatorial optimization problems, producing state-of-the-art algorithms for the problems it was applied to, as demonstrated by the work of Gamboa [3], Rego et al. [4] and Matos and Gamboa [5, 6] and Matos et al. [7, 8].

The rest of the paper is organized as follows. Section 2 describes the SSCFLP. Section 3 presents some of the most relevant solution methods found in the literature to solve this problem. In Sect. 4, we present the proposed algorithm divided by its main components, namely, dual phase, projection method and primal phase. In Sect. 5, we discuss the computational results. Finally, in Sect. 6, conclusions and future work directions are presented.

2 Single Source Capacitated Facility Location Problem

The SSCFLP considers a set J of clients and a set I of candidate locations for opening facilities. Each client j has a demand d_j that must be satisfied by a single facility. Each facility i has a fixed opening cost f_i and a maximum capacity s_i. Satisfying the demand of client j by a facility located in i has an assignment cost (a.k.a., transportation cost) of c_{ij}.

Considering the following decision variables:

$$x_{ij} = \begin{cases} 1, & if\ client\ j\ is\ \text{assigned to facility}\ i, \\ 0, & otherwise. \end{cases}$$

$$y_i = \begin{cases} 1, & if\ facility\ i\ \text{is open}, \\ 0, & otherwise. \end{cases}$$

This problem can be formulated as:

$$SSCFLP = min \sum_{i \in I} f_i y_i + \sum_{i \in I} \sum_{j \in J} c_{ij} x_{ij} \qquad (1)$$

subject to

$$\sum_{i \in I} x_{ij} = 1, \forall j \in J \qquad (2)$$

$$\sum_{j \in J} d_j x_{ij} \leq s_i y_i, \forall i \in I \qquad (3)$$

$$x_{ij} \in \{0, 1\}, \forall i \in I, \forall j \in J \tag{4}$$

$$y_i \in \{0, 1\}, \forall i \in I \tag{5}$$

The objective function (1) is to minimize the total cost of opening facilities and assigning clients to such facilities. The set of constraints (2) guarantee that each client is served by a single facility and the set of constraints (3) ensures that clients are assigned to open facilities and that the total demand assigned to a facility cannot exceed its capacity. Constraints (4) and (5) are integrity constraints.

3   Related Work

Several authors proposed solution methods for solving the SSCFLP, being most of them based on the lagrangean relaxation of the original problem.

Neebe and Rao [9] modelled the SSCFLP as a Set Partitioning Problem that is solved by a column-generating Branch-and-Bound procedure.

Barceló and Casanovas [10] proposed a lagrangean heuristic, dualizing constraints (2), in which the maximum number of open facilities is predefined.

Klincewicz and Luss [11] presented a lagrangean heuristic dualizing the capacity constraints (3). The relaxed problem, which results in an Uncapacitated Facility Location Problem, is solved by the Dual Ascent algorithm proposed by Erlenkotter [12].

Both Sridharan [13] and Pirkul [14] presented lagrangean heuristics relaxing constraints (2). Sridharan [13] used an algorithm based on the resolution of the Single-Source Transportation Problem to obtain feasible solutions on the primal side. Pirkul [14] presented an algorithm to solve a problem equivalent to the SSCFLP, namely the Capacitated Concentrator Location Problem.

Beasley [15] presented a robust framework to develop lagrangean heuristics for various location problems, namely p-Median Location Problem, Uncapacitated Facility Location Problem, CFLP and SSCFLP. The results obtained relaxing constraints (2) and (3) were better than those obtained by Klincewicz and Luss [11] but worse than the ones obtained by Pirkul [14]. The work of Beasley [15] is, usually, referred to justify the choice to relax constraints (2) when implementing lagrangean heuristics for the SSCFLP.

Delmaire et al. [16] proposed four heuristics: Evolutive algorithm, Greedy Randomized Adaptive Search Procedure (GRASP), Simulated Annealing and Tabu Search. The GRASP incorporates three more neighbourhood structures than the other heuristics as it was less demanding in computational resources. In this work, the best results were obtained by the proposed Tabu Search and GRASP. Also, Considering:

$$cmin_j = \min\{c_{ij}, \forall i \in I\} \tag{6}$$

$$\Delta_{ij} = c_{ij} - cmin_j \tag{7}$$

The authors reformulate the objective function (1) and represent the problem as follows:

$$SSCFLP = \sum_{j \in J} cmin_j + min\left(\sum_{i \in I} f_i y_i + \sum_{i \in I}\sum_{j \in J} \Delta_{ij} x_{ij}\right) \tag{8}$$

subject to (2), (3), (4) and (5)

The relative cost (7) of satisfying the demand of client j in facility i with respect to its minimum cost is used in several of the heuristics proposed by Delmaire et al. [16] as a criterion to assign clients.

Delmaire et al. [17] presented a reactive GRASP (RGRASP), a Tabu Search and two hybrid algorithms that combine elements of GRASP, RGRASP and Tabu Search. The RGRASP and Tabu Search were more efficient than the heuristics proposed in the previous work [16], although the RGRASP was more demanding in computational resources when compared with the GRASP. The two hybrid approaches achieved the best results.

Rönnqvist et al. [18] presented a heuristic based on Repeated Matching. Holmberg et al. [19] proposed a Branch-and-Bound that considers a lagrangean relaxation dualizing constraints (2) to obtain a lower bound, and a heuristic based on Repeated Matching for the upper bound.

Hindi and Pienkosz [20] presented a lagrangean heuristic relaxing constraints (2). The upper bounds were obtained through a greedy heuristic based on Maximum Regret and Restricted Neighbourhood Search. This approach proved to be efficient for large-sized instances. Another heuristic that proved to be very efficient for large scale instances was the Very Large Scale Neighbourhood (VLSN) presented by Ahuja et al. [21].

Cortinhal and Captivo [22] proposed a lagrangean heuristic relaxing the set of constraints (2). The solution of the relaxed problem, obtained at each iteration of the subgradient optimization, is subjected to a projection method and improved by a Tabu Search.

Cortinhal and Captivo [23] presented several approaches based on genetic algorithms to solve the SSCFLP. In this study, the authors concluded that genetic algorithms are not an efficient solution method for this problem.

Contreras and Díaz [24] presented a Scatter Search based on its division into five methods as proposed by Laguna and Martí [25]. In the construction phase of the reference set, it is employed a GRASP based on the work of Delmaire et al. [16]. The Scatter Search obtained better results, for small- and medium-sized instances, than the ones presented by Ahuja et al. [21], but was outperformed by the hybrid approaches proposed by Delmaire et al. [16].

Various heuristics based on Ant Colony Optimization (ACO) were presented for the resolution of SSCFLP, as exemplified by the work of Kumweang [26], Chen and Ting [27] and Lina et al. [28].

Yang et al. [29] presented three versions of an exact method based on Cut-and-Solve (see Climer and Zhang [30]).

Guastaroba and Speranza [31] extended the Kernel Search heuristic framework and applied it to the SSCFLP obtaining very good results outperforming most of the best algorithms presented so far. This heuristic is based on the resolution to optimality of a sequence of subproblems that are restricted to a subset of the decision variables. This heuristic relies, as stated by the authors, on the high performance of commercial solvers for Linear Programming and Mixed Integer Linear Programming problems.

Ho [32] presented an Iterated Tabu Search (ITS) heuristic that makes use of randomized neighbourhood sampling and perturbation to obtain diversification in the search.

4 RAMP Algorithm for the SSCFLP

Alternating the search between the dual and the primal side of the problem, the solution method proposed in this paper relies on subgradient optimization to solve the dual problem obtained through the lagrangean relaxation of the SSCFLP.

At each iteration, a solution for the relaxed problem is obtained and projected, through a projection method, to the primal solution space. The solution obtained by the projection method is improved through a Local Search method. If the best lower bound found so far is improved in the current iteration, the improved primal solution is submitted to a Tabu Search to explore a wider neighbourhood.

The basic algorithm is described in Fig. 1, where $\overline{Z}$ and $\underline{Z}$ represents, respectively, the best upper and lower bound found.

---

1. $\underline{Z} \leftarrow 0$
2. $\overline{Z} \leftarrow \sum_{j \in J}(max_i\ c_{ij}) + \sum_{i \in I} f_i$
3. $\lambda_j \leftarrow min_i\ c_{ij}, \forall j \in J$
4. $P \leftarrow \emptyset$
5. $iteration \leftarrow 0$
6. While stopping criteria are not met do
7.   Solve $Z(\lambda)$ with current lagrangean multipliers
8.   Get primal solution Z_c through the projection method
9.   If Z_c is feasible and $Z_c \notin P$ then
10. $P \leftarrow Z_c \cup P$
11.   Improve Z_c through Local Search
12.   If $f(Z(\lambda)) > \underline{Z}$ then
13.   Improve Z_c through Tabu Search
14.   If $f(Z_c) < \overline{Z}$ then
15. $\overline{Z} \leftarrow f(Z_c)$
16.   Update lagrangean multipliers
17.   If $f(Z(\lambda)) > \underline{Z}$ then
18. $\underline{Z} \leftarrow f(Z(\lambda))$
19. $iteration \leftarrow iteration + 1$

Fig. 1. RAMP's algorithm pseudo-code

One of the stopping criteria of this algorithm is the maximum number of iterations that was set to 2000. The other three stopping criteria are related to the dual phase and are described in the next section (see Sect. 4.1).

The next sections present a more detailed description of each main components of this algorithm, namely dual phase, projection method and primal phase.

4.1 Dual Phase

The dual problem, obtained through the lagrangean relaxation of the original problem, is solved through subgradient optimization. Dualizing the set of constraints (2) and given the set of lagrangean multipliers λ_j, $j = 1, \ldots, n$, we obtain following the dual problem:

$$Z(\lambda) = \sum_{j \in J} \lambda_j + min \sum_{i \in I} \left(f_i y_i + \sum_{j \in J} (c_{ij} - \lambda_j) x_{ij} \right) \qquad (9)$$

subject to (3), (4) and (5)

The resolution of the relaxed problem (9) is achieved after solving one knapsack problem for each facility (see Sridharan [13] and Holmberg et al. [19]). To solve the knapsack problems resulting from the lagrangean relaxation of SSCFLP, we used the algorithm proposed by Martello et al. [33]. The code used was made publicly available at http://www.diku.dk/hjemmesider/ansatte/pisinger/codes.htm.

The dual phase starts with the lagrangean multipliers initialized as

$$\lambda_j = \min_i c_{ij}, \forall j \in J \qquad (10)$$

and, at the end of each iteration, updated as follows.

$$\lambda_j = \lambda_j + \Delta \delta_j, \forall j \in J \qquad (11)$$

The subgradient vector δ, where x_{ij} indicates if client j is assigned to facility i in the dual solution, is obtained as follows.

$$\delta_j = 1 - \sum_{i \in I} x_{ij}, \forall j \in J \qquad (12)$$

The step size (Δ), where $\|\delta\|$ represents the norm, is calculated as follows.

$$\Delta = \frac{\pi \left(f\left(\overline{Z}\right) - f(Z(\lambda)) \right)}{\|\delta\|} \qquad (13)$$

The agility parameter (π) is initialized with the value 2 and is halved every 5 consecutive iterations without improving the lower bound. The agility is restarted with the value 2 every 50 iterations.

The dual RAMP algorithm has four stopping criteria, the agility (π) is less than 0.003, the norm ($\|\delta\|$) is equal to 0, the difference between $\overline{Z}$ and $\underline{Z}$ is less than 1 and, as already stated, the maximum number of iterations is reached.

4.2 Projection Method

The projection method, based on the one proposed by Cortinhal and Captivo [22], tries to generate a feasible primal solution considering the current dual solution. Since the dual solution is, in most cases, infeasible, we need to ensure that all clients are served by only one facility and that the capacity of opened facilities is not exceeded in the solution projected to the primal solution space.

Considering that, in the dual solution, the set of clients J is divided into three subsets $J_1 = \left\{ j \in J : \sum_{i \in I} x_{ij} = 1 \right\}$, $J_2 = \left\{ j \in J : \sum_{i \in I} x_{ij} > 1 \right\}$ and $J_3 = \left\{ j \in J : \sum_{i \in I} x_{ij} = 0 \right\}$, we try to generate a primal solution through the following steps:

1. The assignments of the subset J_1 are not changed from the dual to the primal solution.
2. From the facilities in which each client in J_2 is assigned in the dual solution, the client j is assigned, in the primal solution, to the facility i that present the lowest Δ_{ij}.
3. Assign clients from J_3 to the opened facilities in step 1 and 2 using as criterion of choice their lowest Δ_{ij}.
4. Finally, while there are clients not served in J_3 and it is possible to open facilities, open one facility at a time, ordered by their opening cost, and assign the clients not yet served, ordered by their relative cost Δ_{ij}.

Since we ensure that in step 1–4 the facilities capacity is never exceeded, at the end of this method, if all clients are assigned to a facility, the primal solution is feasible.

4.3   Primal Phase

All distinct feasible primal solutions are subjected to a local search phase that explores the solution neighbourhood to improve the solution. If the value of the dual solution is greater than the current best lower bound, the solution improved through the local search is subjected to a Tabu Search to explore a wider neighbourhood.

The local search phase explores, iteratively, five neighbourhood structures. The neighbourhood structures explored in the local search phase are: (1) *Close Facility* that closes one facility and reassigns the clients to the remaining open facilities, (2) *Open Facility* that opens a facility and reassign the clients, (3) *Swap Facilities* that swaps the assignments between facilities, considering that at least one is open, (4) *Shift Client* that performs the reassignment of one client to another facility, and (5) *Swap Clients* that swaps the facility assignments between two clients. To reduce the computational effort of *Swap Clients*, for each client assigned to a facility i, we only try exchanges with clients that have lower Δ_{ij} in that facility. Noteworthy, that in all neighbourhood structures, only the best movement, i.e., the movement that most reduce the value of the objective function, is performed.

The Tabu Search performs the best movement found in *Shift Client* $\cup$ *Swap Clients* even if this move worsens the solution. Unlike in the local search phase, the *Swap Clients* considers all the neighbourhood. The Tabu Search stops after 100 iterations without improving the solution. *Shift Client* and *Swap Clients* use a list of moves considered tabu, represented in the form of (i, j) where i is the location of the facility and j is the client. When a facility assigned to a client changes, the prior assignment (i, j) is considered tabu for a period of time preventing revisiting solutions. The tabu move can, however, be performed if it leads to the best-known solution (aspiration criterion). Also, the tabu state is not permanent, being controlled by a time interval (tabu tenure), e.g., the number of iterations that must elapse for the movement to be removed from the list. In our algorithm, the tabu tenure, as proposed by Cortinhal and Captivo [22], is given by:

$$7 + n \times \frac{number}{All\_number} \tag{14}$$

where n is the number of facilities, $number$ is the number of times the assignment has been made and All_number is the maximum value of $number$ in all assignments made.

5 Computational Results

The proposed algorithm was implemented in C and the tests were run on a computer with Intel Core i7-4800MQ 2.70 GHz with 8 Gb RAM and operating system Ubuntu 15.10.

We compare the results obtained by our heuristic with the ones obtained by Guastaroba and Speranza [31] and by Yang *et al.* [29] for large-sized instances. For each instance, RAMP was executed only once.

To demonstrate the effectiveness of their approaches, Guastaroba and Speranza [31] and Yang *et al.* [29] used the two benchmark datasets of large-sized instances that we will refer to as Yang [29] and TBED1 [34]. The main characteristics of these datasets are presented in Table 1, where the columns show in order of appearance, the dataset name, the number of subsets, the number of facilities and the number of clients.

Table 1. Main features of the datasets.

Name	Subsets	Number of facilities	Number of clients
Yang	4	30 to 80	200 to 400
TBED1	5	300 to 1000	300 to 1500

The results obtained solving Yang and TBED1 datasets are given in Tables 2 and 3, respectively. The content of the columns, in order of appearance, is the following: the subset of instances, the algorithm identifier, the average and worst deviation from the best-known upper bound, and the average computational time (in seconds) needed to solve the subset.

Table 2 presents the computational results for the Yang dataset obtained by the Kernel Search heuristics KS, KS(1) and KS(01) presented by Guastaroba and Speranza [31], and by the Cut-and-Solve algorithm, denoted as CS3, presented by Yang *et al.* [29].

Observing Table 2 it can be noticed that the RAMP approach can obtain, with low computational time, an average deviation from the best-known solution that is under 0.6% for all subsets of the Yang dataset. Also, noteworthy that the worst deviation never exceeded 1.12%. With respect to this dataset, although the proposed RAMP approach provided good quality solutions, obtaining a lower average and worst gap than the ones obtained by KS(01), it is outperformed by the other Kernel Search versions and by the Cut-and-Solve in terms of the quality of solutions. Even though comparisons between CPU time are not possible, since different computers were used, we estimate that the time required by the RAMP approach is lower than the ones required by the other approaches.

Table 2. Computational results for Yang dataset.

Subset	Algorithm	Avg. GAP (%)	Worst GAP (%)	CPU (s)
Y1	KS	0.00	0.00	411.28
	KS(1)	0.27	0.74	186.29
	KS(01)	1.02	2.02	44.27
	CS3	0.00	0.00	220.60
	RAMP	0.38	0.71	3.67
Y2	KS	0.00	0.01	1640.42
	KS(1)	0.33	1.49	1187.59
	KS(01)	0.67	2.96	368.74
	CS3	0.00	0.00	16196.90
	RAMP	0.56	1.12	5.57
Y3	KS	0.00	0.01	597.06
	KS(1)	0.18	0.70	558.08
	KS(01)	1.61	3.24	184.53
	CS3	0.00	0.00	1334.64
	RAMP	0.47	0.82	9.02
Y4	KS	0.00	0.00	1409.11
	KS(1)	0.01	0.04	1149.78
	KS(01)	0.51	1.46	369.76
	CS3	0.00	0.00	7009.52
	RAMP	0.45	0.91	21.96

Table 3 presents the computational results for the TBED1 dataset obtained by Guastaroba and Speranza [31] with algorithms KS, KS(1) and KS(01), as well as the results obtained by the same authors using the commercial solver CPLEX with a time limit of 7200 s.

Table 3. Computational results for TBED1 dataset.

Instances	Algorithm	Avg. GAP (%)	Worst GAP (%)	CPU (s)
TB1	KS	0.56	2.22	2206.96
	KS(1)	0.59	2.29	2110.35
	KS(01)	0.78	2.29	408.21
	CPLEX	0.59	2.23	3159.61
	RAMP	0.95	2.84	30.14
TB2	KS	0.00	0.00	334.70
	KS(1)	0.00	0.00	299.77
	KS(01)	0.21	0.74	186.53
	CPLEX	0.00	0.00	186.67
	RAMP	0.02	0.08	65.82
TB3	KS	0.66	2.04	4190.28
	KS(1)	0.71	2.07	4050.81
	KS(01)	0.79	2.09	673.56
	CPLEX	0.74	2.15	6007.62
	RAMP	1.11	2.40	140.05
TB4	KS	0.90	2.29	5244.69
	KS(1)	0.91	2.34	5169.96
	KS(01)	1.00	2.70	654.17
	CPLEX	1.24	3.35	6865.53
	RAMP	1.38	2.75	412.41
TB5	KS	1.07	3.11	6533.15
	KS(1)	1.02	2.48	6509.26
	KS(01)	1.10	2.67	968.12
	CPLEX	2.44	6.47	7269.60
	RAMP	1.49	2.80	1742.29

The computational results for the TBED1 dataset demonstrate that the RAMP app-roach has obtained very good results, considering the difficulty and the size of these instances. RAMP achieved upper bounds very close to solutions obtained by the Kernel search and by CPLEX with low computational time. Only for TB5 subset, a Kernel search version has presented better results, simultaneously, in all parameters considered: average and worst deviation and computational time.

Notwithstanding that the Kernel Search heuristics presented by Guastaroba and Sper-anza [31] obtained high-quality results and have outperformed most of the heuristics proposed in the literature to solve the SSCFLP, the RAMP approach demonstrated that, with low computational effort, it can achieve good results when dealing with large-scale instances. Also, it must be noted that both the Kernel Search and the Cut-and-Solve approach proposed for solving the SSCFLP rely on commercial solvers.

6 Conclusions

In this paper, we present a new algorithm to solve the Single Source Capacitated Facility Location Problem (SSCFLP) based on the RAMP framework. The SSCFLP considers a set of possible locations for opening facilities and a set of clients with a given demand that must be satisfied by one facility. The problem is to determine the facilities to be opened to satisfy the demand such that total costs are minimized. The proposed algorithm follows the RAMP approach in its simplest version. The dual side is based on the resolution of the lagrangean dual problem through subgradient optimization. At the end of each subgradient iteration, the dual solution is projected to the primal solution space. Then, the primal solution is subjected to an improvement method. The solution method proposed in this paper was compared against state-of-the-art algorithms for solving the SSCFLP and demonstrated to be capable of providing good quality solution with reduced computational times for large-scale instances. The next step in our research will be to strengthen the primal-dual relationship, moving to a more complex RAMP version, a PD-RAMP algorithm, that, we expect, could lead us to achieve better results.

References

1. Current, J., Daskin, M.S., Schilling, D.: Discrete network location models. In: Drezner, Z., Hamacher, H. (eds.) Facility Location; Applications and Theory, pp. 83–120 (2001)
2. Rego, C.: RAMP: a new metaheuristic framework for combinatorial optimization. In: Rego, C., Alidaee, B. (eds.) Metaheuristic Optimization via Memory and Evolution, pp. 441–460. Kluwer Academic Publishers, Boston (2005)
3. Gamboa, D.: Adaptive Memory Algorithms for the Solution of Large Scale Combinatorial Optimization Problems. PhD Thesis, Instituto Superior Técnico, Universidade Técnica de Lisboa (2008). (in Portuguese)
4. Rego, C., Mathew, F., Glover, F.: RAMP for the capacitated minimum spanning tree problem. Ann. Oper. Res. **181**, 661–681 (2010)
5. Matos, T., Gamboa, D.: Dual-RAMP for the capacitated single allocation p-hub location problem. In: 47th International Conference on Computers & Industrial Engineering 2017 (CIE47): How Digital Platforms and Industrial Engineering are Transforming Industry and Services, pp. 696–708. Computers and Industrial Engineering, Lisboa (2017)
6. Matos, T., Gamboa, D.: Dual-RAMP for the capacitated single allocation hub location problem. In: Gervasi, O., et al. (eds.) ICCSA 2017. LNCS, vol. 10405, pp. 696–708. Springer, Cham (2017). https://doi.org/10.1007/978-3-319-62395-5_48
7. Matos, T., Maia, F., Gamboa, D.: Improving traditional dual ascent algorithm for the uncapacitated multiple allocation hub location problem: a RAMP approach. In: Nicosia, G., Pardalos, P., Giuffrida, G., Umeton, R., Sciacca, V. (eds.) The Fourth International Conference on Machine Learning, Optimization, and Data Science, Volterra, Tuscany, Italy, 13–16 September 2018, pp. 243–253. Springer, Italy (2019). https://doi.org/10.1007/978-3-030-13709-0_20
8. Matos, T., Maia, F., Gamboa, D.: A simple dual-RAMP algorithm for the uncapacitated multiple allocation hub location problem. In: Madureira, A.M., Abraham, A., Gandhi, N., Varela, M.L. (eds.) HIS 2018. AISC, vol. 923, pp. 331–339. Springer, Cham (2020). https://doi.org/10.1007/978-3-030-14347-3_32

9. Neebe, A.W., Rao, M.R.: an algorithm for the fixed-charge assigning users to sources problem. J. Oper. Res. Soc. **34**, 1107–1113 (1983)

10. Barceló, J., Casanovas, J.: A heuristic Lagrangian relaxation algorithm for the capacitated plant location problem. Eur. J. Oper. Res. **15**, 212–226 (1984)

11. Klincewicz, J.G., Luss, H.: A Lagrangian relaxation heuristic for capacitated facility location with single-source constraints. J. Oper. Res. Soc. **37**, 495–500 (1986)

12. Erlenkotter, D.: A dual-based procedure for uncapacitated facility location. Oper. Res. **26**, 992–1009 (1978)

13. Sridharan, R.: A Lagrangian heuristic for the capacitated plant location problem with side constraints. J. Oper. Res. Soc. **42**, 579–585 (1991)

14. Pirkul, H.: Efficient algorithms for the capacitated concentrator location problem. Comput. Oper. Res. **14**, 197–208 (1987)

15. Beasley, J.E.: Lagrangean heuristics for location problems. Eur. J. Oper. Res. **65**, 383–399 (1993)

16. Delmaire, H., Díaz, J.A., Fernández, E., Ortega, M.: Comparing new heuristics for the pure integer capacitated plant location problem. Invest. Oper. **8**, 217–242 (1997)

17. Delmaire, H., Díaz, J.A., Fernández, E., Ortega, M.: Reactive GRASP and tabu search based heuristics for the single source capacitated plant location problem. INFOR Inf. Syst. Oper. Res. **37**, 194–225 (1999)

18. Rönnqvist, M., Tragantalerngsak, S., Holt, J.: A repeated matching heuristic for the single-source capacitated facility location problem. Eur. J. Oper. Res. **116**, 51–68 (1999)

19. Holmberg, K., Rönnqvist, M., Yuan, D.: An exact algorithm for the capacitated facility location problems with single sourcing. Eur. J. Oper. Res. **113**, 544–559 (1999)

20. Hindi, K.S., Pienkosz, K.: Efficient solution of large scale, single-source, capacitated plant location problems. J. Oper. Soc. **50**, 268–274 (1999)

21. Ahuja, R.K., Orlin, J.B., Pallottino, S., Scaparra, M.P., Scutellà, M.G.: A multi-exchange heuristic for the single-source capacitated facility location problem. Manage. Sci. **50**, 749–760 (2004)

22. Cortinhal, M.J., Captivo, M.E.: Upper and lower bounds for the single source capacitated location problem. Eur. J. Oper. Res. **151**, 333–351 (2003)

23. Cortinhal, M.J., Captivo, M.E.: Genetic algorithms for the single source capacitated location problem. In: Metaheuristics: Computer Decision-Making, vol. 151, pp. 333–351 (2003)

24. Contreras, I.A., Díaz, J.A.: Scatter search for the single source capacitated facility location problem. Ann. Oper. Res. **157**, 73–89 (2007)

25. Laguna, M., Marti, R.: Scatter Search. Springer, New York (2003). https://doi.org/10.1007/978-1-4615-0337-8

26. Kumweang, K., Kawtummachai, R.: Solving a SSCFLP in a supply chain with ACO. Suranaree J. Sci. **12**, 28–38 (2005)

27. Chen, C.-H., Ting, C.-J.: Combining Lagrangian heuristic and Ant Colony System to solve the Single Source Capacitated Facility Location Problem. Transp. Res. Part E Logistics Transp. Rev. **44**, 1099–1122 (2008)

28. Lina, Y., Xu, S.U.N., Tianhe, C.H.I.: A hybrid ant colony optimization algorithm with local search strategies to solve single source capacitated facility location problem. Eng. Technol. **11**, 1920–1924 (2011)

29. Yang, Z., Chu, F., Chen, H.: A cut-and-solve based algorithm for the single-source capacitated facility location problem. Eur. J. Oper. Res. **221**, 521–532 (2012)

30. Climer, S., Zhang, W.: Cut-and-solve: an iterative search strategy for combinatorial optimization problems. Artif. Intell. **170**, 714–738 (2006)

31. Guastaroba, G., Speranza, M.G.: A heuristic for BILP problems: the Single Source Capacitated Facility Location Problem. Eur. J. Oper. Res. **238**, 438–450 (2014)

32. Ho, S.C.: An iterated tabu search heuristic for the Single Source Capacitated Facility Location Problem. Appl. Soft Comput. **27**, 169–178 (2015)
33. Martello, S., Pisinger, D., Toth, P.: Dynamic programming and strong bounds for the 0-1 knapsack problem. Manage. Sci. **45**, 414–425 (1999)
34. Avella, P., Boccia, M.: A cutting plane algorithm for the capacitated facility location problem. Comput. Optim. Appl. **43**, 39–65 (2009)

A Novel Approach for Solving Large-Scale Bike Sharing Station Planning Problems

Christian Kloimüllner[⊠] and Günther R. Raidl

Institute of Logic and Computation, TU Wien, Favoritenstraße 9–11/19201,
1040 Vienna, Austria
{kloimuellner,raidl}@ac.tuwien.ac.at

Abstract. In large cities all around the world, individual and motorized traffic is still prevalent. This circumstance compromises the quality of living, and moreover, space inside cities for parking individual vehicles for movement is scarce and is becoming even scarcer. Thus, the need for a greener means of transportation and less individual vehicles inside the cities is demanded and rising. An already accepted and established solution possibility to these problems are public bike sharing systems (PBS). Such systems are often freely available to people for commuting within the city and utilize the available space in the city more efficiently than individual vehicles. When building or extending a PBS, a certain optimization goal is to place stations inside a city or a part of it, such that the number of bike trips per time unit is maximized under certain budget constraints. In this context, it is also important to consider rebalancing and maintenance costs as they introduce substantial supplementary costs in addition to the fixed and variable costs when building or extending a PBS. In contrast to the literature, this work introduces a novel approach which is particularly designed to scale well to large real-world instances. Based on our previous work, we propose a multilevel refinement heuristic operating on hierarchically clustered input data. This way, the problem is coarsened until a manageable input size is reached, a solution is derived, and then step by step extended and refined until a valid solution for the whole original problem instance is obtained. As an enhancement to our previous work, we introduce the following extensions. Instead of considering an arbitrary integral number of slots for stations, we now use sets of predefined station configurations. Moreover, a local search is implemented as refinement step in the multilevel refinement heuristic and we now consider real-world input data for the city of Vienna.

Keywords: Bike-Sharing Station Planning Problem · Multilevel refinement · Facility location problem · Mixed integer linear programming

We thank the LOGISTIKUM Steyr, the Austrian Institute of Technology, and Rosinak & Partner for the collaboration on this topic. This work is supported by the Austrian Research Promotion Agency (FFG) under contract 849028.

© Springer Nature Switzerland AG 2020
N. F. Matsatsinis et al. (Eds.): LION 13 2019, LNCS 11968, pp. 184–200, 2020.
https://doi.org/10.1007/978-3-030-38629-0_15

1 Introduction

Nowadays, public bike-sharing systems (PBS) are an essential ingredient for building smart and green cities. In many cities around the world, mostly larger ones, municipalities already implemented PBSs or are thinking about building one. Of course, such systems incur high costs but the advantages are manifold. Bikes do not need so much parking space as e.g., individual cars. Cycling is healthy and it helps to motivate the people to do sports. It is a supplementary means of transportation to public transport, i.e., can solve the last mile problem if a sufficient density of the system is considered. For more information on this topic see also [3].

Most PBSs consist of rental stations which are distributed over a city or an area of a city. Usually, each rental station is equipped with a self service computer terminal. Moreover, each station has a particular number of slots where bikes are "docked" into. This is a mechanism introduced to prevent theft and vandalism of the bikes. Customers can rent bikes at any station of the system and can return them at any station they want. Problematic scenarios arise, when a customer arrives at a station with the intention to rent a bike and there is no bike available, or the even worse case, when she or he aims to return a bike at a target station that has no free slots available. In the latter case, the customer has to look for an alternative station with free slots. These two scenarios can never be fully eliminated, but it the goal of a PBS to minimize such cases because they annoy customers. Much work about rebalancing PBSs is available such as [7,8,13]. But instead of planning rebalancing routes, In this work we concentrate on planning station locations and station configurations for new or existing PBSs, however, also considering the rebalancing as an important factor.

When considering the setup of a PBS, the first aspect is where to build stations of the system such that the usage, i.e., the number of trips during a "usual" day is maximized, and at the same time, a given available budget is considered. Operators of a PBS usually have a particular amount of money available for a given planning horizon. Thus, it is also important to estimate rebalancing and maintenance costs over the whole planning horizon. Most often it is not possible to build an arbitrary number of slots per station but a small amount of possible station configurations is given.

Obviously, there are many factors influencing the decision where to position the stations inside a given area. First of all, there are the commuters and students who most often travel in the morning to their job and in the afternoon to their home. Thus, for instance, large housing complexes, and large train stations play a major role in a PBS network. Also, the time of the day is significant when estimating the demand in a particular area. Especially, due to commuters, a morning and afternoon peak of demands is typically measurable. Other aspects to consider, when estimating demands of a city, are local recreational areas, shops, clubs, bars, etc. Of course, demand also varies with the weather, working day, and special events, but when planning PBS we assume a regular working day with good weather where the demand can be assumed to be highest.

In this work we consider real-world instances of the Bike-Sharing Station Planning Problem (BSSPP) from the city of Vienna. We do not only compute optimal station locations but also consider the rebalancing effort and compute suggestions for initial inventory at the bike-sharing stations. We introduce a refinement step based on local-search techniques and are given possible station configurations, and the prices therefore. Results are shown for up to 4000 prospective station candidates and customer cells.

The remainder of this paper is structured as follows. In Sect. 2, related literature is examined in detail. Then, in Sect. 3, the problem definition is given. The multilevel refinement approach is explained in Sect. 4 and computational results are presented in Sect. 5. The paper ends with conclusions in Sect. 6.

2   Related Work

There already exists some work on the BSSPP, however, most of this previous work only concentrates on small examples or small regions of cities. In contrast, we aim in this work at solving scenarios with thousands of possible locations for stations originating from real-world data.

Most of the previous approaches are not entirely coherent and consider different constraints and optimization goals which makes direct comparisons impossible. Many approaches utilize mixed integer (non-)linear programming (MIP) as core technology. However, for larger instances it is usually impossible to solve such compact models exactly. Nevertheless, MIP technology can also be used inside (meta)heuristic methods.

One of the first approaches was published by Yang et al. [19] in 2010. In their problem formulation, origin/destination pairs with given demands, and potential locations for bike-sharing stations are given. The goal is to find optimal positions for bike-sharing stations but also to decide, where to build bike lanes. The objective is to minimize foot paths, fixed costs for rental stations, bike inventory costs, and a penalty introduced for uncovered demand. The authors solve the problem by a two-phase approach. The first part is a heuristic determining a set of rental stations to be opened and the second part is a kind of solution evaluation by computing a shortest path for origin/destination pairs. The authors illustrate their approach on an example with 11 station candidates.

Lin et al. [10] propose a mixed integer non-linear programming model for the problem formulation from [19]. For solving the mathematical model they used the LINGO solver. The authors show a solution for an instance consisting again of 11 station candidates. Moreover, they provided also a sensitivity analysis with respect to the following parameters: fixed costs for stations, penalty costs for uncovered demands, construction costs for bike lanes, bike riding speed, and the availability rate of bikes at stations.

Martinez et al. [12] propose a hybrid approach consisting of a hourly MIP model which is embedded into a heuristic approach. They aim to maximize the net revenue of the system and also consider rebalancing costs but the latter are not explicitly modeled and only estimated beforehand. The authors aim to solve a real-world problem in Lisbon where they consider a relatively small part of the city with 565 potential station locations. Results for four different scenarios with various parameters are shown.

Lin et al. [11] again use the same problem formulation as given in [10] and state a mixed-integer non-linear programming model for it. As they aim to compute a minimum inventory to fulfill a certain service level, the model is non-linear and they conclude that it is not exactly solvable in practice and thus, propose a heuristic approach for solving the problem. The authors introduce a greedy algorithm which starts by opening all possible station candidates and building all possible bicycle lanes between stations. Then, they alternatively close bike stations and bike lanes that result in a largest cost reduction but are still possible to close with respect to the minimum service level requirement. This procedure is iterated until some termination criterion is met. For evaluating a solution, Dijkstras shortest path algorithm is used to compute paths between all origin/destination pairs. A small test instance with 11 possible station candidates is given and different scenarios are evaluated to provide a sensitivity analysis.

Saharidis et al. [15] propose a pure MIP formulation to the BSSPP in a case study for the city of Athens. They present a time-discretized model in one hour steps and aim to minimize the total walking time of the users of the PBS, and the unfulfilled demand of the system. Demands in the city are estimated by analyzing the usage patterns of the Vélib' system in Paris. The considered instance for the case study is small and only considers 50 prospective candidate stations. The authors have been able to solve the provided instance with CPLEX and considered two case studies: one which assumes that the new PBS will be used heavily by the Athenians, and a second one, which assumes that the PBS is not that popular among the Athenian population.

Hu et al. [6] present a small case study for establishing a PBS along a metro line. Different to most other work they aim at cost minimization, whereas other work aims at maximizing net revenue, fulfilled customer demand or minimizing unfulfilled demand. The MIP model is rather simple, and the only considered constraints in their MIP model are the minimum number of stations that have to be built. The authors show results for different scenarios based on a dataset of ten possible station candidates.

Kloimüllner and Raidl [9] present a novel approach to the BSSPP utilizing hierarchical clustering and present an algorithm based on the multilevel refinement paradigm. They introduced two linear programming (LP) models where one is designed to estimate the maximum satisfied demand and the other is used to estimate rebalancing effort in the prospective system. With these two basic ingredients they propose a multilevel refinement approach where the initialization is solved by a full MIP model and the extension phases are solved by reduced MIP models. However, the authors do not consider station configurations and have not implemented a refinement procedure. Tests have been performed on randomly generated instance data. Straub et al. [16] present a semi-automated planning tool based on the optimization algorithm by Kloimüllner and Raidl and show the development procedure from the requirement analysis to the frontend of the semi-automated planning tool.

There exists also other literature related to this topic. In particular, the problem belongs the class of *facility location problems* [14] and more generally to *hub location problems* [2,4]. Moreover, a related and currently hot topic is the optimal placement of stations in car sharing systems [1]. Gavalas et al. [5], for instance, summarized diverse algorithmic approaches for the design and management of vehicle-sharing systems (e.g., car sharing and PBS).

We conclude, that with the exception of our previous work [9] all other works on combinatorial optimization approaches for designing PBS only consider rather small scenarios. Most previous work accomplish the optimization with compact mathematical models directly approached by a MIP solver. We think that such methods are unsuited for tackling large realistic scenarios with 2,000 cells or more, as such approaches did only solve small problem sizes in the current state-of-the-art. We extend our multilevel refinement heuristic by using station configurations instead of integral slots per station, propose a local search as refinement heuristic in the multilevel refinement paradigm and provide results for instances based on real-world data of the city of Vienna.

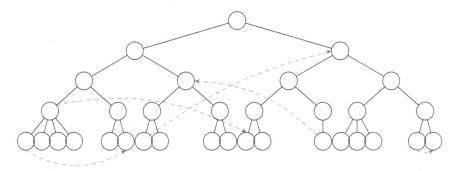

Fig. 1. Example of a hierarchical clustering and the corresponding graph G^t.

3   Problem Definition

We consider a geographical area which is partitioned into discrete cells. Let S denote the set of possible candidate cells where stations may be located, and let V be the set of cells containing some positive demand of prospective customers. Furthermore, let $m = |S|$ and $n = |V|$. In the simplest case $S = V$, in each station with positive customer demand, a station might be located, but this does not necessarily be the case.

As it is meaningful to define cells with about 150×150 m, the input size grows rapidly in larger cities. In particular, it is not meaningful and not practicable anymore to consider the full origin-destination demand matrix. Instead we use a hierarchical abstraction as data structure to be able to have a chance for solving such instances.

This hierarchical clustering is given as a rooted tree where the leaves of the tree correspond to single cells and all other cells correspond to clusters of cells of the geographical area. Let h denote the height of the tree and let $C = C_0 \cup \ldots \cup C_h$ be the set of all nodes of the tree. By C_d we denote all nodes at height $d = 0, \ldots, h$ of the tree. Set C_0 only contains the root node representing the whole considered area and C_h is the set of all original cells contained in the geographical area, i.e., $C_h = V$. For convenience, we define super$(p) \in C$ to return the parent cluster of node $p \in C \setminus \{0\}$, where 0 is the root node, and sub$(p) \subset C$ to return the children of cluster $p \in C \setminus V$.

To model varying demand throughout the day, we consider by $T = \{1, \ldots, \tau\}$ a set of time periods. We derive a weighted directed graph for each time period $t \in T : G^t = (C^t, A^t)$ of the full demand matrix by Algorithm 1. An example of the graph G^t on the hierarchically clustered input data is shown in Fig. 1. A full demand matrix is a matrix where each pair $u \in V, v \in V$ is assigned a prospective demand value which could also be 0, i.e., $d_{u,v}^t = x \mid x \in \mathbb{R}^+, t \in T, u \in V, v \in V$. The demand from node $v \in V$ to cluster $p \in C$ in time period $t \in T$ is denoted by $d_{v,p}^t > 0$. For convenience, we define $V(p) \subseteq V$ to be subset of all leaf nodes rooted under subtree $p \in C \setminus V$. Moreover, the set $C(p)$ contains all nodes which are part of the subtree rooted under p, also including $V(p)$ and p itself. The derivation of the arc set A^t of graph $G^t, \forall t \in T$ can be found in Algorithm 1 having parameters G^{in}, the full demand matrix input graph and θ, a parameter which defines the minimum demand for an arc in the newly constructed graph.

Algorithm 1. Derivation of A^t from original graph $G^{in} = (V^{in}, A^{in})$

Require: graph of full demand matrix $G^{in} = (V^{in}, A^{in}) \mid V^{in} = V^t$, threshold θ
1: $A^t = A^{in}$
2: **for all** $l \in \{h-1, \ldots, 0\}$ **do**
3:   **for all** $c \in C_l$ **do**
4:     **for all** $v \notin V(c)$ **do**
5:       $w^+ = 0, w^- = 0$
6:       **for all** $(v, p) \in A^t \mid p \in C(c) \cap C_{l+1}$ **do**
7:         **if** $d_{v,p} < \theta$ **then**
8:           $w^- = w^- + d_{v,p}^t$
9:           $A^t = A^t \setminus (v, p)$
10:        **else**
11:          $A^t = A^t \cup (v, p)$
12:        **end if**
13:      **end for**
14:      **for all** $(p, v) \in A \mid p \in C(c) \cap C_{l+1}$ **do**
15:        **if** $d_{p,v} < \theta$ **then**
16:          $w^+ = w^+ + d_{p,v}^t$
17:          $A^t = A^t \setminus (p, v)$
18:        **else**
19:          $A^t = A^t \cup (p, v)$
20:        **end if**
21:      **end for**
22:      **if** $w^- > 0$ **then**
23:        $A^t = A^t \cup (v, c)$ with $d_{v,c}^t = w^-$
24:      **end if**
25:      **if** $w^+ > 0$ **then**
26:        $A^t = A^t \cup (c, v)$ with $d_{c,v}^t = w^+$
27:      **end if**
28:    **end for**
29:   **end for**
30: **end for**

To keep this demand graph as sparse as possible in order to be able to solve large instances, we define a set of rules for the demands in the input graph. It is not allowed that there exists demand from a node v to one of its predecessors p, i.e., $d_{v,p}^t > 0 \mid v \in \mathrm{sub}(p)$ is not possible. Self-loops, however, are a special case, and are explicitly allowed to model trips of customers within a cell or cluster. There must not exist arcs with negligible demand, i.e., arcs with $d_{v,p}^t < \theta$ where θ is a predefined threshold, are not allowed. These "low" demands have to be subsumed to a bigger demand at a higher level of the clustering.

Theorem 1. *If there exists already an arc with weight $d_{v,p}^t \geq \theta$, there cannot exist an arc $(v, q) \mid p \in \mathrm{sub}(q) \vee p \in \mathrm{super}(q)$.*

This follows from Algorithm 1:

Proof. Given $(v, p) \in A^t$ with $d_{v,p}^t \geq \theta$ we distinguish the following two cases:

Case 1: Let $(v, q) \in A^t \mid p \in \mathrm{sub}(q)$. Since $p \in \mathrm{sub}(q)$ and $d_{v,p}^t \geq \theta$ no arc (v, q) is generated according to Line 7 of Algorithm 1.

Case 2: Let $(v, q) \in A^t \mid p \in \mathrm{super}(q)$. An arc $(v, p) \in A^t$ can only be generated if $d_{v,q}^t < \theta$ according to Line 7. In this case, however, arc $(v, q) \in A^t$ would be removed according to Line 9 of Algorithm 1. Moreover, if $d_{v,q}^t \geq \theta$, then arc (v, p) is not generated according to Line 7. Thus, only one of these arcs, either (v, p) or (v, q), can exist in the final arc set A^t. □

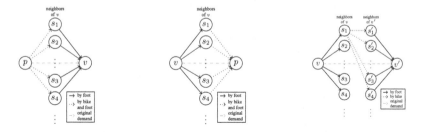

(a) An arc from a cluster node to a leaf node   (b) An arc from a leaf node to a cluster node   (c) An arc from a leaf node to another leaf node

Fig. 2. By modeling the neighbor stations for each customer cell, a separate network is created for every arc of graph G^t.

An important condition that needs to be ensured is that incoming and outgoing demands must be consistent. Therefore, for any $p \in C \setminus V$ the following two conditions must hold: $\sum_{(v,q) \in A^t \mid v \in V(p), q \notin C(p)} d_{v,q}^t \geq \sum_{(q,v) \in A^t \mid q \in C(p), v \notin V(p)} d_{q,v}^t$ and $\sum_{(q,v) \in A^t \mid q \notin C(p), v \in V(p)} d_{q,v}^t \geq \sum_{(v,q) \in A^t \mid v \notin V(p), q \in C(p)} d_{v,q}^t$. The former condition ensures that the total demand originating at the leaves of the subtree rooted at p and leading to a destination outside of the tree is never less than the total incoming demand at all the cells outside the tree originating from some cluster inside the tree. The latter condition provides a symmetric condition for the total incoming demand at all the leaves of the tree. Furthermore, for the root node $p = 0$ both inequalities must hold with equality.

An important fact for public bike-sharing systems (PBS) is that demand of prospective customers may not only be fulfilled by its own cell but also by neighbor cells within a reasonable walking distance. Thus, we define for each leaf node $v \in V$ a set of station cells $S(v) \subseteq S$ which are in the neighborhood of v and with which v's demand can at least be partly fulfilled. To model partly fulfilled demand, we introduce an *attractiveness value* $a_{v,s} \in (0, 1]$, $\forall v \in V$, $s \in S(v)$. This value determines the percentage of demand of customer cell v which can be maximally fulfilled at a station at location s. Note, that if $v = s$, then $a_{v,v} = 1$ will hold. The modeling of these neighbor cells is shown in Fig. 2.

For each cell $s \in S$, a set of possible station configurations $K_s = \{0, \ldots, \gamma_s\}$ is specified. The special configuration 0 always corresponds to the case that no station is built or an existing station is removed. Each other configuration $i \in K_s$ has associated the following values: The number of parking slots is defined by $b_{s,i}^{\text{ps}} \in \mathbb{N}$, the fixed costs, i.e., the costs for constructing the station including the purchase of a corresponding number of bikes are denoted by $b_{s,i}^{\text{cfix}} \geq 0$, and the variable costs, i.e., total maintenance and operating costs over the whole planning horizon, including the maintenance of a respective number of bikes are denoted by $b_{s,i}^{\text{cvar}} \geq 0$. Rebalancing costs are not included in the variable costs and are introduced later on. In case the cell contains an already existing station, $b_{s,i}^{\text{cfix}}$ corresponds to the costs for upgrading/downgrading or removing (if $i = 0$) the station.

Finally, we are given the following global parameters. Parameter $b^{\text{reb}} \geq 0$ represents the average costs for rebalancing a single bike per day over the whole planning horizon, whereas $B_{\max}^{\text{tot}}$ and $B_{\max}^{\text{fix}}$ represent the total maximum costs (budget) and the maximum fixed costs respectively. Basically, when considering two different kind of budgets, possible solution candidates may be excluded which could achieve better solution quality according to the objective function, but having two different budget types is a requirement from practice.

3.1 Solution Representation

A solution $x = (x_s)_{s \in S}$ with $x_s \in K_s$ assigns each station cell $s \in S$ a valid configuration $x_s \in K_s$ (which might also be the "no-station" configuration 0).

3.2 Optimization Goal

The goal is to maximize the total number of trips in the system, i.e., the total demand that can be fulfilled at each day over all time periods, considering a maximum total budget $B_{\max}^{\text{tot}}$ as well as a maximum budget for the overall fixed costs $B_{\max}^{\text{fix}}$ alone.

Let $D(x, t)$ be the total demand fulfilled by solution x in time period $t \in T$, and let $Q_x(s)$ be the required rebalancing effort arising at each station $s \in S \mid x_s \neq 0$ in terms of the number of bikes to be moved to some other station. The calculation of these terms via MIP models was already presented in detail in [9]. The corresponding optimization problem can then be stated as follows.

$$\max \sum_{t \in T} D(x, t) \tag{1}$$

$$\sum_{s \in S} \left(b_{s,x_s}^{\text{cfix}} + b_{s,x_s}^{\text{cvar}} + b^{\text{reb}} \cdot Q_x(s) \right) \leq B_{\max}^{\text{tot}} \tag{2}$$

$$\sum_{s \in S} b_{s,x_s}^{\text{cfix}} \leq B_{\max}^{\text{fix}} \tag{3}$$

$$x_s \in K_s \qquad\qquad s \in S \tag{4}$$

The objective function (1) maximizes the total satisfiable demand for each time period $t \in T$. The left side of Inequality (2) calculates the total costs by summing up the fixed and variable costs resulting from a change in station configurations and the rebalancing costs. Inequality (3) restricts the maximum fixed costs of the new system and in (4) domain definitions for the decision variables are given.

4   Multilevel Refinement Approach

In our opinion, for this type of problem, single construction heuristics based on greedy principles are not the ultimate choice as solution technique and neither is a direct application of classical local-search based metaheuristics, like variable neighborhood search and iterated local search. We think that these strategies are not able to grasp the connections and interactions between all the clusters and leaf nodes in large instances. In this context many local decisions are not the way to go. One has to find a solution method which is able to overlook the complete and complex problem more as a whole.

An intuitive way of achieving this, is to apply the so called *multilevel refinement approach*. Multilevel refinement was originally introduced by Walshaw [17,18]. Initially, it was thought as an additional ingredient to any metaheuristic to improve its solutions. However, we are going to use the approach as the main solution technique. This technique fits to the already given hierarchical

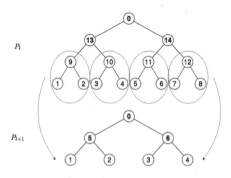

Fig. 3. Example of a coarsening step in the hierarchical clustering of customer cells. Leaf nodes are merged with their parent nodes to form new leaf nodes in the coarsened tree.

clustering of the geographic cells, as we can do the coarsening accordingly, level by level, until we reach a problem size that can be reasonably well solved. Then, we compute an initial solution by the *initialize* procedure and finally, *extend* and *refine* the solution until we obtain a solution to the original input instance. In the following the *coarsen*, *initialize*, *extend*, and *refine* functions are explained in detail.

4.1   Coarsening

In the coarsening, we iteratively merge neighboring clusters into larger clusters according to the already given hierarchical clustering.

Coarsening of Customer Cells: As illustrated in Fig. 3, in the coarsening step, customer cells are merged together into their parent cluster, so that the problem becomes smaller and gets easier solvable.

The outgoing demand of a node p that corresponds to the merging of nodes $V(p)$ is the demand $\sum_{(v,q)\in A^t|v\in V(p),q\notin C(p)} d_{v,q}^t$ and the incoming demand of p is the total demand of nodes $V(p)$, i.e., $\sum_{(q,v)\in A^t|v\in V(p),q\notin C(p)} d_{q,v}^t$.

Coarsening of Station Cells: Merging a set of station cells, each representing possible station configurations with different associated costs, is, however, not as straight-forward. Considering all possible combinations of station configurations appears not meaningful, since the resulting number of these combinations would grow exponentially with the number of merged original station configurations. Furthermore, in particular on higher abstraction levels, individual station configurations do not play a practically significant role anymore. A simpler approximate model for the number of possible parking slots and corresponding costs appears thus reasonable for practice.

We apply the following continuous linear model as approximation for the possibilities at each station cell $s \in S^l$. Let b_s^{ps} be the maximum number of bike parking slots at prospective station s. The possible number of bike parking slots at s can be chosen freely from the (continuous) interval $[0, b_s^{\mathrm{ps}}]$ where the upper bound b_s^{ps} will be chosen as explained in the following. A solution to P^l is a vector containing for each station cell the selected number of bike parking slots, i.e., $x = (x_s)_{s\in S^l}$ with $x_s \in [0, b_s^{\mathrm{ps}}]$. Costs are calculated in dependence of x_s as follows $b_s^{\mathrm{cfix}}(x_s) = b_s^{\mathrm{cfix,a}} \cdot x_s + b_s^{\mathrm{cfix,b}}$ and $b_s^{\mathrm{cvar}}(x_s) = b_s^{\mathrm{cvar,a}} \cdot x_s + b_s^{\mathrm{cvar,b}}$. For original station cells $s \in S^0$ we assume $b_s^{\mathrm{ps}} = \max_{i\in K_s} b_{s,i}^{\mathrm{ps}}$ and the model parameters $b_s^{\mathrm{cfix,a}}$, $b_s^{\mathrm{cfix,b}}$, $b_s^{\mathrm{cvar,a}}$, and $b_s^{\mathrm{cfix,b}}$ are determined as follows:

$$b_{s^{l+1}}^{\mathrm{cfix,a}} = \sum_{s^l \in \mathrm{sub}(s^{l+1})} b_{s^l}^{\mathrm{cfix,a}} \quad \text{and} \quad b_{s^{l+1}}^{\mathrm{cfix,b}} = \frac{1}{|\mathrm{sub}(s^{l+1})|} \sum_{s^l \in \mathrm{sub}(s^{l+1})} b_{s^l}^{\mathrm{cfix,b}} \tag{5}$$

$$b_{s^{l+1}}^{\mathrm{cvar,a}} = \sum_{s^l \in \mathrm{sub}(s^{l+1})} b_{s^l}^{\mathrm{cvar,a}} \quad \text{and} \quad b_{s^{l+1}}^{\mathrm{cvar,b}} = \frac{1}{|\mathrm{sub}(s^{l+1})|} \sum_{s^l \in \mathrm{sub}(s^{l+1})} b_{s^l}^{\mathrm{cvar,b}} \tag{6}$$

The aggregation of station nodes with their approximate cost models is then done as follows. Let $\mathrm{sub}(s^{l+1}) \subseteq S^l$ denote the set of all station cells at level l which are to be aggregated into $s^{l+1} \in S^{l+1}$. The maximum number of parking slots naturally is the sum of the maximum values over all station nodes to be aggregated $b^{\mathrm{ps}}_{s^{l+1}} = \sum_{s^l \in \mathrm{sub}(s^{l+1})} b^{\mathrm{ps}}_{s^l}$.

Attractiveness Values: As we coarsen the problem, we also have to aggregate the attractiveness values of merged stations for the merged customer cells. Hence, we take a weighted average value of all attractiveness values of all respective pairs of customer cells and station cells. In more detail, let $p^l, q^l \in C_{h-l-1}$ and v^{l+1} the node arising from merging all the nodes contained in the subtree rooted at p^l and let s^{l+1} be the resulting station from merging the stations in the subtree rooted at q^l. Then, we compute the attractiveness values as follows

$$
a_{v^{l+1},s^{l+1}} = \begin{cases} 1 & \text{if } v = s \\ \dfrac{\sum_{v \in V(p)} \sum_{s \in V(q)} \left(a_{v,s} \cdot \max_{i \in K_s} \{b^{\mathrm{ps}}_{s,i}\} \right)}{\sum_{v \in V(p)} \sum_{s \in V(q)} \max_{i \in K_s} \{b^{\mathrm{ps}}_{s,i}\}} & \text{if } l = 0, v \neq s \\ \dfrac{\sum_{v^l \in V^l(p^l)} \sum_{s^l \in V^l(q^l)} \left(a_{v^l,s^l} \cdot b^{\mathrm{ps}}_{s^l,l} \right)}{\sum_{v^l \in V^l(p^l)} \sum_{s^l \in V^l(q^l)} b^{\mathrm{ps}}_{s^l,l}} & \text{if } l > 0, v \neq s \end{cases}
\tag{7}
$$

It is impossible to keep the original attractiveness values when coarsening the problem. Of course, we have to find a method to keep deviation to original attractiveness values to a minimum. We take the average between all pairs of customer cells and station cells weighted by the maximum possible parking slots at the particular stations. At level 0 we have to compute this value and find the maximum through all configurations K_s for the station s, and for all other levels we simply take the maximum parking slots as computed by the coarsening of the stations. If the customer cell v and the station cell s refer to the same cell, we set the attractiveness value to 1 which implies that every demand within this cell, i.e., self loops, are always satisfiable.

4.2   Initialization

We have to initialize the solution at some reasonably coarsened level. We solve it via a MIP model when an appropriate level l, where the problem is well solvable, is reached. Parts of the model have been already shown, like calculating fulfillable demands or rebalancing costs in [9] and the optimization goal in Sect. 3.2.

As it is not important and also not meaningful to make decisions about exact station configurations in all levels $l > 0$ we use continuous variables $x^{\mathrm{flex}}_{s,l}$ for computing a completely flexible number of slots per station. The value 0 means no station is going to be built. Note, that variables are also indexed by the level $l \geq 0$ for which we compute the number of slots for the various stations. However, it is important that all stations which are going to be planned have a minimum number of slots which we denote as $x^{\mathrm{flex}}_{\min}$. We need this condition because it is only meaningful to plan a number of slots for a specific station under a particular minimum slot count. If the minimum slot count of some station configuration at level $l = 0$ is greater than the slot count planned at level

$l = 1$ the information obtained through the coarsening and extension steps would be useless since no station can be built in this scenario. Due to this constraint, we introduce additional variables $g_s \in \{0, 1\}$ that decide whether a station is to be built in cell $s \in S$ or not.

The calculation of the demand is done by the linear program $D(x, t)$ defined in [9] but extended with an additional index for modeling the time periods. Rebalancing effort is determined by the linear program $Q_x(s)$ which is also proposed in [9]. For the initialization it is also enriched by an additional index regarding the various prospective station candidates. By $A^{p,q}_{f,l}$ we denote the flow network of nodes/clusters (p, q) in level l, see also Fig. 2 for a visualization of these networks. The set S^l corresponds to a set of possible station candidates at level l. The MIP model for initialization of the solution at the coarsest level $l^{\max}$ is finally defined as follows:

$$\max \quad \sum_{t \in T} \left(\sum_{(v,p) \in A^{t,l^{\max}} | v \in V^{l^{\max}}} \sum_{(v,s) \in A^{v,p}_{f,l^{\max}}} f^{t,v,p}_{v,s} \right) \tag{8}$$

$$\text{s.t.} \quad \text{inequalities from } D(x, t) \text{ hold, see [9].} \tag{9}$$

$$\text{inequalities from } Q_x(s) \text{ hold, see [9].} \tag{10}$$

$$\sum_{s \in S} \left(b^{\text{cfix}} \cdot x^{\text{flex}}_{s,l^{\max}} + b^{\text{cvar}} \cdot x^{\text{flex}}_{s,l^{\max}} + b^{\text{reb}} \cdot \left(\sum_{t \in T} (r^+_{t,s} + r^-_{t,s}) + r^o_s \right) \right)$$
$$\leq B^{\text{tot}}_{\max} \tag{11}$$

$$\sum_{s \in S} b^{\text{cfix}} \cdot x^{\text{flex}}_{s,l^{\max}} \leq B^{\text{fix}}_{\max} \tag{12}$$

$$x^{\text{flex}}_{s,l^{\max}} \geq g_s \cdot x^{\text{flex}}_{\min} \quad \forall s \in S_{l^{\max}} \tag{13}$$

$$x^{\text{flex}}_{s,l^{\max}} \leq g_s \cdot x^{\text{flex}}_{\min} \quad \forall s \in S_{l^{\max}} \tag{14}$$

$$x^{\text{flex}}_{s,l^{\max}} \geq 0 \quad \forall s \in S_{l^{\max}} \tag{15}$$

$$g_s \text{ binary} \quad \forall s \in S_{l^{\max}} \tag{16}$$

$$f^{t,v,p}_{\alpha,\beta} \geq 0 \quad \forall t \in T, (p,q) \in A_{l^{\max}}, (\alpha, \beta) \in A^{p,q}_{f,l^{\max}} \tag{17}$$

$$r^+_{t,s}, r^-_{t,s}, y_{t,s} \geq 0 \quad \forall s \in S_{l^{\max}}, t \in T \tag{18}$$

$$r^o_s \geq 0 \quad \forall s \in S_{l^{\max}} \tag{19}$$

The objective (8) is to maximize the overall prospective demand. All inequalities defined in the maximum flow polytope (9) and all inequalities of the rebalancing polytope (10) must hold. The total budget (11) and the maximum allowed fixed budget (12) have to be respected. Inequalities (13) and (14) are used to ensure that a minimum number of stations is going to be built, if a station should be built in the particular station cell $s \in S$. Domain definitions are given in (15)–(19). All variables are continuous except the $g_s \mid s \in S$ variables which are binary decision variables.

At an appropriate level of the coarsening, the proposed MIP model is able to yield near optimal solutions in a reasonable runtime when solved with a state-of-the-art MIP solver such as Gurobi. As the underlying mathematical model is difficult to solve because of numerical issues and precision, it is useful to give the MIP solver a reasonable optimality gap. The actual optimality gap could vary from problem to problem and should be set according to practical observations. This initial solution is then further extended using the extension algorithm proposed in the next section.

4.3 Extension

The extension is done heuristically as in the lower levels the instance size gets larger and utilizing a MIP formulation becomes time consuming. We use a simple heuristic which is fast as after the extension step the solution is refined using local search.

For each station in the upper level we try to distribute the number of slots over the corresponding children of this cluster node in the lower level. We make use of a priority queue which is sorted according to the highest possible demand which can be fulfilled by each node in the lower level. Thus, we compute the demand LP $D(x,t)$ from [9] by setting the solution vector x to the highest possible slot count so that we know how much demand this node would fulfill in the ideal case. At the end, the new solution for the lower level may be infeasible due to violations of the budget restrictions. If this is the case, we try to iteratively remove stations until we again reach a valid solution. For removal, we choose the station with the lowest satisfied demand. At the end of the algorithm we obtain a valid solution for the lower level.

4.4 Refinement

As the solution evaluation is done by executing two LP models from [9], namely $D(x,t)$ to calculate the demand and $Q_x(s)$ to estimate the rebalancing costs, it is expensive. Therefore, refinement is done by local search, but only a limited amount of time. The following neighborhoods are iterated in the given, static order until a predefined time limit or local optimum has been reached.

Change station: This neighborhood removes a station, i.e., assigns a station the 0 configuration and assigns another customer cell which has currently a 0 configuration some other configuration.

Add station: This neighborhood assigns a station, which has currently the 0 configuration, a new one. Here, also to concentrate on promising stations, the station among the 0 configuration stations is chosen, which has the highest demand.

Change configuration: This neighborhood changes the configuration of an arbitrary station candidate.

We consider only moves that construct valid solutions as it does not make sense to work with infeasible moves/solutions as repairing infeasible solutions consumes too much time due to the expensive evaluation function. Fixed and variable costs are checked before solution evaluation to avoid calling the LP for infeasible solutions. However, it can happen that newly computed solution is still infeasible when considering rebalancing costs. If this is the case, the solution is simply discarded.

5   Computational Results

We derived our test instances[1] from real-world data of the city of Vienna. The whole city was partitioned into cells and a full demand matrix on these cells is given. Different instance sizes have been generated by picking some cells from the whole instance. The clustering on these instances was computed using the Nearest Point Algorithm. Consider cluster p_u and cluster p_v, where $V(p_u) = \{u_1, \ldots, u_n\}, V(p_v) = \{v_1, \ldots, v_m\}$. Among the remaining clusters, the nearest neighbor is chosen as follows: $d_{p_u,p_v} = \min_{u \in V(p_u), v \in V(p_v)} \{dist(u,v)\}$. The graph $G^t = (C^t, A^t)$ was derived by utilizing Algorithm 1. Therefore, we set $\theta = 0.01$. The attractiveness values have also been given as input as a full matrix of attractiveness values between the different cells. However, to have computationally tractable instances we consider only subset of neighbors for a single cell. For instances with 20 and 50 customer cells we consider 10 neighbors for each cell, for instances with 300, 500 and 750 cells, 7 neighbors, for instances with 1,000 and 2,000 cells, 5 neighbors and for instances with 2,500, 3,000, 3,500, and 4,000 cells 3 neighbors. The neighbors with highest attractiveness values are taken. Configurations have been set as follows for all customer cells of the instances: $\forall s \in S$: $K_s = \{(0,0,0), (10, 17500, 10000), (20, 25000, 20000), (40, 30000, 40000)\}$. These tuples represent the number of slots to be built for the particular configuration, its fixed and variable costs. Rebalancing a single bike for a single day has been estimated with 3 € which is $b^{\text{reb}} = 365 \cdot 3 = 1095$ € for a whole year which is equal to the planning horizon. As the visualization component (see later) uses only one time interval, we also use only one time interval for our computational tests and set its duration to 1440 min. We introduced a flexible number of time intervals as it is important for practice. In our first experiments, however, we use only one time interval because of the visualization component. This time interval uses an average prospective user demand over a whole day. We stop the coarsening step when we reached a maximum number of 128 customer cells. We set the time limit for the local search to 15 s. The optimality gap for solving the initial MIP model at the coarsest level is set to five percent.

All algorithms are implemented in C++ and have been compiled with gcc 5.5.0. For solving the LPs and MIPs we used Gurobi 7.0. All experiments were executed as single threads on an Intel Xeon E5540 2.53 GHz Quad Core processor. We executed a single run per instance which seems meaningful for a long-term planning problem.

---

[1]   https://www.ac.tuwien.ac.at/files/resources/instances/bsspp/lion19.bz2.

Table 1. Results for the multilevel refinement heuristic.

	Instance		Multilevel Refinement Heuristic				
#cells	B_{max}^{tot} [€]	B_{max}^{fix} [€]	obj	#coarsen	time [s]	totcost [€]	fixcost [€]
20	100,000.00	80,000.00	0.79	2	167.03	82,598.27	52,500.00
50	200,000.00	160,000.00	45.85	2	3,775.28	200,000.00	122,500.00
300	1,200,000.00	960,000.00	160.18	4	4,635.18	1,200,000.00	752,500.00
500	2,200,000.00	1,760,000.00	451.90	5	8,720.94	2,115,602.79	1,155,000.00
750	3,350,000.00	2,680,000.00	665.62	7	8,588.95	2,428,067.58	1,365,000.00
1,000	4,500,000.00	3,600,000.00	876.77	6	6,532.26	3,343,472.61	1,872,500.00
2,000	9,000,000.00	7,200,000.00	1,125.29	8	11,460.01	6,565,538.53	3,920,000.00
2,500	10,000,000.00	8,000,000.00	1,214.90	8	12,434.51	7,707,661.10	4,462,500.00
3,000	12,000,000.00	9,600,000.00	1,315.58	9	12,464.25	9,321,026.50	5,460,000.00
3,500	15,000,000.00	12,700,000.00	1,435.98	9	13,255.25	11,277,185.57	6,702,500.00
4,000	17,000,000.00	13,600,000.00	1,213.81	9	13,518.10	10,841,911.99	6,510,000.00
		Average	773.33	6.27	8,686.52	3,265,499.73	3,101,000.00

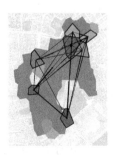

Fig. 4. Solution visualization.

Result of the proposed method are shown in Table 1. For each instance we show the number of customer cells/station candidates (#cells), the maximum total budget (B_{max}^{tot}), the maximum budget for the fixed costs (B_{max}^{fix}), the objective value of the solution, i.e., number of trips (obj), the number of levels to be coarsened in the clustering tree such that we reach our goal, the number of nodes to initialize the solution (#coarsen), the time needed to find the solution (time), the total (totcost) and the fixed costs (fixcost) of the found solution. As shown in Table 1, the approach is able to solve instances derived from real-world data with up to 4,000 customer cells. Interesting is that for the instance with 1,000 customer cells we need less coarsening steps than for the instance with 750 customer cells, but this also strongly depends on the clustering and the depth of the original cluster tree. The second thing which looks interesting is that the fulfilled demand of the instance with 4,000 cells is less than the fulfilled demand for some instances with fewer customer cells. This happens because the refinement, i.e., local search, has fewer iterations for the instance with 4,000 customer cells and the *addStation* neighborhood is often successful. A detailed visualization of the solution found for the instance with 50 customers cells is given in Fig. 4. The visualization tool was written by Markus Straub from the Austrian Institute of Technology[2]. The darker a cell, the more demand is fulfilled and the bolder a line, the more demand is going through this connection.

6   Conclusion

Based on previous work we developed practical as well as algorithmic extensions. We introduced an algorithm which constructs a sparse graph on clustered input data from the original full demand matrix. This drastically reduces the input size of the problem instance. We developed a refinement method based on local search for the multilevel refinement approach which improves the solutions after each extension step in the algorithm. In this work we consider particular station

---

[2]   https://www.ait.ac.at/.

configurations instead of an arbitrary number of slots which is much more realistic as only a full configuration can be bought in practice. This introduces higher complexity in the multilevel refinement approach, which we have successfully shown to work within the algorithm. Moreover, we derived instances which are based on real-world data from the city of Vienna. Results seem reasonable and we were able to solve instances with 4,000 prospective station candidates which demonstrate scalability of the proposed approach. Using the provided visualization the solutions can also be visually verified.

In future work we want to extend our benchmark suite with also other cities like e.g., Linz in Austria. Furthermore, the whole Vienna instance consists of 7,216 prospective stations cells and it is interesting which instance sizes the proposed multilevel refinement can solve. Moreover, it is also interesting to study alternatives and improvements to the extension heuristic and the refinement part of the algorithm. Last but not least tests on small instances can be performed where exact solutions are compared to solutions of the multilevel refinement heuristic.

References

1. Biesinger, B., Hu, B., Stubenschrott, M., Ritzinger, U., Prandtstetter, M.: Optimizing charging station locations for electric car-sharing systems. In: Hu, B., López-Ibáñez, M. (eds.) EvoCOP 2017. LNCS, vol. 10197, pp. 157–172. Springer, Cham (2017). https://doi.org/10.1007/978-3-319-55453-2_11
2. Contreras, I.: Hub location problems. In: Laporte, G., Nickel, S., da Gama, F.S. (eds.) Location Science, pp. 311–344. Springer, Cham (2015). https://doi.org/10.1007/978-3-319-13111-5_12
3. DeMaio, P.: Bike-sharing: history, impacts, models of provision, and future. Public Transp. 12(4), 41–56 (2009)
4. Farahani, R.Z., Hekmatfar, M., Arabani, A.B., Nikbakhsh, E.: Hub location problems: a review of models, classification, solution techniques, and applications. CAIE 64(4), 1096–1109 (2013)
5. Gavalas, D., Konstantopoulos, C., Pantziou, G.: Design and management of vehicle sharing systems: A survey of algorithmic approaches. In: Obaidat, M.S., Nicopolitidis, P. (eds.) Smart Cities and Homes: Key Enabling Technologies, pp. 261–289. Elsevier Science (2016)
6. Hu, S.R., Liu, C.T.: An optimal location model for a bicycle sharing program with truck dispatching consideration. In: IEEE 17th International Conference on Intelligent Transportation Systems (ITSC), pp. 1775–1780. IEEE (2014)
7. Kloimüllner, C., Papazek, P., Hu, B., Raidl, G.R.: Balancing bicycle sharing systems: an approach for the dynamic case. In: Blum, C., Ochoa, G. (eds.) EvoCOP 2014. LNCS, vol. 8600, pp. 73–84. Springer, Heidelberg (2014). https://doi.org/10.1007/978-3-662-44320-0_7
8. Kloimüllner, C., Raidl, G.R.: Full-load route planning for balancing bike sharing systems by logic-based benders decomposition. Networks 69(3), 270–289 (2017)
9. Kloimüllner, C., Raidl, G.R.: Hierarchical clustering and multilevel refinement for the bike-sharing station planning problem. In: Battiti, R., Kvasov, D.E., Sergeyev, Y.D. (eds.) LION 2017. LNCS, vol. 10556, pp. 150–165. Springer, Cham (2017). https://doi.org/10.1007/978-3-319-69404-7_11

10. Lin, J.R., Yang, T.H.: Strategic design of public bicycle sharing systems with service level constraints. Transp. Res. E-Log. **47**(2), 284–294 (2011)
11. Lin, J.R., Yang, T.H., Chang, Y.C.: A hub location inventory model for bicycle sharing system design: formulation and solution. CAIE **65**(1), 77–86 (2013)
12. Martinez, L.M., Caetano, L., Eiró, T., Cruz, F.: An optimisation algorithm to establish the location of stations of a mixed fleet biking system: an application to the city of Lisbon. Procedia Soc. Behav. Sci. **54**, 513–524 (2012)
13. Rainer-Harbach, M., Papazek, P., Hu, B., Raidl, G.R., Kloimüllner, C.: PILOT, GRASP, and VNS approaches for the static balancing of bicycle sharing systems. JOGO **63**(3), 597–629 (2015)
14. ReVelle, C.S., Eiselt, H.A.: Location analysis: a synthesis and survey. EJOR **165**(1), 1–19 (2005)
15. Saharidis, G., Fragkogios, A., Zygouri, E.: A multi-periodic optimization modeling approach for the establishment of a bike sharing network: a case study of the city of Athens. In: Proceedings of the International MultiConference of Engineers and Computer Scientists 2014, vol. II, No. 2210, pp. 1226–1231. LNECS. Newswood Limited (2014)
16. Straub, M., et al.: Semi-automated location planning for urban bike-sharing systems. In: Proceedings of the 7th Transport Research Arena (TRA 2018), pp. 1–10, Vienna, Austria (2018)
17. Walshaw, C.: A multilevel approach to the travelling salesman problem. Oper. Res. **50**(5), 862–877 (2002)
18. Walshaw, C.: Multilevel refinement for combinatorial optimisation problems. Ann. Oper. Res. **131**(1), 325–372 (2004)
19. Yang, T.H., Lin, J.R., Chang, Y.C.: Strategic design of public bicycle sharing systems incorporating with bicycle stocks considerations. In: 40th International Conference on Computers and Industrial Engineering (CIE), pp. 1–6. IEEE (2010)

Asymptotically Optimal Algorithms for the Prize-Collecting Traveling Salesman Problem on Random Inputs

Edward Kh. Gimadi[1,2] and Oxana Tsidulko[1,2(✉)]

[1] Sobolev Institute of Mathematics, Novosibirsk, Russia
[2] Department of Mechanics and Mathematics, Novosibirsk State University, Novosibirsk, Russia
{gimadi,tsidulko}@math.nsc.ru

Abstract. The Prize-Collecting Traveling Salesman Problem is a class of generalizations of the classic Traveling Salesman Problem (TSP) where it is not necessary to visit all the vertices. Given the edge costs and a certain profit associated with each vertex, the goal is to find a route which satisfies maximum collected profit and minimum traveling costs constraints. We show polynomial-time approximation algorithms for two variants of the problem and establish conditions under which the presented algorithms are asymptotically optimal on random inputs.

Keywords: Asymptotically optimal algorithm · Random inputs · NP-hard problem · Prize-collecting TSP

1 Introduction

The Traveling Salesman problem (TSP) is one of the most well-known and widely studied NP-hard combinatorial optimization problems. The classic formulation states as follows: given a complete graph and the costs of edges, the goal is to find the cheapest simple cycle visiting all the vertices. TSP has numerous applications in planning, logistics, routing, scheduling, data mining, clustering [9].

Often in applications one cannot or do not want to visit all vertices of the graph due to different time or budget constrains. For example, the TSP solutions can be used in solving clustering problems [3,12], in this case the vertices represent the objects and the costs of edges represent the measure of similarity between the objects. However, since data is often noisy, there might be outliers that should be discarded from a good solution. One of the natural special cases of the TSP that captures such features is the prize-collecting TSP, where in addition for each vertex v a value $p(v) \in \mathbb{R}$ is given, which represents the prize one gets for visiting vertex v or the penalty one pays for not visiting v [1,2,4,5,11].

Supported by the program of fundamental scientific researches of the SB RAS, project 0314-2019-0014, and by the Ministry of Science and Higher Education of the Russian Federation under the 5-100 Excellence Programme.

N. F. Matsatsinis et al. (Eds.): LION 13 2019, LNCS 11968, pp. 201–207, 2020.
https://doi.org/10.1007/978-3-030-38629-0_16

In the sense of the above clustering example, the vertex prizes might characterize the probability of an object not to be an outlier.

In [10,11] four most common objectives for the prize-collecting problems were mentioned: the *Goemans-Williamson* objective that minimizes the cost of the tour plus a penalty for not visited vertices, the *Net Worth* objective that maximizes the prize of visited vertices minus the cost of used edges, the *Quota* objective that minimizes the cost of a tour containing at least Q vertices, and the *Budget* objective that maximizes the number of vertices in the tour subject to the cost of the tour being at most B. All the problems of this kind are trivially NP-hard.

In this short paper we consider the following cases of the prize-collecting TSP:

Problem 1 (m-Quota TSP).

Input: A complete n-vertex undirected graph $G = (V, E)$, positive integer $m \le n$, costs of edges $c : E \to \mathbb{R}_{\ge 0}$, vertex prizes $p : V \to \mathbb{R}_{\ge 0}$, and a profit $Q \in \mathbb{R}_{\ge 0}$.
Find: A simple cycle H with at least m vertices such that $\sum_{v \in V(H)} p(v) \ge Q$ and $F_1(H) := \sum_{e \in E(H)} c(e) \to \min$.

Problem 2 (m-Net Worth prize-collecting TSP, m-NW-TSP).

Input: A complete n-vertex undirected graph $G = (V, E)$, costs of edges $c : E \to \mathbb{R}_{\ge 0}$, vertex prizes $p : V \to \mathbb{R}_{\ge 0}$, and a positive integer $m \le n$.
Find: A simple cycle H with at least m vertices such that $F_2(H) := \sum_{v \in V(H)} p(v) - \sum_{e \in E(H)} c(e) \to \max$.

A number of previous results [6–8] present analysis of a simple greedy heuristic called the Nearest City algorithm for the classic minimum TSP and prove the conditions under which it gives asymptotically optimal solutions for the TSP on random inputs. Given an n-vertex complete weighted graph, the Nearest City algorithm starts with an arbitrary vertex v_1 and at each step extends the partial chain $C = \{v_1, \ldots, v_k\}$ by adding a new vertex v_{k+1} such that the edge (v_k, v_{k+1}) is the shortest possible. Obviously, the time-complexity of the Nearest City algorithm is $O(n^2)$.

We show how the greedy algorithm can be modified to give asymptotically optimal solutions for the two versions of prize-collecting TSP stated above.

2   Greedy Approximation Algorithms

In this section we present two greedy algorithms that approximately solve the m-Quota TSP and the m-NW-TSP. Algorithms 2.1 and 2.2 for the m-Quota TSP and the m-NW-TSP, respectively, start by sorting the vertices of a given graph G in non-increasing order according to their prizes: $p_1 \ge p_2 \ge \ldots \ge p_n$. Then for each possible value of $\mu : m \le \mu \le n$ algorithms find a μ-vertex subset

$V' \subseteq V$ of the most profit promising vertices and compute an approximate solution H_μ to the minimum TSP in the induced subgraph $G[V']$ by applying the greedy Nearest City algorithm from [6–8]. Finally, the best found solution H_μ, $m \leq \mu \leq n$ is returned as an answer. Since the time complexity of the Nearest City algorithm is $O(n^2)$, it is clear, that both Algorithms 2.1 and 2.2 run in $O(n^3)$ time.

Algorithm 2.1: Algorithm for the m-Quota TSP.

Input: Complete graph $G = (V, E)$, $V = \{1, \ldots, n\}$, positive integer $m \leq n$,
edge costs $c: E \rightarrow \mathbb{R}_{\geq 0}$, vertex prizes $p_1, \ldots, p_n \in \mathbb{R}$, $Q \in \mathbb{R}$.
Output: Cycle H with at least m vertices and at least Q profit.

1 Sort the vertices of G in non-increasing order according to their prizes:
  $p_1 \geq p_2 \geq \ldots \geq p_n$;
2 $m' := \arg\min\{\mu \mid \sum_{i=1}^{\mu} p_i \geq Q \text{ and } \mu \geq m\}$;
3 $F := \infty$;
4 **foreach** $\mu: m' \leq \mu \leq n$ **do**
5 | $V' := \{1, 2, \ldots, \mu\}$;
6 | Using the Nearest City algorithm find an approximate solution H_μ for the
  | minimum TSP on the induced μ-vertex subgraph $G[V']$;
7 | $F := \min\{F, \sum_{e \in E(H_\mu)} c(e)\}$ and $H := \arg\min F$;

8 **return** H.

Algorithm 2.2: Algorithm for the m-NW-TSP.

Input: Complete graph $G = (V, E)$, $V = \{1, \ldots, n\}$, positive integer $m \leq n$,
edge costs $c: E \rightarrow \mathbb{R}_{\geq 0}$, vertex prizes $p_1, \ldots, p_n \in \mathbb{R}$.
Output: Cycle H with at least m vertices.

1 Sort the vertices of G in non-increasing order according to their prizes:
  $p_1 \geq p_2 \geq \ldots \geq p_n$;
2 $F := -\infty$;
3 **foreach** $\mu: m \leq \mu \leq n$ **do**
4 | $V' := \{1, 2, \ldots, \mu\}$;
5 | Using the Nearest City algorithm find an approximate solution H_μ for the
  | minimum TSP on the induced μ-vertex subgraph $G[V']$;
6 | $F := \max\{F, \sum_{i=1}^{m} p_i - \sum_{e \in E(H_\mu)} c(e)\}$ and $H := \arg\max F$;

7 **return** H.

2.1   Analysis of the Algorithms on Random Inputs

In this section for the algorithms presented above we perform probabilistic analysis, which consists in studying successful runs of the algorithms on "typical" instances, instead of the worst-case analysis. To this end, we assume the instances of the considered problems to belong to a certain probability space. Namely, we consider the cases when the input data, that are the edge costs $c(e) \in \mathbb{R}_{\geq 0}$, are identically distributed independent random reals with a uniform UNI(a_n, b_n) distribution function on segment $[a_n, b_n]$, $0 < a_n \leq b_n$; or with a shifted exponential EXP(a_n, a_n) distribution with parameter α_n on interval (a_n, ∞), $a_n > 0$, or a truncated normal NOR(σ_n, a_n) distribution with parameter σ_n on interval (a_n, ∞), $a_n > 0$.

The *performance guarantees* of an approximation algorithm will be defined as follows.

Definition 1. *Let A be an approximation algorithm for some optimization problem X on a graph with n vertices. Let $F_A(I)$ and $OPT(I)$ be the value of the approximate solution found by algorithm A and the optimum on input $I \in X$, correspondingly. Algorithm A is said to have* performance guarantees $\varepsilon_A(n)$ *and* $\delta_A(n)$ *on a set of random inputs of the problem, if*

$$\mathbf{Pr}\left\{\left|\frac{F_A(I) - OPT(I)}{OPT(I)}\right| > \varepsilon_A(n)\right\} \leq \delta_A(n),$$

where $\varepsilon_A(n)$ is the relative error *and $\delta_A(n)$ is the* failure probability *of the algorithm, which is equal to the proportion of cases when the algorithm doesn't hold the relative error $\varepsilon_A(n)$.*

Definition 2. *Algorithm A is called* asymptotically optimal *on a class of instances, if there exist performance guarantees such that $\varepsilon_A(n) \to 0$ and $\delta_A(n) \to 0$ as $n \to \infty$.*

The goal of this section is to present the conditions on the input data under which Algorithms 2.1 and 2.2 are asymptotically optimal. To this end, we will use the following theorem that brings together probabilistic results obtained in [6–8] for the Nearest City algorithm approximately solving the classic minimum TSP.

Theorem 1. *[6–8] Let H^* be an optimal solution for the minimum TSP. In case of the considered random inputs the Nearest City algorithm in $O(n^2)$ time gives a solution H with the following properties:*

$$\mathbf{Pr}\left(na_n \leq \sum_{e \in E(H^*)} c(e) \leq \sum_{e \in E(H)} c(e) \leq (1 + \varepsilon_n)na_n\right) \geq 1 - \delta_n,$$

with relative error $\varepsilon_n = 5\frac{\beta_n/a_n}{n/\ln n}$ and failure probability $\delta_n = n^{-1}$.

The algorithm is asymptotically optimal ($\varepsilon_n \to 0$ and $\delta_n \to 0$ as $n \to \infty$) for

$$\frac{\beta_n}{a_n} = o\left(\frac{n}{\ln n}\right), \quad \text{where} \quad \beta_n = \begin{cases} b_n, & \text{in the case of UNI}(a_n, b_n), \\ \alpha_n, & \text{in the case of EXP}(\alpha_n, a_n), \\ \sigma_n, & \text{in the case of NOR}(\sigma_n, a_n). \end{cases} \quad (1)$$

We now show that Algorithms 2.1 and 2.2 are asymptotically optimal under almost the same conditions as in Theorem 1, if the number m of the vertices that should be visited is a certain proportion of the number n of all the vertices of the graph.

Remark 1. Note that, due to Theorem 1, for each feasible μ-vertex cycle H_μ $(m \leq \mu \leq n)$ found in the inner loop of Algorithm 2.1 (2.2), with probability at least $1 - 1/m$ it holds that

$$\mu a_n \leq \sum_{e \in E(H_\mu)} c(e) \leq \mu a_n(1 + \varepsilon_\mu), \quad \text{where } \varepsilon_\mu \leq \varepsilon_m = 5\frac{\beta_n/a_n}{m/\ln m}.$$

Theorem 2. *Let $m = \rho \cdot n$, where $0 < \rho \leq 1$ is a constant. Then Algorithm 2.1 gives asymptotically optimal solutions for the m-Quota TSP on random input data $UNI(a_n, b_n)$, $EXP(\alpha_n, a_n)$ or $NOR(\sigma_n, a_n)$, if (1) holds.*

Proof. Let H^* and H_μ be an optimal solution and a solution returned by Algorithm 2.1, correspondingly, and let $\mu^* = |V(H^*)|$, $\mu^* \geq m$, and $\mu = |V(H_\mu)|$, $\mu \geq m' \geq m$. Note that all the feasible solutions found in the inner loop of Algorithm 2.1 give at least Q profit, while the number of vertices in any optimal solution is at least m', since m' is the least possible number of vertices that gives profit Q and is not less than m. Since H_μ is the best solution found in the inner loop of Algorithm 2.1, it is not worse than some feasible solution H_{μ^*} with μ^* vertices also found in the inner loop of Algorithm 2.1. Thus, from the obvious lower bound on the cost of an optimal solution and Remark 1, for the objective function $F_1(H)$ it follows that

$$\mu^* a_n \leq F_1(H^*) \leq F_1(H_\mu) \leq F_1(H_{\mu^*}) \leq a_n \mu^*(1 + \varepsilon_{\mu^*})$$

holds with failure probability at most $\delta_{A_1} = 1/m$, and, thus, the relative error of Algorithm 2.1 is

$$\varepsilon_{A_1} = \left| \frac{F_1(H_\mu)}{F_1(H^*)} - 1 \right| \leq \frac{a_n \mu^*(1 + \varepsilon_{\mu^*})}{\mu^* a_n} - 1 = \varepsilon_{\mu^*} \leq \varepsilon_m.$$

Finally, since $m = \rho n$ and (1) holds, relative error ε_{A_1} and failure probability δ_{A_1} of Algorithm 2.1 tend to zero as $n \to \infty$, and, thus, the algorithm is asymptotically optimal. □

Similar analysis can be carried out for the case of m-NW-TSP. We will assume, that the input data has the following property.

Property (∗). *The profit one gains at the vertices of a solution of m-NW-TSP is larger than the cost one pays for traveling the edges.*
That is, the objective function of an optimal solution is positive. Otherwise, one can multiply the objective function by -1 and consider a minimization problem instead of maximization. The analysis of the algorithm in this case would be similar and the relative error will stay the same. The only bad case left is when the objective function is close to zero. To deal with the last case we need to slightly tighten up the conditions of the asymptotic optimality of the algorithm.

Theorem 3. *Let $m = \rho \cdot n$, where $0 < \rho \leq 1$ is a constant. Then Algorithm 2.2 gives asymptotically optimal solutions for the m-NW-TSP on random input data $UNI(a_n, b_n)$, $EXP(\alpha_n, a_n)$ or $NOR(\sigma_n, a_n)$, if*

$$\beta_n/(p - a_n) = o(n/\ln n), \tag{2}$$

where $p = \sum_{i=1}^{n} p_i/n$ is the mean vertex profit and β_n is defined as in (1).

Proof. Let H^* and H_μ be an optimal solution and a solution returned by Algorithm 2.2, respectively, and let μ^* be the number of vertices in H^*, $\mu^* \geq m$, and μ be the number of vertices in H_μ, $\mu \geq m$. Since H_μ is the best solution found in the inner loop of Algorithm 2.2, H_μ is not worse than some feasible solution H_{μ^*} with μ^* vertices found in the inner loop of Algorithm 2.2. Recall that the vertices in the graph are sorted in the non-increasing order of their profits, and denote by P_k the total profit of the first k vertices. Then, for the objective function $F_2(H)$ of m-NW-TSP with failure probability at most $\delta_{A_2} = 1/m$ we have:

$$P_{\mu^*} - \mu^* a_n \geq F_2(H^*) \geq F_2(H_\mu) \geq F_2(H_{\mu^*}) \geq P_{\mu^*} - \mu^* a_n(1 + \varepsilon_{\mu^*}). \tag{3}$$

The first inequality in (3) follows from the fact that P_{μ^*} is the largest possible profit that can be obtained from visiting μ^* vertices and $\mu^* a_n$ is the least possible cost for traveling along a μ^*-vertex cycle. The last inequality in (3) follows from Remark 1 with probability at least $1 - \delta_{A_2}$. Therefore, taking into account Property (*) with the inequalities $\varepsilon_{\mu^*} \leq \varepsilon_m$ and $P_{\mu^*}/\mu^* \geq P_n/n = p$, the relative error of Algorithm 2.2 is:

$$\varepsilon_{A_2} = \left| 1 - \frac{F_2(H_\mu)}{F_2(H^*)} \right| \leq 1 - \frac{P_{\mu^*} - \mu^* a_n(1 + \varepsilon_{\mu^*})}{P_{\mu^*} - \mu^* a_n} \leq \frac{\varepsilon_{\mu^*}}{P_{\mu^*}/(\mu^* a_n) - 1} \leq \frac{\varepsilon_m}{p/a_n - 1}.$$

Finally, since $m = \rho n$ and (2) holds, both ε_{A_2} and δ_{A_2} tend to zero as $n \to \infty$, and, thus, Algorithm 2.2 is asymptotically optimal.   □

Remark 2. It is easy to see that condition $m = n \cdot \rho$ in both Theorems 2 and 3 can be replaced by a weaker one: m is a growing function of n, such that $m \to \infty$ as $n \to \infty$.

3   Conclusion

In this short paper we showed that on certain types of i.i.d. random inputs the greedy approach [6–8], that allowed us to obtain asymptotically optimal solutions for the classic minimum TSP, can be adopted to solve TSP-like problems, in which it is not necessary to visit all n vertices of a given graph.

Namely, we considered two problems m-Quota TSP and m-Net Worth prize-collecting TSP, where in addition to the given edge costs of a graph a certain profit is associated with each vertex, and the goal is to find a cycle which has a large collected profit and small traveling costs. We presented greedy approximation algorithms for these problems on random inputs.

In both problems the sought solution cycle is additionally required to contain at least m vertices, and one of the crucial conditions for the presented algorithms to be asymptotically optimal is m being a growing function of n (for example, it is quite natural for m to be some proportion of n). However, in the probabilistic analysis of the algorithms we mainly made assumptions about the distribution of the edge costs, and not about the vertex prizes. We expect, that assuming both the prizes of vertices and the costs of edges to be i.i.d. random reals, it would be possible to show the asymptotic optimality of the presented algorithms under weaker conditions on the number m of visited vertices.

References

1. Awerbuch, B., Azar, Y., Blum, A., Vempala, S.: New approximation guarantees for minimum-weight k-trees and prize-collecting salesmen. SIAM J. Comput. **28**(1), 254–262 (1999)
2. Bienstock, D., Goemans, M., Simchi-Levi, D., Williamson, D.: A note on the prize collecting traveling salesman problem. Math. Program. **59**, 413–420 (1993)
3. Climer, S., Zhang, W.: Rearrangement clustering: pitfalls, remedies, and applications. JMLR **7**, 919–943 (2006)
4. Dell'Amico, M., Maffioli, F., Sciomachen, A.: A Lagrangian heuristic for the Prize-Collecting Travelling Salesman Problem. Ann. Oper. Res. **81**, 289–305 (1998)
5. Fischetti, M., Toth, P.: An additive approach for the optimal solution of the Prize-Collecting Travelling Salesman Problem. In: Golden, B.L., Assad, A.A. (eds.) Vehicle Routing: Methods and Studies, pp. 319–343 (1988)
6. Gimadi, E.K.: On some probability inequalities for some discrete optimization problems. In: Haasis, H.D., Kopfer, H., Schönberger, J. (eds.) Operations Research Proceedings 2005, pp. 283–289. Springer, Heidelberg (2006). https://doi.org/10.1007/3-540-32539-5_45
7. Gimadi, E.K., Le Gallou, A., Shakhshneyder, A.V.: Probabilistic analysis of an approximation algorithm for the traveling salesman problem on unbounded above instances. J. Appl. Ind. Math. **3**(2), 207–221 (2009)
8. Gimadi, E.K., Khachay, M.Y.: Extremal Problems on the Set of Permutations. Ural Federal University, Ekaterinburg (2016). (in Russian)
9. Gutin, G., Punnen, A.P. (eds.): The Traveling Salesman Problem and Its Variations. Combinatorial Optimization, vol. 12. Springer, Boston (2007). https://doi.org/10.1007/b101971
10. Johnson, D.S., Minkoff, M., Phillips, S.: The prize collecting Steiner tree problem: theory and practice. In: Proceedings of the Eleventh Annual ACM-SIAM Symposium on Discrete Algorithms (SODA), pp 760–769. Society for Industrial and Applied Mathematics (2000)
11. Paul, A., Freund D., Ferber A., Shmoys D.B., Williamson D.P.: Prize-collecting TSP with a budget constraint. In: 25th Annual European Symposium on Algorithms (ESA 2017), Leibniz International Proceedings in Informatics (LIPIcs), pp 62:1–62:14 (2017)
12. Ray, S.S., Bandyopadhyay, S., Pal, S.K.: Gene ordering in partitive clustering using microarray expressions. J. Biosci. **32**(5), 1019–1025 (2007)

An Artificial Bee Colony Algorithm for the Multiobjective Energy Reduction Multi-Depot Vehicle Routing Problem

Emmanouela Rapanaki, Iraklis-Dimitrios Psychas, Magdalene Marinaki[✉],
and Yannis Marinakis

School of Production Engineering and Management,
Technical University of Crete, Chania, Greece
emmarap@hotmail.com, ipsychas102@gmail.com, magda@dssl.tuc.gr,
marinakis@ergasya.tuc.gr

Abstract. Artificial Bee Colony algorithm is a very powerful Swarm
Intelligence Algorithm that has been applied in a number of different
kind of optimization problems since the time that it was published. In
recent years there is a growing number of optimization models that trying to reduce the energy consumption in routing problems. In this paper,
a new variant of Artificial Bee Colony algorithm, the Parallel Multi-Start
Multiobjective Artificial Bee Colony algorithm (PMS-ABC) is proposed
for the solution of a Vehicle Routing Problem variant, the Multiobjective Energy Reduction Multi-Depot Vehicle Routing Problem (MERMD-
VRP). In the formulation four different scenarios are proposed where
the distances between the customers and the depots are either symmetric or asymmetric and the customers have either demand or pickup.
The algorithm is compared with three other multiobjective algorithms,
the Parallel Multi-Start Non-dominated Sorting Differential Evolution
(PMS-NSDE), the Parallel Multi-Start Non-dominated Sorting Particle
Swarm Optimization (PMS-NSPSO) and the Parallel Multi-Start Non-
dominated Sorting Genetic Algorithm II (PMS-NSGA II) in a number
of benchmark instances.

Keywords: Vehicle Routing Problem · Artificial Bee Colony · NSGA
II · NSDE · PSO · VNS

1 Introduction

The **Vehicle Routing Problem** is one of the most famous optimization problems and its main goal is the design of the best routes in order the selected (or
calculated) vehicles to serve the set of customers with the best possible way based
on the selected criteria that each different variant of the problem includes. As
the interesting of the researchers and of the decision makers in industries for the
solution of different variants of Vehicle Routing Problem continuously increases,
a number of more realistic and thus more complicated variants/formulations of

© Springer Nature Switzerland AG 2020
N. F. Matsatsinis et al. (Eds.): LION 13 2019, LNCS 11968, pp. 208–223, 2020.
https://doi.org/10.1007/978-3-030-38629-0_17

the Vehicle Routing Problem has been proposed and a number of more sophisticated algorithms are used for their solutions [26].

The combination of more than one objective functions in the formulation of a Vehicle Routing Problem variant could produce a more realistic problem. Thus, in the recent years there is an increasing number of researches that propose formulations with more than one criteria which are denoted as Multiobjective Vehicle Routing Problems. In this research the formulation of the problem combines more than one depots, the possibility of pickups and deliveries and the simultaneously reduction of the fuel consumption in symmetric and, in the more realistic asymmetric cases. In the first objective function, the total travel time is minimized, while in the second objective function the Fuel Consumption (FC) taking into account the traveled distance and the load of the vehicle is, also, minimized. In symmetric case, except of the load and the traveled distance, we consider that there are no other route parameters or that there are perfect route conditions, while on the other hand in asymmetric case, we take into account parameters of real life, such as weather conditions or uphill and downhill routes.

In recent years a number of evolutionary, swarm intelligence and, in general, nature inspired algorithms have been proposed for the solution of a Vehicle Routing Problem both when one objective or more than one objectives are considered. In this paper a Multiobjective variant of the Artificial Bee Colony algorithm, suitable for Vehicle Routing Problems, the Parallel Multi-Start Multiobjective Artificial Bee Colony Algorithm (PMS-ABC), is proposed for the solution of the Multiobjective Energy Reduction Multi-Depot Vehicle Routing Problem (MEMDVRP). Artificial Bee Colony (ABC) [5] is a nature-inspire meta-heuristic optimization algorithm. It is inspired from the foraging behavior of honey bees. There are two bees species: employed and unemployed foraging bees, which they search for food sources close to their hive. In ABC, first we must find the best parameter vector which minimizes the objective functions. Then, the bees randomly discover a population of initial solution vectors and they move towards better solutions, while they abandon poor solutions. In the PMS-ABC algorithm in addition of the main characteristics of the ABC algorithm we use rank and crowding distance as it will be explained later.

The algorithm is compared with three other evolutionary algorithms, the Parallel Multi-Start Non-dominated Sorting Particle Swarm Optimization (PMS-NSPSO) [21], the Parallel Multi-Start Non-dominated Sorting Differential Evolution (PMS-NSDE) [19] and the Parallel Multi-Start Non-dominated Sorting Genetic Algorithm II (PMS-NSGA II) [20].

This paper is organized into four sections. In the next section, the Multiobjective Energy Reduction Multi-Depot Vehicle Routing Problem is described. In the following section, the proposed algorithm is analyzed in detail, while, in Sect. 4 the computation results are presented. The last section provides concluding remarks and future research.

2   Multiobjective Energy Reduction Multi-Depot Vehicle Routing Problem

In recent years there is a growing number of papers for solving multi-depot vehicle routing problems [15] or energy vehicle routing problems [2,14,24]. There are few papers for the solution of Multi-Depot vehicle routing problem with fuel consumption. One of the first papers about the minimization of fuel consumption was presented by Kuo [11]. Afterwards, other papers were published (Suzuki [25], Li [12], Xiao et al. [27], Niu et al. [16]). In the last years, there are few papers about Multi-Depot Vehicle Routing Problem with Fuel Consumption (Psychas et al. [19–21], Kancharla et al. [4], Pardalos et al. [13]).

In this paper, the first objective function for the four **Multiobjective Route-based Fuel Consumption Vehicle Routing Problems** is the same than the one presented in Psychas et al. [20]. This function is used for the minimization of the **time** needed to travel between two customers or a customer and the depot and the equation is:

$$\min OF1 = \sum_{i=I_1}^{n}\sum_{j=1}^{n}\sum_{\kappa=1}^{m}(t_{ij}^{\kappa}+s_{j}^{\kappa})x_{ij}^{\kappa} \tag{1}$$

where t_{ij}^{κ} is the time needed to visit customer j immediately after customer i using vehicle κ, s_{j}^{κ} is the service time of customer j using vehicle κ, n is the number of nodes, m is the number of homogeneous vehicles and the depots are a subset $\Pi = \{I_1, I_2, ...I_\pi\}$ of the set of the n nodes where denoted by $i = j = I_1, I_2, ...I_\pi$ (π is the number of homogeneous depots). Also, the set of nodes is the following $\{I_1, I_2, ...I_\pi, 2, 3, ..., n\}$.

The second objective function is used for the minimization of the **Route based Fuel Consumption** (RFC) in the case that the vehicle performs only deliveries taking into account real life route parameters (weather conditions or uphills and downhills or driver's behavior). The objective function is described below:

$$\min OF2 = \sum_{h=I_1}^{I_\pi}\sum_{j=2}^{n}\sum_{\kappa=1}^{m}c_{hj}x_{hj}^{\kappa}(1+\frac{y_{hj}^{\kappa}}{Q})r_{hj} + \sum_{i=2}^{n}\sum_{j=I_1}^{n}\sum_{\kappa=1}^{m}c_{ij}x_{ij}^{\kappa}(1+\frac{y_{i-1,i}^{\kappa}-D_i}{Q})r_{ij} \tag{2}$$

with the maximum capacity of the vehicle denoted by Q, the i customer has demand equal to D_i and $D_{I_1} = D_{I_2} = ... = D_{I_\pi} = 0$, x_{ij}^{κ} denotes that the vehicle κ visits customer j immediately after customer i with load y_{ij}^{κ} and $y_{I_1j}^{\kappa} = \sum_{i=I_1}^{n} D_i$ for all vehicles as the vehicle begins with load equal to the summation of the demands of all customers assigned in its route and c_{ij} is the distance from node i to node j. The parameter r_{ij} corresponds to the route parameters from the node i to the node j and it is always a positive number. Due to the fact that it may be $r_{ij} \neq r_{ji}$ the product $c_{ij}r_{ij}$ leads to an asymmetric formulation of the whole problem. If the values of r_{ij} is lower than 1 we consider that the route from i to j is a downhill or the wind is back-wind or the driver drives with smooth shifting. If r_{ij} is larger than 1 we consider that the route from i to j is an uphill or the wind is a head-wind or the driver drives with aggressive shifting. If the $r_{ij} = 1 \forall (i,j)$ that belongs to the route, then, the problem is a symmetric problem. For an analytical presentation of how r_{ij} is calculated please see [19].

The third objective function is used for the minimization of the **Route based Fuel Consumption** (RFC) in the case that the vehicle performs only pick-ups in its route, as mentioned in Psychas et al. [19] and the function is:

$$\min OF3 = \sum_{h=I_1}^{I_\pi}\sum_{j=2}^{n}\sum_{\kappa=1}^{m} c_{hj}x_{hj}^{\kappa}r_{hj} + \sum_{i=2}^{n}\sum_{j=I_1}^{n}\sum_{\kappa=1}^{m} c_{ij}x_{ij}^{\kappa}(1 + \frac{y_{i-1,i}^{\kappa} + D_i}{Q})r_{ij} \quad (3)$$

with r_{ij} is the route parameters as in the OF2 and $y_{I_1 j}^{\kappa} = 0$ for all vehicles as the vehicle begins with empty load. In that case the D_i is the pick-up amount of the customer i. The only difference to the previous functions is that we have more than one depots, where are a subset $\Pi = \{I_1, I_2, ...I_\pi\}$ of the set of the n nodes were denoted by $i = j = I_1, I_2, ...I_\pi$ (π is the number of homogeneous depots). Also, the set of nodes is the following $\{I_1, I_2, ...I_\pi, 2, 3, ..., n\}$. Worth mentioning that each vehicle returns always to the depot where it starts and it does not visits another depot during its travel. Furthermore, there are no transition between the depots (for example from I_1 to I_3). The constraints of the problems are [19]:

$$\sum_{j=I_1}^{n}\sum_{\kappa=1}^{m} x_{ij}^{\kappa} = 1, i = I_1, \cdots, n \quad (4)$$

$$\sum_{i=I_1}^{n}\sum_{\kappa=1}^{m} x_{ij}^{\kappa} = 1, j = I_1, \cdots, n \quad (5)$$

$$\sum_{j=I_1}^{n} x_{ij}^{\kappa} - \sum_{j=I_1}^{n} x_{ji}^{\kappa} = 0, i = I_1, \cdots, n, \kappa = 1, \cdots, m \quad (6)$$

$$\sum_{j=I_1,j\neq i}^{n} y_{ji}^{\kappa} - \sum_{j=I_1,j\neq i}^{n} y_{ij}^{\kappa} = D_i, i = I_1, \cdots, n, \kappa = 1, \cdots, m, \; for \, deliveries \quad (7)$$

$$\sum_{j=I_1,j\neq i}^{n} y_{ij}^{\kappa} - \sum_{j=I_1,j\neq i}^{n} y_{ji}^{\kappa} = D_i, \; i = I_1, \cdots, n, \; \kappa = 1, \cdots, m, \; for \, pick-ups \quad (8)$$

$$Qx_{ij}^{\kappa} \geq y_{ij}^{\kappa}, i, j = I_1, \cdots, n, \kappa = 1, \cdots, m \quad (9)$$

$$\sum_{i,j \in S} x_{ij\kappa} \leq \mid S \mid -1, \; for \, all \; S \subseteq \{2, \cdots, n\}, \kappa = 1, \cdots, m \quad (10)$$

$$x_{ij}^{\kappa} = \begin{cases} 1, & \text{if } (i,j) \text{ belongs to the route} \\ 0, & \text{otherwise} \end{cases} \quad (11)$$

Constraints (4) and (5) represent that each customer must be visited only by one vehicle; constraints (6) ensure that each vehicle that arrives at a node must leave from that node also. Constraints (7) and (8) indicate that the reduced (if it concerns deliveries) or increased (if it concerns pick-ups) load (cargo) of the vehicle after it visits a node is equal to the demand of that node. Constraints (9) are used to limit the maximum load carried by the vehicle and to force y_{ij}^{κ} to be equal to zero when $x_{ij}^{\kappa} = 0$, constraints (10) are the tour eliminates constraints while constraints (11) ensure that only one vehicle will visit each customer. It should be noted that the problems solved in this paper are symmetric (where $r_{ij} = 1 \forall (i,j)$ holds) or asymmetric (where $r_{ij} \neq r_{ji} \forall (i,j)$ holds).

In Table 1, the objectives functions and the r_{ij} parameter are presented for all the problems studied in this paper.

Table 1. Objective functions and r_{ij} for all the problems

	Asymmetric delivery ERVRP	Asymmetric pick-up ERVRP
Objective functions	OF1 and OF2	OF1 and OF3
r_{ij}	$r_{ij} \neq r_{ji}, \forall (i,j)$ that belongs to the route	$r_{ij} \neq r_{ji}, \forall (i,j)$ that belongs to the route
	Symmetric delivery ERVRP	Symmetric pick-up ERVRP
Objective functions	OF1 and OF2	OF1 and OF3
r_{ij}	$r_{ij} = 1, \forall (i,j)$ that belongs to the route	$r_{ij} = 1, \forall (i,j)$ that belongs to the route

3   Parallel Multi-Start Multiobjective Artificial Bee Colony Algorithm (PMS-ABC)

Before we begin to present the Artificial Bee Colony algorithm, it is better to mention a few things about the process of finding bees' food, as it happens in reality. The bees leave the hive to search for food. When they find the food, they usually can not collect it alone and transfer it to the hive. So, they return to the hive unloading the nectar they carried with them. After that, they try to inform the rest bees for the source of food they found by making specific movements that scientists have given the name waggled dance.

Artificial Bee Colony algorithm (PMS-ABC) [5,7] is mainly applied to continuous optimization problems and is based on the process of the waggled dance performed by a bee in the food search process. This is the best known algorithm that simulates behavior of real bees and has been applied in many studies [1,3,5–10,17,23].

In the algorithm there are three subsets of bees: employed, onlookers and scout bees. The employed bees find the food source (possible solution to the problem) from a predetermined set of potential food sources and share this information (bee dance) with the other bees in the hive. The onlookers bees wait in the hive and based on the information they get from the employed bees, they look for a better food source in the neighborhood of food, where the employed bees have indicated to them. Finally, the scout bees are explorers whose food source is over and they seek out randomly a new source of food in the solution space.

In the Parallel Multi-Start Multiobjective Artificial Bee Colony Algorithm (PMS-ABC), initially, a set of food sources (possible solutions) (X) are randomly chosen from scout bees and their available nectar is calculated (value of the objective function). Then, the value of the objective function is calculated and in each food source is assigned a scout bee. Afterwards, the Pareto Front of the initial population is created. In every iteration, the non-dominated solutions are found. Then, we sort the solutions as follows initially the average of the cost values for each objective function is calculated. Then, the Euclidean distance between the average and each of the non-dominated solutions of the solutions is calculated. Finally, the solutions are classified according to their distance from the average.

Scout bees return to the hive and make the bee dance (waggle dance) informing bees that have remained in the beehive, the onlooker bees, where are the food sources. Onlooker bees choose the food source they will visit based on the information they receive about the nectar of each source from the bee dance process. Scout and onlooker bees are placed in the selected sources, matching each bee's source of food. The PMS-ABC algorithm uses the following equation to produce a new food source:

$$x'_{ij} = x_{ij} + rand_2(x_{ij} - x_{kj}), \tag{12}$$

where x'_{ij} is the candidate food source and k is a different from i food source selected from the solutions of the Pareto Front and $rand_2$ is a random number in space $(0,1)$.

As the use of the Eq. (12) for the calculation of the new food sources could produce some inefficient (due to the transformation of the solutions from continuous values (suitable for the equations of ABC) to discrete values (path representation) and vice versa) and dominated from the personal best of each food source and from the global best of the whole solutions, it was decided to add another phase for the calculation of the new positions in the algorithm in order to take advantage of possible good new and old positions in the whole set of food sources. Thus, the solutions of the last two iterations (iteration it and $it+1$) are combined in a new vector and, then, the members of the new vector are sorted using the $rank$ and the $crowding$ $distance$ as in the NSGA II algorithm as it was modified in [18]. The first W food sources of the new vector are the produced solutions (the new food sources) of the iteration $it + 1$. The distribution of the new food sources is performed based on the values of the $rank$ and the $crowding$ $distance$. With this procedure we avoid to add to the next iterations inefficient solutions that will probably be produced using the Eq. (12).

The new W food sources are evaluated by each objective function separately. A Variable Neighborhood Search (VNS) algorithm is applied to the food sources with both the vns_{max} and the $local_{max}$ equal to 10 [18]. The personal best solution of each food source is found by using the following observation. If a solution

214 E. Rapanaki et al.

in iteration $it + 1$ dominates its previous best solution of the iteration it, then, the previous best solution is replaced by the current solution. On the other hand if the previous best solution of dominates the current solution, then, the previous best solution remains the same. Finally, if these two solutions are not dominated between them, then, the previous best solution is not replaced. It should be noted that the non-dominated solutions are not deleted from the Pareto front and, thus, the good solutions will not be disappeared from the population. In the next iterations in order to insert a food source in the Pareto front archive there are two possibilities. First, the food source is non-dominated with respect to the contents of the archive and second it dominates any food source in the archive. In both cases, all the dominated food sources, that already belong to the archive, have to be deleted from the archive. At the end of each iteration, from the non-dominated solutions from all populations the Total Pareto Front is updated considering the non-dominated solutions of the last *initial population*.

The results of the proposed algorithm are compared with the results of the Parallel Multi-Start Non-dominated Sorting Particle Swarm Optimization (PMS-NSPSO) [21], the Parallel Multi-Start Non-dominated Sorting Differential Evolution (PMS-NSDE) algorithm [19] and the Parallel Multi-Start Non-dominated Sorting Genetic Algorithm II (PMS-NSGA II) algorithm [20]. In [21] a number of different versions based on the velocity equation of the Parallel Multi-Start Non-dominated Sorting Particle Swarm Optimization were proposed and tested. The one that performed better than the others was the 3rd version [21] and this is the one that we use for the comparisons in this paper. For all the necessary information about the algorithms PMS-NSPSO, PMS-NSDE and PMS-NSGA II, please see Psychas et al. [19–21].

4 Computational Results

The whole algorithmic approach was implemented in Visual C++ and was tested on a set of benchmark instances. The data were created according Rapanaki et al. [22] research. Also, the algorithms were tested for ten instances for the two objective functions (OF1-OF2 or OF1-OF3) five times [22]. For the comparison of the three algorithms, we use four different evaluation measures: M_k, L, Δ and C. It is preferred the L, M_k and C measures to be as larger as possible and the Δ value to be as smaller as possible, for analytical presentation of the measures please see [19].

Table 2. Results of the first three measures in ten instances for five executions using the proposed algorithm

Multiobjective Asymmetric Delivery Route based Fuel Consumption VRP

	execution 1			execution 2			execution 3			execution 4			execution 5			Average			Best Run		
	L	M_k	Δ	L	M_k	Δ	L	M_k	Δ	L	M_k	Δ	L	M_k	Δ	L	M_k	Δ	L	M_k	Δ
A-B-CD	29	516.02	0.77	25	487.76	0.64	26	537.22	0.69	14	525.10	0.54	32	530.87	0.76	25.2	519.39	0.68	26	537.22	0.69
A-C-BD	27	512.31	0.89	26	513.64	0.69	26	509.60	0.75	23	514.19	0.70	30	523.10	0.94	26.4	514.57	0.80	30	523.10	0.94
A-D-BE	18	522.05	0.67	20	520.56	0.83	22	500.81	0.83	21	490.47	0.79	25	533.70	0.84	21.2	513.52	0.79	25	533.70	0.84
A-E-BD	17	515.82	0.71	23	509.46	0.61	23	517.29	0.75	29	515.53	0.84	30	519.05	0.64	24.4	515.43	0.71	30	519.05	0.64
B-C-AD	23	507.12	0.78	14	452.00	1.05	20	478.85	0.86	20	464.49	0.86	21	494.23	0.78	19.6	479.34	0.87	23	507.12	0.78
B-D-AC	26	514.14	0.69	22	528.87	0.75	21	483.33	0.75	20	512.75	0.70	16	503.56	0.71	21.0	508.53	0.72	26	514.14	0.69
B-E-AD	26	510.33	0.86	21	496.94	0.92	25	470.98	0.74	22	498.24	0.57	22	471.12	0.71	23.2	489.52	0.76	26	510.33	0.86
C-D-AE	30	520.80	0.76	24	509.52	0.82	20	510.73	0.82	25	535.59	0.85	28	545.99	0.86	25.4	524.53	0.82	30	520.80	0.76
C-E-AB	27	518.79	0.91	33	520.96	0.85	20	536.80	0.84	23	500.48	0.69	26	525.11	0.78	25.8	520.43	0.81	33	520.96	0.85
D-E-BC	24	502.93	0.76	23	514.93	0.60	23	516.24	0.76	25	513.47	0.75	21	505.98	0.63	23.2	510.71	0.70	23	514.93	0.60

Multiobjective Symmetric Delivery Route based Fuel Consumption VRP

	execution 1			execution 2			execution 3			execution 4			execution 5			Average			Best Run		
	L	M_k	Δ	L	M_k	Δ	L	M_k	Δ	L	M_k	Δ	L	M_k	Δ	L	M_k	Δ	L	M_k	Δ
A-B	35	603.65	0.74	22	590.99	0.57	24	607.52	0.93	33	603.14	0.87	31	610.84	0.69	29.0	603.23	0.76	31	610.84	0.69
A-C	35	600.87	0.89	32	597.31	0.85	36	602.18	0.70	33	607.11	0.80	25	602.72	0.92	32.2	602.04	0.83	36	602.18	0.70
A-D	28	578.07	0.80	29	576.04	0.80	23	571.98	0.88	34	586.21	0.80	35	610.10	0.88	29.2	584.48	0.84	35	610.10	0.88
A-E	36	596.71	0.74	32	603.25	0.84	36	595.58	0.87	36	598.14	0.81	45	599.44	0.89	35.0	598.62	0.83	36	596.71	0.74
B-C	20	577.17	0.70	25	527.07	0.85	21	521.37	0.82	28	590.73	0.95	26	558.34	0.85	24.0	554.94	0.83	28	590.73	0.95
B-D	29	570.66	0.87	30	598.79	0.93	28	592.98	0.73	26	596.71	0.75	29	587.54	0.76	28.4	589.34	0.81	30	598.79	0.93
B-E	30	601.17	0.98	27	615.58	0.85	31	614.93	0.78	26	606.40	0.69	28	535.69	0.97	28.4	594.76	0.85	31	614.93	0.78
C-D	25	577.24	0.77	36	582.10	0.93	22	595.59	0.79	32	585.57	0.62	30	593.72	0.70	29.0	586.84	0.76	32	585.57	0.62
C-E	37	603.20	0.86	29	616.56	0.68	38	599.04	0.86	31	592.37	0.85	31	608.04	0.74	33.2	603.84	0.80	29	616.56	0.68
D-E	38	595.05	0.76	31	600.07	0.74	28	601.25	0.86	29	616.90	0.79	33	599.06	0.68	31.8	604.43	0.77	33	599.06	0.68

Multiobjective Asymmetric Pick-up Route based Fuel Consumption VRP

	execution 1			execution 2			execution 3			execution 4			execution 5			Average			Best Run		
	L	M_k	Δ	L	M_k	Δ	L	M_k	Δ	L	M_k	Δ	L	M_k	Δ	L	M_k	Δ	L	M_k	Δ
A-B-CD	39	608.17	0.72	31	615.83	0.70	31	592.63	0.75	31	570.85	0.74	23	585.03	0.68	30.2	594.50	0.72	31	615.83	0.70
A-C-BD	33	598.21	0.70	37	595.75	0.79	27	608.82	0.74	18	585.50	0.61	33	609.91	0.73	29.6	599.64	0.72	33	609.91	0.73
A-D-BE	28	593.85	0.78	28	595.78	0.72	31	602.29	0.80	33	614.92	0.85	32	615.39	0.85	30.4	604.45	0.80	33	614.92	0.85
A-E-BD	28	600.35	0.87	32	598.43	0.79	28	603.00	0.72	31	591.92	0.79	35	595.86	0.81	30.8	597.77	0.80	28	603.00	0.72
B-C-AD	25	587.78	0.81	31	585.40	0.76	27	605.47	0.81	26	580.15	0.74	19	595.08	0.84	25.6	590.78	0.79	31	585.40	0.76
B-D-AC	37	570.47	0.72	24	593.56	0.78	22	589.89	0.76	25	602.09	0.67	24	598.23	0.66	24.4	590.85	0.72	25	602.09	0.67
B-E-AD	27	591.80	0.66	34	599.91	0.73	30	602.26	0.85	30	600.24	0.83	26	595.85	0.78	29.4	598.01	0.77	34	599.91	0.73
C-D-AE	29	595.02	0.73	32	590.48	0.79	27	614.45	0.79	28	583.33	0.73	28	583.33	0.73	28.6	598.07	0.72	27	614.45	0.63
C-E-AB	23	599.61	0.80	27	597.77	0.76	29	583.60	0.83	26	600.55	0.87	25	589.61	0.94	26.0	594.23	0.84	27	597.77	0.76
D-E-BC	24	606.77	0.96	25	587.13	0.76	34	571.41	0.78	25	566.57	0.70	30	577.52	0.63	28.0	581.88	0.77	30	577.52	0.63

Multiobjective Symmetric Pick-up Route based Fuel Consumption VRP

	execution 1			execution 2			execution 3			execution 4			execution 5			Average			Best Run		
	L	M_k	Δ	L	M_k	Δ	L	M_k	Δ	L	M_k	Δ	L	M_k	Δ	L	M_k	Δ	L	M_k	Δ
A-B	25	609.69	0.78	25	611.89	0.69	17	600.23	0.72	21	583.23	0.79	19	598.76	0.76	21.4	600.76	0.75	25	611.89	0.69
A-C	25	593.96	0.88	25	601.56	0.76	19	604.92	0.77	18	592.86	0.67	24	604.56	0.86	22.2	599.57	0.79	25	601.56	0.76
A-D	26	582.98	0.59	24	596.21	0.74	22	568.85	0.65	27	581.50	0.62	27	587.25	0.85	25.2	583.36	0.69	26	582.98	0.59
A-E	25	594.05	0.64	27	603.89	0.73	18	595.22	0.77	15	593.26	0.75	19	595.29	0.71	20.8	596.34	0.72	27	603.89	0.72
B-C	24	598.34	0.73	23	590.06	0.79	26	604.70	0.69	25	591.60	0.91	23	590.35	0.66	24.2	595.01	0.76	26	604.70	0.69
B-D	22	588.54	0.69	17	535.37	0.52	20	589.68	0.84	19	602.32	0.71	24	600.10	0.69	20.4	583.20	0.69	24	600.10	0.69
B-E	18	604.22	0.67	17	606.64	0.87	20	605.79	0.70	14	595.38	0.66	19	606.64	0.67	17.6	602.44	0.72	19	606.64	0.67
C-D	23	545.61	0.62	22	606.27	0.57	16	592.48	0.76	16	577.65	0.60	20	594.91	0.51	19.4	583.38	0.61	22	606.27	0.57
C-E	22	586.68	0.73	21	592.58	0.60	23	601.78	0.92	23	619.16	0.65	23	601.53	0.65	22.4	600.35	0.71	23	619.16	0.65
D-E	26	620.04	0.58	22	594.31	0.82	22	608.68	0.66	18	608.79	0.73	32	609.61	0.73	24.0	608.29	0.71	26	620.04	0.58

Table 3. Average results and best runs of all algorithms used in the comparisons.

	Algorithms	Asymmetric delivery FCVRP			Asymmetric pick-up FCVRP		
		L	M_k	Δ	L	M_k	Δ
A-B-CD	PMS-ABC	25.20(26)	519.39(537.22)	0.68(0.69)	30.20(31)	594.50(**615.83**)	0.72(0.70)
	PMS-NSPSO	46.40(53)	598.41(592.84)	0.70(0.68)	47.00(56)	597.42(609.09)	0.66(**0.62**)
	PMS-NSGA II	56.40(**62**)	592.33(598.84)	0.61(0.54)	59.80(**63**)	598.92(608.39)	0.61(0.65)
	PMS-NSDE	50.00(59)	598.17(**604.03**)	0.61(**0.53**)	46.00(49)	598.81(605.87)	0.61(**0.62**)
A-C-BD	PMS-ABC	26.40(30)	514.57(523.10)	0.80(0.94)	29.60(33)	599.64(609.91)	0.72(0.73)
	PMS-NSPSO	47.60(51)	600.33(**611.96**)	0.68(0.68)	47.80(53)	601.70(595.13)	0.64(**0.59**)
	PMS-NSGA II	61.80(**72**)	594.92(602.67)	0.61(0.68)	58.80(**56**)	604.25(**613.08**)	0.59(0.64)
	PMS-NSDE	49.40(56)	594.19(603.86)	0.63(**0.64**)	44.40(50)	597.25(612.67)	0.63(0.62)
A-D-BE	PMS-ABC	21.20(25)	513.52(533.70)	0.79(0.84)	30.40(33)	604.45(614.92)	0.80(0.85)
	PMS-NSPSO	48.80(52)	592.57(**606.60**)	0.64(0.62)	46.00(48)	608.80(**620.73**)	0.66(0.65)
	PMS-NSGA II	54.20(**54**)	601.74(597.52)	0.61(**0.55**)	54.60(**63**)	602.82(610.88)	0.61(0.60)
	PMS-NSDE	46.80(51)	591.33(585.05)	0.68(0.58)	46.20(53)	594.36(608.66)	0.64(**0.58**)
A-E-BD	PMS-ABC	24.40(30)	515.43(519.05)	0.71(0.64)	30.80(28)	597.77(603.00)	0.80(0.72)
	PMS-NSPSO	48.00(**58**)	586.44(595.07)	0.65(0.63)	49.00(**60**)	597.10(610.34)	0.67(0.69)
	PMS-NSGA II	57.20(56)	595.30(**604.81**)	0.58(**0.56**)	56.40(60)	591.73(**615.11**)	0.58(0.53)
	PMS-NSDE	53.80(**58**)	589.49(595.89)	0.66(0.66)	45.40(49)	598.88(578.64)	0.64(**0.52**)
B-C-AD	PMS-ABC	19.60(23)	479.34(507.12)	0.87(0.78)	25.60(31)	590.78(585.40)	0.79(0.76)
	PMS-NSPSO	42.00(45)	587.45(587.39)	0.66(0.69)	43.20(38)	601.86(**616.66**)	0.67(**0.57**)
	PMS-NSGA II	51.20(**64**)	596.11(**602.55**)	0.61(0.60)	53.60(60)	602.63(609.71)	0.63(0.65)
	PMS-NSDE	42.20(47)	593.19(586.51)	0.63(**0.56**)	44.00(47)	596.09(594.45)	0.69(0.61)
B-D-AC	PMS-ABC	21.00(26)	508.53(514.14)	0.72(0.69)	24.40(25)	590.85(602.09)	0.72(0.67)
	PMS-NSPSO	42.00(53)	589.78(591.38)	0.62(0.66)	42.20(**50**)	581.03(589.53)	0.68(0.75)
	PMS-NSGA II	54.60(**63**)	593.43(**618.00**)	0.61(**0.53**)	51.40(46)	587.73(**606.69**)	0.60(**0.51**)
	PMS-NSDE	43.20(48)	591.59(611.63)	0.71(0.72)	44.80(**50**)	592.20(586.71)	0.66(0.58)
B-E-AD	PMS-ABC	23.20(26)	489.52(510.33)	0.76(0.86)	29.40(34)	598.01(599.91)	0.77(0.73)
	PMS-NSPSO	42.00(45)	584.42(543.81)	0.59(**0.52**)	48.00(47)	601.43(611.90)	0.65(0.62)
	PMS-NSGA II	50.80(**57**)	597.52(599.73)	0.56(0.56)	50.20(**55**)	594.89(600.52)	0.60(**0.58**)
	PMS-NSDE	41.80(47)	596.38(**611.30**)	0.66(0.61)	41.00(46)	598.64(**622.57**)	0.66(0.66)
C-D-AE	PMS-ABC	25.40(30)	524.53(520.80)	0.82(0.76)	28.60(27)	598.07(**614.45**)	0.72(0.63)
	PMS-NSPSO	45.00(**57**)	593.92(593.63)	0.67(0.66)	44.00(36)	592.59(602.11)	0.67(**0.55**)
	PMS-NSGA II	53.60(49)	597.24(589.81)	0.58(**0.53**)	51.80(**59**)	597.78(610.64)	0.63(0.59)
	PMS-NSDE	42.40(42)	594.55(**610.84**)	0.63(0.60)	44.40(51)	594.87(591.47)	0.70(0.71)
C-E-AB	PMS-ABC	25.80(33)	520.43(520.96)	0.81(0.85)	26.00(27)	594.23(597.77)	0.84(0.76)
	PMS-NSPSO	45.00(**48**)	586.58(577.86)	0.66(**0.54**)	46.80(41)	588.78(601.76)	0.63(**0.53**)
	PMS-NSGA II	55.00(47)	592.95(**602.86**)	0.61(0.57)	51.00(**65**)	595.58(608.99)	0.65(0.74)
	PMS-NSDE	39.40(47)	592.59(569.13)	0.67(0.63)	47.20(47)	593.76(**612.40**)	0.63(0.57)
D-E-BC	PMS-ABC	23.20(23)	510.71(514.93)	0.70(0.60)	28.00(30)	581.88(577.52)	0.77(**0.63**)
	PMS-NSPSO	42.60(**50**)	571.71(568.50)	0.67(0.69)	39.60(44)	578.42(573.72)	0.60(0.68)
	PMS-NSGA II	43.60(48)	526.74(503.05)	0.62(**0.53**)	50.80(**56**)	581.08(**602.31**)	0.66(0.69)
	PMS-NSDE	42.00(44)	582.50(**589.22**)	0.68(0.64)	43.60(52)	581.38(588.35)	0.60(0.64)

(continued)

Table 3. (*continued*)

	Algorithms	Asymmetric delivery FCVRP			Asymmetric pick-up FCVRP		
		L	M_k	Δ	L	M_k	Δ
		Symmetric delivery FCVRP			Symmetric pick-up FCVRP		
A-B	PMS-ABC	29.00(31)	603.23(**610.84**)	0.76(0.69)	21.40(25)	600.76(611.89)	0.75(0.69)
	PMS-NSPSO	47.80(52)	613.92(610.29)	0.68(0.62)	51.60(54)	607.75(602.86)	0.69(0.62)
	PMS-NSGA II	56.40(**61**)	602.89(603.41)	0.66(0.62)	58.60(**79**)	596.88(605.27)	0.66(**0.60**)
	PMS-NSDE	44.60(45)	605.73(596.04)	0.67(**0.55**)	44.60(46)	602.96(**622.10**)	0.68(0.62)
A-C	PMS-ABC	32.20(36)	602.04(602.18)	0.83(0.70)	22.20(25)	599.57(601.56)	0.79(0.76)
	PMS-NSPSO	50.20(53)	604.15(**615.43**)	0.66(0.67)	47.40(53)	604.96(608.06)	0.68(0.72)
	PMS-NSGA II	62.60(**66**)	609.36(611.80)	0.63(**0.54**)	56.60(**63**)	606.55(**615.56**)	0.64(**0.61**)
	PMS-NSDE	51.40(57)	596.73(602.61)	0.65(0.69)	49.20(44)	603.51(607.51)	0.65(**0.61**)
A-D	PMS-ABC	29.20(35)	584.48(**610.10**)	0.84(0.88)	25.20(26)	583.36(582.98)	0.69(0.59)
	PMS-NSPSO	48.40(54)	577.25(575.63)	0.62(**0.61**)	49.00(58)	586.10(**594.99**)	0.63(0.64)
	PMS-NSGA II	54.40(**57**)	592.85(587.50)	0.67(0.63)	58.60(**66**)	580.64(582.86)	0.66(0.60)
	PMS-NSDE	46.20(40)	581.70(591.46)	0.66(0.67)	47.00(47)	579.02(577.96)	0.66(**0.58**)
A-E	PMS-ABC	35.00(36)	598.62(596.71)	0.83(0.74)	20.80(27)	596.34(603.89)	0.72(0.72)
	PMS-NSPSO	48.00(47)	595.33(602.24)	0.68(0.69)	41.60(46)	592.88(**612.23**)	0.68(0.81)
	PMS-NSGA II	51.20(**51**)	598.85(**608.34**)	0.62(**0.60**)	61.00(**74**)	598.56(598.90)	0.65(0.59)
	PMS-NSDE	44.60(49)	604.13(608.11)	0.68(0.63)	43.80(43)	602.41(605.73)	0.62(**0.58**)
B-C	PMS-ABC	24.00(28)	554.94(590.73)	0.83(0.95)	24.20(26)	595.01(604.70)	0.76(0.69)
	PMS-NSPSO	41.00(51)	589.04(587.03)	0.68(0.62)	47.40(49)	595.11(605.91)	0.67(0.64)
	PMS-NSGA II	58.80(55)	591.49(602.55)	0.65(**0.55**)	56.20(**61**)	597.28(590.74)	0.65(**0.61**)
	PMS-NSDE	49.80(**60**)	589.78(**610.81**)	0.66(0.56)	42.00(49)	589.82(**613.07**)	0.72(0.64)
B-D	PMS-ABC	28.40(30)	589.34(598.79)	0.81(0.93)	20.40(24)	583.20(600.10)	0.69(0.69)
	PMS-NSPSO	43.40(55)	594.80(598.42)	0.65(**0.54**)	43.20(49)	591.07(600.78)	0.64(**0.54**)
	PMS-NSGA II	58.60(**56**)	595.91(**609.20**)	0.64(0.63)	55.80(**54**)	592.55(**608.67**)	0.65(0.66)
	PMS-NSDE	43.20(45)	595.65(595.43)	0.69(0.60)	41.20(**54**)	582.58(600.87)	0.70(0.68)
B-E	PMS-ABC	28.40(31)	594.76(**614.93**)	0.85(0.78)	17.60(19)	602.44(606.64)	0.72(0.67)
	PMS-NSPSO	49.40(52)	606.42(590.66)	0.60(**0.53**)	41.40(40)	604.84(**618.80**)	0.68(0.63)
	PMS-NSGA II	57.60(**59**)	603.48(607.87)	0.61(0.60)	58.60(**63**)	603.78(618.11)	0.63(**0.55**)
	PMS-NSDE	47.80(45)	603.11(609.40)	0.67(0.60)	44.40(46)	580.23(573.63)	0.66(0.58)
C-D	PMS-ABC	29.00(32)	586.84(585.57)	0.76(0.62)	19.40(22)	583.38(**606.27**)	0.61(**0.57**)
	PMS-NSPSO	48.80(46)	584.51(602.85)	0.64(**0.59**)	46.60(56)	586.85(594.19)	0.67(0.73)
	PMS-NSGA II	56.60(**59**)	587.97(**604.68**)	0.63(0.64)	51.20(**57**)	586.43(587.41)	0.59(0.60)
	PMS-NSDE	42.80(51)	577.40(594.58)	0.65(0.61)	46.80(51)	586.32(573.07)	0.66(**0.57**)
C-E	PMS-ABC	33.20(29)	603.84(**616.56**)	0.80(0.68)	22.40(23)	600.35(**619.16**)	0.71(0.65)
	PMS-NSPSO	48.00(51)	598.45(604.81)	0.64(0.65)	43.20(33)	599.71(609.57)	0.67(**0.59**)
	PMS-NSGA II	60.40(**63**)	599.00(592.77)	0.64(**0.58**)	60.40(**61**)	607.48(613.53)	0.65(0.60)
	PMS-NSDE	49.40(53)	594.95(611.88)	0.73(0.75)	52.20(60)	601.83(615.53)	0.67(0.72)
D-E	PMS-ABC	31.80(33)	604.43(599.06)	0.77(0.68)	24.00(26)	608.29(620.04)	0.71(0.58)
	PMS-NSPSO	49.20(**57**)	606.90(**620.98**)	0.71(0.80)	48.00(**55**)	609.46(617.27)	0.69(0.70)
	PMS-NSGA II	60.20(52)	601.63(610.06)	0.67(0.66)	59.00(53)	606.62(617.50)	0.63(**0.57**)
	PMS-NSDE	49.80(46)	604.82(619.96)	0.66(**0.55**)	49.00(54)	615.41(**622.74**)	0.70(0.64)

Generally, based on all Tables (Tables 2, 3, 4, 5, 6 and 7), from the comparison of the four algorithms we conclude that considering the L measure the PMS-NSGA II algorithm performs better than the other three algorithms in 75% of the instances, while the PMS-PSO3 performs better than the other algorithms in 12.5% and PMS-NSDE performs better only in 2.5%. The algorithms have the same percentage in 10% of the instances. Considering the M_k measure,

Table 4. Results of the C measure for the four algorithms in ten instances when the asymmetric delivery problem using objective functions OF1-OF2 is solved

OF1-OF2	Multiobjective asymmetric delivery route based fuel consumption VRP								
A-B-CD	ABC	NSPSO	NSDE	NSGA II	**B-D-AC**	ABC	NSPSO	NSDE	NSGA II
ABC	–	**0.87**	0.69	0.85	ABC	–	**0.66**	0.96	0.83
NSPSO	0	–	0.25	0.82	NSPSO	0.08	–	0.63	0.89
NSDE	0.12	0.55	–	0.90	NSDE	0	0.30	–	0.63
NSGA II	0	0.04	0.08	–	NSGA II	0	0.02	0.17	–
A-C-BD	ABC	NSPSO	NSDE	NSGA II	**B-E-AD**	ABC	NSPSO	NSDE	NSGA II
ABC	–	**0.63**	0.66	0.69	ABC	–	**0.82**	0.38	0.77
NSPSO	0.37	–	0.59	0.78	NSPSO	0.04	–	0.15	0.75
NSDE	0.17	0.39	–	0.83	NSDE	0.23	0.73	–	0.93
NSGA II	0	0.02	0.04	–	NSGA II	0	0.07	0	–
A-D-BE	ABC	NSPSO	NSDE	NSGA II	**C-D-AE**	ABC	NSPSO	NSDE	NSGA II
ABC	–	**0.83**	0.57	0.93	ABC	–	**0.88**	0.76	0.88
NSPSO	0	–	0.16	0.74	NSPSO	0	–	0.38	0.82
NSDE	0.16	0.79	–	0.93	NSDE	0.10	0.65	–	0.82
NSGA II	0	0.15	0.02	–	NSGA II	0	0.07	0.07	–
A-E-BD	ABC	NSPSO	NSDE	NSGA II	**C-E-AB**	ABC	NSPSO	NSDE	NSGA II
ABC	–	**0.76**	0.67	0.88	ABC	–	**0.73**	0.64	0.79
NSPSO	0.03	–	0.21	0.89	NSPSO	0.21	–	0.45	0.79
NSDE	0.10	0.55	–	0.91	NSDE	0.15	0.48	–	0.81
NSGA II	0	0.03	0	–	NSGA II	0.09	0.13	0.13	–
B-C-AD	ABC	NSPSO	NSDE	NSGA II	**D-E-BC**	ABC	NSPSO	NSDE	NSGA II
ABC	–	**0.82**	0.72	0.78	ABC	–	**0.82**	0.68	0.52
NSPSO	0.13	–	0.28	0.86	NSPSO	0.17	–	0.16	0.38
NSDE	0.04	0.38	–	0.94	NSDE	0.30	0.50	–	0.40
NSGA II	0	0.04	0	–	NSGA II	0.52	0.56	0.64	–

PMS-NSGA II algorithm performs better than the other algorithms in 32.5% of the instances, while PMS-NSDE, PMS-NPSO3 and PMS-ABC perform better than the other algorithms in 25%, 22.5% and 20% of the instances, respectively. Also, considering the Δ measure, the PMS-NSGA II algorithm performs better than the other algorithms in 37.5% of the instances, while algorithms PMS-NPSO3, PMS-NSDE and PMS-ABC perform better than the other algorithms in 30%, 22.5% and 2.5% of the instances, respectively. The algorithms have the same percentage in 7.5% of the instances.

Considering the C measure, in Table 4, PMS-ABC algorithm performs better than PMS-NSDE and PMS-NPSO3 algorithms because it performs better in 100% of the instances. Also, PMS-ABC algorithm performs better than PMS-NSGA II algorithm because it performs better in 90% of the instances, while PMS-NSGA II algorithm performs better than PMS-ABC algorithm in 10% of the instances. PMS-ABC algorithm performs slightly better than the other

Table 5. Results of the C measure for the four algorithms in ten instances when the symmetric delivery problem using objective functions OF1-OF2 is solved

OF1-OF2	Multiobjective symmetric delivery route based fuel consumption VRP								
A-B	ABC	NSPSO	NSDE	NSGA II	**B-D**	ABC	NSPSO	NSDE	NSGA II
ABC	–	**0.81**	**0.69**	**0.95**	ABC	–	**0.65**	**0.87**	**0.93**
NSPSO	0.06	–	0.33	**0.97**	NSPSO	0.27	–	**0.67**	**0.96**
NSDE	0.13	**0.42**	–	**0.97**	NSDE	0.07	0.22	–	**0.84**
NSGA II	0.03	0.02	0	–	NSGA II	0	0	0.07	–
A-C	ABC	NSPSO	NSDE	NSGA II	**B-E**	ABC	NSPSO	NSDE	NSGA II
ABC	–	**0.83**	**0.86**	**1.00**	ABC	–	**0.90**	**0.69**	**0.85**
NSPSO	0.11	–	**0.54**	**0.97**	NSPSO	0.03	–	0.38	**0.78**
NSDE	0.08	0.34	–	**0.98**	NSDE	0.19	**0.54**	–	**0.90**
NSGA II	0	0	0	–	NSGA II	0.03	0.10	0.07	–
A-D	ABC	NSPSO	NSDE	NSGA II	**C-D**	ABC	NSPSO	NSDE	NSGA II
ABC	–	**0.70**	**0.60**	**0.84**	ABC	–	**0.87**	**0.90**	**1.00**
NSPSO	0.20	–	0.35	**0.96**	NSPSO	0.03	–	**0.35**	**0.81**
NSDE	0.17	**0.59**	–	**0.82**	NSDE	0.03	0.35	–	**0.90**
NSGA II	0.14	0	0.10	–	NSGA II	0	0.07	0.06	–
A-E	ABC	NSPSO	NSDE	NSGA II	**C-E**	ABC	NSPSO	NSDE	NSGA II
ABC	–	**0.96**	**0.88**	**0.98**	ABC	–	**0.82**	**0.62**	**0.87**
NSPSO	0	–	0.39	**0.92**	NSPSO	0.17	–	0.42	**0.81**
NSDE	0	**0.60**	–	**0.86**	NSDE	0.28	**0.53**	–	**0.86**
NSGA II	0	0.06	0.06	–	NSGA II	0.14	0.20	0.15	–
B-C	ABC	NSPSO	NSDE	NSGA II	**D-E**	ABC	NSPSO	NSDE	NSGA II
ABC	–	**0.82**	**0.82**	**0.82**	ABC	–	**0.70**	**0.50**	**0.96**
NSPSO	0.11	–	0.38	**0.78**	NSPSO	0.15	–	0.35	**0.92**
NSDE	0.11	**0.63**	–	**0.91**	NSDE	0.33	**0.51**	–	**0.98**
NSGA II	0.14	0.12	0.10	–	NSGA II	0.03	0.02	0	–

three algorithms as it performs better in 100% of the instances, based on Tables 5 and 6. In Table 7, the PMS-ABC algorithm performs better than PMS-NPSO3 and PMS-NSGA II algorithms as it performs better in 100% of the instances. Also, PMS-ABC algorithm performs better than PMS-NSGA II algorithm as it performs better in 80% of the instances, while PMS-NSGA II algorithm performs better than PMS-ABC algorithm in 20% of the instances. Finally, in Fig. 1 we conclude that PMS-ABC algorithm dominates the Pareto fronts produced from the other three algorithms. So, we conclude that C measure is the most important measure of all, because only in this measure PMS-ABC algorithm performs better.

According to the average numbers of the results, the PMS-NSGA II algorithm produce Pareto front with more solutions, better distribution than the other algorithms and produce more extend Pareto fronts. Finally, the Pareto fronts produced from PMS-ABC algorithm dominates the Pareto fronts produced from the other three algorithms.

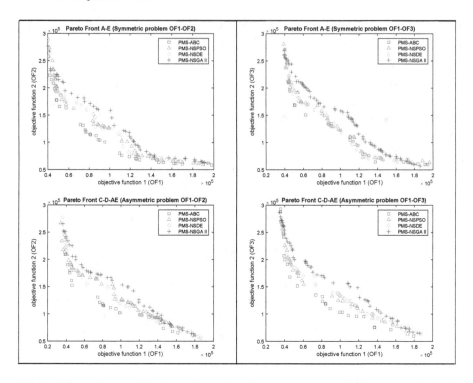

Fig. 1. Pareto fronts of the four algorithms for the instances "A-E" and "C-D-AE".

Table 6. Results of the C measure for the four algorithms in ten instances, when an asymmetric pick-up problem using objective functions OF1-OF3 is solved

OF1-OF3	Multiobjective asymmetric pick-up route based fuel consumption VRP								
A-B-CD	ABC	NSPSO	NSDE	NSGA II	**B-D-AC**	ABC	NSPSO	NSDE	NSGA II
ABC	–	**0.70**	**0.63**	**0.75**	ABC	–	**0.92**	**0.94**	**1.00**
NSPSO	0.13	–	0.41	**0.67**	NSPSO	0.12	–	**0.48**	**0.96**
NSDE	0.26	**0.43**	–	**0.76**	NSDE	0.08	0.38	–	**0.98**
NSGA II	0.16	0.09	0.14	–	NSGA II	0	0	0	–
A-C-BD	ABC	NSPSO	NSDE	NSGA II	**B-E-AD**	ABC	NSPSO	NSDE	NSGA II
ABC	–	**0.70**	**0.50**	**0.88**	ABC	–	**0.70**	**0.78**	**0.84**
NSPSO	0.09	–	0.40	**0.91**	NSPSO	0.12	–	**0.50**	**0.75**
NSDE	0.27	**0.49**	–	**0.82**	NSDE	0.18	0.49	–	**0.84**
NSGA II	0.03	0.02	0.06	–	NSGA II	0.09	0.11	0.11	–

(*continued*)

Table 6. (*continued*)

OF1-OF3	Multiobjective asymmetric pick-up route based fuel consumption VRP							
A-D-BE	ABC	NSPSO	NSDE	NSGA II	**C-D-AE**	ABC	NSPSO	NSDE	NSGA II
ABC	–	**0.73**	**0.64**	**0.89**	ABC	–	**0.83**	**0.92**	**0.88**
NSPSO	0.09	–	0.25	**0.87**	NSPSO	0.07	–	**0.55**	**0.90**
NSDE	0.36	**0.63**	–	**0.87**	NSDE	0.04	0.19	–	**0.83**
NSGA II	0.15	0.08	0.02	–	NSGA II	0.07	0.03	0.02	–
A-E-BD	ABC	NSPSO	NSDE	NSGA II	**C-E-AB**	ABC	NSPSO	NSDE	NSGA II
ABC	–	**0.70**	**0.47**	**0.77**	ABC	–	**0.61**	**0.74**	**0.98**
NSPSO	0.21	–	0.22	**0.80**	NSPSO	0.26	–	**0.40**	**0.85**
NSDE	0.29	**0.65**	–	**0.82**	NSDE	0.19	0.32	–	**0.86**
NSGA II	0.11	0.08	0.10	–	NSGA II	0	0.10	0.04	–
B-C-AD	ABC	NSPSO	NSDE	NSGA II	**D-E-BC**	ABC	NSPSO	NSDE	NSGA II
ABC	–	**0.58**	**0.87**	**0.93**	ABC	–	**0.64**	**0.69**	**0.82**
NSPSO	0.16	–	**0.74**	**0.90**	NSPSO	0.30	–	**0.50**	**0.77**
NSDE	0.06	0.18	–	**0.68**	NSDE	0.17	0.36	–	**0.73**
NSGA II	0.03	0	0.17	–	NSGA II	0.13	0.09	0.17	–

Table 7. Results of the C measure for the four algorithms in ten instances when a symmetric pick-up problem using objective functions OF1-OF3 is solved

OF1-OF3	Multiobjective symmetric pick-up route based fuel consumption VRP							
A-B	ABC	NSPSO	NSDE	NSGA II	**B-D**	ABC	NSPSO	NSDE	NSGA II
ABC	–	**0.54**	**0.43**	**0.95**	ABC	–	**0.71**	**0.59**	**0.87**
NSPSO	0.32	–	**0.48**	**1.00**	NSPSO	0.13	–	0.35	**0.94**
NSDE	0.28	0.31	–	**1.00**	NSDE	0.21	**0.51**	–	**0.89**
NSGA II	0	0	0	–	NSGA II	0.08	0.02	0.02	–
A-C	ABC	NSPSO	NSDE	NSGA II	**B-E**	ABC	NSPSO	NSDE	NSGA II
ABC	–	**0.89**	**0.59**	**0.95**	ABC	–	**0.95**	**0.89**	**0.92**
NSPSO	0.12	–	0.30	**0.84**	NSPSO	0	–	**0.50**	**0.68**
NSDE	0.32	**0.75**	–	**0.90**	NSDE	0	0.43	–	**0.71**
NSGA II	0	0.08	0	–	NSGA II	0	0.15	0.11	–
A-D	ABC	NSPSO	NSDE	NSGA II	**C-D**	ABC	NSPSO	NSDE	NSGA II
ABC	–	**0.78**	**0.49**	**0.94**	ABC	–	**0.52**	**0.73**	**0.84**
NSPSO	0.19	–	0.23	**0.86**	NSPSO	0.23	–	**0.65**	**0.88**
NSDE	0.35	**0.62**	–	**0.94**	NSDE	0.05	0.13	–	**0.81**
NSGA II	0.15	0.07	0.04	–	NSGA II	0.14	0.07	0.14	–
A-E	ABC	NSPSO	NSDE	NSGA II	**C-E**	ABC	NSPSO	NSDE	NSGA II
ABC	–	**0.54**	**0.60**	**0.86**	ABC	–	**0.58**	**0.58**	**0.93**
NSPSO	0.19	–	**0.49**	**0.85**	NSPSO	0.26	–	**0.58**	**0.85**
NSDE	0.19	0.30	–	**0.85**	NSDE	0.43	0.36	–	**0.92**
NSGA II	0.19	0.09	0.12	–	NSGA II	0	0.03	0.03	–
B-C	ABC	NSPSO	NSDE	NSGA II	**D-E**	ABC	NSPSO	NSDE	NSGA II
ABC	–	**0.67**	**0.41**	**0.98**	ABC	–	**0.67**	**0.43**	**0.92**
NSPSO	0.27	–	0.31	**0.95**	NSPSO	0.19	–	0.33	**0.89**
NSDE	0.42	**0.57**	–	**1.00**	NSDE	0.54	**0.49**	–	**0.94**
NSGA II	0	0	0	–	NSGA II	0.04	0.02	0	–

5 Conclusions and Future Research

In this paper, we proposed an algorithm (PMS-ABC) for solving four newly formulated multiobjective fuel consumption multi-depot vehicle routing problems (symmetric and asymmetric pick-up and symmetric and asymmetric delivery cases). The proposed algorithm was compared with other three algorithms, the PMS-NPSO3, the PMS-NSDE and PMS-NSGA II. In general, in the four different problems the PMS-MOCSA algorithm performs slightly better than the other three algorithms in the most measures, as we analyzed in the Computational Results section. As expected, the behavior of the algorithms was slightly different when a symmetric and an asymmetric problem was solved. Our future research will be, mainly, focused on PMS-ABS algorithm in other multiobjective combinatorial optimization problems.

References

1. Baykasoglu, A., Ozbakor, L., Tapkan, P.: Artificial Bee Colony algorithm and its application to generalized assignment problem. In: Chan, F.T.S., Tiwari, M.K., (eds.) Swarm Intelligence, Focus on Ant and Particle Swarm Optimization, pp. 113–144. I-Tech Education and Publishing (2007)
2. Demir, E., Bektaş, T., Laporte, G.: A review of recent research on green road freight transportation. Eur. J. Oper. Res. **237**(3), 775–793 (2014)
3. Hancer, E., Xue, B., Zhang, M., Karaboga, D., Akay, B.: Pareto front feature selection based on artificial bee colony optimization. Inf. Sci. **422**, 462–479 (2017)
4. Kancharla, S., Ramadurai, G.: Incorporating driving cycle based fuel consumption estimation in green vehicle routing problems. Sustain. Cities Soc. **40**, 214–221 (2018)
5. Karaboga, D., Basturk, B.: A powerful and efficient algorithm for numerical function optimization: Artificial Bee Colony (ABC) algorithm. J. Global Optim. **39**, 459–471 (2007)
6. Karaboga, D., Akay, B., Ozturk, C.: Artificial Bee Colony (ABC) optimization algorithm for training feed-forward neural networks. In: Torra, V., Narukawa, Y., Yoshida, Y. (eds.) MDAI 2007. LNCS (LNAI), vol. 4617, pp. 318–329. Springer, Heidelberg (2007). https://doi.org/10.1007/978-3-540-73729-2_30
7. Karaboga, D., Basturk, B.: On the performance of Artificial Bee Colony (ABC) algorithm. Appl. Soft Comput. **8**, 687–697 (2008)
8. Karaboga, D., Akay, B.: A survey: algorithms simulating bee swarm intelligence. Artif. Intell. Rev. (2009). https://doi.org/10.1007/s10462-009-9127-4
9. Karaboga, D., Akay, B.: A comparative study of Artificial Bee Colony algorithm. Appl. Math. Comput. **214**, 108–132 (2009)
10. Karaboga, D., Ozturk, C.: A novel clustering approach: Artificial Bee Colony (ABC) algorithm. Appl. Soft Comput. (2010). https://doi.org/10.1016/j.asoc.2009.12.025
11. Kuo, Y.: Using simulated annealing to minimize fuel consumption for the time-dependent vehicle routing problem. Comput. Ind. Eng. **59**(1), 157–165 (2010)
12. Li, J.: Vehicle routing problem with time windows for reducing fuel consumption. J. Comput. **7**(12), 3020–3027 (2012)

13. Li, J., Wang, R., Li, T., Lu, Z., Pardalos, P.: Benefit analysis of shared depot resources for multi-depot vehicle routing problem with fuel consumption. Transp. Res. Part D Transp. Environ. **59**, 417–432 (2018)
14. Lin, C., Choy, K.L., Ho, G.T.S., Chung, S.H., Lam, H.Y.: Survey of green vehicle routing problem: past and future trends. Expert Syst. Appl. **41**(4), 1118–1138 (2014)
15. Montoya-Torres, J.R., Franco, J.L., Isaza, S.N., Jimenez, H.F., Herazo-Padilla, N.: A literature review on the vehicle routing problem with multiple depots. Comput. Ind. Eng. **79**, 115–129 (2015)
16. Niu, Y., Yang, Z., Chen, P., Xiao, J.: A hybrid tabu search algorithm for a real-world open vehicle routing problem involving fuel consumption constraints. Hindawi (2018). https://doi.org/10.1155/2018/5754908
17. Ozbakir, L., Baykasoglu, A., Tapkan, P.: Bees algorithm for generalized assignment problem. Appl. Math. Comput. **215**, 3782–3795 (2010)
18. Psychas, I.-D., Marinaki, M., Marinakis, Y.: A parallel multi-start NSGA II algorithm for multiobjective energy reduction vehicle routing problem. In: Gaspar-Cunha, A., Henggeler Antunes, C., Coello, C.C. (eds.) EMO 2015. LNCS, vol. 9018, pp. 336–350. Springer, Cham (2015). https://doi.org/10.1007/978-3-319-15934-8_23
19. Psychas, I.D., Marinaki, M., Marinakis, Y., Migdalas, A.: Non-dominated sorting differential evolution algorithm for the minimization of route based fuel consumption multiobjective vehicle routing problems. Energy Syst. **8**, 785–814 (2016)
20. Psychas, I.D., Marinaki, M., Marinakis, Y., Migdalas, A.: Minimizing the fuel consumption of a multiobjective vehicle routing problem using the parallel multi-start NSGA II algorithm. In: Kalyagin, V., Koldanov, P., Pardalos, P. (eds.) Models, Algorithms and Technologies for Network Analysis. Springer Proceedings in Mathematics and Statistics, vol. 156, pp. 69–88. Springer, Cham (2016). https://doi.org/10.1007/978-3-319-29608-1_5
21. Psychas, I.D., Marinaki, M., Marinakis, Y., Migdalas, A.: Parallel multi-start non-dominated sorting particle swarm optimization algorithms for the minimization of the route-based fuel consumption of multiobjective vehicle routing problems. In: Butenko, S., Pardalos, P., Shylo, V. (eds.) Optimization Methods and Applications. Springer Optimization and Its Applications, vol. 130, pp. 425–456. Springer, Cham (2017)
22. Rapanaki, E., Psychas, I.D., Marinaki, M., Marinakis, Y., Migdalas, A.: A clonal selection algorithm for multiobjective energy reduction multi-depot vehicle routing problem. In: Nicosia, G., Pardalos, P., Giuffrida, G., Umeton, R., Sciacca, V. (eds.) LOD 2018. LNCS, vol. 11331, pp. 381–393. Springer, Cham (2019)
23. Sabat, S.L., Udgata, S., Abraham, A.: Artificial Bee Colony algorithm for small signal model parameter extraction of MEFSET. Eng. Appl. Artif. Intell. (2010). https://doi.org/10.1016/j.engappai.2010.01.020
24. Srivastava, S.K.: Green supply-chain management: a state-of the-art literature review. Int. J. Manag. Rev. **9**(1), 53–80 (2007)
25. Suzuki, Y.: A new truck-routing approach for reducing fuel consumption and pollutants emission. Transp. Res. Part D **16**(1), 73–77 (2011)
26. Toth, P., Vigo, D.: Vehicle Routing: Problems, Methods and Applications. MOS-Siam Series on Optimization, 2nd edn. SIAM, Philadelphia (2014)
27. Xiao, Y., Zhao, Q., Kaku, I., Xu, Y.: Development of a fuel consumption optimization model for the capacitated vehicle routing problem. Comput. Oper. Res. **39**(7), 1419–1431 (2012)

PTAS for the Euclidean Capacitated Vehicle Routing Problem with Time Windows

Michael Khachay[1,2,3(✉)] and Yuri Ogorodnikov[1,2]

[1] Krasovsky Institute of Mathematics and Mechanics, Ekaterinburg, Russia
{mkhachay,yogorodnikov}@imm.uran.ru
[2] Ural Federal University, Ekaterinburg, Russia
[3] Omsk State Technical University, Omsk, Russia

Abstract. The Capacitated Vehicle Routing Problem with Time Windows (CVRPTW) is the well-known combinatorial optimization problem having numerous valuable applications in operations research. Unlike the classic CVRP (without time windows constraints), approximability of the CVRPTW (even in the Euclidean plane) in the class of algorithms with theoretical guarantees is much less studied. To the best of our knowledge, the family of such algorithms is exhausted by the Quasi-Polynomial Time Approximation Scheme (QPTAS) proposed by L. Song et al. for the general setting of the planar CVRPTW and two our recent approximation algorithms, which are Efficient Polynomial Time Approximation Schemes (EPTAS) for any fixed capacity q and number p of time windows and remain PTAS for slow-growing dependencies $q = q(n)$ and $p = p(n)$. In this paper, combining the well-known instance decomposition framework by A. Adamaszek et al. and QPTAS by L. Song et al. we propose a novel approximation scheme for the planar CVRPTW, whose running time remains polynomial for the significantly wider range of q and p.

Keywords: Capacitated vehicle routing problem · Time windows · Efficient polynomial time approximation scheme

1 Introduction

The Capacitated Vehicle Routing Problem (CVRP) is the famous combinatorial optimization problem, which was introduced by Dantzig and Ramser in their seminal paper [7] and has a wide range of relevant applications in practice (see, e.g. [28]). In the classic setting of the CVRP, we are given by a finite set of customers and a fleet of vehicles of the same capacity initially located at a depot. The goal is to construct a family of vehicle routes servicing all the customers and having the minimum total transportation cost.

The Capacitated Vehicle Routing Problem with Time Windows (CVRPTW) [19,28] is an extension of the CVRP, where the service of each customer is associated with a specified time interval, called a time window. There are two

© Springer Nature Switzerland AG 2020
N. F. Matsatsinis et al. (Eds.): LION 13 2019, LNCS 11968, pp. 224–230, 2020.
https://doi.org/10.1007/978-3-030-38629-0_18

different types of such intervals. For the former one, called *hard*, each customer should be serviced within its assigned time window exclusively, whilst, in the latter one, *soft*, vehicles can violate any time window constraints for some penalty cost. As a mathematical model, CVRP with hard windows (or, just CVRPTW) is widely employed in low-carbon economy [25], continent-scale distribution of building materials [22], in dial-ride company planning [10] and other practical transportation problems (see, e.g. [24]).

In this paper, we are focused on efficient approximation of the CVRPTW in the class of algorithms with theoretic accuracy and time complexity bounds. Therefore, in the following short literature overview, we intentionally restrict ourselves on recent results of this kind, although the significant progress in the solution of practical instances of the problem by local-search heuristics [12], genetic [29], memetic [6,20], and ant colony algorithms [21] should be mentioned as well.

This short paper is structured as follows. In Sect. 2, we give a short overview of known results. Further, in Sect. 3, we remind the mathematical setting of the CVRPTW. In Sect. 4, we discuss the approximation scheme proposed. For the sake of brevity, we skip all the proofs and technical results postponing them to the forthcoming paper, and concentrate on the main ideas.

2   Related Work

Extending the well-known Traveling Salesman Problem (TSP) [28], the Capacitated Vehicle Routing Problem having the capacity q as a part of the input is strongly NP-hard and remain intractable even in the Euclidean plane [23]. For $q = 2$, the metric CVRP is polynomially solvable, since it can be reduced to the minimum weight perfect matching problem. For any fixed $q \geq 3$, the problem is strongly NP- and APX-hard even for $(1, 2)$-metric. The proof follows from a polynomial time cost preserving reduction from the Partition into Isomorphic Subgraphs [9, Problem GT12].

Approximation results for the Euclidean CVRP date to the seminal paper by Haimovich and Rinnooy Kan [11], where the first PTAS for the CVRP on the plane and capacity $q = o(\log \log n)$ was introduced. Then, in [3] this result was improved to $q = O(\log n / \log \log n)$. Also, they noticed that there can be derived a PTAS for the CVRP problem with $q = \Omega(n)$ from Arora's PTAS for the two-dimensional Euclidean TSP [2].

The ideas proposed by Arora in his paper [2] were used by Das and Mathieu in their Quasi Polynomial Time Approximation Scheme (QPTAS) [8] for the general case of the planar Euclidean CVRP. Their QPTAS finds a $(1 + \varepsilon)$-approximate solution of this problem (for any dependence $q = q(n)$) in time $n^{(\log n)^{O(1/\varepsilon)}}$. Using this QPTAS as a black-box, Adamaszek, Szumaj, and Lingas [1] showed that $(1 + \varepsilon)$-approximate solution can be found in polynomial time for $q \leq 2^{\log^\delta n}$, where $\delta = \delta(\varepsilon)$.

Some part of the aforementioned results found their extension to the case of Euclidean spaces of an arbitrary finite dimension [13,17,18] and several special graphs [4,5].

Unlike CVRP, approximability of the Euclidean CVRPTW is much less investigated. To the best of our knowledge, the family of known approximation algorithms for this problem is exhausted by a quasi-polynomial time approximation scheme (QPTAS) developed in [26,27] for the general case of the problem and an approximation schemes proposed in [15,16] for $\max\{p,q\} = o(\log \log n)$ and $p^3 q^4 = O(\log n)$ respectively, where p is the number of time windows.

In this paper, we propose a novel approximation scheme, having a polynomial time complexity for the faster-growing dependencies $p = p(n)$ and $q = q(n)$.

3   Problem Statement

We consider the simplest single depot case of the CVRPTW in the Euclidean plane with disjoint time windows, which we call in the sequel the planar CVRPTW or just CVRPTW, for the sake of brevity.

The instance of CVRPTW is given by a finite set $X = \{x_1, \ldots, x_n\}$ of *customer* locations (points on the plane), a dedicated location – *depot* y, a capacity q, and an linearly ordered set $\mathcal{T} = \{T_1, \ldots, T_p\}$ of disjoint nonempty intervals—*time windows*. We assume that $T_j \prec T_{j+1}$ holds for any $j \in \{1, \ldots, p-1\} = [p-1]$, i.e. the inequality $t_1 < t_2$ is valid for any moments $t_1 \in T_j$ and $t_2 \in T_{j+1}$. Each customer x_i has a unit demand, which should be serviced by a single *route* during some time window $T = T(x_i) \in \mathcal{T}$.

Any *feasible* route R has the form $R = y, x_{i_1}, \ldots, x_{i_l}, y$, where $l \leq q$ (capacity constraint) and $T(x_{i_j}) \preceq T(x_{i_{j+1}})$ for any $j \in [l-1]$ (time windows constraint), and the *transportation cost* $w(R) = \|y - x_{i_1}\|_2 + \|x_{i_1} - x_{i_2}\|_2 + \ldots + \|x_{i_l} - y\|_2$. The goal is to find a minimum cost collection $U = \{R_1, \ldots, R_b\}$ of feasible routes visiting each customer exactly once.

4   Approximation Scheme

The main idea of our approximation scheme is based on the well-known rounding and instance decomposition framework proposed in [1], quasi-polynomial approximation scheme [27] used as a black-box for finding approximate solutions of the obtained subinstances and our recent technical results [14–16] for the CVRPTW in the Euclidean plane.

Suppose, we are given an instance of the planar CVRPTW and $\varepsilon \in (0,1)$. It is required to find an $(1+\varepsilon)$-approximate solution of the given instance. The proposed scheme consists of several stages, which are described briefly in the subsequent subsections.

4.1 Instance Preprocessing and Accuracy Dependent Rounding

First, we sort the customer locations x_i in decreasing order according to their distances $r_i = \|x_i - y\|_2$ from the depot y and exclude customers with $r_i \leq r_1 \varepsilon / n$, since all of them can be visited for the cost at most $n \cdot 2r_i \leq 2r_1 \varepsilon \leq \varepsilon \, \mathrm{OPT}$, where OPT is an optimum of the given instance. Then, we shift the origin to y and introduce an orthogonal grid in polar coordinates with parameters as follows:

$$\rho_i = \frac{r_1 \varepsilon}{n} \left(1 + \frac{\varepsilon}{q}\right)^i, \ 0 \leq i \leq \left\lceil \log_{1 + \frac{\varepsilon}{q}} \frac{n}{\varepsilon} \right\rceil, \quad \varphi_j = \frac{2\pi j}{s}, \ 0 \leq j \leq s = \left\lceil \frac{2\pi q}{\varepsilon} \right\rceil.$$

Points with polar radii ρ_i and angles φ_j obtained we call *nodes*. To any node ν, we assign p collocated *slots* $\nu(T_1), \dots, \nu(T_p)$, that are "subnodes" induced by time windows T_i. So, after such a transformation we move each customer x_i to the nearest slot $\nu(T(x_i))$.

Lemma 1. *The number of slots is* $N = \Theta \left(\left(\frac{q}{\varepsilon} \right)^2 \log \frac{n}{\varepsilon} \right)$. *Moving each customer to the nearest slot changes the cost of a solution by at most* $\varepsilon \, \mathrm{OPT}$.

As it follows from Lemma 1, in the sequel, we can restrict ourselves to special instances of the planar CVRPTW, where all customers are located at the slots prescribed above. Following Adamaszek et al. [1] we call such instances *rounded*.

4.2 Instance Decomposition

It is easy to verify that, after the rounding, the number of non-empty slots is at most n. To each of these slots, at least one customer is assigned. In the sequel, we show that the total number of the assigned customers can be reduced further.

First, we extend the concept of *non-trivial* routes from [1] to the case of time windows. We call a route R non-trivial if it visits at least two different slots (maybe assigned to the same node). Otherwise, the route R is called trivial.

Lemma 2. *For any rounded CVRPTW, there exists an optimum solution with at most $N \cdot p$ non-trivial routes.*

Due to the capacity constraint, the total number of customers, which can be serviced by non-trivial routes, does not exceed Npq. Since, for the trivial routes, an optimal solution can be constructed trivially as well, we can reduce the number of customers assigned to any slot to at most Npq and consider the rounded instance with at most

$$N^2 p^2 q = O \left(\frac{q^5 p^2}{\varepsilon^4} \log^2 \frac{n}{\varepsilon} \right)$$

customers.

Then, we split the polar grid onto alternating vertical stripes (between fixed polar radii), say *white* and *gray* and decompose the initial instance to smaller subinstances induced by these stripes. Then, we solve any gray subinstance using our modification [15] of the famous Iterative Tour Partition (ITP) [11]. By OPT(w) and ITP(g) denote the optimum value of some white subinstance and the cost of the ITP-based approximate solution for some gray subinstance. The following lemma holds.

Lemma 3. *For any rounded instance of the CVRPTW and any $\varepsilon \in (0,1)$ there exists a partition to at most $O(\log(n/\varepsilon)/(\varepsilon \log(n/\varepsilon)))$ white and gray stripes such that each white subinstance w is defined by at most $poly(p, q, 1/\varepsilon)$ customers and*

$$\sum_w \text{OPT}(w) + \sum_g \text{ITP}(g) \leq (1 + \varepsilon)\text{OPT}.$$

Lemma 3 implies that the initial rounded CVRPTW instance can be decomposed to a collection of much smaller subinstances, which can be solved in parallel.

4.3   Main Result

Applying QPTAS proposed by Song et al. [27] to find approximate solutions for the white subinstances, we obtain our main result

Theorem 1. *For any $\varepsilon \in (0,1)$ there is $\delta = O(\varepsilon)$ such that an $(1 + \varepsilon)$-approximate solution of the CVRPTW in the Euclidean plane can be found in polynomial time provided*

$$\max\{p, q\} = O(2^{\log^\delta n}).$$

5   Conclusion

In this paper, we introduced a new approximation scheme for the Capacitated Vehicle Routing Problem with Time Windows. Due to smart problem decomposition [1] and employing the recent QPTAS proposed in [27] we improve our recent approximation results [15,16] significantly enlarging the range of p and q, for which the CVRPTW in the Euclidean plane can be approximated by PTAS.

In the forthcoming paper, we extend this result to the case of the Euclidean CVRPTW with splittable non-uniform demand.

References

1. Adamaszek, A., Czumaj, A., Lingas, A.: PTAS for k-tour cover problem on the plane rof moderately large values of k. Int. J. Found. Comput. Sci. **21**(06), 893–904 (2010). https://doi.org/10.1142/S0129054110007623
2. Arora, S.: Polynomial time approximation schemes for Euclidean traveling salesman and other geometric problems. J. ACM **45**, 753–782 (1998)
3. Asano, T., Katoh, N., Tamaki, H., Tokuyama, T.: Covering points in the plane by k-tours: towards a polynomial time approximation scheme for general k. In: Proceedings of the Twenty-Ninth Annual ACM Symposium on Theory of Computing, STOC 1997, pp. 275–283. ACM, New York (1997). https://doi.org/10.1145/258533.258602

4. Becker, A., Klein, P.N., Saulpic, D.: A quasi-polynomial-time approximation scheme for vehicle routing on planar and bounded-genus graphs. In: Pruhs, K., Sohler, C. (eds.) 25th Annual European Symposium on Algorithms, ESA 2017, Vienna, Austria, 4–6 September 2017, LIPIcs, vol. 87, pp. 12:1–12:15. Schloss Dagstuhl - Leibniz-Zentrum fuer Informatik (2017). https://doi.org/10.4230/LIPIcs.ESA.2017.12, http://www.dagstuhl.de/dagpub/978-3-95977-049-1

5. Becker, A., Klein, P.N., Saulpic, D.: Polynomial-time approximation schemes for k-center, k-median, and capacitated vehicle routing in bounded highway dimension. In: Azar, Y., Bast, H., Herman, G. (eds.) 26th Annual European Symposium on Algorithms, ESA 2018, August 20–22, 2018, Helsinki, Finland. LIPIcs, vol. 112, pp. 8:1–8:15. Schloss Dagstuhl-Leibniz-Zentrum fuer Informatik (2018). https://doi.org/10.4230/LIPIcs.ESA.2018.8, http://www.dagstuhl.de/dagpub/978-3-95977-081-1

6. Blocho, M., Czech, Z.: A parallel memetic algorithm for the vehicle routing problem with time windows. In: 2013 Eighth International Conference on P2P, Parallel, Grid, Cloud and Internet Computing, pp. 144–151 (2013). https://doi.org/10.1109/3PGCIC.2013.28

7. Dantzig, G., Ramser, J.: The truck dispatching problem. Manage. Sci. **6**, 80–91 (1959)

8. Das, A., Mathieu, C.: A quasipolynomial time approximation scheme for Euclidean capacitated vehicle routing. Algorithmica **73**, 115–142 (2015). https://doi.org/10.1007/s00453-014-9906-4

9. Garey, M.R., Johnson, D.S.: Computers and intractability: a guide to the theory of NP-completeness. W. H. Freeman & Co., New York (1979)

10. Gschwind, T., Irnich, S.: Effective handling of dynamic time windows and its application to solving the dial-a-ride problem. Transp. Sci. **49**(2), 335–354 (2015)

11. Haimovich, M., Rinnooy Kan, A.H.G.: Bounds and heuristics for capacitated routing problems. Math. Oper. Res. **10**(4), 527–542 (1985). https://doi.org/10.1287/moor.10.4.527

12. Hashimoto, H., Yagiura, M.: A path relinking approach with an adaptive mechanism to control parameters for the vehicle routing problem with time windows. In: van Hemert, J., Cotta, C. (eds.) EvoCOP 2008. LNCS, vol. 4972, pp. 254–265. Springer, Heidelberg (2008). https://doi.org/10.1007/978-3-540-78604-7_22

13. Khachai, M.Y., Dubinin, R.D.: Approximability of the vehicle routing problem infinite-dimensional Euclidean spaces. Proc. Steklov Inst. Math. **297**(1), 117–128 (2017). https://doi.org/10.1134/S0081543817050133

14. Khachai, M., Ogorodnikov, Y.: Polynomial time approximation scheme for the capacitated vehicle routing problem with time windows. Trudy instituta matematiki i mekhaniki UrO RAN **24**, 233–246 (2018). https://doi.org/10.21538/0134-4889-2018-24-3-233-246

15. Khachay, M., Ogorodnikov, Y.: Efficient PTAS for the Euclidean CVRP with time windows. In: van der Aalst, W.M.P., et al. (eds.) AIST 2018. LNCS, vol. 11179, pp. 318–328. Springer, Cham (2018). https://doi.org/10.1007/978-3-030-11027-7_30

16. Khachay, M., Ogorodnikov, Y.: Improved polynomial time approximation scheme for capacitated vehicle routing problem with time windows. In: Evtushenko, Y., Jaćimović, M., Khachay, M., Kochetov, Y., Malkova, V., Posypkin, M. (eds.) OPTIMA 2018. CCIS, vol. 974, pp. 155–169. Springer, Cham (2019). https://doi.org/10.1007/978-3-030-10934-9_12

17. Khachay, M., Dubinin, R.: PTAS for the Euclidean capacitated vehicle routing problem in R^d. In: Kochetov, Y., Khachay, M., Beresnev, V., Nurminski, E., Pardalos, P. (eds.) DOOR 2016. LNCS, vol. 9869, pp. 193–205. Springer, Cham (2016). https://doi.org/10.1007/978-3-319-44914-2_16

18. Khachay, M., Zaytseva, H.: Polynomial time approximation scheme for single-depot Euclidean capacitated vehicle routing problem. In: Lu, Z., Kim, D., Wu, W., Li, W., Du, D.-Z. (eds.) COCOA 2015. LNCS, vol. 9486, pp. 178–190. Springer, Cham (2015). https://doi.org/10.1007/978-3-319-26626-8_14

19. Kumar, S., Panneerselvam, R.: A survey on the vehicle routing problem and its variants. Intell. Inf. Manage. 4, 66–74 (2012). https://doi.org/10.4236/iim.2012.43010

20. Nalepa, J., Blocho, M.: Adaptive memetic algorithm for minimizing distance in the vehicle routing problem with time windows. Soft Comput. 20(6), 2309–2327 (2016). https://doi.org/10.1007/s00500-015-1642-4

21. Necula, R., Breaban, M., Raschip, M.: Tackling dynamic vehicle routing problem with time windows by means of ant colony system. In: 2017 IEEE Congress on Evolutionary Computation (CEC), pp. 2480–2487 (2017). https://doi.org/10.1109/CEC.2017.7969606

22. Pace, S., Turky, A., Moser, I., Aleti, A.: Distributing fibre boards: a practical application of the heterogeneous fleet vehicle routing problem with time windows and three-dimensional loading constraints. Procedia Comput. Sci. 51, 2257–2266 (2015). https://doi.org/10.1016/j.procs.2015.05.382. International Conference on Computational Science, ICCS 2015

23. Papadimitriou, C.: Euclidean TSP is NP-complete. Theor. Comput. Sci. 4, 237–244 (1977)

24. Savelsbergh, M., van Woensel, T.: 50th anniversary invited article - city logistics: challenges and opportunities. Transp. Sci. 50(2), 579–590 (2016). https://doi.org/10.1287/trsc.2016.0675

25. Shen, L., Tao, F., Wang, S.: Multi-depot open vehicle routing problem with time windows based on carbon trading. Int. J. Environ. Res. Public Health 15(9), 2025 (2018). https://doi.org/10.3390/ijerph15092025

26. Song, L., Huang, H.: The Euclidean vehicle routing problem with multiple depots and time windows. In: Gao, X., Du, H., Han, M. (eds.) COCOA 2017. LNCS, vol. 10628, pp. 449–456. Springer, Cham (2017). https://doi.org/10.1007/978-3-319-71147-8_31

27. Song, L., Huang, H., Du, H.: Approximation schemes for Euclidean vehicle routing problems with time windows. J. Comb. Optim. 32(4), 1217–1231 (2016). https://doi.org/10.1007/s10878-015-9931-5

28. Toth, P., Vigo, D.: Vehicle Routing: Problems, Methods, and Applications, 2nd edn. MOS-SIAM Series on Optimization, SIAM (2014)

29. Vidal, T., Crainic, T.G., Gendreau, M., Prins, C.: A hybrid genetic algorithm with adaptive diversity management for a large class of vehicle routing problems with time-windows. Comput. Oper. Res. 40(1), 475–489 (2013). https://doi.org/10.1016/j.cor.2012.07.018

On a Cooperative VNS Parallelization Strategy for the Capacitated Vehicle Routing Problem

Panagiotis Kalatzantonakis[ID], Angelo Sifaleras[(✉)][ID], and Nikolaos Samaras[ID]

Department of Applied Informatics, School of Information Sciences,
University of Macedonia, 156 Egnatia Street, 54636 Thessaloniki, Greece
mai18019@uom.edu.gr, {sifalera,samaras}@uom.gr

Abstract. It is generally accepted that cooperation-based strategies in parallel metaheuristics exhibit better performances in contrast with non-cooperative approaches. In this paper, we study how the cooperation between processes affects the performance and solution quality of parallel algorithms. The purpose of this study is to provide researchers with a practical starting point for designing better cooperation strategies in parallel metaheuristics. To achieve that, we propose two parallel models based on the general variable neighborhood search (GVNS) to solve the capacitated vehicle routing problem (CVRP). Both models scan the search space by using multiple search processes in parallel. The first model lacks communication, while on the other hand, the second model follows a strategy based on information exchange. The received solutions are utilized to guide the search. We conduct an experimental study using well-known benchmark instances of the CVRP, in which the usefulness of communication throughout the search process is assessed. The findings confirm that careful design of the cooperation strategy in parallel metaheuristics can yield better results.

Keywords: Parallel metaheuristics · Variable neighborhood search · Cooperation strategies · Vehicle routing problem · Intelligent optimization methods

1 Introduction

Dantzig and Ramser [9] introduced the CVRP, which belongs to the class of routing problems and is a variation of VRP, with additional constraints on the capacities of the vehicles. CVRP is an NP-hard problem with notable impact on the fields of transportation, distribution, and logistics. The fact that most NP-hard problems become intractable for exact methods, mainly when dealing with large instances, has motivated researchers in developing a plethora of approximation algorithms, heuristics, and metaheuristics that provide an optimal, or close to the optimal, solution. The Variable Neighborhood Search (VNS)

© Springer Nature Switzerland AG 2020
N. F. Matsatsinis et al. (Eds.): LION 13 2019, LNCS 11968, pp. 231–239, 2020.
https://doi.org/10.1007/978-3-030-38629-0_19

metaheuristic has been successfully applied for solving many discrete and global optimization problems [5,13].

The purpose of this paper is to present two parallel VNS methods using the general VNS variant to tackle the CVRP, and to examine how the level of cooperation between threads can affect the performance and the quality of the solutions.

The remainder of this paper is organized as follows: In Sect. 2, we present related works, in which the impact of communication in parallel algorithms is analyzed. In Sect. 3, we present two parallel VNS models for the solution of the CVRP. In Sect. 4, we present the summary of the findings for the models. Finally, conclusions and prospects are summarized in Sect. 5.

2 Related Work

Recently, parallelization processing methods are increasingly being used in meta-heuristics, due to the broadly available multicore processors and distributed computing environments. Contributions focused on the communication strategies are sparse. In [6], Crainic focused on different cooperation-based strategies. In this study, Crainic found that approaches based on asynchronous exchanges of information and the formulation of new knowledge out of exchanged data improve the global guidance of the search and display extraordinary performances. The author noticed that, low level communication schemes are particularly attractive when neighborhoods or populations are large, or the neighbor or individual evaluation is costly. Those low level schemes were classified in Crainic taxonomy with the 1st dimension marked as 1-control. Crainic taxonomy is discussed in Sect. 3.3.

The articles assessed for the literature review relate to cooperative parallel metaheuristics that, are based on the VNS algorithm and have been applied on several problems. Table 1 sums the related works. The authors of these works focus on the effectiveness of the proposed cooperative models rather than the reasoning for selecting the cooperation strategy.

Table 1. Parallel VNS metaheuristic applied in several problems

Related work	Metaheuristic algorithm	Problem
García-López et. al. [10]	Parallel VNS	P-median
Crainic et al. [7]	Parallel VNS	P-median
Aydin and Sevkli [2]	Parallel VNS	Job shop scheduling
Polacek et al. [11]	Parallel VNS	MDVRPTW
Coelho et al. [4]	Parallel VNS	SVRPDSP
Polat O. [12]	Parallel VNS	VRPDP
Antoniadis et. al. [1]	Parallel VNS	Inventory optimization

3 Information Exchange Between Parallel Models

It is generally accepted that, adding cooperation to parallel algorithms provides a critical boost to create solutions of the highest quality. In order to study the effect of communication between the threads, we created two parallel GVNS models. The Savings Algorithm of Clarke and Wright [3] was used to construct the initial solution for the two models. Both models are using an identical neighborhood structure, consisting of three widely used inter and intra-route operators, i.e., 2-opt (Intra-route), Swap (Inter-route), and Relocate (Inter-route). In order for the two models to have the same resources in the search for a solution, a single thread was used to play the role of the solution warehouse.

3.1 Parallel GVNS - Managed Information Exchange Model

In this model, Clarke and Wright's algorithm provides the initial solution to all the threads, except for one, which will assume the role of the solution warehouse. Communication is asynchronous and dynamically determined by each process. The threads begin their search in the solution space and if a thread finds a better solution then, and only then, communication between the current thread and the solution warehouse is initiated. The thread passes the solution to the warehouse/manager process. At that point, a check is taking place. If a better solution exists in the solution warehouse, then it is adopted by the thread, and the search continues.

The communication schema used in the managed information exchange model is novel and its purpose is to create a sparse communication graph. The target is to maintain an equilibrium between exploration and exploitation phases. As shown in Fig. 1, while three solutions are generated, one gets rejected by the

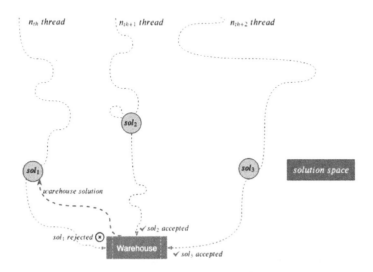

Fig. 1. Three solutions were passed to the warehouse. sol_1 was rejected and the first thread adopted a better solution (sol_3) from the warehouse (blue arrow). (Color figure online)

warehouse and the thread that passed that solution adopts a new, better solution from the warehouse.

The novelty in the cooperative model resides in the fact that, not only no broadcasting takes place, but also the information exchange between a process and the solution warehouse happens at irregular intervals. Each process dynamically determines those intervals, and even when they occur, the thread might not adopt the available solution from the solution warehouse.

Solution adoption by the warehouse, much like communication initiation is being controlled by each individual process and can be configured to filter-out solutions based on several criteria. The algorithm of this model is shown in Algorithm 1.

Algorithm 1. Pseudo code - Managed information exchange model

---

```
// fr: Current best solution route
// kmax: Shake k parameter
// timelimit: The time limit stopping criteria
input  : fr, kmax, timelimit, initial_solution

while true do
    t ← CpuTime()
    fr' ← Shake(fr, kmax)
    fr'' ← VND(fr', t, timelimit)
    // When a new solution has been found communicate with warehouse
    if thread_solution < thread_current_best then
        thread_current_best ← thread_solution
        if thread_current_best < solution_warehouse then
            // Give solution to warehouse and continue
            solution_warehouse ← thread_current_best
        end
        else
            // Get new solution from solution warehouse and restart.
            thread_current_best, thread_solution ← solution_warehouse
        end
    end
    if t > timelimit then
        break
    end
    if solution_warehouse = optimum then
        break
    end
end
```

3.2 Parallel GVNS - A Model with Isolated Processes

This non-cooperative model, as shown in Fig. 2, uses an island-based design where every thread runs the GVNS wholly isolated. All threads utilize identical search procedures. Once the primary solution is produced using the Clarke and Wright algorithm, it is used as a starting point by each thread.

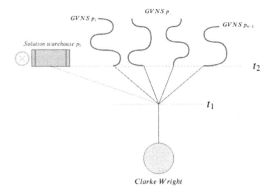

Fig. 2. The Clarke and Wright algorithm generates a solution that is passed to all processes. Best solutions are stored in solution warehouse, but never broadcasted.

Each thread works autonomously, and their paths deviate particularly when the shaking procedure takes place. When the stopping criteria have been met, then all the threads terminate. The best solution is then picked among the list of best solutions. The algorithm of this model is shown in Algorithm 2.

Algorithm 2. Pseudo code - GVNS thread, Isolated model

---

input : $fr, kmax, timelimit$

while $true$ **do**
  $t \leftarrow CpuTime()$
  $fr' \leftarrow Shake(fr, kmax)$
  $fr'' \leftarrow VND(fr', t, timelimit)$
  if $t > timelimit$ **then**
  | break
  end
  // if optimum value exists
  if $best\_value = optimum$ **then**
  | break
  end
end

3.3   Model Classification

Crainic and Hail [8] suggested three dimensions to classify parallel metaheuristic strategies:

- 1st dimension: Search control cardinality
- 2nd dimension: Search control and communication
- 3rd dimension: Search strategies

According to this taxonomy, the proposed models can be classified as follows:

– The non cooperative model fits into the pC/RS/SPSS classification.
– The cooperative model fits into pC/C/SPSS classification.

The first dimension of this taxonomy defines how the search process for new solutions is controlled; pc stands for poly-Control meaning that, there is more than one process that controls the search operation.

In our case, each single thread has its control for the search operation. The second dimension defines how the information between processes is exchanged. RS stands for rigid synchronization, meaning that little or no information exchange takes place when we use the non-cooperative model. The second dimension in the cooperative model is classified as "Collegial", thus we extract and adopt only the best solutions when information exchange occurs. The third dimension refers to how new information is created, and the diversity of searches involved. Both models are classified as "SPSS" that stands for "Same initial Point, Same search Strategy". This makes sense since all the threads use Clarke and Wright as an initial solution and all the threads have an identical neighborhood structure.

4   Computational Experiments

This section presents the results of the computational experiments carried out to ascertain the performance of the two parallel GVNS models. The practical relevance of the communication strategy is presented and analyzed.

All the algorithms were implemented in Python 3.7. The experiments were conducted on an Intel Core i9 7940X CPU (3.50 GHz) and 32 GB RAM at 3333 MHz. Both models have a single termination criterion; the test is repeated until a certain number of GVNS iterations is met. The two parallel models were tested with the following iterations: 5, 10, 20, 30, 40, 50, 100, 200, and 300. All tests were repeated ten times.

The computational tests were carried out on instances from the X set from the CVRP library [14]. The test set is composed of a subset of the X set containing 6 instances (X-n110-k13, X-n143-k7, X-n153-k22, X-n256-k16, X-n261-k13 and X-n280-k17). Every instance in the X set was generated by Uchoa with specific characteristics. From the computational effort associated with the instance characteristics and the size of the neighborhood to be explored, the instances can be categorized into the following three groups:

(a) Easy (X-n110-k13),
(b) Medium (X-n143-k7, X-n153-k22)
(c) Hard (X-n261-k13, X-n256-k16, X-n280-k17)

In the results showed in Table 2, we can observe essential differences among the compared methods. The isolated model appears to be much faster but the cooperative model provides solutions with better quality. "SpI" stands for "seconds per 1 GVNS iteration". The solution manager never broadcasts the best solution in the cooperative model, in order to minimize the communication overhead. In spite of this fact, the isolated method is 21.108% faster. This can be explained to a large extent by the fact that, communication and solution comparison deprives the search procedure of some CPU cycles. Even though the isolated method is much faster and has a more intense diversification phase, communication between processes appears to yield better results.

Table 2. Comparison of the two GVNS parallel variants

	Isolated scheme	Managed scheme
Mean error	10.487%	10.078%
Median error	11.460%	11.471%
SpI	4.982	6.315

When we focus on the instances, based on their characteristics and grouped by the computational effort required by the CPU to complete one GVNS iteration, an interesting pattern emerges. As shown in Table 3, information sharing outperforms isolation in hard instances. When the search space is small (easier instances), the non-cooperative method yields better results. Thus, information sharing seems to outperform the independent search method and constitutes a valuable strategy for tackling hard instances when setting small iteration count as a stopping criterion. As shown in Fig. 3, after several GVNS iterations, the two methods don't display essential differences.

Table 3. Model performance based on computational effort

	Isolated scheme	Managed scheme
Mean error - (easy)	4.310%	4.955%
Mean error - (medium)	11.488%	11.423%
Mean error - (hard)	9.131	8.434%
SpI - (easy)	1.234	1.652
SpI - (medium)	5.000	6.153
SpI - (hard)	18.750	20.689

In order to support our findings, we applied a Friedman test to the performance results collected by the two models executed across the same set of instances and obtained a p-value equal to zero showing that, there is enough statistical evidence to consider the two algorithms different. We consider a common significance level $\alpha = 0.05$ as the threshold for rejecting the null hypothesis.

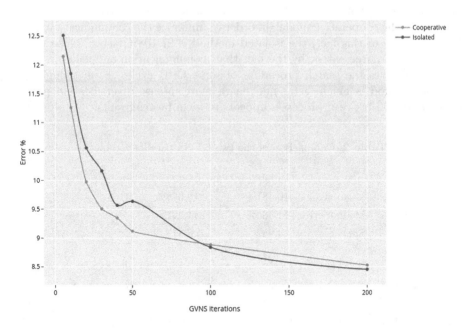

Fig. 3. Performance of the two models at 5, 10, 20, 30, 40, 50, 100 and 200 GVNS iterations

5   Conclusions

In this paper, we proposed two models for the parallelization of the variable neighborhood search for the efficient solution of CVRP. Our goal was to study how the communication between processes affects the performance and the solution quality of parallel algorithms. Well known instances were used in order to compare and analyze the effect of the cooperation strategies between the two parallel metaheuristic models.

Cooperation strategy can have a decisive influence on the quality of the solutions. There is a strong indication that cooperation yields better results over hard instances, whereas in small solution spaces isolation appears to be the best strategy. The timing of communication also appears to play a role since no communication near the end of the search yields better results.

Future studies may include the use of filters to better guide the solution adoption and smarter memory-based strategies to provide better solutions.

Acknowledgements. The second author has been funded by the University of Macedonia Research Committee as part of the "Principal Research 2019" funding scheme (ID 81307).

References

1. Antoniadis, N., Sifaleras, A.: A hybrid CPU-GPU parallelization scheme of variable neighborhood search for inventory optimization problems. Electron. Notes Discrete Math. **58**, 47–54 (2017)
2. Aydin, M.E., Sevkli, M.: Sequential and parallel variable neighborhood search algorithms for job shop scheduling. In: Xhafa, F., Abraham, A. (eds.) Metaheuristics for Scheduling in Industrial and Manufacturing Applications. SCI, vol. 128, pp. 125–144. Springer, Heidelberg (2008). https://doi.org/10.1007/978-3-540-78985-7_6
3. Clarke, G., Wright, J.W.: Scheduling of vehicles from a central depot to a number of delivery points. Oper. Res. **12**(4), 568–581 (1964)
4. Coelho, I.M., Ochi, L.S., Munhoz, P.L.A., Souza, M.J.F., Farias, R., Bentes, C.: The single vehicle routing problem with deliveries and selective pickups in a CPU-GPU heterogeneous environment. In: 14th IEEE International Conference on High Performance Computing and Communication & 9th IEEE International Conference on Embedded Software and Systems (HPCC-ICESS), pp. 1606–1611. IEEE (2012)
5. Coelho, V.N., Santos, H.G., Coelho, I.M., Oliveira, T.A., Penna, P.H.V., Souza, M.J.F., Sifaleras, A.: 5th international conference on variable neighborhood search (ICVNS'17). Electron. Notes Discrete Math. **66**, 1–5 (2018)
6. Crainic, T.: Parallel metaheuristics and cooperative search. In: Gendreau, M., Potvin, J.Y. (eds.) Handbook of Metaheuristics. ISOR, vol. 272, pp. 419–451. Springer, Cham (2018). https://doi.org/10.1007/978-3-319-91086-4_13
7. Crainic, T.G., Gendreau, M., Hansen, P., Mladenović, N.: Cooperative parallel variable neighborhood search for the p-median. J. Heuristics **10**(3), 293–314 (2004)
8. Crainic, T.G., Hail, N.: Parallel metaheuristics applications. Parallel Metaheuristics: New Class Algorithms **47**, 447–494 (2005)
9. Dantzig, G.B., Ramser, J.H.: The truck dispatching problem. Manag. Sci. **6**(1), 80–91 (1959)
10. García-López, F., Melián-Batista, B., Moreno-Pérez, J.A., Moreno-Vega, J.M.: The parallel variable neighborhood search for the p-median problem. J. Heuristics **8**(3), 375–388 (2002)
11. Polacek, M., Benkner, S., Doerner, K.F., Hartl, R.F.: A cooperative and adaptive variable neighborhood search for the multi depot vehicle routing problem with time windows. Bus. Res. **1**(2), 207–218 (2008)
12. Polat, O.: A parallel variable neighborhood search for the vehicle routing problem with divisible deliveries and pickups. Comput. Oper. Res. **85**, 71–86 (2017)
13. Sifaleras, A., Salhi, S., Brimberg, J. (eds.): ICVNS 2018. LNCS, vol. 11328. Springer, Cham (2019). https://doi.org/10.1007/978-3-030-15843-9
14. Uchoa, E., Pecin, D., Pessoa, A., Poggi, M., Vidal, T., Subramanian, A.: New benchmark instances for the capacitated vehicle routing problem. Eur. J. Oper. Res. **257**(3), 845–858 (2017)

A Simple Dual-RAMP Algorithm
for the Capacitated Facility Location Problem

Telmo Matos$^{(\boxtimes)}$ (iD), Óscar Oliveira$^{(\boxtimes)}$ (iD), and Dorabela Gamboa$^{(\boxtimes)}$ (iD)

CIICESI – Center for Research and Innovation in Business Sciences and Information Systems,
School of Technology and Management, Polytechnic of Porto, Porto, Portugal
{tsm,oao,dgamboa}@estg.ipp.pt

Abstract. Facility Location embodies a class of problems concerned with locating a set of facilities to serve a geographically distributed population of customers at minimum cost. We address the classical Capacitated Facility Location Problem (CFLP) in which the assignment of facilities to customers must ensure enough facility capacity and all the customers must be served. This is a well-known NP-hard problem in combinatorial optimization that has been extensively studied in the literature. Due to the difficulty of the problem, significant research efforts have been devoted to developing advanced heuristic methods aimed at finding high-quality solutions in reasonable computational times. We propose a Relaxation Adaptive Memory Programming (RAMP) approach for the CFLP. Our method combines lagrangean subgradient search with an improvement method to explore primal-dual relationships to create advanced memory structures that integrate information from both primal and dual solution spaces. The algorithm was tested on the standard ORLIB dataset and on other very large-scale instances for the CFLP. Our approach efficiently found the optimal solution for all ORLIB instances and very competitive results for the large-scale ones. Comparisons with current best-performing algorithms for the CFLP show that our RAMP algorithm exhibits excellent results.

Keywords: RAMP · Facility Location · Adaptive memory programming · Lagrangean Relaxation

1 Introduction

The Capacitated Facility Location Problem (CFLP) is a well-known combinatorial optimization problem belonging to the class of the NP-hard problems [1]. The CFLP can be formulated as follows:

$$\min \sum_{i=1}^{m} \sum_{j=1}^{n} D_j C_{ij} x_{ij} + \sum_{i=1}^{m} F_i y_i \tag{1}$$

$$\text{s.t.} \sum_{i=1}^{m} x_{ij} = 1, \quad j = 1, \ldots, n \tag{2}$$

$$\sum_{j=1}^{n} D_j x_{ij} \leq S_i y_i, \quad i = 1, \ldots, m \tag{3}$$

© Springer Nature Switzerland AG 2020
N. F. Matsatsinis et al. (Eds.): LION 13 2019, LNCS 11968, pp. 240–252, 2020.
https://doi.org/10.1007/978-3-030-38629-0_20

$$x_{ij} \geq 0, \quad j = 1, \ldots, n \quad i = 1, \ldots, m \tag{4}$$

$$y_i \in \{0, 1\}, \quad i = 1, \ldots, m \tag{5}$$

where m represents the number of possible locations to open a facility and n the number of customers to be served. S_i indicates the capacity of facility i and F_i the fixed cost for opening that facility. D_j represents the demand of client j and C_{ij} the unit shipment cost between facility i and customer j. The variable x_{ij} denotes the amount (scaled to be fractional) shipped from facility i to customer j and y_i indicates whether facility i is open or not. The objective is to locate a set of facilities in such a way that the sum of the costs for opening those facilities and the transportation costs for serving all customers is minimized. Given a set of open facilities ($y_i = 1, i \in I$), the CFLP has the particularity of becoming a Transportation Problem (TP) that can be solved in polynomial time. The TP can be formulated as:

$$\min \sum_{i=1}^{m} \sum_{j=1}^{n} D_j C_{ij} x_{ij} \tag{6}$$

$$\text{s.t.} \sum_{j=1}^{n} D_j x_{ij} \leq S_i, \quad i = 1, \ldots, m \tag{7}$$

(2) and (4)

Where C_{ij} is the unit shipment cost from facility i to customer j, $D_j x_{ij}$ is the amount sent from facility i to customer j, S_i is the availability of facility i and D_j is the demand of customer j. The objective is to determine an optimal transportation scheme between the facilities and customers so that the transportation costs are minimized. Also, if we eliminate the capacity of each facility (eliminating in Eq. (3) variable S_i) and set $D_j = 1$ in the same equation and in the objective function, we get the uncapacitated variant of the problem (UFLP) also widely studied in the literature.

2   Related Work

The CFLP has been extensively studied over the past 50 years, resulting in many algorithmic approaches based on exact and heuristic methods. For the best of our knowledge, the first heuristic for the CFLP is due to Jacobsen [2] who extended Kuehn and Hamburger's [3] heuristic originally proposed for the uncapacitated variant of the problem (UFLP). This heuristic has two phases: an ADD method that starts with all facilities closed and tries to open facilities that produce the highest total cost reduction. This phase ends when no more facilities contribute to the reduction of the total cost. The second phase consists of a local search method in which two facilities (one opened and one closed) alternate their status if this operation reduces the total cost.

Relaxation techniques are frequently used to solve the CFLP. Cornuejols [4] proposed several algorithms based on mathematic relaxation to solve large CFLP and proposed a comparison between several types of relaxations. Guignard and Spielberg [5] presented a Dual Ascent procedure for the CFLP, initially proposed by Bilde and Krarup [6] and Erlenkotter [7] for solving the UFLP. The main algorithm starts by obtaining a linear programming relaxation of the original problem to produce a lower bound. Starting with a dual solution, the dual ascent method iteratively, tries to improve that solution. The procedure stops when no more improvements can be achieved and then the dual solution is projected to the primal solutions space. Finally, the application of a method to solve the transportation problem is applied to the primal solution.

Avella et al. [8] proposed a Lagrangean Relaxation for the CFLP, that selects a subset of "promising" variables, defines the core problem, and uses an exact (Branch-and-Cut) algorithm to solve it. By relaxing the demand constraints and using subgradient optimization to maximize the dual problem, Avella et al. [8] manage to solve very large instances. Lorena [9] and Beasley [10] obtained good upper and lower bounds based on relaxation techniques, obtaining good results for the CFLP.

Sridharan [11] proposed the use of cross decomposition, initially presented by Van Roy [12], in which the main idea is the exploration of primal and dual subproblems into a single decomposition procedure to produce tight lower and upper bounds. Bornstein [13] proposed an ADD/DROP algorithm (that extends the ADD procedure seen earlier) and uses the DROP method to close facilities if this improves the objective function. Bornstein [14] also proposed the use of reduced tests and dominance criteria to define priorities to change the status of facilities (open-close or close-open) and also use simulated annealing for the set of facilities who have not yet defined their status.

Several exact algorithms were also introduced to solve the CFLP. Lai et al. [15] proposed the Benders Decomposition, initially presented by Benders [16] and use Genetic Algorithms (GA) [17] for the main problem. The algorithm in Lai et al. [15] starts by dividing the original problem into subproblems easier to solve. The original problem is relaxed and the constraints and variables that compose the relaxed problem are divided and solved (using the GA) separately alternating between the divided problem and the original problem.

Metaheuristics like Tabu Search (TS) [18, 19] and Greedy Randomized Adaptive Search Procedure (GRASP) [20] have many applications in optimization problems with high-quality results.

Sun [21], proposed a TS for the CFLP obtaining state-of-the-art results. Sun uses flexible memory structures to search in solutions space regions that can be promising. These space regions are kept in long-term memory to be intensively explored. Sun's TS uses diversification and intensification strategies, and a local search method based on changing the status of facilities (open-close or close-open). When a set of opened facilities is found (feasibility is met), then a Network Flow algorithm is used, based on Kennington [22], to solve the transportation problem. Silva [23] proposed a hybrid GRASP with different parameters (simple and reactive GRASP) obtaining good results in reasonable computational times.

Guastaroba and Speranza [24] proposed a kernel search algorithm to solve the CFLP that consists in a heuristic framework based on the idea of identifying subsets of variables and solving a sequence of MILP (mixed integer linear programming) problems, each of them, restrained to one of the identified subsets of associated variables. More information about the kernel search method and its applications, please refer to [24, 25].

Firefly (with GA) by Rahmani and Mirhassani, the Ant Colony by Venables and Moscardini [26] and Bee Colony [27, 28] are other recent metaheuristics based on nature observation that produced good results for the CFLP.

3 RAMP Algorithm for the CFLP

Proposed by Rego [29] in 2005, the Relaxation Adaptive Memory Programming (RAMP) metaheuristic framework combines fundamental principles of mathematical relaxation with concepts of adaptive memory programming techniques, covering primal and dual solutions spaces, with the objective of incorporating information obtained by both sides of the problem.

The RAMP method allows different levels of sophistication, depending on the intensification desired for the primal or dual search. At the first level of sophistication (Dual-RAMP), this framework explores more intensively the dual side, restricting the primal side interaction to the projection of dual solutions to the primal solutions space and to the improvement of these solutions. Higher levels of sophistication (PD-RAMP) allow a more intensive exploration of the primal side, incorporating the simple level, the Dual-RAMP, with more complex memory structures.

Several combinatorial optimization problems have already been solved by RAMP applications, producing excellent results, in some cases with new best-known solutions. Some examples of RAMP approaches with different levels of sophistication are the capacitated minimum spanning tree [30], the linear ordering problem [31], or the hub location problem [32–34], among others.

We propose a Dual-RAMP algorithm for the CFLP that uses a Lagrangean Relaxation [35] on the dual side and a simple improvement method based on tabu search on the primal side. The local search is based on the classical ADD/DROP neighborhood structure [13].

3.1 The Dual Method

The Dual Method proposed in this algorithm relies on the exploration on the dual side with subgradient optimization to solve the dual problem obtained by the Lagrangean Relaxation, based on Daskin [35].

At each iteration of the subgradient optimization, a solution for the relaxed problem is obtained. This solution is projected to the primal solution space through a Projection Method and an Improvement Method tries to improve it. Specifically, if we relax constraints (2), we obtain the following optimization problem:

$$\min \sum_{i=1}^{m} \sum_{j=1}^{n} D_j C_{ij} x_{ij} + \sum_{i=1}^{m} F_i y_i + \sum_{j=1}^{n} \lambda_j \left(1 - \sum_{i=1}^{m} x_{ij} \right) \quad (8)$$

$$Z(\lambda) = \sum_{i=1}^{m} F_i y_i + \sum_{i=1}^{m} \sum_{j=1}^{n} (D_j C_{ij} - \lambda_j) x_{ij} + \sum_{j=1}^{n} \lambda_j \quad (9)$$

$$\text{s.t.} \sum_{j=1}^{n} x_{ij} \leq y_i, \quad i = 1, \dots, m \quad (10)$$

$$\sum_{i=1}^{m} S_i y_i \geq \sum_{j=1}^{n} D_j \quad (11)$$

(3), (4) and (5)

Constraints (10) and (11) have been added to strengthen the linear relaxation and to ensure that the total amount of the capacity that is selected is sufficient to serve all the demand, respectively. The objective is to minimize the lagrangean function (9) with respect to the primal decision variables (y_i and x_{ij}) and to maximize the same function over the lagrangean multipliers λ_j. Maximizing the function over λ_j can be done using subgradient optimization, starting with $\lambda_j = min_i C_{ij}, \forall j \in J$.

The solution of the relaxed problem is achieved after solving to optimality the m continuous knapsack problems, one for each facility as we show below (problem given by Eqs. 12 to 14). To solve the m continuous knapsack problems resulting from the lagrangean relaxation of the CFLP, we use the greedy algorithm proposed by Daskin [35] (we will have one such problem for every candidate location and V_i will be the objective value of locating at candidate facility i. This quantity will always be non-positive and is given by Eq. 12).

$$\text{min } V_i = \sum\nolimits_{j=1}^{n} (D_j C_{ij} - \lambda_j) x_{ij} \tag{12}$$

$$\text{s.t. } \sum\nolimits_{j=1}^{n} D_j x_{ij} \leq S_i \tag{13}$$

$$0 \leq x_{ij} \leq 1, \ \forall j \in n \tag{14}$$

We would like to select facilities with the highest negative coefficients in the objective function (12) and with small demands in order to consume relatively little capacity in constraint (13). Hence, it is considered the following ratio for each demand node:

$$r_{ij} = \frac{\left(D_j C_{ij} - \lambda_j\right)}{D_j} \tag{15}$$

Variable r_{ij} is the amount of the demand j which contributes to the objective function given above, divided by the demand node j. This amount will be the proportion at which each demand node j increases the objective function per unit of the capacity. Since x_{ij} (the assignment variables) can be fractional, we can solve the problem given by Eqs. (12)–(14) to optimality by using the greedy algorithm proposed by Daskin [35].

Once all the V_i values have been computed, we can find values for the location variables y_i in the lagrangean problem, by solving the following optimization problem:

$$\text{min } \sum\nolimits_{i=1}^{m} (F_i + V_i) y_i \tag{16}$$

$$\sum\nolimits_{i=1}^{m} S_i y_i \geq \sum\nolimits_{j=1}^{n} D_j \tag{17}$$

$$y_i \in \{0, 1\} \ \forall i \in m \tag{18}$$

The objective function (16) states that the contribution of candidate facility i to the lagrangean objective function will be equal to the fixed cost of locating at candidate facility i plus the assignment variables if we locate at candidate facility i. If we relax the integrality constraint (18) and replace it by the constraint, $0 \leq y_i \leq 1, \forall i \in I$, we could solve the problem given by Eqs. (16–18) with the same greedy algorithm we used for problem above (12–14). This optimization problem is quite simple, therefore we chose to solve it using the CPLEX solver, obtaining the decision variables of the optimal location and ensuring we have sufficient opened facilities, as we will see in the Projection Method. After we get the values for the decision variables x_{ij} we are ready to compute the remaining subgradient parameters.

The agility parameter π, is initialized with the value of 2, and every three consecutive failures to improve the lower bound, it is divided by 2, in order to decrease the step size (Δ). The agility parameter is restarted every 30 iterations. The step size (Δ) is computed as $\Delta = \frac{\pi(Z_{UB} - Z(\lambda))}{||\delta||}$, where $Z(\lambda)$ is an upper bound to the original problem and $Z(\lambda)$ is the best lower bound found so far. The j-th component of the subgradient (δ) is $\delta_j = 1 - \sum_{i=1}^{m} x_{ij}$. Finally, to determine the set of λ multipliers maximizing the Lagrangean function $Z(\lambda)$, the subgradient method requires generating a new sequence of multipliers (one for each iteration of the Lagrangean Relaxation) that is determined by $\lambda_j^{iter+1} = \lambda_j^{iter} + \Delta \delta_j$.

3.2 Projection Method

Once we obtain the lower bound, $Z(\lambda)$, then the dual solution is projected to the primal solutions space by solving the Transportation Problem (TP) of the decision variables $y_i = 1$ for any $i \in I$, and adding to the TP's objective function value the fixed costs of the selected facilities i, thus obtaining the upper bound (Z_{UB}).

If we get an infeasible solution (when solving the optimization problem (16–18) we choose to set $y_i = 1$ when and only when the coefficient ($F_i + V_i$) is not positive. For more information, please refer to [35]), then a simple procedure makes the solution feasible by opening sufficient facilities to serve all customers. This is accomplished by selecting facilities by descending order of setup cost and opening facilities until all demand is met. For solving the TP, we used the commercial solver CPLEX 12.6 to obtain the optimal solution for this integer programming problem and retrieve the values of the assignment variables x_{ij}.

3.3 Primal Method

After the Projection Method obtains a primal feasible solution, the Dual-RAMP algorithm uses it as the initial solution for starting a simple improvement method based on tabu search that uses the classical ADD/DROP [8] local search neighborhood structures.

The ADD and DROP neighborhood structures are used one at a time and a move can only be applied if it is not tabu (the type of move, add or drop, and the location of the facility is stored in the tabu list). The choice for selecting a facility for ADD and/or DROP is based on choosing a facility by decreasing (DROP case) or increasing (ADD case) order of the total sum allocation for all customers. This choice will penalize facilities that contribute to the allocation cost growth and favor facilities that could make a reduction on this cost. If all moves are tabu then the first item in the tabu list is removed to allow new moves.

Solving many TP is very expensive in terms of computational time, so we chose to apply the first improvement criteria, which means that we apply a ADD or DROP move whenever it improves the current solution in a given neighborhood. The algorithm alternates the search between the dual side and the primal side until one of the four stopping criteria is achieved:

- The agility parameter is less than 0.005;
- The norm reaches the value 0;
- The maximum number of iterations is reached;
- The absolute difference between the upper bound (Z_{UB}) and lower bound ($Z(\lambda)$) is less than 1.

4   Computational Results

The performance of the proposed Dual-RAMP algorithm was evaluated on a standard set of benchmark instances. The first one is the well-known OR-Library data set (http://people.brunel.ac.uk/~mastjjb/jeb/orlib/capinfo.html) proposed by Beasley [36]. This set has 49 instances with known optimal solutions, which have the following sizes (facilities × customers): 16 × 50, 25 × 50 and 50 × 50 for the small instances and 100 × 1000 for the large ones.

The second set of instances (TBED1) was presented by Avella and Boccia [37] (http://www.ing.unisannio.it/boccia) and contains 100 instances also with known optimal solutions. In this set, the small instances have the following sizes: 300 × 300, 300 × 1500, 500 × 500 and 700 × 700.

The large instances have 1000 × 1000 facilities and customers. The last set of instances (TESTBED A, B and C) contains the larger ones. Avella et al. [8] proposed TESTBED A and B (http://wpage.unina.it/sforza/test) and TESTBED C was introduced by Guastaroba and Speranza [24] (http://www-c.eco.unibs.it/~guastaro/InstancesCFLP.html). This set is composed of 800 × 4400, 1000 × 1000, 1000 × 4000, 1200 × 3000 and 2000 × 2000 instances (445 in total), with different costs structures and different facility capacities. To scale these capacities the authors used a ratio with the following values: 1.1, 1.5, 2, 3, 5 and 10. Each of the five subsets shown in Table 1 contains six groups (of five instances), each with a specific ratio (for example, in TESTBEDA, 5 out of the 30 instances belonging to the 800 × 4400 subset have a ratio of 1.1, the next 5 have a ratio of 1.5, and so on). For these large instances (TESTBED A, B and C) no optimal solution is known (or it is not proven yet).

Table 1 summarizes the data sets considered in our computational experiments, where instances belonging to the same data set are divided into subsets according to their size.

The algorithm was coded in C programming language and run on an Intel Pentium I7 2.40 GHz with 8 GB RAM under Ubuntu operating system (only one processor was used) using CPLEX 12.6 to solve the Transportation Problem. To compare our Dual-RAMP algorithm with the state-of-the-art approaches for the CFLP we show, in Tables 2, 3 and 4, the results reported for these algorithms on the data sets previously described. All tables below presents the average computational time for each of the different algorithms. As

Table 1. A summary of the instances to be considered.

Data set	# Instances	# Facilities	# Customers
OR-Library			
1	13	16	50
2	12	25	50
3	12	50	50
4	12	100	1000
TBED1			
1	20	300	300
2	20	300	1500
3	20	500	500
4	20	700	700
5	20	1000	1000
TESTBED A/B/C			
1	30 (TESTBED B = 25)	800	4400
2	30	1000	1000
3	30	1000	4000
4	30	1200	3000
5	30	2000	2000

all authors use different machines to process their algorithms, it is not possible to make a direct comparison regarding this parameter. The comparison between algorithms needs different tables, since some authors do not provide results for all instances.

The second table (Table 2) shows the average results for the OR-Library instances divided by small and large. Some specific instances belonging to this group are also presented in Table 2 to be possible to compare with one specific algorithm. The Dual-RAMP algorithm is compared with Sun's Tabu Search (TS) algorithm [21], Beasley's lagrangean relaxation heuristic (LR) [38], Guastaroba and Speranza's Kernel Search (KS) procedure [24], Hybrid Approach (Bee Algorithm with Mixed Integer Programming – HA) proposed by Cabrera *et al.* [28] and with a Branch-and-Cut-and-Price, next denoted as AB, proposed by Avella and Boccia [37]. The column GAP is the gap computed as $\frac{(UB-Z^*)}{Z^*} * 100$ (Z^* is the optimal solution) and CPU is the computational time in seconds needed to achieve the best UB (upper bound).

Regarding the OR-Library instances, our algorithm found optimal solutions for all 49 instances under 113 s. In particular, in large instances, the RAMP achieved all optimal solutions in less than 216 s, which is very fast considering the size of these problems. For possible comparison with the Bee algorithm (HA), we included instances capa8000, capb8000 and capc5000. Despite this algorithm and the KS achieved all optimal solution, it seems that (this is only an assumption. A direct comparison would be possible if we

Table 2. Comparison table for the ORLIB instances.

ORLIB	KS		TS		LR		HA		AB		Dual-RAMP	
	GAP	CPU	GAP	CPU	GAP	CPU	GAP	CPU	GAP	CPU	GAP	CPU
Small	0.00	0.63	0.00	0.24	0.02	1.49	–	–	–	–	**0.00**	**10.18**
Large	0.00	2158.83	0.07	48.63	0.21	75.79	–	–	0.00	415.79	**0.00**	**215.64**
Average	0.00	1079.73	0.04	24.44	0.12	38.64	–	–	–	–	**0.00**	**112.91**
CAPA 8000	0.00	3604.88	–	–	0.24	73.65	0.00	367.74	–	–	**0.00**	**85.43**
CAPB 8000	0.00	1999.73	–	–	0.40	131.84	0.00	317.47	0.00	319.26	**0.00**	**99.92**
CAPC 5000	0.00	2621.94	–	–	0.00	105.74	0.00	129.34	0.00	577.30	**0.00**	**210.85**

process all algorithms on the same machines) it takes much more computational time on average than RAMP.

The next table (Table 3) shows the results for the TBED1 where instances are separated into subsets according to their size instances and the proposed algorithm is compared with Kernel Search [24] (KS) and with Avella and Boccia [37] (AB). For two instances of this type, $1000 \times 1000-11$ and $1000 \times 1000-12$, the optimality has not been proven. We used the best upper bound given by the [37]. The column Gap was computed as above, and CPU is the computational time in seconds needed to achieve the best UB.

Table 3. Comparison table for the TBED1 instances.

TBED1	KS		AB		Dual-RAMP	
	GAP	CPU	GAP	CPU	GAP	CPU
300×300	0.00	57.82	0.00	327.99	**0.02**	**96.52**
300×1500	0.00	68.68	0.00	807.07	**0.05**	**674.82**
500×500	0.00	225.66	0.00	1518.56	**0.06**	**286.10**
700×700	0.00	795.80	0.00	6569.55	**0.06**	**709.15**
1000×1000	0.00	1745.87	0.00	46895.14	**0.22**	**627.67**
Average	0.00	578.77	0.00	11223.66	**0.08**	**478.85**

For TBED1 instances, our approach didn't achieve all optimal solutions, but it achieved near optimal solutions in low computational time. Comparing with KS algorithm, for size 300×300, 500×500, 700×700 achieved, in average, quality solutions less or equal than 0.06% deviation from optimal solutions in less than 710 s. For the large ones, size 1000×1000, RAMP algorithm achieved in average solutions under 0.22% deviation from optimal in very reasonable computation time.

The next table (Table 4) displays the results for the instances TESTBED A, B and C. Since the optimal solution is not known, we compute the "GAP" column as $\frac{(UB-LB)}{LB} *$ 100. The "CPU" column continues to show the computational time (in seconds) needed to achieve the best UB. The lower bound (LB) values were obtained with our lagrangean relaxation, because we found very good lower bounds. Since the original instances for the TESTBED C are not available, we used the instances provided by the authors of KS, that where generated as described in Avella et al. [39]. Since the instances are not the same, we do not show the results for the AB algorithm on TESTBED C.

Table 4. Comparison table for the TESTBED A, B and C instances.

TESTBED A	KS		AB		**Dual-RAMP**	
	GAP	CPU	GAP	CPU	**GAP**	**CPU**
800 × 4400	0.33	1349.87	0.71	209.27	**0.31**	**1610.40**
1000 × 1000	0.10	336.64	0.57	75.37	**0.23**	**152.66**
1000 × 4000	0.27	1539.77	0.67	304.45	**0.46**	**1714.60**
1200 × 3000	0.18	1570.96	0.63	150.43	**0.32**	**1336.42**
2000 × 2000	0.07	1382.68	0.51	165.40	**0.24**	**964.11**
Average	0.19	1235.99	0.62	180.98	**0.31**	**1155.64**
TESTBED B						
800 × 4400	0.33	1497.19	1.98	113.22	**0.90**	**4034.01**
1000 × 1000	0.34	1409.52	0.60	48.03	**0.90**	**477.24**
1000 × 4000	0.34	1519.93	1.21	129.85	**1.04**	**4270.93**
1200 × 3000	0.36	1727.18	0.78	108.78	**0.74**	**2956.66**
2000 × 2000	0.40	2073.38	0.96	161.25	**1.12**	**2588.78**
Average	0.35	1645.44	1.10	112.23	**0.92**	**2865.53**
TESTBED C						
800 × 4400	1.51	265.86	–	–	**0.09**	**6121.58**
1000 × 1000	3.57	1358.44	–	–	**0.91**	**2069.96**
1000 × 4000	2.09	465.63	–	–	**0.17**	**7934.40**
1200 × 3000	2.94	1001.21	–	–	**0.28**	**7481.75**
2000 × 2000	4.65	1833.15	–	–	**1.17**	**11391.92**
Average	2.95	984.86	–	–	**0.52**	**6999.92**

For these very large instances, the Dual-RAMP algorithm shows its robustness by being very consistent with the previous results, since it achieved good quality solutions in reasonable computational times. Dual-RAMP produced very competitive results for TESTBED A and B and proved to be very effective for TESTBED C despite needing much higher computational times. Dual-RAMP achieved a 0.52% average Gap, a much better value than the 2.95% obtained by the KS approach.

In summary, our Dual-RAMP algorithm proved to be a robust approach for the solution of the CFLP, by effectively solving the best-known (small, large and very large) instances available in the literature.

5 Conclusions

This paper describes a Dual-RAMP algorithm for the CFLP that competes with the best-known algorithms for the solution of this problem. Despite the fact that only the first level of sophistication of the RAMP method was implemented, the proposed algorithm managed to produce excellent results for the complete testbed in reasonable computational times, and even introduced ten new best-known solutions. We conjecture that our algorithm owes its advantage to the premise that a judicious exploration of primal-dual relationships provides an effective interplay between intensification and diversification that is absent in search methods confined to the primal solution space. The consistent encouraging results obtained by RAMP applications to hard combinatorial optimization problems certainly invite further studies in the application of the method. In particular, the impressive results reported by the present study strongly suggest extending our RAMP approach to the solution of other facility location problems.

References

1. Current, J., Daskin, M., Schilling, D.: Discrete Network Location Models. Springer-Verlag, Berlin (2001)
2. Jacobsen, S.K.: Heuristics for the capacitated plant location model. Eur. J. Oper. Res. **12**, 253–261 (1983)
3. Kuehn, A., Hamburger, M.: A heuristic program for locating warehouses. Manage. Sci. **9**, 643–666 (1963)
4. Cornuéjols, G., Sridharan, R., Thizy, J.: A comparison of heuristics and relaxations for the capacitated plant location problem. Eur. J. Oper. Res. **50**, 280–297 (1991)
5. Guignard, M., Spielberg, K.: A direct dual method for the mixed plant location problem with some side constraints. Math. Program. **17**, 198–228 (1979)
6. Bilde, O., Krarup, J.: Sharp lower bounds and efficient algorithms for the simple plant location problem. In: Annals of Discrete Mathematics, pp. 79–97 (1977)
7. Erlenkotter, D.: A dual-based procedure for uncapacitated facility location. Oper. Res. **26**, 992–1009 (1978)
8. Avella, P., Boccia, M., Sforza, A., Vasil'ev, I.: An effective heuristic for large-scale capacitated facility location problems. J. Heuristics **15**, 597–615 (2008)
9. Lorena, L., Senne, E.: Improving traditional subgradient scheme for Lagrangean relaxation: an application to location problems. Int. J. Math. Algorithms **1**, 133–151 (1999)
10. Beasley, J.E.: Lagrangean heuristics for location problems. Eur. J. Oper. Res. **65**, 383–399 (1993)
11. Sridharan, R.: The capacitated plant location problem. Eur. J. Oper. Res. **87**, 203–213 (1995)
12. Van Roy, T.: A cross decomposition algorithm for capacitated facility location. Eur. J. Oper. Res. **34**, 145–163 (1986)
13. Bornstein, C.T.: An ADD/DROP procedure for the capacitated plant location problem. Pesqui. Operacional **24**, 151–162 (2003)

14. Bornstein, C.T.C., Azlan, H.H.B.: The use of reduction tests and simulated annealing for the capacitated plant location problem. Locat. Sci. **6**, 67–81 (1998)
15. Lai, M.-C., Sohn, H., Tseng, T.-L. (Bill), Chiang, C.: A hybrid algorithm for capacitated plant location problem. Expert Syst. Appl. **37** 8599–8605 (2010)
16. Benders, J.F.: Partitioning procedures for solving mixed-variables programming problems. Numer. Math. **4**, 238–252 (1962)
17. Sastry, K., Goldberg, D.E., Kendall, G.: Genetic algorithms. In: Search Methodologies, pp. 93–117. Springer US, Boston (2014)
18. Glover, F., Laguna, M.: Tabu Search (1997)
19. Glover, F.: Tabu search—part I. ORSA J. Comput. **1**, 190–206 (1989)
20. Feo, T., Resende, M.: Greedy randomized adaptive search procedures. J. Glob. Optim. **134**, 109–134 (1995)
21. Sun, M.: A tabu search heuristic procedure for the capacitated facility location problem. J. Heuristics **18**, 91–118 (2012)
22. Kennington, J.L., Helgason, R.V.: Algorithms for network programming. John Wiley & Sons Inc., New York (1980)
23. Ronaldo Silva, V.F.: Um Algoritmo GRASP Híbrido para o Problema de Localização Capacitada de Custo Fixo, (2007)
24. Guastaroba, G., Speranza, M.G.: Kernel search for the capacitated facility location problem. J. Heuristics **18**, 877–917 (2012)
25. Guastaroba, G., Speranza, M.G.: Kernel search: an application to the index tracking problem. Eur. J. Oper. Res. **217**, 54–68 (2012)
26. Venables, H., Moscardini, A.: Ant based heuristics for the capacitated fixed charge location problem. In: Dorigo, M., Birattari, M., Blum, C., Clerc, M., Stützle, T., Winfield, A.F.T. (eds.) ANTS 2008. LNCS, vol. 5217, pp. 235–242. Springer, Heidelberg (2008). https://doi.org/10.1007/978-3-540-87527-7_22
27. Levanova, T., Tkachuk, E.: Development of a bee colony optimization algorithm for the capacitated plant location problem. In: II International Conference Optimization and Applications (OPTIMA-2011), pp. 153–156, Petrovac, Montenegro (2011)
28. Cabrera, G., Cabrera, E., Soto, R., Rubio, L.J.M., Crawford, B., Paredes, F.: A hybrid approach using an artificial bee algorithm with mixed integer programming applied to a large-scale capacitated facility location problem. Math. Probl. Eng. **2012**, 14 (2012)
29. Rego, C.: RAMP: a new metaheuristic framework for combinatorial optimization. In: Rego, C., Alidaee, B. (eds.) Metaheuristic Optimization via Memory and Evolution: Tabu Search and Scatter Search, pp. 441–460. Kluwer Academic Publishers (2005)
30. Rego, C., Mathew, F., Glover, F.: RAMP for the capacitated minimum spanning tree problem. Ann. Oper. Res. **181**, 661–681 (2010)
31. Gamboa, D.: Adaptive Memory Algorithms for the Solution of Large Scale Combinatorial Optimization Problems, Ph.D. Thesis (in Portuguese) (2008)
32. Matos, T., Gamboa, D.: Dual-RAMP for the capacitated single allocation hub location problem. In: Gervasi, O., et al. (eds.) ICCSA 2017, Part II. LNCS, vol. 10405, pp. 696–708. Springer, Cham (2017). https://doi.org/10.1007/978-3-319-62395-5_48
33. Matos, T., Maia, F., Gamboa, D.: Improving traditional dual ascent algorithm for the uncapacitated multiple allocation hub location problem: a RAMP approach. In: Nicosia, G., Pardalos, P., Giuffrida, G., Umeton, R., Sciacca, V. (eds.) LOD 2018. LNCS, vol. 11331, pp. 243–253. Springer, Cham (2019)
34. Matos, T., Maia, F., Gamboa, D.: A simple dual-RAMP algorithm for the uncapacitated multiple allocation hub location problem. In: 18th International Conference on Hybrid Intelligent Systems (HIS 2018) in Portugal, 2018, pp. 331–339 (2020)
35. Daskin, M.M.S.: Network and Discrete Location: Models, Algorithms, and Applications. Wiley, New York (1995)

36. Beasley, J.: OR-library: distributing test problems by electronic mail. J. Oper. Res. Soc. **65**, 1069–1072 (1990)
37. Avella, P., Boccia, M.: A cutting plane algorithm for the capacitated facility location problem. Comput. Optim. Appl. **43**, 39–65 (2009)
38. Beasley, J.E.: An algorithm for solving large capacitated warehouse location problems. Eur. J. Oper. Res. **33**, 314–325 (1988)
39. Avella, P., Boccia, M., Sforza, A., Vasilev, I.: An effective heuristic for large-scale capacitated facility location problems - draft available from the authors upon request. J. Heuristics **15**, 597–615 (2006)

A Novel Solution Encoding in the Differential Evolution Algorithm for Optimizing Tourist Trip Design Problems

Dimitra Trachanatzi[✉], Manousos Rigakis, Andromachi Taxidou,
Magdalene Marinaki, Yannis Marinakis, and Nikolaos Matsatsinis

Technical University of Crete, School of Production Engineering and Management
University Campus, Chania, Crete, Greece
{dtrachanatzi,mrigakis,ataxidou}@isc.tuc.gr, magda@dssl.tuc.gr,
{marinakis,nikos}@ergasya.tuc.gr

Abstract. In this paper, a tourist trip design problem is simulated by the Capacitated Team Orienteering Problem (CTOP). The objective of the CTOP is to form feasible solution, as a set of itineraries, that represent a sequence visit of nodes, that maximize the total prize collected from them. Each itinerary is constrained by the vehicle capacity and the total travelled time. The proposed algorithmic framework, the Distance Related Differential Algorithm (DRDE), is a combination of the widely-known Differential Evolution algorithm (DE) and a novel encoding/decoding process, namely the *Distance Related* (DR). The process is based on the representation of the solution vector by the Euclidean Distance of the included nodes and offers a data-oriented approach to apply the original DE to a discrete optimization problem, such as the CTOP. The efficiency of the proposed algorithm is demonstrated over computational experiments.

Keywords: Capacitated Team Orienteering Problem · Differential Evolution Algorithm · DR solution encoding

1 Introduction

Over the years, the tourism industry promotes products that offer customize solutions with respect to their customers' preferences, and their time and budget limitations. Following this trend, the scientific community has turned to the modelling and optimization of problems that simulate a tourist trip, bounded by several constraints. The objective of these problems is to simulate the design

This research is co-financed by Greece and the European Union (European Social Fund-ESF) through the Operational Programme Human Resources Development, Education and Lifelong Learning in the context of the project "Strengthening Human Resources Research Potential via Doctorate Research" (MIS-5000432), implemented by the State Scholarships Foundation (IKY).

© Springer Nature Switzerland AG 2020
N. F. Matsatsinis et al. (Eds.): LION 13 2019, LNCS 11968, pp. 253–267, 2020.
https://doi.org/10.1007/978-3-030-38629-0_21

of an itinerary that respects the customer's (tourist) requirements. A tourist trip design problem can be represented by an Orienteering Problem (OP), firstly introduced in 1987, by Golden et al. [12]. The objective of the OP is to maximize the total score, collected from nodes that are included in a route, considering the formulation of a shortest path. In contrast to the original Travelling Salesman Problem (TSP), the OP takes into account the realistic issue that a common tourist trip imposes, i.e it is impossible to visit all the POIs in a city due to budget or time limitations [22]. Thus, a specific set of nodes has to be selected from all the available nodes, that is sufficient to form a feasible itinerary by maximizing the collected score. The extension of the OP to a number of vehicles is the Team Orienteering Problem (TOP), initially introduced as the Multiple Tour Maximum Collection Problem by Butt and Cavalier in 1994 [5]. Since then, several heuristics, meta-heuristics and exact algorithms were proposed for the solution of the TOP, while it has been widely used to model tourist trip design problems [11,13]. As an OP, in TOP it is not obligatory to visit all the available nodes, and by including a homogeneous fleet, each vehicle follows a route starting from an initial node (depot), up to a final node, with respect to time travel constraints.

In this paper we focus on the Capacitated Team Orienteering Problem (CTOP) as a variant of the aforementioned TOP, that was introduced by Archetti et al. [1]. The CTOP differs from TOP as follows: (a) all vehicles should return to the initial node (depot), and (b) an extra constraint exists that facilitates the capacity of the vehicle in each route. Initially, for the solution of the CTOP, Archetti et al. [1] proposed three algorithmic schemes, a Variable Neighborhood Search (VNS) algorithm and two variants of the Tabu Seach algorithm, the *TabuFeasible* (TSF) and the *TabuAdmissible* (TSA). In their approach, all the nodes are grouped and arranged into routes, and then, the most profitable routes are chosen to form a solution, while unfeasible solutions are also considered to further explore the search space. Moreover local search and hump mechanisms were employed to escape from local optima, and the computational results show that the heuristics obtain good results within a reasonable amount of time. The unsolved benchmark instances in their work, were latter solved to optimality by Archetti et al. in 2013 [2], utilizing a branch-and-price algorithm. In addition, in 2013, Tarantilis et al. [20] presented a hierarchical bi-level search framework, namely a Bi-level Filter-and-Fan method, for the solution of the CTOP. Their algorithm, in the first phase aims to optimal node selection, in terms of score, by employing a Tabu Search scheme and the Filter-and-Fan search, while in the second phase the travel time minimization is performed by a Variable Neighborhood Descent algorithm. The proposed method is tested on the benchmark instances introduced in [1], and on a new set of lager instances. The result shown the efficiency and effectiveness of their approach as 18 best known solutions were obtained.

Also, in 2013, Luo et al. [15] described the ADaptive Ejection Pool with Toggle-Rule Diversification (ADEPT-RD) for the solution of the CTOP. In their method, at each iteration, nodes with high priority are inserted into a solution based on a scheme that switches between two ejection pool priority rules, and subsequently, a local search technique is performed to restore the feasibility of the solution. Compared to [1], the approach of Luo et al. managed to obtain outperforming solution in terms of both quality and execution time. In 2016, Ben-Said et al. [3] presented an Adaptive Iterative Destruction/Construction Heuristic (AIDCH), that is based on two adaptive mechanisms, the best insertion and the perturbation, following an adaptive parametrization according to the solution progress. Their algorithm showed high quality results with a competitive computational time in the solution of the CTOP benchmark instances proposed in [1]. The same authors, in 2018 [4], proposed a Variable Space Search (VSS) heuristic, combining the exploration and exploitation that the Greedy Randomized Adaptive Search Procedure (GRASP) and the Evolutionary Local Search (ELS), respectively offer. Their algorithmic framework operates in two search spaces, the unfeasible, i.e. the *giant tour* space, and the feasible, i.e. the *route* space. Moreover, to enhance their solutions, in the *route* search space, a Tabu Search and a Simulated Annealing are employed. The described method results in two variant the VSS-Tabu and the VSS-SA, that outperform the computational results found in the literature , by obtaining 74 and 77 new best solutions on the large CTOP benchmark instances [20], respectively.

Based on the above brief literature review, there is no publication that studies the solution of the CTOP utilizing the Differential Evolution (DE) algorithm. The DE is a stochastic, real-parameter optimization algorithm and was fist proposed by Storn and Price, in 1997, [19]. Based on the literature, the DE has been employed for the solution of several VRP problems such as: the Capacitated Vehicle Routing Problem (CVRP) [10,21]; the Multi-Depot Vehicle Routing Problem (MDVRP) [14]; the Open Vehicle Routing Problem (OVRP) [6]; and the Vehicle Routing Problem with Simultaneous Pickups and Deliveries and Time Windows (VRP-SPDTW) [16]. Because of the continuous nature of DE, researchers develop encoding and decoding methods, in order to be able to apply it to discrete optimization problems, e.g. the VRP. Thus, in this paper we present a novel encoding and decoding scheme for the solution of the Capacitated Team Orienteering Problem, that is incorporated to a DE algorithmic scheme, enhanced by multiple local search procedures. The encoding/decoding method will be referred as *Distance Related*.

The rest of the paper is organized as follows: Sect. 2 provides the definition and mathematical formulation of the CTOP; Sect. 3 presents the original DE algorithm; in Sect. 4 the proposed *Distance Related* method is described in detail; in Sect. 5 the proposed algorithmic framework *"DRDE"* is presented; while Sects. 6 and 7 provide the computational experiments, their analysis and the drawn conclusions of this study.

2 Capacitated Team Orienteering Problem

The Capacitated Team Orienteering Problem can be defined on a complete undirected graph $G = (V, A)$ where $V = \{1, \cdots, N\}$ is the set of nodes and

$A = \{(i,j)|i,j \in V\}$ defines the set of arcs. Every node i, included in set V, is associated with a prize value p_i and a demand value D_i. Moreover, for each pair of nodes (i,j), the corresponding travel time (cost), is denoted by t_{ij}. According to the mathematical formulation of the problem, M number of routes has to be created, corresponding to a homogeneous fleet of vehicles. Such, each vehicle, with Q maximum carrying capacity, follows an itinerary, starting from, and returning to the depot, without exceeding $Tmax$ time units. The objective of the CTOP is the maximization of the total prize, collected from the nodes that are included in the feasible itineraries. Due to problem restrictions, i.e. vehicle capacity and total travel duration, it is not possible to formulate a feasible solution including all nodes of set V. Respecting the above description, the CTOP is mathematically formulated as follows, based on the three-index vehicle flow formulation presented by Tarantilis et al. [20]. The decision variables used, are:

- $y_{id} = 1$ if node i $(i = 1, \ldots, N)$ is included in route d $(d = 1, \ldots, M)$, $y_{id} = 0$ otherwise.
- $x_{ijd} = 1$ if the arc i, j $(i, j = 1, \ldots, N)$ is included in route d $(d = 1, \ldots, M)$, $x_{ijd} = 0$ otherwise. The symmetry of the problem ensures that $t_{ij} = t_{ji}$, thus, only the variables x_{ijd} for $i < j$ are defined.

The mathematical formulation of CTOP is depicted below.

$$Maximize \sum_{i=2}^{N} \sum_{d=1}^{M} p_i y_{id} \tag{1}$$

s.t.

$$\sum_{j=2}^{N} \sum_{d=1}^{M} x_{1jd} = \sum_{i=1}^{N} \sum_{d=1}^{M} x_{i1d} = M \tag{2}$$

$$\sum_{d=1}^{M} y_{id} \leq 1, \quad \forall i = 2, \ldots, N \tag{3}$$

$$\sum_{j \in N} x_{ijd} = \sum_{j \in N} x_{jid} = y_{id}, \quad \forall i = 2, \ldots, N; \forall d = 1, \ldots, M \tag{4}$$

$$\sum_{i \in N} \sum_{j \in N} t_{ij} x_{ijd} \leq T_{max}, \quad \forall d = 1, \ldots, M \tag{5}$$

$$\sum_{i \in N} D_i x_{ijd} \leq Q, \quad \forall d = 1, \ldots, M \tag{6}$$

$$\sum_{i \in S} \sum_{ij \in S} x_{ijd} \leq |S| - 1, \quad \forall S \subseteq V, |S| \geq 2; \forall d = 1, \ldots, M \tag{7}$$

$$x_{ijd}, y_{id} \in \{0, 1\}, 1 \leq i < j \leq N; d = 1, \ldots, M \tag{8}$$

Equation 1 express the objective function i.e. the total profit maximization. Constraint 2 ensures that M vehicles leave and return to the depot, node 1. Constraints 3 impose that each node is allowed to be included only once in a route.

The continuity of the route is facilitated by constraints 4, that is the same vehicle d should depart from the same node if that node is visited. Constraints 5, 6 ensure the feasibility of each route in the solution in respect of time duration and vehicle capacity. Constraints 7 prohibit sub-tour formulation, while Constraints 8 impose binary restrictions to the decision variables.

3   Differential Evolution Algorithm

Differential Evolution (DE) was designed for continuous-optimization problems, as a population-based search method, which includes processes of Evolutionary Algorithms (EAs), such as mutation, crossover and selection. One of the DE advantages is the small number of control parameters, the population size (NP), the mutation rate (F) and the crossover rate (Cr). The main idea is the perturbation of a population of vectors through a number of generations, that incorporates vector differences and recombination. Initially, similar to EAs, a randomly disturbed population of NP individuals is generated. Each one is a D-dimensional real vector x_{ij}, where $i \in \{1, \cdots, NP\}$ and $j \in \{1, \cdots, D\}$. The first evolutionary process that takes place in every generation is the mutation. During mutation, three vectors are randomly chosen, a *base* vector $(i_1 \neq i)$ and two others $(i \neq i_1 \neq i_2 \neq i_3)$. The difference of x_{i_2} and x_{i_3} is amplified by the mutation rate F, which is a real, constant value between 0 and 2. The scaled difference is added to the base vector, see Eq. 9, in order to form the mutant vector $v_{ij}(t)$, for each target vector of the population, in t generation, i.e. for each individual.

$$v_{ij}(t+1) = x_{i_1 j}(t) + F * (x_{i_2 j}(t) - x_{i_3 j}(t)) \tag{9}$$

Afterwards, the crossover process occurs, which is a recombination of each target and its corresponding mutant and generates the trial vector. There are two common kinds of crossover methods, exponential and binomial. The binomial crossover will be used in this research and it is implemented as follows, for every target vector in the population. Through Eq. 10, it is determined what parameters will be inherited to the trial vector from the mutant and what from the target vector. The crossover is controlled by the Cr parameter and its value is decided by the user, within the range [0,1]. For each parameter j, a random number ϕ and a random index j_{rand} are generated such as $\phi, j_{rand} \in [0, 1]$. If the random number ϕ is less or equal to Cr, or if the parameter's index equals to j_{rand}, the trial vector inherits the corresponding element from the mutant vector and otherwise from the target vector. The j_{rand} ensures that at least one parameter will be forwarded to the trial vector from the mutant vector.

$$u_{ij}(t+1) = \begin{cases} v_{ij}(t+1), & \text{if } \phi \leq C_r \text{ or } j = j_{rand} \\ x_{ij}(t), & \text{otherwise} \end{cases} \tag{10}$$

Both processes, mutation and crossover, increase the diversity of the population and thereby they carry out the exploration phase of the search. In order to involve an exploitation phase of the search and to retain the size of the population, a selection procedure is performed. Thus, subsequent to crossover process, one of the correlated vectors, target and trial, have to remain in the population and the other one has to be discarded. Differential evolution uses a greedy technique as selection and the vector with the highest fitness value will survive over the other and will be included in the next generation's population, see Eq. 11.

$$x_{ij}(t+1) = \begin{cases} u_{ij}(t+1), & \text{if } f(x_{ij}(t)) \leq f(u_{ij}(t+1)) \\ x_{ij}(t), & \text{otherwise} \end{cases} \tag{11}$$

A number of variations to the basic Differential Evolution algorithm have been developed over the years. Different DE mutation strategies have a general notation $DE/x/y/z$, where x is cited to the way that a target vector is selected, y is the number of difference vectors considered for perturbation of x and z refers to the crossover scheme (i.e. exp: exponential; bin: binomial).For a recent theoretical discussion focused on the DE algorithm, one can referred to [9] and [18]. The presented research adopts the $DE/best/1/bin$ variation, thus the vector to be mutated is the best individual vector with the best fitness value, including one difference vector and the binomial crossover is implemented. Although, the crossover part of the adopted framework is enriched, as described in Sect. 5.

4   The Proposed *Distance Related* Solution Encoding/Decoding Scheme

First, the representation of a feasible CTOP solution need to be reported. Inspired by several relating studies, i.e meta-heuristic optimization approaches for the VRP, a sequence representation of nodes in integer values (ordinal number encoding) is adopted. Thus, a feasible depot-returning vehicle itinerary starts from node 1, a permutation of integer values follows (showing the sequence that the nodes should be visited) and finally, the node 1 concludes it, e.g. [1, 4, 6, 7, 2, 1]. A complete feasible solution consists of M routes in the same vector such as: *Solution Vector* [1, 4, 6, 7, 2, 1, 3, 10, 5, 1]. From this small example, one may observe that the solution consists of two routes; that the duplication of node one is successive order it omitted; and based on the selective nature of CTOP, not every node in the set {1...10} is included into the solution.

Second, it is crucial to align the solution representation with the scheme and requirements of the optimization algorithm. As described above, the DE operates in the continues search space and based on Eq. 9, in order to form the mutant vector, element-wise operations have to occur between the vectors chosen among the population of solutions. Thus, although the node sequence representation, described above, is convenient for proving feasibility and calculating the objective function value, it is not the appropriate form to apply the equations of mutation. Consequently, the integer values of the solution vector should be transformed to continuous that the DE is able to handle. In the following, we present methods found in the literature, addressing this issue.

- Assuming that the mutation equation is directly applied to the ordinal encoded vectors, the mutant vector occurs, consisting of floating numbers. Subsequently, the Integer order criterion (IOR) [6] can be utilized. According to IOR the largest floating number corresponds to the largest node ordinal number N, the lowest to depot and the rest may be deduced by analogy. For example the mutant vector $[-7.1, 1.3, -5.6, 2.5, -3.7, 0, 3.3, 5.4]$ is transformed to $[1, 5, 2, 6, 3, 4, 7, 8]$.
- The same issue occurs during the movement equations of the Particle Swarm Optimization Algorithm (PSO), and as an example in [23], the integer part of an element in the vector represents the vehicle. Thus, the same integer part represents the node in the same route, while the fractional part represents the sequence of the customer in the vehicle, following the IOR logic.
- Another technique that has been applied to the solution of the flow shop scheduling problem with PSO [17] is to use transformation functions based on real number encoding. The proposed equations to encode and decode the solution vector are the Eqs. 12 and 13, respectively, where f is a large scaling factor.

$$z_i = -1 + z_i' * f * (length_{vector})/(10^3 - 1) \qquad (12)$$

$$z_i' = (1 + z_i * (10^3 - 1)/(f * (length_{vector}) \qquad (13)$$

Moreover, after the completion of a decoding process, all of the aforementioned methods require a feasible control, to check and correct the transformed vector accordingly. These methods are applied to problem where the complete set of nodes must be included in a feasible solution, while in CTOP that is not the case. Thus, the *Distance Related* (DR) solution encoding/decoding scheme is proposed. The fundamental element of the process is the Euclidean Distance of successive nodes in a route. The novelty of the DR is that utilizes floating values that directly represent a sequence of nodes, and that provides a data-orientated transformation scheme, without requiring auxiliary operators. A simple example is illustrated in Table 1, where the distance representation of the vector express the Euclidean Distance e.g. the distance between the depot and node 12 is 12.04 units. When all the solution vectors in the population have been expressed accordingly the mutation equation (Eq. 9) is applied and the corresponding mutant vector emerges. Subsequently, the crossover operation have to take place, as described in the following section. Nevertheless, the crossover results in a trial vector that consists of floating numbers. At this stage, the decoding procedure has to be employed. Based on the *Distance Related* scheme, the floating values are transformed to integer values, that align to the

Table 1. DR: encoding and mutation example

Solution vector	1	12	39	17	1	48	6	1
Distance representation	Inf	12.04	6.71	9.22	22.02	9.43	15.13	14.14
Mutant vector	0	7.96	7.41	15.79	22.02	11.10	15.63	10.59

representation of a nodes' sequence. The following steps are included in the decoding process:

1. The decoded trial vector is initialized with node 1, $x_{i,1}(t+1)$: [1].
2. Then, we seek the pr nodes that are not included in $x_{i,:}(t+1)$, and their distance from the last node included is in close proximity to the corresponding value of the trial vector $u_{i,j}(t+1)$.
3. Among the selected nodes, the node with the maximum prize value is inserted into the $x_{i,:}(t+1)$, in the position j, with respect to the constraints of the CTOP.
4. The process continues for all the values included in the trial vector $u_{i,:}(t+1)$.

As an instance, decoding the mutant vector presented in Table 1, the first node to be inserted in the first route (after the depot), will be the one with the maximum prize, among the pr nodes with distance from depot, i.e $t_{1,j}$, close to 7.96. Based on the data depicted in Table 2, with $pr = 6$, node 48 should be inserted in the route subsequent to the depot. In the second step, the process is repeated with respect to node 48, i.e $t_{48,j}$, and the distance 7.41 (the next element in the mutant vector), hence node 19 should be positioned after node 48. The rest of the process may be deduced by analogy, while, in case of an unfeasible insertion the route should be completed and node 1 should be forced into the solution, accordingly.

Table 2. Example of the decoding process based on the *distance related* method

The first decoding step						The second decoding step					
Nodes to be considered:	28	13	48	33	7	12	Nodes to be considered:	5	19	13	18	1	49
$t_{1,j}$	8.00	8.06	**9.43**	10.00	11.45	12.04	$t_{48,j}$	7.81	**8.06**	6	9.22	9.43	9.89
Prize:	16	16	**32**	9	15	23	Prize:	6	**31**	16	2	0	5

5 Distance Related Differential Evolution Algorithm (DRDE)

In this section, the proposed algorithmic framework, namely the Distance Related Differential Evolution Algorithm (DRDE) is described, in detail. The DRDE follows the scheme of the original DE, while it incorporates the novel *Distance Related* procedure, presented in the previous section, along with local search techniques, oriented to the requirements and to the goal of the CTOP.

Following the DRDE framework, depicted in Algorithm 1, the first step is the formulation of the initial population, according to the *Initial solution* heuristic technique (Algorithm 2). The goal of the initial process is to create NP feasible solutions, that show diversity and, respectively, good solution quality.

Algorithm 1. Distance Related Differential Evolution Algorithm

Define: size population NP, number of generations L, F & Cr
Initialize target population via *Initial solution*, see Algorithm 2
repeat
    for each $x_{i,:}$, $i = 1 : NP$ **do**
        Update F & Cr
        Compute the objective function value $f(x_{i,:})$
        Find $x_{best,:}$: max f & $x_{rand1,:}$, $x_{rand2,:}$
        Express chosen solutions by distance representation: $dist_{i,:}$, see Section 3
        Apply mutation operator: mutant vector $v_{i,:}$, see Equation 9
        Apply crossover operator:
        if $rand \leq 0.5$ **then**
            Apply binomial operator: trial vector $u_{i,:}$, see Equation 10
        else
            Apply $bi - route$ operator:
            Find index of $x_{i,:} == 1$: separate M routes
            for each route $d = 1 : M$ **do**
                if $rand \leq Cr$ **then**
                    $u_{i,route_d} \leftarrow v_{i,route_d}$
                else
                    $u_{i,route_d} \leftarrow dist_{i,route_d}$
                end if
            end for
        end if
        Update pr
        Apply decoding process on $u_{i,:}$, see Section 3
        Apply *Exchange nodes* : $1 - 1$ on $u_{i,:}$
        Apply *Exchange nodes* : $2 - 1$ on $u_{i,:}$
        Apply *Remove node* on $u_{i,:}$
        Apply *Reinforcement* on $u_{i,:}$
        Calculate objective function $f(u_{i,:})$
        if $f(u_{i,:}) \geq f(x_{i,:})$ **then**
            $x_{i,:} \leftarrow u_{i,:}$
        end if
    end for
until L generations are reached

These characteristics are achieved by the random selection of the M initial nodes and, subsequently by selecting elite nodes (in terms of prize) to augment the solution, since the objective of the CTOP is the maximization of the total collected prize from the nodes. Continuing with the process, the distance representation of the solutions is adopted, as described in Sect. 4. Then, the mutation operator is applied and the mutant vectors $v_{i,:}$ are created. Although, the utilization of the binomial crossover has been mentioned above, the crossover operator applied is enriched, denoted by $bi - route$ crossover. Based on a random number generator, the binomial crossover is used, alternatively to the $bi - route$. The objective of the later operator, is to crossover complete routes (as partial solution) between the target and the mutant solution vector. Preserving the distance elements of a complete route, could aid the DRDE to inherit efficient parts of a solution to the next generations, while the decoding process gains in effectiveness.

Across the proposed algorithmic framework, nodes have to be inserted in the most efficient position in a route (index of the solution vector), with respect to the corresponding time travel needed. Therefore, the *Savings* criterion of Clarke and Wright [7], is employed. Regardless of whether the most efficient position is to be determinate among the hole solution (M routes) or along a partial solution (d route), the process is identical. For each position j to be examined, Eq. 14, and, hence, the minimum $s_{k,j}$ corresponds to the position that node k should be inserted.

$$s_{k,j} = t_{route(j-1),k} + t_{route(j+1),k} - t_{route(j-1),route(j+1)} \qquad (14)$$

Algorithm 2. *Initial solution*

Randomly select M initial nodes to create the $u_{i,:}$
$u_{i,:} = [1, n_1, 1, n_2, 1, \cdots, 1, n_M, 1]$
Find all nodes n not included in the solution vector $u_{i,:}$
$S \cup S + n$
Sort S in descent order based on prize
repeat
    Find rand node n in $S' \leftarrow 0.2 * length(S)$
    Calculate s of n, considering the hole solution $u_{i,:}$
    Find position k: min $s_{n,:}$
    $u'_{i,:} \leftarrow u_{i,:}$
    Insert n in $u'_{i,k}$
    Check feasibility of $u'_{i,:}$
    Update $u'_{i,:}$ accordingly
    Update $u_{i,:} \leftarrow u'_{i,:}$ accordingly
until *itermax* number of iterations is reached

Furthermore, three local search techniques are sequentially utilized, i.e. the *Exchange nodes* : $1 - 1$, the *Exchange nodes* : $2 - 1$, the *Remove node* and the *Reinforcement*. The goal of the *Exchange nodes* : $1 - 1$ heuristic technique is to replace a node in the trial solution vector by an unused one, with greater prize value. Moreover, it facilitated that the later node is inserted in the most efficient position in the vector. Expanding this logic, the *Exchange nodes* : $2 - 1$ aids to the total prize maximization of the solution vector, by replacing two nodes by an unused one, when the prize of the later node exceeds the total prize of the replaced ones. The *Remove node* technique is applied to remove nodes from the trial solution vector, that have minimum contribution in maximizing the total prize of the solution i.e. the nodes with the minimum prize. Such, the node removal corresponds to a reduction in the total travel time, and that creates the opportunity for other nodes with greater prize to be inserted, subsequently, in the solution via *Reinforcement*.

6 Computational Experiments

6.1 Parametrization and Benchmark Instances

To evaluate the proposed Distance Related Differential Evolution Algorithm in the solution of the Capacitataed Team Orienteering Problem, the benchmark instances proposed by Archetti et al. [1] were used. These instances are variations of the 10 original benchmark instances for the Capacitated VRP, presented by Christofides et al. [8], while the number of nodes varies between 51 and 200. To align with the formulation of the CTOP, the prize element has been included, as such, each node i is associated to a prize value p_i, defined as $(0.5 + h) * D_i$, where h is a random number uniformly generated in the interval $[0, 1]$. Furthermore, 12 variants for each of the 10 Christofides' instances have been created, by modifying the fleet size M, the maximum vehicle capacity Q and the maximum allowed travel time T_{max}, grouped in 10 distinctive sets.

The Differential Evolution algorithm, requires only a few control variables, as mentioned above: the L, NP, F, and Cr. Additionally, the DRDE requires the following control variables: the pr in the decoding DR process, as the number of nodes to be considered (see Sect. 4), and the $itermax$, as the number of local search iterations. The values of these parameters are summarized in Table 3, deduced through experimental algorithm executions over several representative instances. To enhance the exploration ability of the DRDE, in each generation, for each solution, the parameters F, Cr and pr, are updated randomly, within their predetermined range, respectively.

Table 3. Control variables

L	NP	F	Cr	pr	$Itermax$
500	$1.2 * N$	$rand[0.6 : 0.7]$	$rand[0.8 : 0.9]$	$rand[0.01 : 0.04] * N$	50

6.2 Results

Tables 4 and 5 present the results of the conducted computational experiments. The name of each tested benchmark instance is presented in the first and sixth column of each table, expressed as: *Set indicator* $- N - Q - T_{max}$. The second and seventh column of each table holds the best solution found in the literature, denoted by *Best*, considering the studies of [1–4,15,20]. The best obtained objective function value, among five algorithmic executions of the proposed DRDE, is presented in columns three and eight, denoted by z_{best}. Subsequently, the relative percentage error rpe per instance i, is reported, calculated by Eq. 15. Finally, columns five and ten, depict the average relative percentage error over the five algorithmic executions, denoted by $arpe$.

$$rpe = \frac{(Best_i - z_{best_i})}{Best_i}\%$$

(15)

Table 4. Computational results

Instance name	DRDE				Instance name	DRDE			
	Best	z_{best}	$rpe(\%)$	$arpe(\%)$		Best	z_{best}	$rpe(\%)$	$arpe(\%)$
03-101-2-100-100	277	277	0.00	0.00	08-101-2-100-100	277	277	0.00	0.00
03-101-2-200-200	536	536	0.00	0.00	08-101-2-200-230	536	536	0.00	0.00
03-101-2-50-50	133	133	0.00	0.00	08-101-2-50-50	133	133	0.00	0.15
03-101-2-75-75	208	208	0.00	0.00	08-101-2-75-75	208	208	0.00	0.10
03-101-2-100-100	277	277	0.00	0.00	08-101-2-100-100	277	277	0.00	0.00
03-101-2-200-200	536	536	0.00	0.00	08-101-2-200-230	536	536	0.00	0.00
03-101-2-50-50	133	133	0.00	0.00	08-101-2-50-50	133	133	0.00	0.15
03-101-2-75-75	208	208	0.00	0.00	08-101-2-75-75	208	208	0.00	0.10
03-101-3-100-100	408	408	0.00	0.00	08-101-3-100-100	408	408	0.00	0.15
03-101-3-200-200	762	760	0.26	0.42	08-101-3-200-230	762	759	0.39	0.55
03-101-3-50-50	198	198	0.00	0.00	08-101-3-50-50	198	198	0.00	0.00
03-101-3-75-75	307	307	0.00	0.00	08-101-3-75-75	307	307	0.00	0.20
03-101-4-100-100	532	530	0.38	0.49	08-101-4-100-100	532	530	0.38	0.45
03-101-4-200-200	950	945	0.53	0.53	08-101-4-200-230	950	948	0.21	0.53
03-101-4-50-50	260	260	0.00	0.00	08-101-4-50-50	260	260	0.00	0.08
03-101-4-75-75	403	403	0.00	0.00	08-101-4-75-75	403	403	0.00	0.15
06-51-2-100-100	252	252	0.00	0.00	09-151-2-100-100	279	279	0.00	0.00
06-51-2-160-200	403	403	0.00	0.00	09-151-2-200-200	548	546	0.36	0.62
06-51-2-50-50	121	121	0.00	0.00	09-151-2-50-50	137	137	0.00	0.29
06-51-2-75-75	183	183	0.00	0.00	09-151-2-75-75	210	210	0.00	0.00
06-51-3-100-100	369	369	0.00	0.00	09-151-3-100-100	415	414	0.24	0.63
06-51-3-160-200	565	565	0.00	0.00	09-151-3-200-200	797	795	0.25	0.60
06-51-3-50-50	177	177	0.00	0.00	09-151-3-50-50	201	201	0.00	0.00
06-51-3-75-75	269	269	0.00	0.00	09-151-3-75-75	312	312	0.00	0.13
06-51-4-100-100	482	482	0.00	0.00	09-151-4-100-100	546	542	0.73	0.84
06-51-4-160-200	683	683	0.00	0.00	09-151-4-200-200	1033	1030	0.29	0.52
06-51-4-50-50	222	222	0.00	0.00	09-151-4-50-50	262	262	0.00	0.00
06-51-4-75-75	349	349	0.00	0.00	09-151-4-75-75	408	407	0.25	0.34
07-76-2-100-100	266	266	0.00	0.00	10-200-2-100-100	282	281	0.35	0.43
07-76-2-140-160	377	377	0.00	0.00	10-200-2-200-200	556	554	0.36	0.43
07-76-2-50-50	126	126	0.00	0.00	10-200-2-50-50	134	134	0.00	0.15
07-76-2-75-75	193	193	0.00	0.10	10-200-2-75-75	208	208	0.00	0.10
07-76-3-100-100	397	397	0.00	0.00	10-200-3-100-100	418	415	0.72	0.86
07-76-3-140-160	548	548	0.00	0.15	10-200-3-200-200	816	813	0.37	0.44
07-76-3-50-50	187	187	0.00	0.00	10-200-3-50-50	200	200	0.00	0.20
07-76-3-75-75	287	287	0.00	0.00	10-200-3-75-75	311	311	0.00	0.13
07-76-4-100-100	521	521	0.00	0.12	10-200-4-100-100	553	550	0.54	0.83
07-76-4-140-160	707	707	0.00	0.14	10-200-4-200-200	1064	1060	0.38	0.47
07-76-4-50-50	240	240	0.00	0.08	10-200-4-50-50	265	265	0.00	0.15
07-76-4-75-75	378	378	0.00	0.11	10-200-4-75-75	411	407	0.97	1.12

Summarizing, the computational results presented in this Section, the DRDE reached the best known results for 88 instances among the 120 explored. For 24 instances, the reported *rpe* does not exceed the value of 0.39%, while, in only 8 benchmark instances the algorithm showed deviation from the best known solution inside the range [0.52%,1.24%]. Examining each benchmark set, for all the instances in sets 6 and 7, with 51 and 76 nodes, respectively, the maximum known prize is achieved by the proposed algorithm. With respect to sets 03, 08

Table 5. Computational results (continued)

Instance name	DRDE				Instance name	DRDE			
	$Best$	z_{best}	$rpe(\%)$	$arpe(\%)$		$Best$	z_{best}	$rpe(\%)$	$arpe(\%)$
13-121-2-100-100	253	253	0.00	0.00	15-151-2-100-100	282	282	282	0.00
13-121-2-200-720	513	513	0.00	0.00	15-151-2-200-200	550	550	550	0.00
13-121-2-50-50	134	134	0.00	0.00	15-151-2-50-50	134	134	134	0.00
13-121-2-75-75	193	193	0.00	0.00	15-151-2-75-75	211	211	211	0.00
13-121-3-100-100	344	344	0.00	0.00	15-151-3-100-100	417	418	417	0.24
13-121-3-200-720	727	727	0.00	0.06	15-151-3-200-200	800	802	800	0.25
13-121-3-50-50	193	193	0.00	0.31	15-151-3-50-50	200	200	200	0.20
13-121-3-75-75	265	265	0.00	0.30	15-151-3-75-75	315	315	315	0.13
13-121-4-100-100	419	416	0.72	0.81	15-151-4-100-100	548	549	548	0.18
13-121-4-200-720	908	908	0.00	0.04	15-151-4-200-200	1020	1031	1020	1.07
13-121-4-50-50	243	243	0.00	0.25	15-151-4-50-50	266	266	266	0.15
13-121-4-75-75	323	323	0.00	0.06	15-151-4-75-75	415	415	415	0.14
14-101-2-100-100	271	271	0.00	0.00	16-200-2-100-100	285	285	0.00	0.00
14-101-2-200-1040	534	534	0.00	0.00	16-200-2-200-200	558	557	0.18	0.47
14-101-2-50-50	124	124	0.00	0.00	16-200-2-50-50	137	137	0.00	0.00
14-101-2-75-75	190	190	0.00	0.00	16-200-2-75-75	212	212	0.00	0.09
14-101-3-100-100	399	398	0.25	0.35	16-200-3-100-100	423	423	0.00	0.00
14-101-3-200-1040	770	770	0.00	0.00	16-200-3-200-200	822	820	0.24	0.24
14-101-3-50-50	184	184	0.00	0.00	16-200-3-50-50	203	203	0.00	0.10
14-101-3-75-75	279	279	0.00	0.00	16-200-3-75-75	317	317	0.00	0.19
14-101-4-100-100	525	523	0.38	0.46	16-200-4-100-100	558	556	0.36	0.36
14-101-4-200-1040	975	975	0.00	0.00	16-200-4-200-200	1073	1071	0.19	0.19
14-101-4-50-50	241	238	1.24	1.41	16-200-4-50-50	269	269	0.00	0.07
14-101-4-75-75	366	365	0.27	0.27	16-200-4-75-75	420	420	0.00	0.10

and 14, that contain instances with 101 nodes, 9, 9 and 8 best known solution were found (among the twelve instances in each set), with average rpe per set, 0.10%, 0.08% and 0.18%, respectively. In set 13, only in one instance the z_{best} deviates from the best known solution. Considering the sets with 151 nodes, in set 09 and 15, 6 and 8 best solution were achieved, with average rpe per set, 0.18% and 0.14%, respectively. Finally, sets 10 and 16 include the largest instances with 200 nodes, and the DRDE obtained 5 and 8 best solution, with average rpe per set, 0.31% and 0.08%, respectively.

7 Conclusions

In this paper, two main issues are addressed, and the first one is the solution of the Capacitated Team Orienteering Problem, a selective Vehicle Routing Problem variant to simulate the modelling and optimization of tourist trip design problems. The objective of the CTOP is formatting a set of depot-returning itineraries, that maximize the total prize collected from the visited nodes, adhering to constraints related to the capacity of the vehicles and to the allowed total travel time. For the solution of the CTOP, a hybrid Differential Evolution (DE) algorithm, is proposed, namely the DRDE. According to the conducted experimental results, the DRDE has obtained the best known solution for the majority of the benchmark instances tested, while the maximum occurred relative percentage error, does not exceed the 1.24%. Duo to space limitations, execution time, heuristic element contribution and parametrization analysis, are omitted from this paper, but the corresponding analysis will be included in a future work that will extend the proposed methodology. The second issue addressed, in this paper, is the utilization of the DE algorithm for optimizing the CTOP. The main consideration is that the DE requires a continuous representation of the solution vectors, i.e. floating numbers, to apply the mutation equation. Hence, the *Distance Related* (DR) encoding/decoding method is introduced. The novelty, of the DR is the utilization of the Euclidean Distance, and as such, in the encoding step the solution vectors are represented by the distance of each consecutive pair of nodes. Subsequently, in the decoding step, the new continues values of the trial vector, are associated to pairs of nodes with approximate distance value, while nodes with greater prize value are prioritized, enhancing the quality of the new solutions. The main advantage of this encoding/decoding scheme is that it is data- and problem-oriented, offering flexibility to the solution process. Moreover, the same scheme could be applied for all the Vehicle Routing Problems and for their solution by meta-heuristic and nature inspired algorithms, such as by the Particle Swarm Optimization and the Firefly algorithm, that utilize equation in the continuous solution space. Further exploration of effective combinations of the DR and other meta-heuristic algorithms is recommended, and will be included in a future study.

References

1. Archetti, C., Feillet, D., Hertz, A., Speranza, M.G.: The capacitated team orienteering and profitable tour problems. J. Oper. Res. Soc. **60**(6), 831–842 (2009)
2. Archetti, C., Bianchessi, N., Speranza, M.G.: Optimal solutions for routing problems with profits. Discrete Appl. Math. **161**(4–5), 547–557 (2013)
3. Ben-Said, A., El-Hajj, R., Moukrim, A.: An adaptive heuristic for the capacitated team orienteering problem. IFAC-PapersOnLine **49**(12), 1662–1666 (2016)
4. Ben-Said, A., El-Hajj, R., Moukrim, A.: A variable space search heuristic for the capacitated team orienteering problem. J. Heuristics **25**(2), 273–303 (2018)
5. Butt, S.E., Cavalier, T.M.: A heuristic for the multiple tour maximum collection problem. Comput. Oper. Res. **21**(1), 101–111 (1994)
6. Cao, E., Lai, M., Yang, H.: Open vehicle routing problem with demand uncertainty and its robust strategies. Expert Syst. Appl. **41**(7), 3569–3575 (2014)
7. Clarke, G., Wright, J.W.: Scheduling of vehicles from a central depot to a number of delivery points. Oper. Res. **12**(4), 568–581 (1964)
8. Christofides, N.: The vehicle routing problem. In: Christofides, N., Mingozzi, A., Toth, P., Sandi, C. (eds.) Combinatorial Optimization (1979)
9. Das, S., Mullick, S.S., Suganthan, P.N.: Recent advances in differential evolution-an updated survey. Swarm Evol. Comput. **27**, 1–30 (2016)

10. Dechampai, D., Tanwanichkul, L., Sethanan, K., Pitakaso, R.: A differential evolution algorithm for the capacitated VRP with flexibility of mixing pickup and delivery services and the maximum duration of a route in poultry industry. J. Intell. Manuf. **28**(6), 1357–1376 (2017)
11. Gavalas, D., Konstantopoulos, C., Mastakas, K., Pantziou, G.: A survey on algorithmic approaches for solving tourist trip design problems. J. Heuristics **20**(3), 291–328 (2014)
12. Golden, B.L., Levy, L., Vohra, R.: The orienteering problem. Nav. Res. Logist. (NRL) **34**(3), 307–318 (1987)
13. Gunawan, A., Lau, H.C., Vansteenwegen, P.: Orienteering problem: a survey of recent variants, solution approaches and applications. Eur. J. Oper. Res. **255**(2), 315–332 (2016)
14. Kunnapapdeelert, S., Kachitvichyanukul, V.: Modified DE algorithms for solving multi-depot vehicle routing problem with multiple pickup and delivery requests. In: Kachitvichyanukul, V., Sethanan, K., Golinska- Dawson, P. (eds.) Toward Sustainable Operations of Supply Chain and Logistics Systems. EcoProduction (Environmental Issues in Logistics and Manufacturing, pp. 361–373. Springer, Cham (2015). https://doi.org/10.1007/978-3-319-19006-8_25
15. Luo, Z., Cheang, B., Lim, A., Zhu, W.: An adaptive ejection pool with toggle-rule diversification approach for the capacitated team orienteering problem. Eur. J. Oper. Res. **229**(3), 673–682 (2013)
16. Mingyong, L., Erbao, C.: An improved differential evolution algorithm for vehicle routing problem with simultaneous pickups and deliveries and time windows. Eng. Appl. Artif. Intell. **23**(2), 188–195 (2010)
17. Onwubolu, G., Davendra, D.: Scheduling flow shops using differential evolution algorithm. Eur. J. Oper. Res. **171**(2), 674–692 (2006)
18. Opara, K.R., Arabas, J.: Differential evolution: a survey of theoretical analyses. Swarm Evol. Comput. **44**, 546–558 (2019)
19. Storn, R., Price, K.: Differential evolution-a simple and efficient heuristic for global optimization over continuous spaces. J. Global Optim. **11**(4), 341–359 (1997)
20. Tarantilis, C.D., Stavropoulou, F., Repoussis, P.P.: The capacitated team orienteering problem: a bi-level filter-and-fan method. Eur. J. Oper. Res. **224**(1), 65–78 (2013)
21. Teoh, B.E., Ponnambalam, S.G., Kanagaraj, G.: Differential evolution algorithm with local search for capacitated vehicle routing problem. Int. J. Bio-Inspired Comput. **7**(5), 321–342 (2015)
22. Vansteenwegen, P., Souffriau, W., Berghe, G.V., Van Oudheusden, D.: The city trip planner: an expert system for tourists. Expert Syst. Appl. **38**(6), 6540–6546 (2011)
23. Wang, W., Wu, B., Zhao, Y., Feng, D.: Particle swarm optimization for open vehicle routing problem. In: Huang, D.-S., Li, K., Irwin, G.W. (eds.) ICIC 2006, Part II. LNCS (LNAI), vol. 4114, pp. 999–1007. Springer, Heidelberg (2006). https://doi.org/10.1007/978-3-540-37275-2_126

Sonar Inspired Optimization in Energy Problems Related to Load and Emission Dispatch

Alexandros Tzanetos$^{(\boxtimes)}$ ⓘ and Georgios Dounias

Management and Decision Engineering Laboratory, Department of Financial and Management
Engineering, School of Engineering,
University of the Aegean, Kountouriotou 41, 82100 Chios, Greece
{atzanetos,g.dounias}@aegean.gr

Abstract. One of the upcoming categories of Computational Intelligence (CI) is meta-heuristic schemes, which derive their intelligence from strategies that are met in nature, namely Nature Inspired Algorithms. These algorithms are used in various optimization problems because of their ability to cope with multi-objective problems and solve difficult constraint optimization problems. In this work, the performance of Sonar Inspired Optimization (SIO) is tested in a non-smooth, non-convex multi-objective Energy problem, namely the Economic Emissions Load Dispatch (EELD) problem. The research hypothesis was that this new nature-inspired method would provide better solutions because of its mechanisms. The algorithm manages to deal with constraints, namely Valve-point Effect and Multi-fuel Operation, and produces only feasible solutions, which satisfy power demand and operating limits of the system examined. Also, with a lot less number of agents manages to be very competitive against other meta-heuristics, such as hybrid schemes and established nature inspired algorithms. Furthermore, the proposed scheme outperforms several methods derived from literature.

Keywords: Meta-heuristics · Sonar Inspired Optimization · Nature Inspired Algorithms · Load Dispatch · Economic Emissions Load Dispatch · Constrained optimization

1 Introduction

In previous works, Sonar Inspired Optimization (SIO) has proven to be quite challenging comparing to other meta-heuristics in real world applications. Recently, SIO managed to provide better or equally good solutions in constrained [1] and high-dimensional [2] problems, which belong in the wider field of engineering optimization problems. A question arising from the abovementioned works was if this method could cope with a challenging problem with multiple objectives.

The aim of this paper is to measure the performance of SIO in a more difficult constraint optimization problem i.e. a non-smooth, non-convex multi-objective energy problem, namely the Economic Emissions Load Dispatch (EELD) problem. SIO is at

© Springer Nature Switzerland AG 2020
N. F. Matsatsinis et al. (Eds.): LION 13 2019, LNCS 11968, pp. 268–283, 2020.
https://doi.org/10.1007/978-3-030-38629-0_22

first applied in the three single-objective problems (minimization of fuel, emissions and power losses respectively), which consist the multi-objective problem. Also, three alterations of the minimization of fuel cost are considered in this paper, where the basic Economic Dispatch (ED) problem is constrained by valve-point effects, multi-fuel operation and network power losses. In addition, SIO is implemented to solve the multi-objective problem of all above mentioned objectives.

A 10-unit test system, which has 2000 MW demand, is examined. SIO managed to outperform a number of compared competitive methodologies and it has also been capable of solving the multi-objective problem with the usage of a single objective function, while other schemes use various multi-objective optimization methods. What is more, SIO was found to outperform five known optimization methods.

This paper is organized as follows; in Sect. 1 a short introduction of the topic is done. In Sect. 2, a brief literature review on the application of nature inspired meta-heuristics in energy problems is presented. In Sect. 3, the algorithmic formulation is discussed. In Sect. 4, the mathematic formulation of the problems tackled in this work is presented. In Sect. 5, the experimental results are included and in Sect. 6, future work is discussed.

2 Literature Review

The different Power system optimization problems have been presented in detail in a recent broad-based survey on articles using hybrid bio-inspired Computational Intelligent (CI) techniques [3]. Categories of these problems are Economic Dispatch (ED), Optimal Power Flow (OPF), Load Forecasting, etc. In this paper, SIO copes with the known problem of Economic Dispatch. This problem consists of four constraints; Real power balance and Real power operating limits, which are strict limitations, and also, Valve-point effects and Multi-fuel operation. The last two constraints make the problem really challenging, when they are taken into consideration. The problem's objectives and constraints are explained in detail in [4].

In literature, various algorithms have been implemented to solve the Economic Dispatch problem. Nature inspired schemes were also developed to solve this problem. Biogeography-based optimization (BBO) [5], Flower Pollination Algorithm (FPA) [6] and Grey Wolf Optimizer (GWO) [7] are some of these schemes. Established algorithms, such as Particle Swarm Optimization (PSO) [8–10] and Differential Evolution [11], are among the most used methods. Recently, a hybrid method consisting of these two algorithms [12] was proposed to deal with ED problem. This method's results are used to compare the corresponding results of SIO in this paper.

Other versions of the Economic Dispatch problem can be found in literature; such as Environmental/economic power dispatch [9, 11] and Combined Economic Emission Dispatch (CEED) [6]. Furthermore, a lot hybrid nature inspired algorithms have been presented to solve EELD, many of which are mentioned in [3].

Usually different test systems are used for experimentation, consisting of 6, 10, 15 and 40 generating units. In this work, a 10-unit system is taken into consideration to measure the performance of SIO.

3 Energy Problems

The problem of Economic Dispatch (ED) has many alterations. The basic problem (BP) consists of the minimization of total fuel cost and the emissions, while trying to minimize also the real power loss and meet the power demand of the system. In literature, these objectives are solved separately or with various formulations, like weighted sum method and Pareto-optimal solution.

Table 1. Characteristics of each problem case.

Problem - Cases		Objective	VPE	MFO	NL
I	I.I	Minimization of fuel cost	–	–	–
	I.II	Minimization of fuel cost	✓	✓	–
	I.III	Minimization of fuel cost	✓	✓	✓
II		Minimization of emissions	–	–	✓
III		Minimization of power losses	–	–	–
IV		Multi-objective	–	–	–

In this paper, the ED problem is solved both separately (Problem I, II and III) and by combining the goals in one single objective function (Problem IV). In Problem I the objective is the minimization of fuel cost. This Problem is divided in three Cases, as it is shown in Table 1; in Case I.I no constraints are taken into consideration, while in Cases I.II and I.III Valve-point effects (VPE) and Multi-fuel operation (MFO) limit the solution space of the problem. In Case I.III, Network Losses (NL) which are related to active power losses of the examined system, are also implemented. By considering the extra constraints such as VPE and MFO, the proposed problem becomes a non-smooth and non-convex optimization problem.

In Problem II the objective is the minimization of emissions, while in Problem III the objective is the minimization of power losses.

In Problem IV, all three objectives are taken into consideration. The minimization of fuel cost, emissions and power losses consist the Economic Emissions Load Dispatch problem (EELD) [13]. Therefore, the problem can be characterized as non-smooth, non-convex and multi-objective. However, in this Problem Case the constraints of Valve-point effects and Multi-fuel operation are not taken into consideration.

The examined system in this study is a 10-unit test system derived from literature [12]. The power demand of the system is 2000 MW in all Problem Cases. All coefficients and power operating limits of the system examined in this work are presented in Tables A1 and A2 in Appendix.

3.1 Objective Functions Consisting the Multi-objective Problem

Minimizing the Total Fuel Cost. According to [4], the initial generators cost layout is presented by quadratic function as a common form for cost function in conventional ED studies as is shown in (1).

$$FC(P_{G_i}) = \sum_{i=1}^{N_G} \left(a_i + b_i P_{G_i} + c_i P_{G_i}^2 \right) \tag{1}$$

where, $a_i\,(\$/h)$, $b_i\,(\$/hMW)$ and $c_i\,(\$/hMW^2)$ are the fuel cost coefficients for the i-th thermal unit. N_G is the number of generators in the system. P_{G_i} is the active power generation of the i-th thermal unit and $FC(P_{G_i})$ is the total fuel cost function for singe-fuel generation scheme, here measured in $\$/h$.

Minimizing the Emission. Power plants produce different kinds of emissions, such as SO_x, NO_x[1] etc., along with the electricity generated. The objective of (2), which is used in literature [3], is the reduction of these emissions:

$$EC(P_{G_i}) = \sum_{i=1}^{N_G} \left(\alpha_i + \beta_i P_{G_i} + \gamma_i P_{G_i}^2 + \xi_i e^{\lambda_i P_{G_i}} \right) \tag{2}$$

where, $\gamma_i(ton/h)$, $\beta_i(ton/hMW)$, $\alpha_i(ton/hMW^2)$, $\xi_i(ton/h)$ and $\lambda_i(1/MW)$ are emission coefficients for the i-th thermal unit. N_G is the number of generators. P_{G_i} is the active power generated from i-th unit and $EC(P_{G_i})$ is the summation of all emission types, measured in ton/h.

Minimizing the Real Power Losses. It is understandable that a generating unit exhibits loss of power when it's working. This power loss can be formulated as (3).

$$PL(P_{G_i}) = \sum_{i=1}^{N_G} \sum_{j=1}^{N_G} P_{G_i} B_{ij} P_{G_j} + \sum_{i=1}^{N_G} B_{oi} P_{G_i} + B_{oo} \tag{3}$$

where, B_{ij} is the ij-th element of the loss coefficient square matrix. B_{oi} is the i-th element of loss coefficient vector. B_{oo} is the loss coefficient constant and finally $PL(P_{G_i})$ is the total network losses, measured in MW. N_G is the number of generation units and P_{G_i} is the active power generated by i-th unit.

3.2 Constraints of the Problem

Valve-Point Effects. Several control valves are employed in power generation units for controlling the flow of steam in order to adjust the output power. There is a sudden increase in power losses whenever one of the valves gets opened leading to addition of several ripples in the cost function which is called Valve-Point Effect (VPE). Figure 1 presents the comparison between the cost function with and without VPE [10].

According to [14], a cost function is obtained based on the ripple curve for more accurate modeling. This curve contains higher order nonlinearity and discontinuity due

---

[1] SO_x stands for Sulphur Oxides and NO_x for Nitrogen Oxides.

to the valve point effect and should be refined by a sine function. The quadratic cost function defined in Eq. (1) above is altered because of the valve-point effect as follows:

$$C_{VPE}(P_{G_i}) = \sum_{i=1}^{N_G} \left(a_i P_{G_i}^2 + b_i P_{G_i} + c_i P_{G_i}^2 \right) + \left| d_i \sin\left(e_i \left(P_{G_i}^{min} - P_{G_i} \right) \right) \right| \quad (4)$$

where, a_i, b_i, c_i, d_i and e_i are the fuel cost coefficients for i-th thermal unit, considering Valve-point Effect (VPE). N_G is number of generators. P_{G_i} is the active power generated of i-th generator and $C_{VPE}(P_{G_i})$ is the total fuel cost considering Valve-point Effect (VPE), here measured in $/h$.

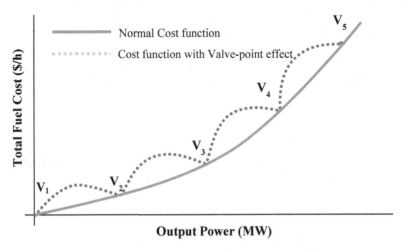

Fig. 1. Cost curve with and without valve-point effects

Multi-fuel Operation. Various types of fuels are used usually in power plants different kinds of units. The cost objective function turns into a piecewise polynomial function by considering the different fuels, as illustrated in Fig. 2.

Therefore based on Fig. 2, (4) cannot represent the generation cost anymore and cost function must be rewritten as a multidisciplinary function as follows [12]:

$$C_{VPE\_MFO}(P_{G_i}) = \sum_{i=1}^{N_G} \left(a_{i,k} P_{G_i}^2 + b_{i,k} P_{G_i} + c_{i,k} P_{G_i}^2 \right) + \left| d_{i,k} \sin\left(e_{i,k} \left(P_{G_i}^{min} - P_{G_i} \right) \right) \right| \quad (5)$$

subject to $P_{G_{i,k}}^{min} \leq P_{G_i} \leq P_{G_{i,k}}^{max}$, for fuel option $k = 1, 2, \ldots, N_F$.

In Eq. (5), $a_{i,k}, b_{i,k}, c_{i,k}, d_{i,k}$ and $e_{i,k}$ are the cost function coefficients of the i-th generation unit, when fuel type k is used. P_{G_i} is the active power produced by the i-th generation unit, while $P_{G_{i,k}}^{min}$ and $P_{G_{i,k}}^{max}$ are its corresponding minimum and maximum limits, when operating with fuel type k. N_G and N_F are the number of generators and available fuels, respectively. Finally, $C_{VPE\_MFO}(P_{G_i})$ represents the total fuel cost considering Valve-point effects (VPE) and Multi-fuel operation (MFO), here measured in $/h$.

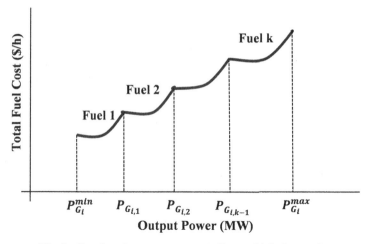

Fig. 2. Cost function curve corresponding multi-fuel operation

Power Balance Constraint. The power balance constraint [4] is based on the principle of equilibrium between total system generation and total system loads (PD) and losses (PL).

$$\sum_{i=1}^{N_G} P_{G_i} = P_D + P_L \tag{6}$$

where, P_D is the system's total power demand (MW), P_L is the total real power transmission losses of the network (MW) and N_G is the number of generators.

Generation Constraint. The net output power of each generation unit is limited by its lower and upper bounds i.e. Power Operating Limits [4].

$$P_{G_i}^{min} \leq P_{G_i} \leq P_{G_i}^{max} \tag{7}$$

where, P_{G_i} is the active power generated by i-th generation unit, while $P_{G_i}^{min}$ and $P_{G_i}^{max}$ are unit's minimum and maximum operating limits, respectively.

4   Sonar Inspired Optimization

Sonar Inspired Optimization (SIO) is a meta-heuristic Nature Inspired Algorithm, which has been presented in [15]. Various applications of the algorithm have been done, including engineering design problems [1], resource leveling [2] etc., attesting that SIO is a good optimization tool in constrained optimization problems. Given that fact, in this work, SIO is applied on energy problems, in which there are various constrains that make these problems non-smooth and non-convex.

In the algorithmic formulation, we consider each agent $X_i = \{P_{G_1}, P_{G_2}, P_{G_3}, \ldots, P_{G_n}\}$ as a solution containing the active Power generation P_{G_i} for each unit i, where

$i \in 1, 2, \ldots, N$ and N being the maximum number of agents, while n is the maximum number of generating units.

At start, the position of the agents is initialized somewhere in the solution space; the easiest way to do that is with random way via the normal distribution function, as it is shown in (8). In this paper, the problem's variables are limited by the Power Operating Limits discussed in the previous section, so the function used is:

$$P_{G_i} = P_{G_i}^{min} + \left(P_{G_i}^{max} - P_{G_i}^{min} \right) \cdot rand \tag{8}$$

where $rand$ is a random number between $(0, 1)$, $P_{G_i}^{min}$ and $P_{G_i}^{max}$ are the lower and upper operating limits of each unit i, respectively.

Sonar Inspired Optimization

Initialization of agents' position
Initialize effective radius and intensity for each agent
While *stopping criteria not met*
    If *counter = checkpoint*
        Relocate the agent
        Recalculate intensity and radius
    End
    Update radius for every agent
    Calculate Intensity for every agent
    While *full_scan = false*
        Update the rotation angle in every dimension of the problem
        Calculate fitness of possible new position
        Save the best so far for each agent in the current scan
    End
    Update best position and fitness
    Update intensity and acoustic power output for every agent
End

Fig. 3. Pseudocode of the proposed Sonar Inspired Algorithm (SIO)

Also, the initial values of effective radius and intensity for each agent are calculated. Then, while the stopping criteria are not met, each agent is updating its position and in each step (iteration) a sonar-alike scan is taking place in its surrounding solution space. This mechanism is the main novelty of this scheme, because each agent checks more possible positions in each step, in opposition with other meta-heuristics where only one check per iteration is done. The scanning space is limited by the effective radius and the number of possible positions that will be investigated is based on agent's current fitness. All agents are sorted based on their fitness and according to the sub-group into which they belong the maximum rotation angle is altered. In this work, six sub-groups have been used, given the values of maximum rotation angle as follows:

$$\vec{a}^{\circ} = [5\ 10\ 20\ 30\ 40\ 50]$$

which is the vector presented in [1] reversed.

The algorithm is explained in detail in [1]. A representative pseudocode of the algorithm is shown in Fig. 3.

4.1 Dealing with Constraints

Among other novel concepts, Sonar Inspired Optimization implements a correction mechanism to relocate solutions which exceed their limits. This mechanism is useful when facing with the Generation Constraint of each unit, because each solution is transformed in a feasible one. Therefore, the algorithm does not waste computational time or power into non-feasible solutions. The correction mechanism is described in (9):

$$P_{G_i} = P_{G_i}^{min} + \left(P_{G_i}^{max} - P_{G_i}^{min} \right) \cdot \cos\left(x_i^d \right) \tag{9}$$

in order to fulfil the relation (7).

This mechanism is also applied when an agent has done multiple steps (iterations) without improving its fitness. The critical number of steps is defined by Eq. (10):

$$checkpoint = scans \cdot \mu \tag{10}$$

where μ is the equivalent of generations without environmental change of genetic algorithm. In this paper, the value of μ is set to 0.05, leaving each agent enough time (50 steps) to evolve its solution. If a smaller value was used (e.g. 0.01 as in [1]), then each agent would have only a few (10 in the example) iterations to improve his solution before relocating. On the other hand, a very big value of μ will led the algorithm to be trapped in local optima.

Furthermore, a penalty function is implemented to deal with the power balance constraint. Solutions that violate this limitation take a penalty, namely the Power Balance Penalty, which is calculated as:

$$PBP = \left| P_D - \sum_{i=1}^{N_G} P_{G_i} \right| \tag{11}$$

where, P_D is the system's total power demand (MW), P_{G_i} is the active power generated by i-th generation unit and N_G is the number of generators.

4.2 Objective Functions

Each agent's fitness is calculated following the corresponding objective function based on the problem case (Table 2):

Table 2. Modified objective functions for each problem.

Objective function	Problem	Case	
$fit_i = FC(P_{G_i}) + 10^3 \cdot PBP$	I	I	(12)
$fit_i = C_{VPE\_MFO}(P_{G_i}) + 10^1 \cdot PBP$		II	(13)
$fit_i = C_{VPE\_MFO}(P_{G_i}) + 10^3 \cdot PBP + PL(P_{G_i})$		III	(14)
$fit_i = EC(P_{G_i}) + 10^4 \cdot PBP + 10^2 \cdot PL(P_{G_i})$	II		(15)
$fit_i = PL(P_{G_i}) + PBP$	III		(16)
$fit_i = FC(P_{G_i}) + EC(P_{G_i}) + 10^5 \cdot PBP + 10^2 \cdot PL(P_{G_i})$	IV		(17)

The above modified objective functions have been constructed in order to take all constraints into consideration for each problem. This approach seems similar to the weighted sum method; however, the multiplied values have been chosen to make all magnitudes of the same size.

5   Experimental Results

All experiments were conducted using Matlab R2015a on a 4 GB, 3.6 GHz Intel Core i7 Windows 10 Pro. The parameters used in experiments are shown in Table 3 below, while the corresponding parameters of the comparing schemes are also included. For every problem, 40 independent runs were done to measure the statistical performance of the algorithm. The results are compared with the corresponding results obtained by two hybrid nature-inspired schemes, namely Fuzzy Based Hybrid Particle Swarm Optimization-Differential Evolution and Hybrid Particle Swarm Optimization-Differential Evolution, and the established Particle Swarm Optimization algorithm, which achieved the best results in recent work [12] dealing with the same problem. What is more, a number of standard methods like adaptive Hopfield Neural Networks with bias adjustment method, Evolutionary Programming, Hierarchical Structure Method, Non-dominated Sorting Genetic Algorithm II and Strength Pareto Evolutionary Algorithm 2, are used as benchmarks in the problem cases that numerical results are available. Works are underway to hybridize SIO to maximize its performance. The research assumption of this paper was that SIO will be challenging in this problem without being hybridized. Also, the drawbacks of SIO, which will be arisen in the experimental process, will be used as a factor on which method would be the fittest for SIO to perform better in this problem. The parameters of comparing schemes are given in detail in [12].

Table 3. Parameters used in each scheme.

Parameter	SIO	FBHPSO-DE	HPSO-DE	PSO
Population	50	300	300	300
Generations	1000	1000	1000	1000
μ	0,05	–	–	–

Maybe, the important detail that can someone observe in Table 3 is that SIO implements significantly fewer agents. Although generally the more agents there are, the higher is the probability of finding the optimal solution, in SIO this is not the case. The main novelty of this algorithm is the multitude of generated points around each agent, which provide a wider search of the solution space, while the number of agents can remain the same.

Four different variations of the ED problem are tested in this work. In Sect. 2, these problems and their constraints were presented. Problem I is divided in three cases; Case I.I where no constraints are taken into consideration and Cases I.II and I.III where Valve-point effects (VPE) and Multi-fuel operation (MFO) limit the solution space of the problem. In Case I.III, Network Losses (NL) are also taken into consideration. By considering the extra constraints such as VPE and MFO, the proposed problem becomes a non-smooth and non-convex optimization problem.

In Problem II the objective is the minimization of emissions, while in Problem III the objective is the minimization of power losses. In Problem IV, all three objectives are taken into consideration.

Table 4. Power operating limits for the 10-unit system.

Unit	1	2	3	4	5	6	7	8	9	10
$P_{G_i}^{min}$	10	20	47	20	50	70	60	70	135	150
$P_{G_i}^{max}$	55	80	120	130	160	240	300	340	470	470

The data used in all problems are based on a 10-unit system. In Table 4, the Power Operation Limits for each unit of the system are presented. The cost coefficients considering VPE and considering multi-fuel cost coefficients with VPE, emission coefficients and loss coefficients of the system are given in the Appendix, according to [16].

In Table 5, the dispatch outcomes of SIO for each problem are presented. In each occasion, the objective of the corresponding problem is denoted with bold.

The results of SIO are compared with the corresponding results of two hybrid schemes consisting of PSO and DE (FBHPSO-DE and HPSO-DE), and the results of PSO. All comparing results are included in [12]. As it can be seen in Table 6, SIO outperforms the other schemes in two cases of Problem I, where the objective is the minimization of fuel cost. In both Cases I.II and I.III, where constraints were added, the performance of the algorithm was equal (Case I.II) or superior (Case I.III) to the other schemes. For Case I.II, where SIO didn't manage to overcome the result of FBHPSO, the provided results were concentrated very close to the best performance achieved, as it can be seen in Fig. 4 in the upper histogram.

Also, as benchmark, the results of classic methods are used, such as adaptive Hopfield Neural Networks with bias adjustment method (aHNN) [17], Evolutionary Programming (EP) [18], Hierarchical Structure Method (HSM) [19] for Problem Case I.II and Non-dominated Sorting Genetic Algorithm II (NSGA II) [11] and Strength Pareto Evolutionary Algorithm 2 (SPEA 2) for Problem II [11].

Table 5. Dispatch outcomes of the proposed algorithm for all problems.

Control variables	Problem I			Problem II	Problem III	Problem IV
	Case I.I	Case I.II	Case I.III			
Load of system (MW)	2000	2700	2700	2000	2000	2000
P_{G_1}	54.1063	217.2060	222.8991	54.8694	46.1135	51.5205
P_{G_2}	78.2936	212.6991	211.7118	73.4562	61.2648	79.3155
P_{G_3}	111.7484	287.0443	273.7992	90.7497	118.6768	87.6005
P_{G_4}	76.3568	239.1194	239.6446	100.1967	128.2961	83.0033
P_{G_5}	71.8717	270.5148	277.7134	137.7294	159.4203	103.8117
P_{G_6}	71.4365	240.9868	242.8708	221.6755	237.5198	141.1924
P_{G_7}	294.0659	292.5837	295.6449	255.1432	293.6623	286.3655
P_{G_8}	329.2705	237.8868	239.0937	290.3642	318.8012	306.7620
P_{G_9}	449.5824	425.0959	422.5314	391.4659	407.1916	440.4434
$P_{G_{10}}$	463.2645	276.8645	274.1013	384.3498	229.0538	419.9852
Total P_G (MW)	2000.00	2700.00	2829.20	2000.00	2000.00	2000.00
Emission (lb/h)	–	–	–	3746.3581	–	3950.12
Power loss (MW)	–	–	129.1903	75.2671	**72.5526**	78.2415
Total cost $(\$/h)$	**106130.50**	**624.2600**	**624.2703**	–	–	**107220.20**

Table 6. Comparison of SIO with other schemes in Problems I, II and III.

Algorithms	Problem I			Problem II	Problem III
	Case I.I	Case I.II	Case I.III		
SIO	**106130.50**	624.2600	**624.2703**	**3746.3581**	72.5526
FBHPSO-DE	106170.09	**623.8224**	700.0118	3920.4819	**68.7708**
HPSO-DE	–	624.1034	–	3921.6618	69.1269
PSO	–	627.6902	–	3929.4431	76.8925
aHNN	–	626.24	–	–	–
EP	–	626.26	–	–	–
HSM	–	625.18	–	–	–
NSGA II	–	–	–	4130.20	–
SPEA 2	–	–	–	4109.10	–

In Problem II, where the objective is the minimization of emissions, SIO provided a very improved solution compared with the solutions of the other schemes. At last, in Problem III, the objective was to minimize the power losses of the system. SIO didn't manage to overcome all three comparing algorithms.

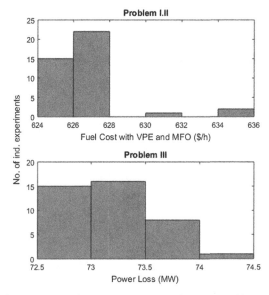

Fig. 4. Histograms showing the performance of SIO in Problems I.II and III

In Problem IV, or in other words the multi-objective problem, SIO provided a very improved solution both in terms of fuel cost and emissions. However, power losses were slightly worse than the best solution derived by FBHPSO-DE (Table 7).

Table 7. Comparison of SIO with FBHPSO-DE in the multi-objective problem (Problem IV).

Parameter	SIO	FBHPSO-DE
Total cost $(\$/h)$	107220.20	115193.4416
Emission (lb/h)	3950.12	5225.0893
Power loss (MW)	78.2415	77.7879

It must be denoted that in FBHPSO-DE the Best Compromise Solution [12] is concluded, meaning that the best solution presented is derived by selecting the best value from the values that arise from running the hybrid scheme with different objectives for each criterion. On the other hand, SIO managed to find the best solution for all three objectives using one objective function, as it is shown in (12)–(17).

Important information derived from Fig. 5 is the iteration where the comparing solution of FBHPSO-DE was surpassed. In Problems I.III and II, the solutions provided by SIO were superior to the corresponding of FBHPSO-DE from the start of the algorithmic process. For the Problem I.I, the algorithm approached the best solution of FBHPSO-DE in the first 250 iterations and then, overcame it after the 500 iteration.

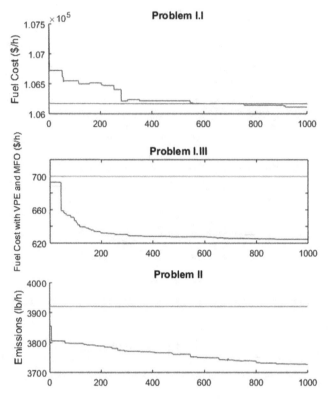

Fig. 5. Evolution of the best solution for Problems I.I, I.III and II

6 Future Work

In this work, a recently presented nature inspired meta-heuristic algorithm is applied on energy problems related to load and emission dispatch. Experimentation has been done on a 10-unit power system. Sonar Inspired Optimization (SIO) performance is measured in variations of the Economic Dispatch (ED) problem implementing Valve-point Effect (VPE) and Multi-fuel Operation (MFO).

SIO overcame a number of known methods and managed to outperform the comparing nature-inspired schemes in most of the cases, while using a lot fewer agents. In Problem Case I.II, the results of the proposed algorithm were statistically equal to these of the best comparing scheme. In Problem III, SIO didn't manage to overcome all other schemes. However, its equally good performance made us think that hybridizing SIO, the new scheme will overcome the known best solution of other schemes. Results are collated with corresponding findings of two hybrid algorithms, consisting of PSO and DE, and the established PSO algorithm. Known optimization methods, namely adaptive Hopfield Neural Networks with bias adjustment method, Evolutionary Programming, Hierarchical Structure Method, Non-dominated Sorting Genetic Algorithm II and Strength Pareto Evolutionary Algorithm 2 were used as benchmarks. What is more, SIO dealt with the multi-objective problem without the usage of any multi-objective optimization technique. The solution provided in the multi-objective problem was better than the corresponding one of FBHPSO-DE in terms of cost and emissions.

SIO is proven to be very challenging and also provides good solutions without checking non-feasible solutions, saving valuable computational time and power. In future research, SIO will be applied on larger systems and in real systems to improve its mechanisms and to provide a good decision tool for energy dispatch problems.

Appendix

Table A1. System data of units with single-fuel cost coefficients considering VPE and emission coefficients.

Unit	a_i	b_i	c_i	d_i	e_i	α_i	β_i	γ_i	ξ_i	λ_i
1	1000.403	40.5407	0.12951	33	0.0174	360.0012	−3.9864	0.04702	0.25475	0.01234
2	950.606	39.5804	0.10908	25	0.0178	350.0056	−3.9524	0.04652	0.25475	0.01234
3	900.705	36.5104	0.12511	32	0.0162	330.0056	−3.9023	0.04652	0.25163	0.01215
4	800.705	39.5104	0.12111	30	0.0168	330.0056	−3.9023	0.04652	0.25163	0.01215
5	756.799	38.5390	0.15247	30	0.0148	13.8593	0.3277	0.00420	0.24970	0.01200
6	451.325	46.1592	0.10587	20	0.0163	13.8593	0.3277	0.00420	0.24970	0.01200
7	1243.531	38.3055	0.03546	20	0.0152	40.2669	−0.5455	0.00680	0.24800	0.01290
8	1049.998	40.3965	0.02803	30	0.0128	40.2669	−0.5455	0.00680	0.24990	0.01203
9	1658.569	36.3278	0.02111	60	0.0136	42.8955	−0.5112	0.00460	0.25470	0.01234
10	1356.659	38.2704	0.01799	40	0.0141	42.8955	−0.5112	0.00460	0.25470	0.01234

The loss coefficients matrix values:

$$B_{ij} = \begin{bmatrix}
0.000049 & 0.000014 & 0.000015 & 0.000015 & 0.000016 & 0.000017 & 0.000017 & 0.000018 & 0.000019 & 0.000020 \\
0.000014 & 0.000045 & 0.000016 & 0.000016 & 0.000017 & 0.000015 & 0.000015 & 0.000016 & 0.000018 & 0.000018 \\
0.000015 & 0.000016 & 0.000039 & 0.000010 & 0.000012 & 0.000012 & 0.000014 & 0.000014 & 0.000016 & 0.000016 \\
0.000015 & 0.000016 & 0.000010 & 0.000040 & 0.000014 & 0.000010 & 0.000011 & 0.000012 & 0.000014 & 0.000015 \\
0.000016 & 0.000017 & 0.000012 & 0.000014 & 0.000035 & 0.000011 & 0.000013 & 0.000013 & 0.000015 & 0.000016 \\
0.000017 & 0.000015 & 0.000012 & 0.000010 & 0.000011 & 0.000036 & 0.000012 & 0.000012 & 0.000014 & 0.000015 \\
0.000017 & 0.000015 & 0.000014 & 0.000011 & 0.000013 & 0.000012 & 0.000038 & 0.000016 & 0.000016 & 0.000018 \\
0.000018 & 0.000016 & 0.000014 & 0.000012 & 0.000013 & 0.000012 & 0.000016 & 0.000040 & 0.000015 & 0.000016 \\
0.000019 & 0.000018 & 0.000016 & 0.000014 & 0.000015 & 0.000014 & 0.000016 & 0.000015 & 0.000042 & 0.000019 \\
0.000020 & 0.000018 & 0.000016 & 0.000015 & 0.000016 & 0.000015 & 0.000018 & 0.000016 & 0.000019 & 0.000044
\end{bmatrix}$$

$B_{oi} = 0$
$B_{oo} = 0$

Table A2. System data of units considering multi-fuel cost coefficients with VPE.

Unit	Generation $P_{G_i}^{min}$	F_1	P_1	F_2	P_2	F_3	$P_{G_i}^{max}$	Fuel type	a_i	b_i	c_i	d_i	e_i
1	100		196		–		250	1	26.97	−0.3975	0.002176	0.02697	−3.975
		1		2		–		2	21.13	−0.3059	0.001861	0.02113	−3.059
2	50		114		157		230	1	118.4	−1.269	0.004194	0.1184	−12.69
		2		3		1		2	1.865	−0.03988	0.001138	0.001865	−0.3988
								3	13.65	−0.1980	0.001620	0.01365	−1.980
3	200		332		388		500	1	39.79	−0.3116	0.001457	0.03979	−3.116
		1		3		2		2	−59.14	0.4864	0.00001176	−0.05914	4.864
								3	−2.875	0.03389	0.0008035	−0.002876	0.3389
4	99		138		200		265	1	1.983	−0.03114	0.001049	0.001983	−0.3114
		1		2		3		2	52.85	−0.6348	0.002758	0.05285	−6.348
								3	266.8	−2.338	0.005935	0.2668	−23.38
5	190		338		407		490	1	13.92	−0.08733	0.001066	0.01392	−0.8733
		1		2		3		2	99.76	−0.5206	0.001597	0.09976	−5.206
								3	−53.99	0.4462	0.0001498	−0.05399	4.462
6	85		138		200		265	1	52.85	−0.6348	0.002758	0.05285	−6.348
		2		1		3		2	1.983	−0.03114	0.001049	0.001983	−0.3114
								3	266.8	−2.338	0.005935	0.2668	−23.38
7	200		331		391		500	1	18.93	−0.1325	0.001107	0.01893	−1.325
		1		2		3		2	43.77	−0.2267	0.001165	0.04377	−2.267
								3	−43.35	0.3559	0.0002454	−0.04335	3.559
8	99		138		200		265	1	1.983	−0.03114	0.001049	0.001983	−0.3114
		1		2		3		2	52.85	−0.6348	0.002758	0.05285	−6.348
								3	266.8	−2.338	0.005935	0.2668	−23.38
9	130		213		370		440	1	88.53	−0.5675	0.001554	0.08853	−5.675
		3		1		3		2	15.30	−0.04514	0.007033	0.01423	−0.1817
								3	14.23	−0.01817	0.0006121	0.01423	−0.1817
10	200		362		407		490	1	13.97	−0.09938	0.001102	0.01397	−0.9938
		1		3		2		2	−61.13	0.5084	0.00004164	−0.06113	5.084
								3	46.71	−0.2024	0.001137	0.04671	−2.024

References

1. Tzanetos, A., Dounias, G.: Sonar inspired optimization (SIO) in engineering applications. Evol. Syst. 1–9 (2018). https://link.springer.com/article/10.1007/s12530-018-9250-z#article-info
2. Tzanetos, A., Kyriklidis, C., Papamichail, A., Dimoulakis, A., Dounias, G.: A nature inspired metaheuristic for optimal leveling of resources in project management. In: Proceedings of the 10th Hellenic Conference on Artificial Intelligence. ACM, July 2018
3. Rahman, I., Mohamad-Saleh, J.: Hybrid bio-inspired computational intelligence techniques for solving power system optimization problems: a comprehensive survey. Appl. Soft Comput. **69**, 72–130 (2018)

4. Das, S., Suganthan, P.N.: Problem definitions and evaluation criteria for CEC 2011 competition on testing evolutionary algorithms on real world optimization problems. Jadavpur University, Nanyang Technological University, Kolkata (2010)
5. Bhattacharya, A., Chattopadhyay, P.K.: Biogeography-based optimization for different economic load dispatch problems. IEEE Trans. Power Syst. **25**(2), 1064–1077 (2010)
6. Abdelaziz, A., Ali, E., Elazim, S.: Implementation of flower pollination algorithm for solving economic load dispatch and combined economic emission dispatch problems in power systems. Energy **101**, 506–518 (2016)
7. Pradhan, M., Roy, P., Pal, T.: Grey wolf optimization applied to economic load dispatch problems. Int. J. Electr. Power Energy Syst. **83**, 325–334 (2016)
8. Zafar, H., Chowdhury, A., Panigrahi, B.K.: Solution of economic load dispatch problem using Lbest-particle swarm optimization with dynamically varying sub-swarms. In: Panigrahi, B.K., Suganthan, P.N., Das, S., Satapathy, S.C. (eds.) SEMCCO 2011. LNCS, vol. 7076, pp. 191–198. Springer, Heidelberg (2011). https://doi.org/10.1007/978-3-642-27172-4_24
9. Wang, L., Singh, C.: Environmental/economic power dispatch using a fuzzified multi-objective particle swarm optimization algorithm. Electr. Power Syst. Res. **77**(12), 1654–1664 (2007)
10. Chakraborty, S., Senjyu, T., Yona, A., Saber, A.Y., Funabashi, T.: Solving economic load dispatch problem with valve-point effects using a hybrid quantum mechanics inspired particle swarm optimisation. IET Gener. Transm. Distrib. **5**(10), 1042–1052 (2011)
11. Basu, M.: Economic environmental dispatch using multi-objective differential evolution. Appl. Soft Comput. **11**(2), 2845–2853 (2011)
12. Naderi, E., Azizivahed, A., Narimani, H., Fathi, M., Narimani, M.: A comprehensive study of practical economic dispatch problems by a new hybrid evolutionary algorithm. Appl. Soft Comput. **61**, 1186–1206 (2017)
13. Apostolopoulos, T., Vlachos, A.: Application of the firefly algorithm for solving the economic emissions load dispatch problem. Int. J. Comb. **2011** (2010). https://www.hindawi.com/journals/ijcom/2011/523806/cta/
14. Coelho, L., Mariani, V.: Combining of chaotic differential evolution and quadratic programming for economic dispatch optimization with valve-point effect. IEEE Trans. Power Syst. **21**(2), 989–996 (2006)
15. Tzanetos, A., Dounias, G.: A new metaheuristic method for optimization: sonar inspired optimization. In: Boracchi, G., Iliadis, L., Jayne, C., Likas, A. (eds.) EANN 2017. CCIS, vol. 744, pp. 417–428. Springer, Cham (2017). https://doi.org/10.1007/978-3-319-65172-9_35
16. Chiang, C.L.: Improved genetic algorithm for power economic dispatch of units with valve-point effects and multiple fuels. IEEE Trans. Power Syst. **20**(4), 1690–1699 (2005)
17. Lee, K., Sode-Yome, A., Park, J.: Adaptive Hopfield neural networks for economic load dispatch. IEEE Trans. Power Syst. **13**(2), 519–526 (1998)
18. Jayabarathi, T., Sadasivam, G.: Evolutionary programming-based economic dispatch for units with multiple fuel options. Eur. Trans. Electr. Power **10**(3), 167–170 (2000)
19. Lin, C., Viviani, G.: Hierarchical economic dispatch for piecewise quadratic cost functions. IEEE Trans. Power Appar. Syst. **6**, 1170–1175 (1984)

A New Hybrid Firefly – Genetic Algorithm for the Optimal Product Line Design Problem

Konstantinos Zervoudakis$^{(\boxtimes)}$ (iD), Stelios Tsafarakis (iD),
and Sovatzidi Paraskevi-Panagiota

School of Production Engineering and Management,
Technical University of Crete, Chania, Greece
{kzervoudakis,tsafarakis}@isc.tuc.gr

Abstract. The optimal product line design is one of the most critical decisions for a firm to stay competitive, since it is related to the sustainability and profitability of a company. It is classified as an NP-hard problem since no algorithm can certify in polynomial time that the optimum it identifies is the overall optimum of the problem. The focus of this research is to propose a new hybrid optimization method (FAGA) combining Firefly algorithm (FA) and Genetic algorithm (GA). The proposed hybrid method is applied to the product line design problem and its performance is compared to those of previous approaches, like genetic algorithm (GA) and simulated annealing (SA), by using both actual and artificial consumer-related data preferences for specific products. The comparison results demonstrate that the proposed hybrid method is superior to both genetic algorithm and simulated annealing in terms of accuracy, efficiency and convergence speed.

Keywords: Product line design · Hybridization · Firefly algorithm · Genetic algorithm

1 Introduction

To stay competitive, now that the economic environment is becoming more competitive than ever, firms have to face difficulties, such as market globalization and shorter product life cycles. As a result, firms must introduce new products or redesign existing ones to maintain their sustainability and profitability, even though such processes can be uncertain and expensive. An illuminating example is the commercial failure of the Edsel model, which cost Ford $350 million [1]. To avoid such risks, managers estimate the potential success of a new product concept before producing it. These kinds of estimations have consisted of a wide area of research in quantitative marketing for about 40 years, known as the optimal product design problem and is usually formulated in the context of conjoint analysis [2]. In a product line, several related products are designed. The problem is classified as an NP-hard combinatorial optimization problem, since no algorithm can certify in polynomial time that the optimum it identifies is the overall optimum of the problem [3]. Numerous optimization algorithms have been applied to the problem in previous attempts of solving it. The most important attempts are dynamic

© Springer Nature Switzerland AG 2020
N. F. Matsatsinis et al. (Eds.): LION 13 2019, LNCS 11968, pp. 284–297, 2020.
https://doi.org/10.1007/978-3-030-38629-0_23

programming [4], beam search [5], genetic algorithms [6], Lagrangian relaxation with branch and bound [7], simulated annealing [8] and particle swarm optimization [9]. In this paper, we solve the optimal product line design problem with the use of a new hybrid optimization method (FAGA) combining firefly algorithm (FA) [10] and Genetic algorithm (GA) [11]. Previous attempts of successful FA and GA hybridizations are reported in the literature [12–14]. The proposed hybrid method is applied to the product line design problem and its performance is compared to those of previous approaches, like genetic algorithm (GA) [11] and simulated annealing (SA) [15], using both actual and artificial consumer-related data preferences for specific products.

The rest of the paper is organized into 6 sections as follows: Sect. 2 provides a brief description of the optimal product line design problem, while in Sect. 3 the FA and GA are described. In Sect. 4, we describe the problem formulation and we explain the algorithmic structure of the proposed hybrid FAGA method. In Sect. 5 we evaluate the effectiveness of FAGA through a comparison of its performance with the methods that Belloni et al. [8] compared, like GA and SA. Finally, Sect. 6 provides an overview of the main conclusions of the study and future research areas are suggested.

2 The Optimal Product Line Design Problem

The optimal product line design problem refers to the problem where individual firms want to introduce a new "optimal" product line, to optimize a specific objective (usually the objective is the maximization of market share or profit). A product is usually presented as a set of attributes (characteristics), each one having specific levels. A tablet for example, could consist of the following attributes: screen, camera, processor and memory, which have, respectively, the levels $8''$ or $10''$, 8 MP, 16 MP or 24 MP, quad core or octa-core, 16 GB or 32 GB etc.

Customers' selection varies according to the levels of the attributes that they prefer. A writer, for instance, most probably prefers a larger screen, whereas an average person may choose a smaller one as it is more convenient in terms of mobility. In order for companies to design and sell new products, they should be aware of consumer habits. This is achieved through market research and surveys. Customer data give important information to firms, since through them, the attributes of their products are linked with individual customer preferences. Conjoint analysis [2], is a widely known method used to achieve this task. Through this method, individual part-worth estimates that associate the perceived utility value of each customer with each level of the product's attributes, are generated. The utility value of a product usually corresponds to the sum of the part-worths of the corresponding attribute levels (linear-additive part-worth model). Through them, choice probabilities are easily calculated, and the expected market share is estimated. Consequently, optimization algorithms can be applied using the part-worths data to find a product line that optimizes a criterion set by the firm.

However, in real world problems, products are characterized by numerous attributes and corresponding levels. Consequently, the existence of many different product lines makes the managerial task of selecting the appropriate combination of attribute levels very difficult.

Consider, for example, a company that produces tablets and plans to introduce three new ones. If each tablet consists of 8 attributes, with the first attribute taking two different levels (screen 8″ or 10″) and the rest of them taking 4 different levels, the number of possible tablet profiles is 32,768, whereas for designing a line of three tablets the number of candidate solutions is more than 3×10^{13}. Kohli and Krishnamurti [16] proved that the share of choices problem (market share maximization) for a single-product design is an NP-hard problem, which means that it is impossible to completely explore the solution space in real time. Numerous optimization algorithms have been applied to the problem, providing good, near-optimal solutions. A review of them can be found in Belloni et al. [8].

3 Firefly Algorithm and Genetic Algorithm

3.1 Firefly Algorithm

Firefly Algorithm (FA) was developed in 2008 by Yang [10], and it is based on the flashing patterns and behavior of fireflies. The FA is used to solve both continuous and discrete optimization problems. Two very important factors in the firefly algorithm are the light intensity (I) and the attractiveness (β). The light intensity of a firefly is determined by the landscape of the objective function to be optimized. The attractiveness is proportional to the brightness and they both decrease as the distance between two fireflies increases. As the fireflies search for better solutions, their movements are updated based on their current position, their attractiveness, and a randomization term. For any two fireflies, the less bright one will move towards the brighter one. If no one is brighter, they move randomly.

3.2 Genetic Algorithm

Genetic Algorithm (GA) was introduced in the 1960s by Holland and further analyzed by Goldberg in 1989 [11]. GA is a global search optimization technique that imitates processes from natural evolution, based on the survival and reproduction of the fittest. In GA, solutions that are usually coded as binary or integer strings called chromosomes, evolve over the iterations through genetic operations like crossover and mutation. The solutions are evaluated using an objective function. Should the new solutions turn out to be better than the old ones, they will replace the worse ones in the next generation.

4 The Proposed Approach

In this section, the formulation of the product line design problem is described, and the algorithmic structure of the proposed hybrid approach is explained. All the algorithms have been programmed with the use of the MATLAB platform. The simulations have been carried out on a i5 3.3 GHz desktop computer, with 8 GB of RAM and a 64-bit operating system.

4.1 Problem Formulation

Following Belloni et al. [8], the first problem that we used is an actual product line design problem faced by a manufacturer of bags, named Timbuk2. A conjoint study that focused on price and nine binary product features was conducted by a group of academic researchers who cooperated with the company. Additional details are reported in Toubia et al. [17].

In this problem, our goal is to maximize the predicted earnings by introducing a product line consisting of five bags to the market, compared to three already-known competitive products.

After a market survey, we know the consumer preferences on the ten characteristics of each bag, as well as the cost of each feature. The first feature is the price that can take seven different levels ($70, $75, $80, $85, $90, $95, $100) while the remaining nine are yes/no values (exists/does not exist). Table 1 represents a solution vector of a random bag's characteristics.

Table 1. Example of a solution vector of a random bag's characteristics.

Attributes	1	2	3	4	5	6	7	8	9	10
Features	3	0	1	0	1	0	0	1	0	1

The possible products that can be designed by the combinations of these features are 3,584, of which 4.9×10^{15} different product lines consisting of five products can be created. For each one of these 3,584 products, the profit of selling a backpack is equal to its sale price, minus the cost of its features and a fixed construction cost of $35. The marginal cost of each feature for the construction of a backpack as well as their average part-worths of each feature are presented in Table 2.

Table 2. Incremental marginal cost and average part-worths of each feature.

Feature	Average part-worth	Incremental marginal cost ($)
$5 Price increase	−7.6	−5.00
Large size	17.9	3.50
Red color (not black)	−36.0	0.00
School logo	9.0	2.00
Handle	37.7	3.50
Gadget holder	5.2	3.00
Cell phone holder	5.5	3.00
Mesh pocket	9.7	2.00
Velcro flap	18.2	3.50
Reinforcing boot	24.4	4.50

For each one of the 324 consumers, the preferred product from both our own product line and the competitive one is identified. Then the two corresponding utilities are compared, and the final choice of each consumer is obtained. Even though there are various models that simulate how consumers will make their final choice, in this problem, we assume that the first choice or maximum utility model is in place. As a result, a consumer will buy the product with the highest utility, ignoring any stochastic parameter that may occur during the purchase.

Finally, by summing up the earnings from the sales of each bag, we calculate the earnings of the product line. As a result, the earnings of the product line is the objective function of the problem.

4.2   Proposed Hybrid Approach (FAGA) for the Product Line Design Problem

In this subsection we describe the proposed hybrid approach (FAGA), based on FA and GA, by combining some of the advantages of both algorithms. The hybrid approach takes N individuals (chromosomes or fireflies) that are randomly generated. The N individuals are sorted from best to worst according to their light intensity I, which is equal to the objective function's value. During the iterations, the light intensity I of two fireflies, is compared. Should firefly j have a brighter light than firefly i, genetic crossover for the two fireflies is applied. On the contrary, if firefly i has a brighter light than firefly j, the genetic mutation is applied in both fireflies. As a result, in both ways, two new solutions are generated. The two new solutions replace the old ones in the iteration process and their light intensity is equal to the average light intensity of the parents used. As a result, the proposed hybrid FAGA approach introduces a different way of selecting individuals for crossover and mutation in genetic algorithm by using the main algorithmic structure of firefly algorithm. Just like most metaheuristics, FAGA runs until a stopping criterion is met (e.g. maximum number of iterations, maximum number of function evaluations etc.).

FAGA pseudo-code. A pseudo-code of the FAGA is presented in the following.

Objective function f(x), x=$(x_1, ... , x_d)^T$
Initialize the firefly population $x_i(i = 1,2, ... , N)$
Evaluate solutions, update light intensity I and sort of fireflies
Do While *stopping criteria are not met*
    keep the old solutions
    for i ← 2:N do
        j ← 1:i-1
        if I_j> I_i then
            Use genetic crossover for the two fireflies
            Update light intensity as $I_{i,j}=(I_i+I_j)/2$
        else
            Use genetic mutation in both fireflies
        end
    end
    evaluate new solutions and update light intensity I
    merge old and new solutions and sort of fireflies
    keep the N best solutions so far
end while
Postprocess results and visualization.

4.3 FAGA Configuration

FAGA first creates a set of random solutions, which are evaluated by calculating the earnings of each product line. Each solution will be represented as a table of *bags* × *attributes*. Then, an iterating process begins until a stopping criterion is met.

During crossover, the genetic information of two parents are combined to generate new offspring. A crossover point on both parents' chromosomes is picked randomly. Values before and after of that point are swapped between the two parent chromosomes. This results in two offspring, each carrying some genetic information from both parents. Figure 1 shows an example of crossover.

Parent 1
4 0 0 0 1 0 1 0 1 0
1 0 0 0 0 0 0 1 1 0
0 1 1 1 0 0 1 1 1 1
6 0 0 1 1 0 0 1 1 1
4 0 1 1 0 0 0 0 1 1
Parent 2
3 0 0 0 0 0 0 0 0 1
2 0 0 1 1 1 1 0 1 0
4 0 1 1 1 1 0 0 0 1
5 1 0 1 0 0 0 0 0 1
2 1 0 0 1 1 0 0 1 1
Offspring 1
3 0 0 0 0 0 0 0 0 1
2 0 0 1 1 1 1 0 1 0
4 0 1 1 1 1 1 1 1 1
6 0 0 1 1 0 0 1 1 1
4 0 1 1 0 0 0 0 1 1
Offspring 2
4 0 0 0 1 0 1 0 1 0
1 0 0 0 0 0 0 1 1 0
0 1 1 1 0 0 0 0 0 1
5 1 0 1 0 0 0 0 0 1
2 1 0 0 1 1 0 0 1 1

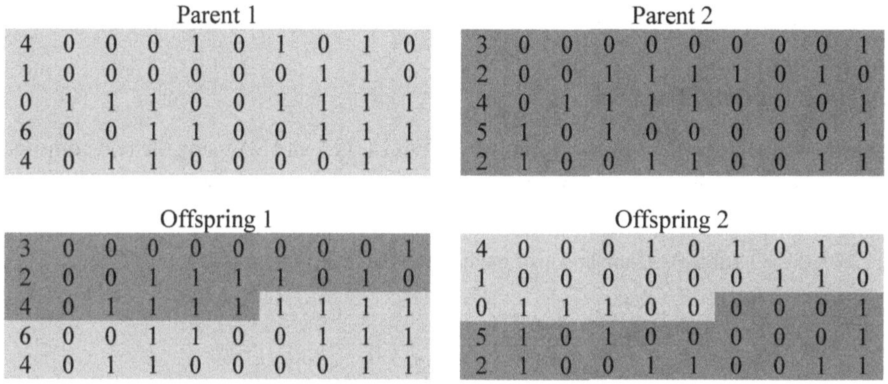

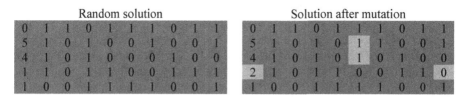

Fig. 1. Example of crossover, using 3 × 6 cell as a crossover point.

Mutation is the second way FAGA explores the search space. It can introduce traits not in the original population and prevents the algorithm from converging too fast. Mutation alters one or more gene values in a chromosome from its initial state. An example of mutation is presented below (see: Fig. 2).

Random solution
0 1 1 0 1 1 1 0 1 1
5 1 0 1 0 0 1 0 0 1
4 1 0 1 0 0 0 1 0 0
1 1 0 1 1 0 0 1 1 1
1 0 0 1 1 1 1 0 0 1
Solution after mutation
0 1 1 0 1 1 1 0 1 1
5 1 0 1 0 1 1 0 0 1
4 1 0 1 0 1 0 1 0 0
2 1 0 1 1 0 0 1 1 0
1 0 0 1 1 1 1 0 0 1

Fig. 2. Example of mutation.

Parameter Settings. To determine the best parameter settings for the FAGA, several experiments have been carried out. We found that a population value of n = 50 to 120 is sufficient. Therefore, we used a fixed population of n = 100 for all simulations of FAGA. The mutation rate used, was set to 0.05, just like Belloni et al. [8] used in genetic algorithm for the product line design problem.

5   A Comparison of FAGA with Genetic Algorithm (GA) and Simulated Annealing (SA)

In this section, the proposed method is compared to those of the most successful methods used by Belloni et al. [8]. Particularly, FAGA's performance will be compared with the performance of genetic algorithm (GA) [11] and simulated annealing (SA) [15]. The choice of parameters values for each one of the comparing algorithms, was made according to Belloni et al. [8]. At first, we use the real conjoint data set. Then, we test whether the model we have developed is affected by errors in consumer preference measurements, and whether the model can be generalized by changing the size of the problem. In our simulations, each algorithm runs until 70,000 function evaluations are reached, each one performing 50 replications. Each run of FAGA takes less than 7 s.

5.1   Real Conjoint Data Set

Table 3 shows the results of the comparison of FAGA, GA and SA using the real conjoint data, while Fig. 3 shows the convergence characteristic curves of them.

Table 3. Comparison of methods on the real conjoint data set.

Statistics	FAGA	GA	SA
Best	**12226**	**12226**	11781.5
Worst	**12032.5**	11895.5	10867
Average	**12121.38**	12083.62	11140.82
Median	**12055.5**	**12055.5**	11140
Std.	**8.3549E+01**	1.1229E+02	1.9132E+02

The comparison results above indicate the superiority of FAGA over the two other algorithms. Particularly, FAGA's objective function value has an average value of 12121.38, a standard deviation of 8.3549E+01 and a median value of 12055.5 in the range of 12032.5 to 12226. On the contrary, GA's objective function value has an average value of 12083.62, a standard deviation of 1.1229E+02 and a median value of 12055.5 in the range of 11895.5 to 12226. Finally, SA's objective function value has an average value of 11140.82, a standard deviation of 1.9132E+02 and a median value of 11140 in the range of 10867 to 11781.5. Figure 3 shows that the proposed hybrid method converges faster than GA and SA.

5.2   Robustness Testing

The accuracy of the results does not only depend on the efficiency, effectiveness and accuracy of the method used to solve the problem, but also on other factors. One of the most important factors is whether the estimation of consumer preferences, according to the conjoint analysis, as well as the rating of part worths which corresponds to each

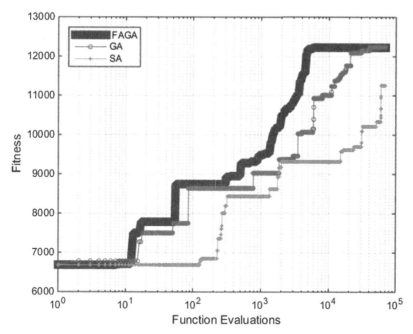

Fig. 3. Convergence characteristic curves of FAGA, GA and SA.

feature are accurate or not. In addition, there are methods to measure whether these estimations correspond to actual consumer preferences. To test the robustness of the methods in the presence of a measurement error, we repeated our analysis after perturbing the original part-worth estimates. These perturbations were accomplished by adding a (simulated) error to the part-worths:

$$u'_{i,j} = u_{i,j} + \varepsilon_{i,j} \qquad (1)$$

where $u_{i,j}$ is the original part-worth for respondent i on product feature j, $\varepsilon_{i,j}$ is a zero-mean, independent normal error term which works differently across customers or attribute levels, and $u'_{i,j}$ is the perturbed part-worth.

Following Belloni et al. [8], we run each algorithm for 100 runs, obtaining 100 sets of perturbed part-worths for each respondent. The perturbation terms are treated as measurement error. Under this interpretation, the original part-worths that we analyzed in Table 3, is the only one set of "true" part-worths. However, we assume that the researcher can only observe the part-worths that are subject to measurement error. The results are reported in Table 4.

From the results obtained above, we note that the proposed hybrid FAGA method is the less affected by the measurement error of consumer preferences.

5.3   Results Using Simulated Data

In this subsection, we also test the performance of FAGA comparing to the performance of GA [11] and SA [15], using simulated data. In our simulated problems, a product line

Table 4. Comparison of methods under measurement error.

Statistics	FAGA	GA	SA
Best	**12223.5**	11986.5	11275.5
Worst	**11834**	10903.5	10162
Average	**12021.76**	11589.51	10782.06
Median	**12032.5**	11630.5	10806.25
Std.	**9.55935E+01**	2.24975E+02	2.90559E+02

consisting of 3 or 4 products, each composed of 3, 5, or 7 attributes that can take on 2, 3, 5, or 8 different levels, is designed. We assume that the customers are 50 or 100 for each case. We select 12 different problem sizes and 10 different problems for each size are generated, for a total of 120 simulated data sets. According to Belloni et al. [8], this kind of problem sizes are chosen, because Lagrangian relaxation [18] finds an optimal solution in a reasonable amount of time.

According to Belloni et al. [8], SA is extremely accurate, since it identifies the same solutions with the Lagrangian relaxation for all problem sizes, while GA identifies the 99.9%. However, to make accurate comparisons and to measure the accuracy of FAGA, we run all the methods again, using the same configuration used in the real data comparison for all the three algorithms. Each algorithm runs until 70,000 function evaluations are reached, like before. Then we compare the results to the overall optimum results. The summary results are shown in Table 5.

Table 5. Comparison of methods using simulated data.

Statistics	FAGA	GA	SA
Average performance	**98.95%**	98.94%	95.44%

From the comparison results above, we notice that FAGA's average performance dominates that of GA and SA. Obviously, all 3 algorithms need more function evaluations to locate more accurate results. However, FAGA still converges faster than the other three.

To further comparison of the models, we increase specific dimensions of the above sets. As a result, larger sizes of problems are created, as shown in Table 6. The comparison of the methods is presented in Table 7. The results will be compared according to FAGA's results, which was the most successful algorithm comparing to GA and SA, according to Table 5.

As we can see, FAGA is still superior to GA and SA. We now increase the number of products for all sets by one. This creates a set of problems as presented in Table 8. The results are presented in Table 9.

For once more, FAGA's results are better than those of GA and SA. We now increase the number of attributes for all sets by one. This creates a set of problems as presented in Table 10. The results are presented in Table 11.

Table 6. Problem sizes with increased dimensions.

Customers	Attributes	Levels	Products
50	3	5	5
100	3	5	5
50	8	3	4
100	8	3	4
50	9	2	4
100	9	2	4
50	3	9	3
100	3	9	3
50	5	5	6
100	5	5	6
50	9	3	3
100	9	3	3

Table 7. Comparison of methods performing on the problems presented in Table 6.

Statistics	FAGA	GA	SA
Average performance	100%	99.73%	93.29%

Table 8. Problem sizes with increased products by one.

Customers	Attributes	Levels	Products
50	3	5	6
100	3	5	6
50	8	3	5
100	8	3	5
50	9	2	5
100	9	2	5
50	3	9	4
100	3	9	4
50	5	5	7
100	5	5	7
50	9	3	4
100	9	3	4

Table 9. Comparison of methods performing on the problems presented in Table 8.

Statistics	FAGA	GA	SA
Average performance	100%	99.89%	90.64%

Table 10. Problem sizes with increased attributes.

Customers	Attributes	Levels	Products
50	4	5	5
100	4	5	5
50	9	3	4
100	9	3	4
50	10	2	4
100	10	2	4
50	4	9	3
100	4	9	3
50	6	5	6
100	6	5	6
50	10	3	3
100	10	3	3

Table 11. Comparison of methods performing on the problems presented in Table 10.

Statistics	FAGA	GA	SA
Average performance	100%	99.55%	90.99%

Finally, we increase the levels of all products by one. The new created set of problems is presented in Table 12 and the results are shown in Table 13.

Table 12. Problem sizes with increased levels.

Customers	Attributes	Levels	Products
50	3	6	5
100	3	6	5
50	8	4	4
100	8	4	4
50	9	3	4
100	9	3	4
50	3	10	3
100	3	10	3
50	5	6	6
100	5	6	6
50	9	4	3
100	9	4	3

Table 13. Comparison of methods performing on the problems presented in Table 12.

Statistics	FAGA	GA	SA
Average performance	100	99.89%	89.81%

FAGA is still superior to GA and SA when increasing the levels of attributes. Moreover, FAGA is becoming more and more superior to GA as the size of the problem is increased.

6 Conclusions

In this paper, a hybrid optimization algorithm (FAGA) is proposed for solving the product line design problem. The proposed FAGA approach is based on hybridization of firefly algorithm (FA) [10] and genetic algorithm (GA) [11] and introduces a different way of selecting individuals for crossover and mutation in genetic algorithm by using the main algorithmic structure of firefly algorithm. FAGA's results are compared to those of 2 high quality optimization algorithms. The results reveal that the proposed FAGA method has superior performance to those of both GA and SA. It has a higher probability of finding a global optimum using less function evaluations and a better ability to deal with multimodality comparing to GA and SA. The convergence behavior of the FAGA is also exceptional, since most of the times it reaches the best overall solution in the early iterations. It provides a robust model and it has the potential to be generalized in large scale product line design problems of well-known firms with complex product lines. Moreover, while changing the dimension of the problem we noticed that FAGA is superior over GA and SA in terms of accuracy and efficiency. As already mentioned, FAGA is more and more superior to GA as we increase the size of the problem, regardless of the growing dimension. Finally, FAGA is an easy hybrid approach to be coded. Particularly, it needs only two initial parameters (population size and mutation rate).

Future research may include the application of the proposed algorithm to other engineering and industrial optimization problems. Furthermore, we intend to propose new approaches based on hybridization of swarm intelligence and evolutionary algorithms.

Acknowledgments. This research is co-financed by Greece and the European Union (European Social Fund- ESF) through the Operational Programme «Human Resources Development, Education and Lifelong Learning» in the context of the project "Strengthening Human Resources Research Potential via Doctorate Research" (MIS-5000432), implemented by the State Scholarships Foundation (ΙΚΥ).

Ευρωπαϊκή Ένωση
European Social Fund

**Operational Programme
Human Resources Development,
Education and Lifelong Learning**

Co-financed by Greece and the European Union

References

1. Kotler, P., Armstrong, G.: Principles of Marketing. Pearson/Prentice Hall, Upper Saddle River (2012)
2. Luce, R.D., Tukey, J.W.: Simultaneous conjoint measurement: a new type of fundamental measurement. J. Math. Psychol. **1**, 1–27 (1964)
3. Papadimitriou, C.H., Steiglitz, K.: Combinatorial Optimization: Algorithms and Complexity. Prentice Hall, Upper Saddle River (1982)
4. Kohli, R., Sukumar, R.: Heuristics for product-line design using conjoint analysis. Manage. Sci. **36**, 1464–1478 (1990)
5. Nair, S.K., Thakur, L.S., Wen, K.-W.: Near optimal solutions for product line design and selection: beam search heuristics. Manage. Sci. **41**, 767–785 (1995)
6. Balakrishnan, P.V., Gupta, R., Jacob, V.S.: Development of hybrid genetic algorithms for product line designs. IEEE Trans. Syst. Man, Cybern. Part B Cybern. **34**, 468–483 (2004)
7. Camm, J.D., Cochran, J.J., Curry, D.J., Kannan, S.: Conjoint optimization: an exact branch-and-bound algorithm for the share-of-choice problem. Manage. Sci. **52**, 435–447 (2006)
8. Belloni, A., Freund, R., Selove, M., Simester, D.: Optimizing product line designs: efficient methods and comparisons. Manage. Sci. **54**, 1544–1552 (2008)
9. Tsafarakis, S., Marinakis, Y., Matsatsinis, N.: Particle swarm optimization for optimal product line design. Int. J. Res. Mark. **28**, 13–22 (2011)
10. Yang, X.-S.: Nature-Inspired Metaheuristic Algorithms. Luniver Press, Bristol (2008)
11. Goldberg, D.E.: Genetic Algorithms in Search, Optimization, and Machine Learning. Addison-Wesley, Reading (1989)
12. Elkhechafi, M., Hachimi, H., Elkettani, Y.: A new hybrid firefly with genetic algorithm for global optimization. Int. J. Manag. Appl. Sci. **3**, 47–51 (2017)
13. Masouleh, M.F., Kazemi, M.A.A., Alborzi, M., Eshlaghy, A.T.: Engineering, technology & applied science research. ETASR (2016)

14. Rahmani, A., MirHassani, S.A.: A hybrid Firefly-Genetic Algorithm for the capacitated facility location problem. Inf. Sci. (Ny) **283**, 70–78 (2014)
15. Kirkpatrick, S., Gelatt, C.D., Vecchi, M.P.: Optimization by simulated annealing. Science **220**, 671–680 (1983)
16. Kohli, R., Krishnamurti, R.: Optimal product design using conjoint analysis: computational complexity and algorithms. Eur. J. Oper. Res. **40**, 186–195 (1989)
17. Toubia, O., Simester, D.I., Hauser, J.R., Dahan, E.: Fast polyhedral adaptive conjoint estimation. Mark. Sci. **22**, 273–303 (2003)
18. Fisher, M.L.: The Lagrangian relaxation method for solving integer programming problems. Manage. Sci. **27**, 1–18 (1981)

Assessing Simulated Annealing with Variable Neighborhoods

Eduardo Lalla-Ruiz[1]([✉]), Leonard Heilig[2], and Stefan Voß[2]

[1] Department of Business Information Systems and Industrial Engineering,
University of Twente, Enschede, The Netherlands
e.a.lalla@utwente.nl
[2] Institute of Information Systems, University of Hamburg, Hamburg, Germany
{leonard.heilig,stefan.voss}@uni-hamburg.de

Abstract. Simulated annealing (SA) is a well-known metaheuristic commonly used to solve a great variety of $\mathcal{NP}$-hard problems such as the quadratic assignment problem (QAP). As commonly known, the choice and size of neighborhoods can have a considerable impact on the performance of SA. In this work, we investigate and propose a SA variant that considers variable neighborhood structures driven by the state of the search. In the computational experiments, we assess the contribution of this SA variant in comparison with the state-of-the-art SA for the QAP applied to printed circuit boards and conclude that our approach is able to report better solutions by means of short computational times.

1 Introduction

This paper proposes and assesses the incorporation of variable neighborhoods into SA driven by the state of the search. From a practical standpoint, the basic idea for this strategy was initially proposed in [7] for a vehicle routing application in order to obtain better solutions than those provided by a standard SA. At a methodological level, the contribution of our paper is to evaluate this strategy in a more general setting with standard problem instances.

In the quest of showing the impact of the algorithmic enhancement regarding a known method, i.e., SA, we exemplify by means of the quadratic assignment problem (QAP). The QAP is an $\mathcal{NP}$-hard combinatorial optimization problem introduced by Koopmans and Beckman [10] that have received a lot of attention due to its numerous applications. In the QAP, we are given a set of facilities denoted as $\mathcal{F} = \{1, 2, \ldots, n\}$ and a set of locations denoted as $\mathcal{L} = \{1, 2, \ldots, n\}$. Each pair of facilities, $(i, j) \in \mathcal{F}$, requires a certain flow, i.e., $f_{ij} \geq 0$. The distance between the locations $k, l \in \mathcal{L}$ is denoted as $d_{kl} \geq 0$. It should be mentioned that the flows and distances are symmetric (*i.e.*, $f_{ij} = f_{ji}, \forall i, j \in \mathcal{F}$ and $d_{kl} = d_{lk}, \forall k, l \in \mathcal{L}$) and the flow/distance between a given facility/location and itself is zero (*i.e.*, $f_{ii} = 0, \forall i \in \mathcal{F}$ and $d_{kk} = 0, \forall k \in \mathcal{L}$). Its objective is to minimize the cost derived from the distance and flows among facilities. Duman et al. [5] present a practical application of the QAP for sequencing placement

© Springer Nature Switzerland AG 2020
N. F. Matsatsinis et al. (Eds.): LION 13 2019, LNCS 11968, pp. 298–303, 2020.
https://doi.org/10.1007/978-3-030-38629-0_24

and configuration printed circuit boards (PCB) and extensively analyze the use of SA for addressing the problem. The QAP is formally expressed as follows

$$minimize \sum_{i=1}^{n} \sum_{j=1}^{n} f_{ij} d_{\phi(i)\phi(j)}, \tag{1}$$

where ϕ is a solution belonging to the set composed of all the feasible permutations, denoted as $\mathcal{S}_n$, such that $\phi : \mathcal{F} \rightarrow \mathcal{L}$. The cost associated to assign facility i to location $\phi(i)$ and facility j to facility $\phi(j)$ is, according to Eq. (1), $f_{ij} d_{\phi(i)\phi(j)}$. In addition, let us denote as $f(\phi)$ the objective function value of solution $\phi \in \mathcal{S}_n$. Drezner et al. [4] review the applicability of widespread metaheuristics from the literature to address the QAP. The interested reader is referred to the detailed survey provided by Loiola et al. [13].

We use the same problem instances of [6] and the best state-of-the-art SA [5] to properly compare our proposed approach. From the computational experiments, we conclude that our approach provides a better performance compared to the SA when using a single neighborhood structure.

The remainder of this paper is structured as follows. First, we review some related works in Sect. 2. The proposed variable SA algorithm is presented in Sect. 3. In Sect. 4, we report the results of the computational experiments. The paper ends with some conclusions and an outlook.

2   Related Works

In general terms, Cheh et al. [2] studied the effect that neighborhood structures have on SA. On the other hand, Ogbu and Smith [14] showed the benefit of using larger neighborhoods within SA. Henderson et al. [9] review the impact that the choice of neighborhoods has on SA and indicate that the efficiency of SA is highly influenced by the neighborhood selection.

Besides Heilig et al. [7], some authors have investigated ideas to control the neighborhood structure during the search. Xu and Qu [17] investigate the use of variable neighborhoods within an evolutionary multi-objective SA (EMOSA) for solving multicast routing problems. Using multiple neighborhood structures specifically designed for each objective significantly improves the performance of the SA. Ying et al. [19] propose an SA algorithm with variable neighborhoods and define additional parameters to control the random selection of the neighborhood structure. While the performance of the algorithm depends on a good configuration of the parameters, requiring additional experiments. The proposed approach was able to find new best-known solutions for the cell formation problem.

Instead of defining a parameter, Rodriguez-Cristerna and Torres-Jimenez [15] use only two neighborhood structures and select them with uniform probabilities. Some authors investigate ideas to adjust the size of neighborhood structures, such as by means of a non-uniform mutation operator for monotonously decreasing the neighborhood size [16] or by using a circle-directed mutation as done in [12].

Other than in previous works, we propose a dynamic neighborhood variation where the neighborhood structures are changed depending on the success of finding better solutions.

Furthermore, SA is used in some works to extend the acceptance criterion of the variable neighborhood search (VNS) for accepting also non-improving solutions under certain conditions (see, e.g., [3,8,11,18]).

3   Variable Neighborhood Simulated Annealing Algorithm

In order to evaluate the contribution of variable neighborhood within SA, the best state-of-the-art SA proposed for solving the QAP-PCB [5] is used as a base template. For extending it, we include the novel incorporation of neighborhood variation in lines 5 to 14, where a parameter k is introduced for regulating the change of neighborhood structures.

Algorithm 1. SA with variable neighborhood structures (SA-VN)

Require: $Temp_{min}$, α, β, r_{max}

1: $S \leftarrow$ generate initial solution at random
2: $Temp \leftarrow f_{obj}(S)\alpha$; $k \leftarrow 1$
3: **while** ($Temp_{min} < Temp$ and $it \leq it_{max}$) **do**
4:      **for** ($r = 1$ to r_{max}) **do**
5:          Generate a solution $S' \in \mathcal{N}^k(\phi)(S)$
6:          Calculate $\Delta_{S,S'} = f_{obj}(S') - f_{obj}(S)$
7:          **if** ($\Delta_{S,S'} \leq 0$) **then**
8:              $S \leftarrow S'$
9:              $k \leftarrow 1$
10:             Update best solution S_{best} if applicable
11:         **else**
12:             $S \leftarrow S'$ with probability $e^{-\Delta/Temp}$
13:             $k + +$
14:         **end if**
15:         $Temp = Temp \cdot \beta$
16:         $it + +$
17:     **end for**
18: **end while**
19: Return S_{best}

We apply the *swap* neighborhood structure. That is, given a solution, $\phi \in \mathcal{S}_n$, the swap neighborhood, $\mathcal{N}^1(\phi) = \{\phi \circ (i,j) : 1 \leq i,j \leq n, i \neq j\}$, performs the transposition (i,j) by swapping the two relevant locations assigned to the indexes, i and j, respectively. Moreover, we define $\mathcal{N}^2(\phi)$ and $\mathcal{N}^3(\phi)$ as the application of swap consecutively two and three times, respectively.

4 Computational Results

The computational experiments were executed over the instances proposed by Duman *et al.* [6] which were generated considering the QAP-PCB. In each case, 30 executions of our SA-VN were carried out for each problem instance. The algorithms were executed on a computer equipped with an Intel i7-7700HQ and 16 GB of RAM. In order to properly evaluate the contribution of our approach, we have used the SA and its best parameters as provided in [5] with a maximum number of $|n|^{3.5}$ solutions.

Table 1. Comparison among SA-VN algorithms for the QAP-PCB instances. Best values in bold.

Instance (size)	SA-VN (k = 3)				SA-VN (k = 2)			
	Min	Avg.	Max.	t (ms)	Min	Avg.	Max.	t (ms)
B1 (58)	**1066**	1084.33	1132	3281.47	**1066**	1088.33	1126	3302.47
B2 (54)	**754**	771.067	798	2865.37	756	775.4	796	2852.4
B3 (52)	**730**	749.667	784	2666.57	732	749.867	776	2653.37
B4 (50)	**1450**	1456	1482	1651.27	**1450**	1466.13	1502	1643.07
B5 (48)	**752**	765.133	802	1534.33	754	767.067	814	1525.43
B6 (49)	**1388**	1400.33	1436	1601.3	1392	1397.07	1438	1597.5
B7 (47)	1350	1362.6	1394	1474.2	**1348**	1360.6	1374	1503.83
B8 (40)	**714**	725.067	740	491.533	718	726.333	738	497.667
	1025.5	1039.28	1071	1945.75	1027	1041.35	1070.5	1946.97

Table 2. Comparison among SA algorithms using one neighborhood structure for the QAP-PCB instances. Best values in bold.

Instance (size)	SA_1				SA_2				SA_3			
	Min	Avg.	Max.	t (ms)	Min	Avg.	Max.	t (ms)	Min	Avg.	Max.	t (ms)
B1(58)	1070	1087.73	1130	3319	**1066**	1084.13	1126	3323.57	1068	1092	1130	3345.23
B2 (54)	758	778.4	814	2850.1	756	775.533	814	2894.47	760	774.60	796	2938.67
B3 (52)	**730**	753.4	780	2667.2	734	757.733	784	2701.37	**730**	752.67	804	2744.90
B4 (50)	**1450**	1458.4	1488	1628.9	**1450**	1462.47	1484	1664.33	**1450**	1463.73	1514	1696.17
B5 (48)	**752**	764.533	804	1512.07	754	766.533	802	1533.57	754	764.20	808	1556.43
B6 (49)	1390	1399.93	1438	1589.63	**1388**	1397.53	1444	1589.5	1390	1399.20	1438	1638.93
B7 (47)	1350	1362.07	1386	1454.9	**1348**	1360	1378	1472.3	**1348**	1364.40	1398	1512.27
B8 (40)	718	724.867	736	487.1	716	724.667	734	492.133	716	725.20	736	505.87
	1027.25	1041.17	1072	1938.61	1026.5	1041.08	1070.75	1958.9	1027	1042	1078	1992.31

Table 1 shows the comparison between SA with variable neighborhoods considering two, i.e., SA-VN (k = 2) and three neighborhoods, i.e., SA-VN (k = 3).

Moreover, the best SA proposed by [5] considering individually all the used neighborhoods is compared in Table 2. For each problem instance, the performances of the algorithms in terms of average objective value (Avg.), best objective value (Min.), the worst objective value (Max.), and the computational time (t (ms)) of all the executions in milliseconds are reported.

From the results, it can be seen that all algorithms require similar computational times. The strategy of including variable neighborhoods permits to obtain more best-known solutions. Although there is not a relevant difference in terms of the worst values, the average performance is enhanced when more neighborhoods are considered in SA-VN. Moreover, there is a relevant performance benefit when the number of neighborhoods increases.

5   Conclusions

In this work, a novel simulated annealing with variable neighborhoods changing along the search is proposed for solving the quadratic assignment problem. This new SA approach includes alternating neighborhoods when there is no improvement or the probability of acceptance does not permit a worsening movement. It is noticeable from the numerical experiments that the proposed algorithm exhibits a better performance within similar time frames as the standard SA. The promising results encourage to further explore this research direction. The results also go in line with those shown in [7] where the inclusion of variable neighborhoods leads to an overall improvement of the SA search framework.

As future work, we aim at extending and analyzing the performance of SA-VN on other QAP instances such as those from the QAPLIB [1] as well as other optimization problems. Moreover, we aim to add a look ahead component like known from the pilot method.

References

1. Burkard, R.E., Karisch, S.E., Rendl, F.: QAPLIB-a quadratic assignment problem library. J. Global Optim. **10**(4), 391–403 (1997)
2. Cheh, K.M., Goldberg, J.B., Askin, R.G.: A note on the effect of neighborhood structure in simulated annealing. Comput. Oper. Res. **18**(6), 537–547 (1991)
3. Chen, P., Huang, H.K., Dong, X.Y.: Iterated variable neighborhood descent algorithm for the capacitated vehicle routing problem. Expert Syst. Appl. **37**(2), 1620–1627 (2010)
4. Drezner, Z., Hahn, P.M., Taillard, E.D.: Recent advances for the quadratic assignment problem with special emphasis on instances that are difficult for metaheuristic methods. Ann. Oper. Res. **139**(1), 65–94 (2005)
5. Duman, E., Or, I.: The quadratic assignment problem in the context of the printed circuit board assembly process. Comput. Oper. Res. **34**(1), 163–179 (2007)
6. Duman, E., Uysal, M., Alkaya, A.F.: Migrating birds optimization: a new metaheuristic approach and its performance on quadratic assignment problem. Inf. Sci. **217**, 65–77 (2012)

7. Heilig, L., Lalla-Ruiz, E., Voß, S.: port-IO: an integrative mobile cloud platform for real-time inter-terminal truck routing optimization. Flex. Serv. Manuf. J. **29**(3), 504–534 (2017)
8. Hemmelmayr, V.C., Doerner, K.F., Hartl, R.F.: A variable neighborhood search heuristic for periodic routing problems. Eur. J. Oper. Res. **195**(3), 791–802 (2009)
9. Henderson, D., Jacobson, S.H., Johnson, A.W.: The theory and practice of simulated annealing. In: Glover, F.W., Kochenberger, G.A. (eds.) Handbook of Metaheuristics, pp. 287–319. Springer, Boston (2003). https://doi.org/10.1007/0-306-48056-5_10
10. Koopmans, T.C., Beckman, M.: Assignment problems and the location of economic activities. Econometrica **25**, 53–76 (1957)
11. Kuo, Y., Wang, C.C.: A variable neighborhood search for the multi-depot vehicle routing problem with loading cost. Expert Syst. Appl. **39**(8), 6949–6954 (2012)
12. Lin, Y., Bian, Z., Liu, X.: Developing a dynamic neighborhood structure for an adaptive hybrid simulated annealing-tabu search algorithm to solve the symmetrical traveling salesman problem. Appl. Soft Comput. **49**, 937–952 (2016)
13. Loiola, E.M., Maia de Abreu, N.M., Boaventura-Netto, P.O., Hahn, P., Querido, T.: A survey for the quadratic assignment problem. Eur. J. Oper. Res. **176**(2), 657–690 (2007)
14. Ogbu, F., Smith, D.K.: The application of the simulated annealing algorithm to the solution of the n/m/Cmax flowshop problem. Comput. Oper. Res. **17**(3), 243–253 (1990)
15. Rodriguez-Cristerna, A., Torres-Jimenez, J.: A simulated annealing with variable neighborhood search approach to construct mixed covering arrays. Electron. Notes Discret. Math. **39**, 249–256 (2012)
16. Xinchao, Z.: Simulated annealing algorithm with adaptive neighborhood. Appl. Soft Comput. **11**(2), 1827–1836 (2011)
17. Xu, Y., Qu, R.: Solving multi-objective multicast routing problems by evolutionary multi-objective simulated annealing algorithms with variable neighbourhoods. J. Oper. Res. Soc. **62**(2), 313–325 (2011)
18. Xu, Y., Wang, L., Yang, Y.: A new variable neighborhood search algorithm for the multi depot heterogeneous vehicle routing problem with time windows. Electron. Notes Discret. Math. **39**, 289–296 (2012)
19. Ying, K.C., Lin, S.W., Lu, C.C.: Cell formation using a simulated annealing algorithm with variable neighbourhood. Eur. J. Ind. Eng. **5**(1), 22–42 (2010)

Evolving Gaussian Process Kernels for Translation Editing Effort Estimation

Ibai Roman[1(✉)] , Roberto Santana[1] , Alexander Mendiburu[1] ,
and Jose A. Lozano[1,2]

[1] University of the Basque Country (UPV/EHU), San Sebastian, Spain
{ibai.roman,roberto.santana,alexander.mendiburu,ja.lozano}@ehu.eus
[2] Basque Center for Applied Mathematics (BCAM), Bilbao, Spain

Abstract. In many Natural Language Processing problems the combination of machine learning and optimization techniques is essential. One of these problems is estimating the effort required to improve, under direct human supervision, a text that has been translated using a machine translation method. Recent developments in this area have shown that Gaussian Processes can be accurate for post-editing effort prediction. However, the Gaussian Process kernel has to be chosen in advance, and this choice influences the quality of the prediction. In this paper, we propose a Genetic Programming algorithm to evolve kernels for Gaussian Processes. We show that the combination of evolutionary optimization and Gaussian Processes removes the need for a-priori specification of the kernel choice, and achieves predictions that, in many cases, outperform those obtained with fixed kernels.

Keywords: Evolutionary search · Gaussian Processes · Genetic Programming · Kernel selection · Quality Estimation

1 Introduction

Gaussian Processes (GP) [31] have been extensively applied for function approximation. A GP is a model that lies on strong Bayesian inference foundations and can be updated when new evidence is available. In comparison to other regression methods, a GP provides not only a prediction of a given function but also estimates the uncertainty of the predictions. A GP requires a kernel function to be defined and adjusts its hyperparameters to the data. Usually, the kernel function is specified a-priori, and the search for the hyperparameters is posed as an optimization process. This optimization is an essential component of a GP model since it highly influences the quality of the approximation.

Within Natural Language Processing (NLP) literature, GPs have been applied for text classification [28], modeling periodic distributions of words over time [30], emotion or sentiment classification [1] and Quality Estimation (QE) [7,35]. In these works, the choice of the GP kernel is made a-priori. For instance, for QE regression the most common kernel is the Squared Exponential (SE) or

N. F. Matsatsinis et al. (Eds.): LION 13 2019, LNCS 11968, pp. 304–318, 2020.
https://doi.org/10.1007/978-3-030-38629-0_25

Radial Basis Function (RBF) kernel [7,35]. For modeling text periodicities, the periodic (PER) and the Periodic Spikes (PS) kernels have been proposed as a sensible way to capture the periodicities of the function [30]. In addition to SE, two Matern kernels, Matern 32 (M32) and Matern 52 (M52), were evaluated in [1] for emotion analysis. There is a repertoire of kernel functions available in the literature [11,15,31], and selecting the best kernel for each problem requires an expert knowledge of the domain.

In this paper we propose to simultaneously optimize the kernel function itself, along with its hyperparameters. For this purpose, we propose the use of Genetic Programming (GenProg) [20]. In this work, kernels are encoded as a set of genes and optimized though an evolutionary process. As opposed to other proposals in the GP literature, we learn the initial kernels from scratch, without seeding human-designed kernels, allowing us to derive kernels that are not constrained by the prior knowledge while at the same time being optimized for the desired objective.

We focus on a practical NLP regression problem that is related to automatic translation of texts. In this domain, post-editing work is frequently required, and an estimation of the cost of the editing process (in terms of time, effort, and editing distance) is essential. In [37], Specia investigated the question of translation QE from different perspectives. In particular, a number of metrics describing translation quality were proposed and Support Vector Machine (SVM) regression models [40] were used to predict them from a set of pre-defined features.

Instead of manually defining the features, our approach relies on sentence embeddings, vector representations of the source and the automatically translated texts to predict the post-editing effort that leads to the final text. This allows a fully automated approach, where there is no need to extract the features from the sentences, nor learn the kernel function. We investigate several methods to construct the sentence embeddings along with the evolution of the GP kernels, measuring the joint effect that these questions have in QE performance.

The remainder of the paper is organized as follows: The next section introduces the main concepts related to GP regression. Section 3 presents the addressed problem in the context of QA. The GenProg approach to evolve kernel functions is presented in Sect. 4 and a review on related work is provided. We describe the experimental framework used to validate our algorithm, along with the numerical results in Sect. 5. The conclusions of the paper and discussion of future work are presented in Sect. 6.

2   Gaussian Process Regression

A GP is a stochastic process, defined by a collection of random variables, any finite number of which have a joint Gaussian distribution [31]. A GP can be interpreted as a distribution over functions, and each sample of a GP as a function.

GPs can be completely defined by a mean function $m(\mathbf{x})$ and a covariance function, which depends on a kernel $k(\mathbf{x}, \mathbf{x}')$. Given that, a GP can be expressed as follows:

$$f(\mathbf{x}) \sim GP(m(\mathbf{x}), k(\mathbf{x}, \mathbf{x}')) \tag{1}$$

where we assume that $\mathbf{x} \in \mathbb{R}^d$. We also consider an a-priori equal-to-zero mean function $(m(\mathbf{x}) = 0)$ to focus on the kernel search problem.

A GP can be used for regression by obtaining its posterior distribution given some (training) data. The joint distribution between the training outputs $\mathbf{f} = (f_1, f_2, ..., f_n)$ (where $f_i \in \mathbb{R}$, $i \in \{1, ..., n\}$ and $n \in \mathbb{N}$) and the test outputs $\mathbf{f}_* = (f_{n+1}, f_{n+2}, ..., f_{n+n_*})$ is given by:

$$\begin{bmatrix} \mathbf{f} \\ \mathbf{f}_* \end{bmatrix} \sim \mathcal{N}\left(0, \begin{bmatrix} K(X, X) & K(X, X_*) \\ K(X_*, X) & K(X_*, X_*) \end{bmatrix} \right) \tag{2}$$

where $\mathcal{N}(\mu, \Sigma)$ is a multivariate Gaussian distribution, $X = (\mathbf{x}_1, \mathbf{x}_2, ..., \mathbf{x}_n)$ ($\mathbf{x}_i \in \mathbb{R}^d$, $i \in \{1, ..., n\}$ and $n \in \mathbb{N}$) corresponds to the training inputs and $X_* = (\mathbf{x}_{n+1}, ..., \mathbf{x}_{n+n_*})$ to the test inputs. $K(X, X_*)$ denotes the $n \times n_*$ matrix of the covariances evaluated for all the (X, X_*) pairs.

The predictive Gaussian distribution can be found by obtaining the conditional distribution given the training data and the test inputs:

$$\mathbf{f}_* | X_*, X, \mathbf{f} \sim \mathcal{N}(\hat{M}(X_*), \hat{K}(X_*, X_*))$$
$$\hat{M}(X_*) = K(X_*, X)K(X, X)^{-1}\mathbf{f} \tag{3}$$
$$\hat{K}(X_*, X_*) = K(X_*, X_*) - K(X_*, X)K(X, X)^{-1}K(X, X_*)$$

2.1 Kernel Functions

GP models use a kernel to define the covariance between any two function values [11]:

$$cov\left(f(\mathbf{x}), f(\mathbf{x}')\right) = k(\mathbf{x}, \mathbf{x}') \tag{4}$$

The kernel functions used in GPs are positive-definite kernels. According to Mercer's Theorem [24], any PSD kernel can be represented as an inner product in some Hilbert Space.

The best known kernels in GP literature are translation invariant, often referred to as stationary kernels. Among them, we focus on isotropic kernels, where the covariance function depends on the norm:

$$k(\mathbf{x}, \mathbf{x}') = \hat{k}(r) \text{ where } r = \frac{1}{\theta_l} \|\mathbf{x} - \mathbf{x}'\| \tag{5}$$

where θ_l is the length scale hyperparameter and $\hat{k}$ a function that guarantees that the kernel is PSD.

Table 1 shows four well-known kernels that have been previously applied to NLP applications [1, 3, 30, 37]. The SE kernel, described as k_{SE} in the table, is known to capture the smoothness property of the objective functions. Matern class kernels, are denoted as $\hat{k}_{M32}$ and $\hat{k}_{M52}$ in the table, while the Periodic kernel is shown as $\hat{k}_{PER}$.

Table 1. Well-known kernel functions. θ_0 and θ_p are the kernel hyperparameters called amplitude and period respectively.

Kernel function expressions	
Squared Exp.	$\hat{k}_{SE}(r) = \theta_0^2 \, exp\left(-\frac{1}{2}r^2\right)$
Matern 32	$\hat{k}_{M32}(r) = \theta_0^2 \left(1 + \sqrt{3}r\right) exp\left(-\sqrt{3}r\right)$
Matern 52	$\hat{k}_{M52}(r) = \theta_0^2 \left(1 + \sqrt{5}r + \frac{5}{3}r^2\right) exp\left(-\sqrt{5}r\right)$
Periodic	$\hat{k}_{PER}(r) = \theta_0^2 \exp\left(-\frac{2\sin^2(\pi r)}{\theta_p^2}\right)$

2.2   Hyperparameter Optimization

The choice of the kernel function and its hyperparameters has a critical influence on the behavior of the model, and it is crucial to achieve good results in NLP applications of GPs [1]. This selection has been usually made by choosing one kernel a-priori, and then adjusting the hyperparameters of the kernel function so to optimize a given metric for the data. The most common approach is to find the hyperparameter set that maximizes the log marginal likelihood (LML):

$$log\, p\,(\mathbf{f}|X, \boldsymbol{\theta}, K) = -\frac{1}{2}\mathbf{f}^T K(X,X)^{-1}\mathbf{f} - \frac{1}{2}log\, |K(X,X)| - \frac{n}{2}\, log\, 2\pi \quad (6)$$

where $\boldsymbol{\theta}$ is the set of hyperparameters of the kernel and n is the length of X.

3   Quality Estimation and Feature Extraction

Following [37], we use three different metrics for QE. We assume the existence of an original text in the source language which is divided into sentences. For each sentence, an automatic translation to the target language is available. These translations are the subject of post-editing work. The metrics considered are, *post-editing effort, Human Translation Edit Rate* (HTER) and *post-edit time*.

In [37], translators were asked to post-edit each sentence and score the *post-editing effort* according to the following options:

1. Requires complete retranslation.
2. Requires some retranslation, but post-editing is still quicker than retranslation.
3. Very little post-editing needed.
4. Fit for purpose.

Another metric used to evaluate the translation quality is the edit distance between the automatic translation and its post-edited version. This is computed using the *Human Translation Edit Rate* (HTER) [36]. HTER is defined as $HTER = \frac{e}{pe_w}$, where e is the number of edits, which can be standard insertion, deletion and substitution of single words, and shifting of word sequences. pe_w is the number of words in the sentence.

Finally, the *post-edit time* was computed in [37] using the average number of seconds required by two translators to post-edit each word in the sentence. It is the number of seconds to post-edit the sentence, normalized by the number of words in that sentence.

For more details on the ways these metrics were defined, [37] can be consulted.

3.1   Sentence Embeddings

There are a number of approaches to extract relevant information for QE [5]. In this domain, feature engineering to obtain informative features can be very labor-intensive [38]. In [37], 80 shallow and machine-translation independent features were extracted from the source sentences, their corresponding translations, and monolingual and parallel corpora. A selected set of these features, comprising only 17 features, is used in [38]. Syntactic information represented in tree fragments that contain complete grammar rules are used as feature representation in [2].

Here, we focus on sentence embeddings [25], a common text representation for NLP tasks, although less investigated for QE. We use word embedding dictionaries for the source and target languages (before post-edition). For each sentence, in each language, we compute the embedding representation of all the words. Words missing in the dictionary are assigned a zero-vector representation. In each corpus, for each sentence, a function that combines the embedding representations of all the words in a single vector (e.g., mean of the word vectors) is computed to generate the sentence embedding. Finally, for each pair of original and translated sentences, their sentence embeddings are concatenated to produce the vector representation that is used for QE.

We examine two questions related to the embedding representation. The embedding dimension and the way sentence embeddings are computed. In addition to the commonly applied *mean* function, which computes the mean of all word embeddings in a sentence, we use another two other functions: the maximum (*max*) of all word embeddings (maximum value of each embedding component across all word embeddings), and the standard deviation (*std*) of word embeddings. We hypothesize that variations in the words as captured by the maximum and standard deviation may provide a clue as to the difficulty of the translation.

4   Kernel Function Search

In this work, we automatically search for new kernel functions in order to better predict the translation effort. Specifically, our goal is to find the optimal $\hat{k}(r)$ as in Eq. (5).

To guide this search, we propose using Genetic Programming (GenProg) [20]. In GenProg, computer programs are encoded as solutions using an evolvable representation. At each iteration, solutions are recombined and selected according to the program performance, until an optimal solution is found or a stop criterion is satisfied. In our case, the GP kernel function is the program that is encoded into an expression tree and its performance is evaluated in the context of the translation quality evaluation task.

We define a strongly-typed grammar [26] that specifies the possible combinations these kernels can be composed of, by breaking down the mathematical expressions present in the well-known kernels of Table 1 into basic operations (multiplication, square root, ...).

To randomly generate kernel expression trees that conform the initial population, we propose a grow method based on the work done in [20]. The initial expression tress are created by a recursive process where, at each step, a random terminal or operator is added.

GenProg also needs perturbation or variation methods to be defined in order to modify previous solutions to obtain new ones. A crossover operator, which combines two kernel functions to generate a new one that keeps some of the features of its parents, and a mutation operator, which introduces slight modifications to the original kernel to obtain a new individual are used. We propose a crossover operator that randomly selects a subtree from each kernel and combines them with the sum or the product operator. Furthermore, the mutation operator replaces, shrinks or inserts an elementary mathematical expression at a random position in the tree in a type-safe manner.

As in [12], we use the Bayesian Information Criterion (BIC) [33] as a quality metric for each kernel. This is the fitness function of our GenProg algorithm. As can be seen in Eq. (7), this metric is similar to the LML, but a regularization term that penalizes the complexity of the kernels is added:

$$BIC(k_i) = -2\,log\,p\,(\mathbf{f}|X, k_i, \boldsymbol{\theta}_{i,best}) + q\,log\,n \qquad (7)$$

where q is the number of hyperparameters of the kernel and n is the number of data points in X. $\boldsymbol{\theta}_{i,best}$ is the best hyperparameter set for the kernel k_i according to a given metric.

In contrast to other GenProg applications, the solutions in our approach do not encode all the necessary information to be evaluated. The optimal values of the hyperparameters, according to the LML, have to be determined. Thus, the performance of the solutions depends on the results of the hyperparameter optimization. Both search procedures, the selection of the best hyperparameters for each kernel and the selection of the best kernel given these hyperparameters, are illustrated in Fig. 1.

In this paper, the hyperparameters are optimized by means of *Powell*'s local search algorithm [29]. As this algorithm is not bounded, the search space has to be constrained by penalizing non-feasible hyperparameter sets. Besides, as the function to optimize might be multi-modal, a multi-start approach was used, performing a random restart every time the stopping criteria of *Powell*'s algorithm are met, and getting the best overall result. During this hyperparameter search, a maximum number of 150 evaluations of the LML were allowed.

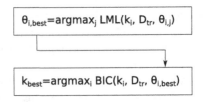

Fig. 1. Two nested search procedures: the selection of the best hyperparameters for each kernel is made according to LML and the selection of the best kernel according to the BIC.

Note that, as a result of the inclusion of the randomized restarts, the hyperparameters found for a certain kernel in two independent evaluations may not be the same. In fact, this implies that the fitness function optimized by the GP algorithm is stochastic.

4.1   Related Work

GPs are particularly suited to model uncertainty in the predictions and allow accurate prediction under noisy conditions. As such, there are diverse scenarios in which GP can be applied to NLP tasks [6]. In [35], they are used for feature selection for QE. Another property of GP, the possibility of extending them to model correlated tasks using multi-task kernels, is used in [7] to jointly learn a series of annotator and metadata specific models.

The most frequently used kernel for NLP tasks is the SE kernel. However, other kernels have shown a better performance than SE in specific tasks. In [30], frequencies of tweets are modeled using GP kernels specifically suited to capture periodicities. In the same paper, the PER and the PS kernels are shown to outperform non-periodic kernels and capture different periodic patterns. In [1], three different kernels are compared: the SE and two Matern kernels. The Matern kernels are reported to produce better results than SE. In addition to numerical kernels, structural-kernels (e.g., tree-kernels) have been also combined with GP. In [2], they are applied to emotion analysis and quality estimation.

A common characteristic of GP applications to NLP is that the choice of the GP kernel has to be made a-priori and does not depend on the quality of the function approximation. The focus is placed on hyperparameter optimization. However, research on evolutionary algorithms has shown that it is also possible to explore the space of kernel functions beyond the hyperparameter optimization. In the GP literature this has been done by combining known kernels [12,21,23]. Kernels have also been evolved for Support Vector Machines (SVMs) [9,10,14,18, 19,39] and Relevance Vector Machines (RVMs) [4]. Some of the SVM approaches are also based on combining the well-known kernels [10,39], although in some other works the kernels are learned from simple mathematical expressions [9,14, 18,19].

In contrast to other works in the GP field, our approach is to evolve kernels from scratch. This method does not rely on previously proposed kernels, and thus, new kernels may naturally arise.

Word embeddings [25] are extensively applied to NLP tasks [22]. The usual approach when combining embeddings from words in a sentence is to compute the average. This is the procedure used in [1], where 100-dimensional Glove embeddings [27] are the representation of choice for mapping texts to emotion scores using GP regression. Word-embeddings have been also used together with GenProg in [32], but with the goal of solving the analogy task problem. This is a completely different problem and the solution presented in [32] does not consider the optimization of GPs or kernels.

5   Experiments

The objectives of the experimentation are twofold. On the one hand, we would like to know if the evolution of GP kernels can improve the results of well-known stationary kernels in translation post-editing effort estimation though sentence embeddings. On the other hand, we would also like to investigate the best solution to aggregate the word vectors to conform these sentence embeddings.

5.1   Datasets and Embeddings

Our experiments consist of learning a GP regressor based on some combination of source and target embeddings, and use this combined representation to predict a particular metric. To carry out these experiment, we use the datasets originally proposed in [37]:

1. en-es news-test2010: First 1000 English news sentences and their translations into Spanish using Moses software. 800 sentences were used for training and 200 for testing.
2. fr-en news-test2009: 2525 French news sentences and their Moses translations into English. 800 sentences were used for training and the remaining ones for testing.

Punctuations were removed from the sentences, the text was tokenized, and also case ignored. For the "HTER" metric, an equal cost was used for all edits [37].

For each source language, we created two embedding representations with different vector dimensions. For the first dataset, we use English Glove embeddings [27] of dimensions 50 and 300. To represent sentences in the target language, we used Spanish embeddings of size 100 computed from the Spanish CoNLL17 corpus and available from the NLPL word embeddings repository[1]. For the second dataset, we used French embeddings of sizes 300 and 52. The

---

[1] http://vectors.nlpl.eu/repository/.

300-dimensional embeddings[2] were trained on Common Crawl and Wikipedia using fastText [16]. The 52-dimensional embeddings[3] were trained on tweets [8]. The 100-dimensional embeddings for the target language were those provided by Glove.

5.2 Experimental Set-Up

Learning a GP regressor implies optimizing the hyperparameters of the kernel. Since the local optimizer used for that optimization is a stochastic process, we run 30 executions of the fitting process using the training data. For traditional Gaussian kernels, this amounts to learning 30 different hyperparameter configurations of the same kernel (e.g., of SE). For the evolved kernels, this means obtaining 30 different kernels. Each regressor is then used to predict the corresponding metric on test data. Finally, for each kernel, the quality of the prediction is measured by computing the root mean squared error (RMSE) between the known true metric values and the predictions. In order to compare the different variants of the algorithms, we analyze the distribution of the RMSE and the mean RMSE values.

5.3 Results

In our first experiment, we compare the classical kernel models to the evolved kernels introduced in this paper. We also evaluate the effect of using embeddings of different dimensions to represent the source language text. Figure 2 shows the results. In the figure, *orig_vw_len* refers to the dimensions of the embeddings of the source data.

The first remarkable fact in Fig. 2 is that GenProg kernels show a similar variance comparing to the classical kernels. This can be a sign of convergence at least in performance. On the other hand, it can be seen that the dimension of the source embeddings has a higher effect in the quality of the predictions than the type of kernel used. This is particularly evident for the *fr-en news-test2009* dataset for which all distribution shapes are clearly asymmetric, with embeddings of size 52 producing lower errors. It is important to take into consideration that, for this dataset, embeddings are different not only in terms of dimension but they also have been actually produced using different corpora and methods. This is not the case for *en-es news-test2010* dataset that uses 50- and 300-dimensional Glove embeddings. This fact may explain why differences for this dataset, particularly for the "time" and "HTER" metrics, are so small.

In terms of the class of kernels used, differences are only visible for the *fr-en news-test2009* and the "score" metric, for which the use of evolved kernels produces better predictions with a higher probability. The improvement over the Matern 52 and SE kernels is particularly clear. The results of the second experiment are summarized in Table 2.

---

[2] https://fasttext.cc/docs/en/crawl-vectors.html.
[3] https://www.spinningbytes.com/resources/wordembeddings/.

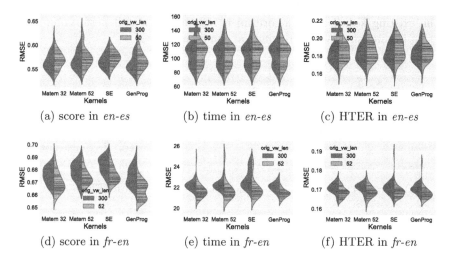

(a) score in *en-es*   (b) time in *en-es*   (c) HTER in *en-es*

(d) score in *fr-en*   (e) time in *fr-en*   (f) HTER in *fr-en*

Fig. 2. Comparison of the kernel models. For all kernels, the *mean* sentence embedding has been used.

Table 2. Comparison between the GP models (mean RMSE from 30 executions). The best results are shown in bold.

	Matern 32		Matern 52		SE		GenProg	
en-es	50	300	50	300	50	300	50	300
score	0.5686	0.5655	0.5731	0.5705	0.5742	0.5678	0.5679	**0.5609**
time	106.3459	105.5226	106.1876	**105.3945**	106.4699	105.4241	106.6932	105.4191
HTER	0.1851	**0.1827**	0.1857	0.1830	0.1872	0.1837	0.1846	0.1835
fr-en	52	300	52	300	52	300	52	300
score	0.6703	0.6755	0.6741	0.6786	0.6758	0.6834	**0.6638**	0.6748
time	21.7082	22.1893	21.8140	22.2194	21.9587	22.8207	**21.4360**	22.0969
HTER	0.1686	0.1708	0.1691	0.1710	0.1687	0.1714	**0.1684**	0.1706

We conducted a statistical test to assess the existence of significant differences among the kernels. For each dataset and metric, we applied Friedman's test [13] and we found significant differences in most comparisons (p-values can be seen in the figures). Then, for each configuration, we applied a post-hoc test based on Friedman's test, and adjusted its results with Shaffer's correction [34].

The results of the statistical tests are shown in Fig. 3. The results confirm a coherent pattern where GenProg is the best performing kernel for most of the configurations. Particularly, in *fr-en news-test2009* dataset, it achieves significantly better results that the classical kernels for the "time" metric. However, according to this test, it can also be appreciated that for the rest of the configurations the differences between GenProg and M32 are not significant. In evaluating these results it is important to take into account that the kernels produced by GenProg have been generated completely from scratch, with no prior knowledge of the existing kernels. The algorithm is able evolve a well performing kernel starting from elementary mathematical components.

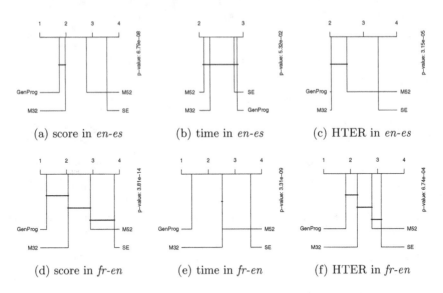

(a) score in *en-es*   (b) time in *en-es*   (c) HTER in *en-es*

(d) score in *fr-en*   (e) time in *fr-en*   (f) HTER in *fr-en*

Fig. 3. Critical difference diagrams. The kernels are ordered following the results in their ranking. The metrics with no significant differences among them are matched with a straight line. On the top, the results for the *en-es news-test2010* dataset with 50-dimensional word-vectors can be found. On the bottom of the figure, the results for *fr-en news-test2009* with 52-dimensional word-vectors are shown. For all kernels, the *mean* sentence embedding has been used.

Another important question is whether the different ways to compute sentence embeddings influences the quality of the prediction. In the next experiment, we investigate this issue. Figure 4 shows the violin plots [17] for RMSE as computed in the test set with the *max*, *mean*, and *std* functions of the word-vectors used. In these experiments, the 300-dimensional embeddings have been used. Each violin plot represents a histogram smoothened using a kernel density with Normal kernel. RMSE values for each of the 30 executions are shown as black vars.

The analysis of Fig. 4 reveals that there are not major differences in sentence embeddings for the "HTER" metric. However, for the "score" metric, *max* embeddings produce smaller errors in the predictions than the other two sentence embedding functions for the two datasets. This effect is more pronounced for the time metric in the *fr-en news-test2009* dataset, for which the *max* embeddings produce better predictions.

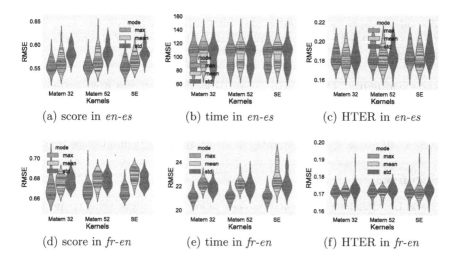

Fig. 4. Comparison of the three sentence embeddings: *max*, *mean*, and *std* on the two datasets, and three effort estimation metrics. For all kernels, 300-dimensional embeddings have been used.

6   Conclusions

Quality estimation of automatic translation has grown in interest in recent years. There are many factors that influence the quality of the final estimation. In particular, the type of text representation used and the class of regressor models are very relevant questions. In this paper, using different QE metrics, we have investigated how the joint determination of these factors influences the final QE. We have focused on GP models, evaluating evolutionary generated kernels, which can be more flexible than classical kernels.

Our results show that GenProg is the best performing kernel for most of the configurations, although in most cases no significant differences were found. For one particular metric, the "score" metric, evolved kernels produce better results than simpler classical models on average in both datasets. Moreover, in *fr-en news-test2009* dataset, significant differences with the classical kernels were found for the "time" metric. If the effort needed to train the model is not an issue, the GenProg approach is the recommended option. Alternatively, classical kernels, particularly the M32 kernel, are a good choice if a faster prediction is needed.

In terms of the sentence embeddings used, *max* embedding showed a slight advantage over extensively used *mean* embeddings. However, this effect seems to depend on the particular metric approximated since for the "HTER" metric we did not observe any difference between the sentence embeddings used. This may indicate that word embeddings in general are not a good feature representation for "HTER". The choice of the dimensionality of word embeddings produced a more marked effect in the quality of the predictions.

As future work we consider the further evaluation of the GP kernels using other features and more sophisticated approaches to compute sentence embeddings.

Acknowledgments. The research presented in this paper has been partially supported by the Basque Government (ELKARTEK programs), and Spanish Ministry of Economy and Competitiveness MINECO (project TIN2016-78365-R). Jose A. Lozano is also supported by BERC 2018–2021 (Basque Government), and Severo Ochoa Program SEV-2017-0718 (Spanish Ministry of Economy, Industry and Competitiveness).

References

1. Beck, D.: Modelling representation noise in emotion analysis using Gaussian processes. In: Proceedings of the Eighth International Joint Conference on Natural Language Processing (Volume 2: Short Papers), vol. 2, pp. 140–145 (2017)
2. Beck, D., Cohn, T., Hardmeier, C., Specia, L.: Learning structural kernels for natural language processing. CoRR abs/1508.02131 (2015). http://arxiv.org/abs/1508.02131
3. Beck, D.E.: Gaussian processes for text regression. Ph.D. thesis, University of Sheffield (2017)
4. Bing, W., Wen-qiong, Z., Ling, C., Jia-hong, L.: A GP-based kernel construction and optimization method for RVM. In: 2010 the 2nd International Conference on Computer and Automation Engineering (ICCAE), vol. 4, pp. 419–423, February 2010. https://doi.org/10.1109/ICCAE.2010.5451646
5. Callison-Burch, C., Koehn, P., Monz, C., Zaidan, O.F.: Findings of the 2011 workshop on statistical machine translation. In: Proceedings of the Sixth Workshop on Statistical Machine Translation, pp. 22–64. Association for Computational Linguistics (2011)
6. Cohn, T., Preotiuc-Pietro, D., Lawrence, N.: Gaussian processes for natural language processing. In: Proceedings of the 52nd Annual Meeting of the Association for Computational Linguistics: Tutorials, pp. 1–3 (2014)
7. Cohn, T., Specia, L.: Modelling annotator bias with multi-task Gaussian processes: an application to machine translation quality estimation. In: Proceedings of the 51st Annual Meeting of the Association for Computational Linguistics (Volume 1: Long Papers), vol. 1, pp. 32–42 (2013)
8. Deriu, J., et al.: Leveraging large amounts of weakly supervised data for multi-language sentiment classification. In: Proceedings of the 26th International Conference on World Wide Web, pp. 1045–1052. International World Wide Web Conferences Steering Committee (2017)
9. Diosan, L., Rogozan, A., Pecuchet, J.P.: Evolving kernel functions for SVMs by genetic programming. In: Sixth International Conference on Machine Learning and Applications (ICMLA 2007), pp. 19–24 (2007). https://doi.org/10.1109/ICMLA.2007.70
10. Dioşan, L., Rogozan, A., Pecuchet, J.P.: Improving classification performance of support vector machine by genetically optimising kernel shape and hyper-parameters. Appl. Intell. **36**(2), 280–294 (2012). https://doi.org/10.1007/s10489-010-0260-1. https://link.springer.com/article/10.1007/s10489-010-0260-1
11. Duvenaud, D.: Automatic model construction with Gaussian processes. Thesis, University of Cambridge (2014). http://www.repository.cam.ac.uk/handle/1810/247281

12. Duvenaud, D., Lloyd, J., Grosse, R., Tenenbaum, J., Zoubin, G.: Structure discovery in nonparametric regression through compositional kernel search. In: Proceedings of the 30th International Conference on Machine Learning, pp. 1166–1174 (2013). http://jmlr.org/proceedings/papers/v28/duvenaud13.html
13. Friedman, M.: The use of ranks to avoid the assumption of normality implicit in the analysis of variance. J. Am. Stat. Assoc. **32**(200), 675–701 (1937)
14. Gagné, C., Schoenauer, M., Sebag, M., Tomassini, M.: Genetic programming for kernel-based learning with co-evolving subsets selection. In: Runarsson, T.P., et al. (eds.) PPSN 2006. LNCS, vol. 4193, pp. 1008–1017. Springer, Heidelberg (2006). https://doi.org/10.1007/11844297_102
15. Genton, M.G.: Classes of kernels for machine learning: a statistics perspective. J. Mach. Learn. Res. **2**, 299–312 (2002). http://dl.acm.org/citation.cfm?id=944790.944815
16. Grave, E., Bojanowski, P., Gupta, P., Joulin, A., Mikolov, T.: Learning word vectors for 157 languages. CoRR abs/1802.06893 (2018). http://arxiv.org/abs/1802.06893
17. Hintze, J.L., Nelson, R.D.: Violin plots: a box plot-density trace synergism. Am. Stat. **52**(2), 181–184 (1998)
18. Howley, T., Madden, M.G.: An evolutionary approach to automatic kernel construction. In: Kollias, S., Stafylopatis, A., Duch, W., Oja, E. (eds.) ICANN 2006. LNCS, vol. 4132, pp. 417–426. Springer, Heidelberg (2006). https://doi.org/10.1007/11840930_43
19. Koch, P., Bischl, B., Flasch, O., Bartz-Beielstein, T., Weihs, C., Konen, W.: Tuning and evolution of support vector kernels. Evol. Intell. **5**(3), 153–170 (2012). https://doi.org/10.1007/s12065-012-0073-8. https://link.springer.com/article/10.1007/s12065-012-0073-8
20. Koza, J.R.: Genetic Programming: On the Programming of Computers by Means of Natural Selection. MIT Press, Cambridge (1992). google-Books-ID: Bhtxo60BV0EC
21. Kronberger, G., Kommenda, M.: Evolution of covariance functions for gaussian process regression using genetic programming. In: Moreno-Díaz, R., Pichler, F., Quesada-Arencibia, A. (eds.) EUROCAST 2013. LNCS, vol. 8111, pp. 308–315. Springer, Heidelberg (2013). https://doi.org/10.1007/978-3-642-53856-8_39
22. Lampos, V., Zou, B., Cox, I.J.: Enhancing feature selection using word embeddings: the case of flu surveillance. In: Proceedings of the 26th International Conference on World Wide Web, pp. 695–704 (2017)
23. Lloyd, J.R., Duvenaud, D., Grosse, R., Tenenbaum, J., Ghahramani, Z.: Automatic construction and natural-language description of nonparametric regression models. In: Twenty-Eighth AAAI Conference on Artificial Intelligence, June 2014. https://www.aaai.org/ocs/index.php/AAAI/AAAI14/paper/view/8240
24. Mercer, J.: XVI. Functions of positive and negative type, and their connection the theory of integral equations. Phil. Trans. R. Soc. Lond. A **209**(441–458), 415–446 (1909). https://doi.org/10.1098/rsta.1909.0016. http://rsta.royalsocietypublishing.org/content/209/441-458/415
25. Mikolov, T., Sutskever, I., Chen, K., Corrado, G.S., Dean, J.: Distributed representations of words and phrases and their compositionality. In: Advances in Neural Information Processing Systems, pp. 3111–3119 (2013)
26. Montana, D.J.: Strongly typed genetic programming. Evol. Comput. **3**(2), 199–230 (1995). https://doi.org/10.1162/evco.1995.3.2.199

27. Pennington, J., Socher, R., Manning, C.D.: GloVe: global vectors for word representation. In: Empirical Methods in Natural Language Processing (EMNLP), vol. 14, pp. 1532–1543 (2014)
28. Polajnar, T., Rogers, S., Girolami, M.: Protein interaction detection in sentences via Gaussian processes: a preliminary evaluation. Int. J. Data Min. Bioinf. **5**(1), 52–72 (2011)
29. Powell, M.J.D.: An efficient method for finding the minimum of a function of several variables without calculating derivatives. Comput. J. **7**(2), 155–162 (1964). https://doi.org/10.1093/comjnl/7.2.155. http://comjnl.oxfordjournals.org/content/7/2/155
30. Preoţiuc-Pietro, D., Cohn, T.: A temporal model of text periodicities using Gaussian processes. In: Proceedings of the 2013 Conference on Empirical Methods in Natural Language Processing, pp. 977–988 (2013)
31. Rasmussen, C.E., Williams, C.K.: Gaussian Processes for Machine Learning. MIT Press, Cambridge (2006)
32. Santana, R.: Reproducing and learning new algebraic operations on word embeddings using genetic programming. CoRR abs/1702.05624 (2017). https://arxiv.org/abs/1702.05624v
33. Schwarz, G.: Estimating the dimension of a model. Ann. Stat. **6**(2), 461–464 (1978). https://doi.org/10.1214/aos/1176344136. https://projecteuclid.org/euclid.aos/1176344136
34. Shaffer, J.P.: Modified sequentially rejective multiple test procedures. J. Am. Stat. Assoc. (2012). https://amstat.tandfonline.com/doi/abs/10.1080/01621459.1986.10478341
35. Shah, K., Cohn, T., Specia, L.: An investigation on the effectiveness of features for translation quality estimation. In: Proceedings of the Machine Translation Summit, vol. 14, pp. 167–174. Citeseer (2013)
36. Snover, M., Dorr, B., Schwartz, R., Micciulla, L., Makhoul, J.: A study of translation edit rate with targeted human annotation. In: Proceedings of Association for Machine Translation in the Americas, vol. 200 (2006)
37. Specia, L.: Exploiting objective annotations for measuring translation post-editing effort. In: Proceedings of the 15th Conference of the European Association for Machine Translation, pp. 73–80 (2011)
38. Specia, L., Shah, K., Souza, J.G., Cohn, T.: QuEst-a translation quality estimation framework. In: Proceedings of the 51st Annual Meeting of the Association for Computational Linguistics: System Demonstrations, pp. 79–84 (2013)
39. Sullivan, K.M., Luke, S.: Evolving kernels for support vector machine classification. In: Proceedings of the 9th Annual Conference on Genetic and Evolutionary Computation, GECCO 2007, pp. 1702–1707. ACM, New York (2007). https://doi.org/10.1145/1276958.1277292. http://doi.acm.org/10.1145/1276958.1277292
40. Vapnik, V.: The Nature of Statistical Learning Theory. Springer-Verlag, New York (2000). https://doi.org/10.1007/978-1-4757-3264-1

Predicting the Execution Time of the Interior Point Method for Solving Linear Programming Problems Using Artificial Neural Networks

Sophia Voulgaropoulou[1] ⓘ, Nikolaos Samaras[1](✉) ⓘ, and Nikolaos Ploskas[2] ⓘ

[1] Department of Applied Informatics, School of Information Sciences,
University of Macedonia, 54636 Thessaloniki, Greece
svoulgaropoulou@uom.edu.gr, samaras@uom.gr
[2] Department of Electrical and Computer Engineering, Faculty of Engineering,
University of Western Macedonia, 50100 Kozani, Greece
nploskas@uowm.gr

Abstract. Deciding upon which algorithm would be the most efficient for a given set of linear programming problems is a significant step in linear programming solvers. CPLEX Optimizer supports primal and dual variants of the simplex algorithm and the interior point method. In this paper, we examine a prediction model using artificial neural works for the performance of CPLEX's interior point method on a set of benchmark linear programming problems (netlib, kennington, Mészáros, Mittelmann). Our study consists of the measurement of the execution time needed for the solution of 295 linear programming problems. Specific characteristics of the linear programming problems are examined, such as the number of constraints and variables, the nonzero elements of the constraint matrix and the right-hand side, and the rank of the constraint matrix of the linear programming problems. The purpose of our study is to identify a model, which could be used for prediction of the algorithm's efficiency on linear programming problems of similar structure. This model can be used prior to the execution of the interior point method in order to estimate its execution time. Experimental results show a good fit of our model both on the training and test set, with the coefficient of determination value at 78% and 72%, respectively.

Keywords: Linear programming · Interior point method · CPLEX Optimizer · Artificial neural network

1 Introduction

Various algorithms exist for the solution of linear programming problems. When it comes to the selection of the most appropriate algorithm to use for a set of problems, the question still remains: how to select the most efficient method,

ⓒ Springer Nature Switzerland AG 2020
N. F. Matsatsinis et al. (Eds.): LION 13 2019, LNCS 11968, pp. 319–324, 2020.
https://doi.org/10.1007/978-3-030-38629-0_26

in terms of the execution time? CPLEX Optimizer [5] includes several high-performance linear programming algorithms, supporting, among other methods, primal and dual variants of the simplex algorithm, as well as the interior point method. In this study, we are exploring a first approach towards creating a prediction model for the efficiency of the interior point method. Contrary to the simplex algorithm, the interior point method reaches a best solution by traversing the interior of the feasible region and not by walking along the edges of its surface to identify extreme points with gradually greater objective values [6]. Interior point methods are superior to the simplex algorithm for certain classes of linear programming problems. Extensive computational studies shows that interior point methods outperform the simplex algorithm for large scale linear programming problems [1]. The simplex algorithm tends to perform poorly on large-scale degenerate linear programming problems [10]. Mehrotra [8] presented a practical algorithm for solving linear programming problems in 1992. This algorithm remains the basis of most current linear programming solvers. A thorough description of the interior point method would exceed the scope of this short paper, therefore, more information about its steps and functionality can be found in [3, 10, 12].

2   Empirical Results

2.1   Data

For the purpose of our computational study, 295 benchmark linear programming problems were used from the netlib (25), kennington (13), Mészáros (217), and Mittelmann (40) libraries. The problems were solved with CPLEX's 12.6.1 [5] interior point method and the respective execution time, needed for their solution, was recorded for each problem. The linear programming problem characteristics which were examined in this study and set as input in our model are the following:

- m: the number of constraints
- n: the number of variables
- $nnzA$: the number of nonzero elements of the constraint matrix
- $nnzb$: the number of nonzero elements of the right-hand side vector
- $rankA$: the rank of the constraint matrix.

The execution time was set as the output of our model. Apart from the above characteristics, we examined also the number of variables in the problems after adding slack variables, the density of the problem, the data length (bit length), required in order to represent integer data, as well as the norm of the constraint matrix. However, these characteristics did not contribute to the creation of our models. For this set of problems, the lower and upper values in number of constraints, variables, nonzero elements of the constraint matrix and right-hand side vector, are given respectively in Table 1 below. Each minimum and maximum value may be related to different LPs, since these values are provided as reference in this paper and they are not necessarily linked to the same LP.

Table 1. Lower and Upper values in examined LP characteristics

	Minimum	Maximum
Constraints	25	986,069
Variables	882	15,000,000
Nonzero elements of constraint matrix	3108	30,000,000
Nonzero elements of right-hand side vector	0	512,209

2.2 Neural Network Model

The neural network model presented in this study has been generated using the scikit-learn toolkit [9]. While testing several statistical environments, we found scikit-learn to be the most suitable for the purpose of our analysis, since it supports numerous methods of supervised and unsupervised learning, model selection and evaluation and transformation of data. The algorithm used for the generation of our model is the Multi-layer Perceptron (MLP). MLP is a supervised learning algorithm that can learn a function $f(\Delta) : R^x \rightarrow R^o$ by training on a dataset, where x is the number of dimensions for input and o is the number of dimensions for output. Given a set of input features and an output target, MLP can learn a nonlinear function approximator for either classification or regression. We also experimented with other supervised learning methods available in scikit-learn toolkit, such as linear regression, lasso regression, ridge regression and decision trees, but we obtained the best results using the MLP method.

The greatest challenge during our analysis was related to the number of hidden layers, which had to be set while testing our models, along with the activation function we had to choose. It was noted that the more hidden layers we have in our models, the worse results we eventually get. The activation function is very significant since it converts an input signal of the last hidden layer to an output signal for the next layer. Commonly used activation functions are the hyperbolic tan function (*tanh*), the logistic sigmoid function (*logistic*) and the rectified linear unit function (*relu*). Such activation functions were tested while repeated and extensive testing was also performed on the number of hidden layers. One more aspect that was taken into consideration through extensive testing, was the scale of our data. The input parameters of a model may have different scales, which makes it difficult for the examined problems to be modeled. Scaling and normalizing the original data was a significant step we took while generating our models. Evaluating the metrics of each model, we formed our model using the parameters shown in Table 2 below. Apart from the activation function and the number of hidden layers, the solver selected for weight optimization is *LBFGS*, an optimizer in the family of quasi-Newton methods. *LBFGS* uses a weighted linear summation to transform the input values of previous layers to output values for the next layer. Tolerance value refers to the tolerance for the optimization. For example, if, upon a certain number of iterations, we fail to decrease the training loss or to increase the validation score by at least a value equal to tolerance, convergence is considered to be reached and training stops. The alpha value refers to the L2 penalty (regularization term) parameter, while the maximum number of iterations indicates that the solver will iterate until convergence (determined by tolerance value) or this number of iterations.

Table 2. Model parameters for the MLP method

Algorithm	MLPRegressor
Hidden layer sizes	20
Activation function	relu
Solver	lbfgs
Alpha value	$1e{-}5$
Maximum iterations	1,000
Tolerance	0.0001

The ratio between the training and test set was chosen to be 70 to 30. To evaluate the performance of our model, certain metrics were taken into consideration. The R-squared (R^2, coefficient of determination) provides an estimate of the strength of the relationship between a regression model and the dependent variable (output), since it defines how well a statistical model fits the examined data (i.e., the bigger its value is, the better fitting the model has). The Root Mean Square Error (RMSE) is the standard deviation of the residuals (prediction errors) and measures their spread around the line of best fit. RMSE is considered as a measure of accuracy, usually used for comparison of different models generated for a particular dataset. This metric has non-negative values and a value of 0, although impossible to achieve in practice, would indicate a perfect fit to the data. In general, a lower RMSE value is always better than a higher one. It is important to note that this metric is not used between different datasets, as it is scale-dependent [4], thus comparisons of different types of data would not be valid since the measure depends on the scale of the examined dataset numbers. We also include the mean absolute error that measures the average magnitude of the errors in a set of predictions, without considering their direction, and the median absolute error that is insensitive to outliers [2,7,11]. Table 3 presents the results of the neural network. For the training set, the model achieved an RMSE value of 123.32 and an R^2 value of 0.78, while for the test set the model achieved an RMSE value of 296.73 and an R^2 value of 0.72. Taking into account the variability in the features of the 295 linear programming models and the metrics' values, it is shown that our model can explain the data reasonably well. As an example, an R^2 value of 1 would indicate a perfect fit of the data, so the current R^2 value proves goodness of fit of our model. Further below, a graphical representation of the comparison between the metrics measured for some of the models we tested is presented in Figs. 1 and 2 (different number of hidden layers and different activation functions, respectively). The term *units* as shown in Fig. 1 below, refer to the actual layers of the model.

Table 3. MLP model for the interior point method execution time

	Training set	Test set
Root Mean squared error	123.32	296.73
Absolute Mean error	54.31	97.54
Absolute Median error	7.12	9.49
R^2	0.78	0.72

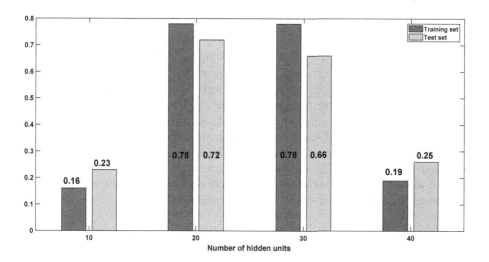

Fig. 1. Tuning the number of hidden layers

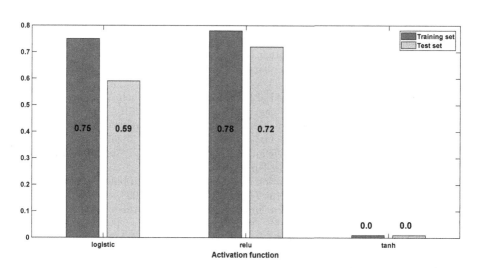

Fig. 2. Tuning the activation function

3   Conclusions

An important step in solving linear programming problems is the selection of the most efficient algorithm. Most linear programming solvers have a heuristic procedure to select the most suitable algorithm based on the characteristic of the input linear programming problem. In this paper, we experiment with the use of a neural network for predicting the execution time of CPLEX's interior point method. Experimental results show that the model can achieve an R^2 value of 78% for the training set and 72% for the test set. Taking into account the variability in the features of the benchmark linear programming models we examined and the metrics used for the comparison of the generated models, the current model proves to have a good fit on the data and thus, can be used for further prediction of the algorithm's efficiency. In future work, we plan to build prediction models for the primal and dual simplex algorithm and experiment with various supervised learning methods. These models will be formed after a further investigation on the above mentioned model parameters (different number of hidden layers, number of neurons, activation functions, scaling and normalization techniques). Building accurate models for the prediction of the execution of the primal simplex algorithm, the dual simplex algorithm, and the interior point method will lead a linear programming solver to select the most efficient algorithm for a given linear programming problem. That would lead to great savings in solving linear programming problems.

References

1. Bixby, E.R.: Implementing the simplex method: the initial basis. ORSA J. Comput. **4**, 267–284 (1992). https://doi.org/10.1287/ijoc.4.3.267
2. Draper, N.R., Smith, H.: Applied Regression Analysis, 3rd edn. Wiley, Hoboken (1998)
3. Gondzio, J.: Interior point methods 25 years later. Eur. J. Oper. Res. **218**(3), 587–601 (2012). https://doi.org/10.1016/j.ejor.2011.09.017
4. Hyndman, R., Koehler, A.: Another look at measures of forecast accuracy. Int. J. Forecast. **22**(4), 679–688 (2006). https://doi.org/10.1016/j.ijforecast.2006.03.001
5. IBM ILOG CPLEX: Cplex 12.6.0 user manual (2017). http://www-01.ibm.com/support/knowledgecenter/SSSA5P_12.6.1/ilog.odms.studio.help/Optimization_Studio/topics/COS_home.html?lang=en. Accessed 15 Feb 2019
6. Karmarkar, N.: A new polynomial-time algorithm for linear programming. Combinatorica **4**(4), 373–395 (1984). https://doi.org/10.1007/BF02579150
7. Kutner, M.H., Neter, J., Nachtsheim, C.J., Wasserman, W.: Applied Linear Statistical Models, 5th edn. McGraw-Hill, New York (2004)
8. Mehrotra, S.: On the implementation of a primal-dual interior point method. SIAM J. Optim. **2**, 575–601 (1992). https://doi.org/10.1137/0802028
9. Pedregosa, F., et al.: Scikit-learn: machine learning in Python. J. Mach. Learn. Res. **12**, 2825–2830 (2011)
10. Ploskas, N., Samaras, N.: Linear Programming Using MATLAB. Springer, Cham (2017). https://doi.org/10.1007/978-3-319-65919-0
11. Rao, C.R.: Coefficient of determination. In: Linear Statistical Inference and its Applications, 2nd edn. Wiley, New York (1973)
12. Wright, S.: Primal-Dual Interior-Point Methods, vol. 54. SIAM, Philadelphia (1997)

A SAT Approach for Finding
Sup-Transition-Minors

Benedikt Klocker[✉], Herbert Fleischner, and Günther R. Raidl

Institute of Logic and Computation, TU Wien,
Favoritenstraße 9-11/192-01, 1040 Vienna, Austria
{klocker,fleischner,raidl}@ac.tuwien.ac.at

Abstract. The cycle double cover conjecture is a famous longstanding
unsolved conjecture in graph theory. It is related and can be reduced to
the compatible circuit decomposition problem. Recently Fleischner et al.
(2018) provided a sufficient condition for a compatible circuit decomposition, which is called SUD-K_5-minor freeness. In a previous work we developed an abstract mathematical model for finding SUD-K_5-minors and
based on the model a mixed integer linear program (MIP). In this work
we propose a respective boolean satisfiability (SAT) model and compare
it with the MIP model in computational tests. Non-trivial symmetry
breaking constraints are proposed, which improve the solving times of
both models considerably. Compared to the MIP model the SAT approach performs significantly better. We use the faster algorithm to further test graphs of graph theoretic interest and were able to get new
insights. Among other results we found snarks with 30 and 32 vertices
that do not contain a perfect pseudo-matching, that is a spanning subgraph consisting of K_2 and $K_{1,3}$ components, whose contraction leads to
a SUD-K_5-minor free graph.

Keywords: Transition minor · Cycle double cover · Compatible
circuit decomposition · SAT

1 Introduction

The famous cycle double cover (CDC) conjecture states that every bridgeless
graph has a cycle double cover, which is a collection of cycles such that every edge
of the graph is part of exactly two cycles. It was originally posed by Szekeres [13]
and Seymour [11] over 40 years ago and is still unsolved. As Jaeger shows in
[5], the CDC conjecture can be reduced to the consideration of a special class
of graphs called snarks by considering a minimum counter example. There are
multiple similar definitions of snarks, we will use the one from Jaeger [5]: A *snark*
is a simple cyclically 4-edge-connected cubic graph with chromatic index four. A
cyclically 4-edge-connected graph is a graph that has no 4-edge cut after whose

This work is supported by the Austrian Science Fund (FWF) under grant P27615 and
the Vienna Graduate School on Computational Optimization, grant W1260.

N. F. Matsatsinis et al. (Eds.): LION 13 2019, LNCS 11968, pp. 325–341, 2020.
https://doi.org/10.1007/978-3-030-38629-0_27

removal at least two components contain a cycle. Although snarks are simple, we consider in general undirected multigraphs without loops.

A problem related to the CDC conjecture is the compatible circuit decomposition (CCD) problem. It is formulated on a *transitioned graph* (G, T), which is a graph G together with a set of transitions T. A *transition* consists of a vertex and two incident edges. We write $T(v)$ for the set of all transitions at vertex v. A transition system has to satisfy that the transitions in $T(v)$ are edge-disjoint. A *compatible circuit decomposition* of a transitioned graph is a collection of circuits such that each edge of the graph is part of exactly one circuit and each circuit does not contain any pair of edges of a transition. The CCD problem asks if a given transitioned graph contains a compatible circuit decomposition. To see the connection between the CDC conjecture and the CCD problem we consider a cubic graph C, for example a snark. A *perfect pseudo-matching (PPM)* of C is a subgraph spanning C whose connected components are either two vertices connected by an edge, i.e. the K_2, or one vertex together with its three incident edges and its three neighbors, i.e. the $K_{1,3}$ which we also call *claw*. Given a PPM of C we can define now a transitioned graph (G, T) by contracting all edges of the PPM. We define a transition in T for each pair of adjacent edges in G that remain after the contraction, see Fig. 1 for an illustration.

Fig. 1. Example contraction of parts of a PPM. The edges of the PPM getting contracted are drawn dashed. The transitions in the resulting graph are represented by a vee ($\vee$) between the two edges of the transition.

Note that the contracted graph may contain loops, but we can ignore them since they are not relevant in the context of circuit decompositions. If we contract a PPM of a snark there are no loops since a snark is simple and has no triangles. As described in [8] if the constructed transitioned graph (G, T) contains a CCD one can construct a CDC in the original graph C. Already in 1980 Fleischner [3] proved that every transitioned graph (G, T) where G is 2-connected and planar contains a CCD. This result was then improved in 2000 by Fan and Zhang [2] who showed that if G is 2-connected and K_5-minor free it must contain a CCD. Those two sufficient conditions for the existence of a CCD are only based on the structure of G and do not consider the transition system T. Recently Fleischner et al. [4] generalized the minor term to transitioned graphs and proved that if (G, T) is 2-connected and SUD-K_5-minor free it must contain a CCD. For the definition of a SUD-K_5-minor we refer to [4] or [8].

Because of the complex nature of the definition of a SUD-K_5-minor it is nontrivial to check if a graph contains a SUD-K_5-minor. In a previous work [8] we generalized the problem of SUD-K_5-minor containment by allowing to replace the K_5 by any 4-regular graph H. Formally, given a transitioned graph $(G, \mathcal{T})$ and a 4-regular completely transitioned graph $(H, \mathcal{S})$ the decision problem *Existence of Sup-Transition-Minors (ESTM)* asks if $(G, \mathcal{T})$ has a sup-$(H, \mathcal{S})$-transition minor. A graph is *completely transitioned* if every edge is part of two transitions. A transitioned graph $(G, \mathcal{T})$ has a sup-$(H, \mathcal{S})$-transition minor if it has a H-minor where every vertex w of H corresponds to a subgraph C_w of G that has a cut vertex v_w. The cut vertex must split C_w in at least two components such that there is a transition at v_w whose edges are part of one of those components C_w^1 and all other edges incident to v_w are part of other components. Furthermore, the transition at v_w must correspond to a transition of H at w such that the two edges of the transition in H are connected to the component C_w^1. For a formal definition of a sup-$(H, \mathcal{S})$-transition minor we refer to [8].

The mathematical model developed in [8] for deciding the ESTM allowed to derive a mixed integer linear program (MIP) model, which could be solved for small graphs, yielding interesting graph theoretic results. In this work we present a more powerful boolean satisfiability (SAT) formulation for the mathematical model developed in [8], which allows addressing significantly larger graphs. To improve the solving times of the MIP as well as the SAT model we propose a non-trivial symmetry breaking based on graph automorphisms of the two input graphs $(G, \mathcal{T})$ and $(H, \mathcal{S})$. The idea of breaking symmetries using automorphism groups has been studied in a general context, see e.g. [1], and in problem-specific contexts, see e.g. [7]. We extend the definition of automorphisms to transitioned graphs and propose problem specific symmetry breaking constraints based on a vertex mapping between the two input graphs $(G, \mathcal{T})$ and $(H, \mathcal{S})$.

Using the new SAT model, which outperforms the MIP model significantly, together with the symmetry breaking constraints we were able to check for all snarks with up to 32 vertices if they contain a PPM whose contraction is SUD-K_5-minor free. Within those tests we were able to find snarks that do not contain such a PPM. This result answers the previously open question that the notion of SUD-K_5-minor freeness in the context of contractions of PPMs in snarks is not enough to prove the CDC conjecture.

In the following section we present a SAT model for finding a sup-$(H, \mathcal{S})$-transition minor. Then we discuss symmetry breaking constraints, which can be used in the MIP and in the SAT model, in Sect. 3. Section 4 gives computational results for the new SAT model in comparison to the MIP model. Also we show the impact of the symmetry breaking. Finally, we will conclude and propose some future work in Sect. 5.

1.1 Terminology and Notation

As already mentioned, when referring to a graph we mean here an undirected multigraph without loops, unless otherwise stated. We denote a graph by $G = (V, E, r)$ with a vertex set V, an edge set E and a function r that maps an edge $e \in E$ to the set of its two end vertices $\{v_1, v_2\}$. If $r(e) = \{v_1, v_2\}$ we also write $e = v_1 v_2$. Note that $e = v_1 v_2$ and $e' = v_1 v_2$ does not imply $e = e'$ since we can have parallel edges. For a vertex $v \in V$ we write $E(v) = \{e \in E \mid v \in r(e)\}$ for the set of all incident edges and $N(v) = \{v' \in V \mid \exists e \in E : e = vv'\}$ for the set of all neighbors.

We represent a transition in a transitioned graph by a pair $T = (v, \{e_1, e_2\})$ where the first element is a vertex and the second element a set of two edges which are incident to the vertex. Furthermore, we use projections $\pi_1(T) = v$ and $\pi_2(T) = \{e_1, e_2\}$ to denote the vertex and the edge set of a transition.

For a partial function $\alpha : A \nrightarrow B$ and a subset $X \subseteq A$ we write $\alpha[X] = \{b \in B : \exists a \in X : b = \alpha(a)\}$ for the image of X under α. Furthermore, if $Y \subseteq B$ we write $\alpha^{-1}[Y] = \{a \in A : \alpha(a) \in Y\}$ for the preimage of Y under α. We also abbreviate the notation in case of only one element by $\alpha[a] = \alpha[\{a\}]$ and $\alpha^{-1}[b] = \alpha^{-1}[\{b\}]$ for $a \in A$ and $b \in B$.

2   The SAT Model

In this section we present a SAT model for checking if a given transitioned graph $(G, \mathcal{T})$ contains a sup-$(H, \mathcal{S})$-transition minor for a given completely transitioned 4-regular graph $(H, \mathcal{S})$. For the formal definition of a sup-$(H, \mathcal{S})$-transition minor see [8].

In the following we first repeat the mathematical model developed in [8] on which the SAT model will be based. The model will use simple trees C_w^i with vertices in G for which we will use the following notation: $E_w^i := \{e \in E(G) \mid r(e) \in E(C_w^i)\}$. The model is defined as finding

1. a partial surjective function $\varphi \colon V(G) \nrightarrow V(H)$,
2. a partial injective and surjective function $\kappa \colon E(G) \nrightarrow E(H)$,
3. a partial injective function $\theta \colon E(G) \nrightarrow V(H)$,
4. $\forall w \in V(H)$ a pair (T_w, S_w) of transitions with $T_w \in \mathcal{T}$ and $S_w \in \mathcal{S}(w)$, and
5. $\forall w \in V(H)$ two simple trees C_w^1 and C_w^2 with $V(C_w^i) \subseteq V(G)$ for $i = 1, 2$,

such that

$$E(C_w^i) \subseteq r_G[E(G)] \qquad\qquad \forall w \in V(H), \forall i \in \{1, 2\} \quad (1)$$

$$\kappa(e) = f \Rightarrow \varphi[r_G(e)] = r_H(f) \qquad\qquad \forall e \in E(G), \forall f \in E(H) \quad (2)$$

$$V(C_w^1) \cup V(C_w^2) = \varphi^{-1}[w] \qquad\qquad \forall w \in V(H) \quad (3)$$

$$\{\pi_1(T_w)\} = V(C_w^1) \cap V(C_w^2) \qquad\qquad \forall w \in V(H) \quad (4)$$

$$\pi_2(T_w) \subseteq \kappa^{-1}[\pi_2(S_w)] \cup \theta^{-1}[w] \cup E_w^1 \qquad\qquad \forall w \in V(H) \quad (5)$$

$$(\kappa^{-1}[\pi_2(S_w)] \cap E(\pi_1(T_w))) \cup \theta^{-1}[w] \subseteq \pi_2(T_w) \qquad\qquad \forall w \in V(H) \quad (6)$$

$$e \in \mathrm{dom}(\kappa) \wedge \kappa(e) \in \pi_2(S_w) \Rightarrow r_G(e) \cap V(C_w^1) \neq \emptyset \qquad \forall w \in V(H), \forall e \in E(G) \quad (7)$$

$$\begin{aligned} e \in \mathrm{dom}(\kappa) \wedge \kappa(e) \in E(w) \setminus \pi_2(S_w) \\ \Rightarrow r_G(e) \cap V(C_w^2) \neq \emptyset \end{aligned} \qquad \forall w \in V(H), \forall e \in E(G) \quad (8)$$

$$\begin{aligned} v \in V(C_w^1) \setminus \{\pi_1(T_w)\} \wedge \deg_{C_w^1}(v) = 1 \wedge \\ v \notin \bigcup r_G[\theta^{-1}[w]] \Rightarrow E(v) \cap \kappa^{-1}[\pi_2(S_w)] \neq \emptyset \end{aligned} \qquad \forall w \in V(H), \forall v \in V(G) \quad (9)$$

$$E_{C_w^1}(\pi_1(T_w)) \subseteq r_G[\pi_2(T_w)] \qquad\qquad \forall w \in V(H) \qquad (10)$$

$$\theta(e) = w \Rightarrow r_G(e) \subseteq V(C_w^1) \qquad\qquad \forall e \in E(G), \forall w \in V(H) \qquad (11)$$

$$\theta(e) = w \Rightarrow r_G(e) \notin E(C_w^1) \qquad\qquad \forall e \in E(G), \forall w \in V(H) \qquad (12)$$

holds.

In [8] we proved that the feasibility of this model is equivalent to the existence of a sup-$(H, \mathcal{S})$-transition minor in $(G, \mathcal{T})$. Most constraints of this model can more or less directly be translated into SAT clauses. One critical aspect is how to model the tree C_w^i for $w \in V(H)$ and $i \in \{1, 2\}$. Constraints (3) and (4) ensure that the subgraph C_w formed by C_w^1 and C_w^2 together is a tree and all trees C_w are disjoint for $w \in V(H)$. Combining all trees C_w for $w \in V(H)$ we obtain a forest and for each of the trees C_w we define a unique root $\pi_1(T_w)$ by (4). When modeling the forest in a directed fashion, we then only have to take care to avoid any cycles. There are different techniques in literature to model acyclicity in SAT models. Some of those techniques are summarized in [6]. We will use the approach based on a transitive closure for ensuring acyclicity in our model. Our SAT model uses the following variables:

- x_v^w for $v \in V(G)$, $w \in V(H)$ represents $\varphi(v) = w$,
- y_e^f for $e \in E(G)$, $f \in E(H)$ represents $\kappa(e) = f$,
- z_e^w for $e \in E(G)$, $w \in V(H)$ represents $\theta(e) = w$,
- a_T^w for $w \in V(H)$, $T \in \mathcal{T}$ represents $T = T_w$,
- b_S^w for $w \in V(H)$, $S \in \mathcal{S}(w)$ represents $S = S_w$,
- $o_v^{i,w}$ for $v \in V(G)$, $w \in V(H)$, $i \in \{1,2\}$ represents $v \in V(C_w^i)$,
- $p_a^{i,w}$ for $a \in A(G)$, $w \in V(H)$, $i \in \{1,2\}$ represents $a \in E(C_w^i)$,
- t_{v_1,v_2} for $v_1, v_2 \in V(G)$ is the transitive closure relation of all $p_a^{i,w}$ variables.

The trees C_w^i are modeled as a directed rooted out-trees and the variables $p_a^{i,w}$ decide which directed arcs are part of the tree. Set $A(G)$ is the set of all directed arcs of edges in G when eliminating parallel edges. So for every pair of adjacent vertices in G there are two arcs in opposite direction in $A(G)$. We write $A^{\text{in}}(v)$ for the ingoing arcs at v and $A^{\text{out}}(v)$ for the outgoing arcs at v. In the following we list all constraints of our SAT model. For simplicity, we will present the constraints in the form of propositional logic formulas. To transform them into clauses we use De Morgan's law and the distributive property. One alternative would be to use Tseitin transformations [14], although for the constraints we will present the number of resulting clauses using the naive transformation is still small and therefore this is not needed. In the following we will use for a given $v \in V(G)$, $w \in V(H)$, and $i \in \{1, 2\}$

$$\text{oneIn}(v, i, w) := \Bigg(\bigvee_{a \in A^{\text{in}}(v)} p_a^{i,w} \Bigg) \wedge \bigwedge_{\substack{a_1, a_2 \in A^{\text{in}}(v) \\ a_1 \neq a_2}} \left(\neg p_{a_1}^{i,w} \vee \neg p_{a_2}^{i,w} \right).$$

The basic structures as defined in the mathematical model are expressed by

$$\neg(x_v^{w_1} \wedge x_v^{w_2}) \qquad\qquad \forall v \in V(G), \forall w_1, w_2 \in V(H), w_1 \neq w_2 \qquad (13)$$

$$\bigvee_{v \in V(G)} x_v^w \qquad\qquad \forall w \in V(H) \qquad (14)$$

$$\neg(y_e^{f_1} \wedge y_e^{f_2}) \qquad\qquad \forall e \in E(G), \forall f_1, f_2 \in E(H), f_1 \neq f_2 \qquad (15)$$

$$\neg(y_{e_1}^{f} \wedge y_{e_2}^{f}) \qquad\qquad \forall e_1, e_2 \in E(G), e_1 \neq e_2, \forall f \in E(H) \qquad (16)$$

$$\bigvee_{e \in E(G)} y_e^f \qquad\qquad \forall f \in E(H) \qquad (17)$$

$$\neg(z_e^{w_1} \wedge z_e^{w_2}) \qquad\qquad \forall e \in E(G), \forall w_1, w_2 \in V(H), w_1 \neq w_2 \qquad (18)$$

$$\neg(z_{e_1}^{w} \wedge z_{e_2}^{w}) \qquad\qquad \forall e_1, e_2 \in E(G), e_1 \neq e_2, \forall w \in V(H) \qquad (19)$$

$$\neg(a_{T_1}^{w} \wedge a_{T_2}^{w}) \qquad\qquad \forall w \in V(H), \forall T_1, T_2 \in \mathcal{T}, T_1 \neq T_2 \qquad (20)$$

$$\bigvee_{T \in \mathcal{T}} a_T^w \qquad\qquad \forall w \in V(H) \qquad (21)$$

$$\neg(a_T^{w_1} \wedge a_T^{w_2}) \qquad\qquad \forall w_1, w_2 \in V(H), w_1 \neq w_2, \forall T \in \mathcal{T} \qquad (22)$$

$$\neg(b_{S_1}^{w} \wedge b_{S_2}^{w}) \qquad\qquad \forall w \in V(H), \forall S_1, S_2 \in \mathcal{S}(w), S_1 \neq S_2 \qquad (23)$$

$$\bigvee_{S \in \mathcal{S}(w)} b_S^w \qquad\qquad \forall w \in V(H) \qquad (24)$$

$$o_v^{i,w} \to \bigvee_{T \in \mathcal{T}(v)} a_T^w \vee \text{oneIn}(v, i, w) \qquad \forall v \in V(G), w \in V(H), i \in \{1,2\} \qquad (25)$$

$$\bigvee_{T \in \mathcal{T}(v)} a_T^w \to o_v^{i,w} \wedge \bigwedge_{a \in A^{\text{in}}(v)} \neg p_a^{i,w} \qquad \forall v \in V(G), w \in V(H), i \in \{1,2\} \qquad (26)$$

$$\neg o_v^{i,w} \to \bigwedge_{a \in A^{\text{in}}(v) \cup A^{\text{out}}(v)} \neg p_a^{i,w} \qquad \forall v \in V(G), w \in V(H), i \in \{1,2\} \qquad (27)$$

$$\neg(t_{v_1,v_2} \wedge t_{v_2,v_1}) \qquad\qquad \forall a = (v_1, v_2) \in A(G) \qquad (28)$$

$$t_{v_1,v_2} \wedge t_{v_2,v_3} \to t_{v_1,v_3} \qquad\qquad \forall a = (v_1, v_2) \in A(G), v_3 \in V(G) \qquad (29)$$

$$\bigvee_{w \in V(H), i \in \{1,2\}} p_a^{i,w} \to t_{v_1,v_2} \qquad\qquad \forall a = (v_1, v_2) \in A(G). \qquad (30)$$

Constraints (13), (15), (18), (20), and (23) ensure that φ, κ, θ, $w \mapsto T_w$, and $w \mapsto S_w$ are partial functions with the special restriction that $S_w \in \mathcal{S}(w)$. Furthermore, constraints (14) and (17) enforce that φ and κ are surjective. On the other hand, constraints (16), (19), and (22) ensure that κ, θ, and $w \mapsto T_w$ are injective. Note that the mathematical model does not state directly that $w \mapsto T_w$ should be injective, but it does indirectly by constraints (3) and (4). Additionally, constraints (21) and (24) guarantee that there exists a T_w and a S_w for each $w \in V(H)$.

Constraints (25)–(27) characterize three types of vertices in G. The root vertices of the trees C_w^i, which are defined by the vertices of the transitions T_w by (4), do not have any ingoing arcs in C_w^i. Other vertices in C_w^i that are not roots have exactly one ingoing arc in C_w^i and vertices that are not in C_w^i have no ingoing or outgoing arc in C_w^i. Last but not least, constraints (28)–(30) ensure that the trees C_w^i have no cycles by using the transitive closure variables t_{v_1,v_2} similarly as it is described in [6]. Instead of having just one variable, which represents if a directed edge is part of the forest, we use in our case the disjunction $\bigvee_{w \in V(H), i \in \{1,2,\}} p_a^{i,w}$ for an arc a. With this we ensured all structural properties formulated in the mathematical model. What is left is to model constraints (1)–(12) which is achieved by

$$y_e^f \to (x_{v_1}^{w_1} \wedge x_{v_2}^{w_2}) \vee (x_{v_2}^{w_1} \wedge x_{v_1}^{w_2}) \qquad \begin{array}{c} \forall e = v_1, v_2 \in E(G), \\ \forall f = w_1 w_2 \in E(H) \end{array} \qquad (31)$$

$$o_v^{1,w} \vee o_v^{2,w} \leftrightarrow x_v^w \qquad \forall v \in V(G), \forall w \in V(H) \qquad (32)$$

$$\bigvee_{T \in \mathcal{T}(v)} a_T^w \leftrightarrow o_v^{1,w} \wedge o_v^{2,w} \qquad \forall v \in V(G), \forall w \in V(H) \qquad (33)$$

$$\bigvee_{\substack{T \in \mathcal{T} \\ e \in \pi_2(T)}} a_T^w \to \bigvee_{S \in \mathcal{S}(w)} \left(b_S^w \wedge \bigvee_{f \in \pi_2(S)} y_e^f \right) \qquad \begin{array}{c} \forall e = v_1 v_2 \in E(G), \\ \forall w \in V(H) \end{array} \qquad (34)$$
$$\vee z_e^w \vee p_{(v_1,v_2)}^{1,w} \vee p_{(v_2,v_1)}^{1,w}$$

$$a_T^w \wedge b_S^w \to \neg \bigvee_{f \in \pi_2(S)} y_e^f \qquad \begin{array}{c} \forall w \in V(H), \forall S \in \mathcal{S}(w), \\ \forall T \in \mathcal{T}, \forall e \in E(\pi_1(T)) \setminus \pi_2(T) \end{array} \qquad (35)$$

$$a_T^w \to \neg z_e^w \qquad \begin{array}{c} \forall w \in V(H), \\ \forall T \in \mathcal{T}, \forall e \in E(\pi_1(T)) \setminus \pi_2(T) \end{array} \qquad (36)$$

$$b_S^w \wedge \bigvee_{f \in \pi_2(S)} y_e^f \to o_{v_1}^{1,w} \vee o_{v_2}^{1,w} \qquad \begin{array}{c} \forall w \in V(H), \forall S \in \mathcal{S}(w), \\ \forall e = v_1 v_2 \in E(G) \end{array} \qquad (37)$$

$$\left(b_S^w \wedge \bigvee_{f \in E(w) \setminus \pi_2(S)} y_e^f \right) \to o_{v_1}^{2,w} \vee o_{v_2}^{2,w} \qquad \begin{array}{c} \forall w \in V(H), \forall S \in \mathcal{S}(w), \\ \forall e = v_1 v_2 \in E(G) \end{array} \qquad (38)$$

$$b_S^w \wedge o_v^{1,w} \wedge \bigwedge_{v' \in N(v)} \neg p_{(v,v')}^{1,w} \wedge \bigwedge_{e \in E(v)} \neg z_e^w$$
$$\to \left(\bigvee_{e \in E(v), f \in \pi_2(S)} y_e^f \vee o_v^{2,w} \right) \qquad \begin{array}{c} \forall w \in V(H), \forall S \in \mathcal{S}(w), \\ \forall v \in V(G) \end{array} \qquad (39)$$

$$a_T^w \to \neg p_{(\pi_1(T),v)}^{1,w} \wedge \neg p_{(v,\pi_1(T))}^{1,w} \qquad \begin{array}{c} \forall w \in V(H), \forall T \in \mathcal{T}, \\ \forall v \in N(\pi_1(T)) \setminus \bigcup r_G[\pi_2(T)] \end{array} \qquad (40)$$

$$z_e^w \to (o_{v_1}^{1,w} \wedge o_{v_2}^{1,w}) \qquad \forall w \in V(H), \forall e = v_1 v_2 \in E(G) \qquad (41)$$

$$z_e^w \to (\neg p_{(v_1,v_2)}^{1,w} \wedge \neg p_{(v_2,v_1)}^{1,w}) \qquad \forall w \in V(H), \forall e = v_1 v_2 \in E(G). \qquad (42)$$

Constraints (1) are already satisfied implicitly and constraints (2)–(5) are realized by constraints (31)–(34) respectively. Furthermore, constraints (6) are guaranteed by (35) and (36). All the other constraints (7)–(12) are modeled via (37)–(42) respectively.

By using our SAT model we can develop an algorithm that checks for a given snark if it contains a PPM whose contraction leads to a planar, a K_5-minor free, a SUD-K_5-minor free, or a CCD-containing graph. The algorithm enumerates all PPMs iteratively by ordering the vertices of the snark and always trying to add all possible edges or claws to the pseudo-matching that contain the smallest not yet visited vertex of the snark. Then it checks for each generated PPM if its contraction leads to a planar graph. If it does not find such a matching it checks for K_5-minor free contractions, if this also is not the case it checks for SUD-K_5-minor free contractions and otherwise it checks for CCD-containing contractions. Using this algorithm one can specify for each snark the type of the strongest matching found for this snark.

3   Symmetry Breaking

The input graphs G and H, especially H, often have symmetries leading to symmetric solutions in our model. To avoid those we analyze the structure of the symmetries in G and H and incorporate symmetry breaking constraints into our model.

To formalize the concept of symmetries in transitioned graphs we extend the definition of homomorphisms on graphs to transitioned graphs. We are only interested in vertex symmetries and therefore a homomorphism between two multigraphs is for us a vertex mapping which preserves the vertex adjacency relation with the correct number of edges, e.g. if there are k edges between two vertices then there must be exactly k edges between the images of those vertices. If we would want to also eliminate edge symmetries this would lead to more complex symmetry breaking constraints and would only help in cases where there are a lot of parallel edges.

We can extend the definition of homomorphisms to transitioned graphs by enforcing that it also preserves transitions. That means that the homomorphism $f : V(G) \mapsto V(H)$ between $(G, \mathcal{T})$ and $(H, \mathcal{S})$, which is a vertex mapping, induces a mapping between the edges $g : E(G) \mapsto E(H)$ according to the end vertex relation of f, i.e

$$f[r(e)] = r(g(e)) \quad \forall e \in E(G)$$

and transitions are preserved, i.e.

$$(f(v), \{g(e_1), g(e_2)\}) \in \mathcal{S} \quad \forall T = (v, \{e_1, e_2\}) \in \mathcal{T}.$$

By using our extended definition of homomorphisms we can define isomorphisms as bijective functions which are a homomorphism in both directions. Furthermore, we can define automorphisms as isomorphisms of a graph to itself and can consider the automorphism group of a transitioned graph $(G, \mathcal{T})$.

Given input graphs $(G, \mathcal{T})$ and $(H, \mathcal{S})$ we can use automorphisms to transform feasible solutions into other feasible solutions. More formally for any feasible solution and any pair of automorphisms $f \in \mathrm{Aut}(G, \mathcal{T})$ and $g \in \mathrm{Aut}(H, \mathcal{S})$ of which at least one is not the identity we can construct another feasible solution by replacing all vertices in G according to f and all vertices in H according to g. Since f and g preserve all edges and all transitions this is sufficient to get a new feasible solution.

Next we propose an approach how to eliminate some of those symmetries. Let S be a feasible solution with vertex mapping $\varphi : V(G) \twoheadrightarrow V(H)$. We assume that $V(G)$ and $V(H)$ are totally ordered sets. We can define for any pair of automorphisms $f \in \text{Aut}(G,T)$ and $g \in \text{Aut}(H,S)$ a sequence $\alpha^{f,g} := (x_w^{f,g})_{w \in V(H)}$ by

$$\alpha_w^{f,g} := \min f[\varphi^{-1}[g(w)]]$$

which is well-defined since φ is surjective. The sequence $\alpha^{f,g}$ contains the smallest vertex of each preimage of φ after applying the automorphisms f and g to the solution. The idea is to enforce that $\alpha := \alpha^{\text{id}_{V(G)},\text{id}_{V(H)}}$ is lexicographically minimal compared to all $\alpha^{f,g}$ for all pairs of automorphisms $f \in \text{Aut}(G,T)$ and $g \in \text{Aut}(H,S)$, i.e.

$$\alpha \leq_{\text{lex}} \alpha^{f,g} \quad \forall f \in \text{Aut}(G,T), \forall g \in \text{Aut}(H,S). \tag{43}$$

Note that there may be multiple different feasible solutions with the same sequence α and therefore this only eliminates some symmetries. Such different solutions with the same α may differ in the mapped edges or transitions, or differ in vertices in G that are not mapped by φ or are not the smallest vertices of the preimages of φ. But if H is simple this restriction eliminates all symmetries occurring only in H, i.e. if we only apply an automorphism in $\text{Aut}(H,S) \setminus \{\text{id}_{V(H)}\}$ to a feasible solution satisfying (43) the resulting solution will not satisfy (43). To formalize (43) in such a way that it can be modeled in a MIP or a SAT formulation we have to expand the definition of a lexicographical ordering. Condition (43) is equivalent to

$$\forall w \in V(H), \forall f \in \text{Aut}(G,T), \forall g \in \text{Aut}(H,S) :$$
$$\alpha_w \leq \alpha_w^{f,g} \vee \exists w' < w : \alpha_{w'} < \alpha_{w'}^{f,g}$$
$$\Leftrightarrow (\forall v < \alpha_w : f(v) \notin \varphi^{-1}[g(w)]) \vee (\exists w' < w : \forall v \leq \alpha_w : f(v) \notin \varphi^{-1}[g(w')]).$$

This constraint is still complicated and results in a lot of constraints in SAT or MIP models. To avoid bloating the models we consider only the variant for the smallest vertex $w_0 := \min(V(H))$ of H. Then the condition can be simplified using orbits.

Definition 1. *Let f be an automorphism on a transitioned graph (G,T). The set*

$$\text{orb}(v) := \{v' \in V \mid \exists f \in \text{Aut}(G,T) : f(v) = v'\}$$

is called the orbit *of $v \in V$. Orbits are the equivalence classes of the equivalence relation corresponding to $\text{Aut}(G,T)$ in which two vertices are equivalent if there exists an automorphism mapping one vertex to the other.*

Using the definition of orbits we can simplify our condition for the special case w_0

$$f(v) \notin \varphi^{-1}[g(w_0)] \; \forall v < \alpha_{w_0}, \forall f \in \text{Aut}(G, \mathcal{T}), \forall g \in \text{Aut}(H, \mathcal{S})$$
$$\Leftrightarrow v' \notin \varphi^{-1}[w'] \; \forall v < \alpha_{w_0}, \forall v' \in \text{orb}(v), \forall w' \in \text{orb}(w_0)$$
$$\Leftrightarrow \varphi(v') = w' \rightarrow \alpha_{w_0} \leq v \; \forall v \in V(G), \forall v' \in \text{orb}(v), \forall w' \in \text{orb}(w_0)$$
$$\Leftrightarrow \varphi(v') = w' \rightarrow \exists v'' \leq v : \varphi(v'') = w_0 \; \forall v' \in V(G), \forall v \in \text{orb}(v'), \forall w' \in \text{orb}(w_0)$$
$$\Leftrightarrow \varphi(v) = w \rightarrow \exists v' \leq \min \text{orb}(v) : \varphi(v') = w_0 \; \forall v \in V(G), \forall w \in \text{orb}(w_0). \quad (44)$$

Another specialization of (43) is if we only consider automorphisms on H, i.e. fix $f = id_{V(G)}$. In this case we simply have $\alpha_w^g := \alpha_w^{id_{V(G)}, g} = \min \varphi^{-1}[g(w)] = \alpha_{g(w)}$, i.e. the α values are simple permutations of each other based on g. Therefore the symmetry breaking condition holds if and only if

$$(\alpha_w)_{w \in V(H)} \leq_{\text{lex}} (\alpha_{g(w)})_{w \in V(H)} \quad \forall g \in \text{Aut}(H, \mathcal{S}). \quad (45)$$

Note that $(\alpha_w)_{w \in V(H)} \leq_{\text{lex}} (\alpha_{g(w)})_{w \in V(H)}$ if and only if for the first vertex w for which $\alpha_w \neq \alpha_{g(w)}$, $\alpha_w < \alpha_{g(w)}$ holds. Since all values in α_w are different we know that $\alpha_w = \alpha_{g(w)}$ if and only if $w = g(w)$. Therefore, if w is the first value where they are different this implies that g fixes all $w' < w$, i.e. $g(w') = w'$ for all $w' < w$.

Definition 2. Let $S \subseteq V(G)$, then the stabilizer of $\text{Aut}(H, \mathcal{S})$ with respect to S is defined by $\text{Aut}_S(H, \mathcal{S}) := \{g \in \text{Aut}(H, \mathcal{S}) \mid \forall s \in S : g(s) = s\}$, which is a subgroup of $\text{Aut}(H, \mathcal{S})$. We can again define stabilizer orbits according to the automorphisms in the stabilizer, i.e. $\text{orb}_S(v) := \{v' \in V \mid \exists f \in \text{Aut}_S(G, \mathcal{T}) : f(v) = v'\}$.

With this definition we can reformulate (45) in the following way:

$$\alpha_w < \alpha_{g(w)} \quad \forall w \in V(H), \forall g \in \text{Aut}_{\{w' \in V(H) : w' < w\}}(H, \mathcal{S}) : g(w) \neq w$$
$$\Leftrightarrow \alpha_w < \alpha_{w''} \quad \forall w \in V(H), \forall w'' \in \text{orb}_{\{w' \in V(H) : w' < w\}}(w) \setminus \{w\}.$$

The condition $\alpha_w < \alpha_{w''}$ can be expressed such that the statement is equivalent to

$$\varphi(v) = w'' \rightarrow \exists v' < v : \varphi(v') = w \qquad \begin{array}{c} \forall v \in V(G), \forall w \in V(H), \\ \forall w'' \in \text{orb}_{\{w' \in V(H) : w' < w\}}(w) \setminus \{w\}. \end{array} \quad (46)$$

To model constraints (44) and (46) we use the inequalities

$$x_v^w \leq \sum_{v' \leq \min \text{orb}(v)} x_{v'}^{w_0} \qquad \forall v \in V(G), \forall w \in \text{orb}(w_0) \quad (47)$$

$$x_v^{w''} \leq \sum_{v' < v} x_{v'}^w \qquad \begin{array}{c} \forall v \in V(G), \forall w \in V(H), \\ \forall w'' \in \text{orb}_{\{w' \in V(H) : w' < w\}}(w) \setminus \{w\} \end{array} \quad (48)$$

for the MIP model and the constraints

$$x_v^w \rightarrow \bigvee_{v' \leq \min \operatorname{orb}(v)} x_{v'}^{w_0} \qquad \forall v \in V(G), \forall w \in \operatorname{orb}(w_0) \qquad (49)$$

$$x_v^{w''} \rightarrow \bigvee_{v' < v} x_{v'}^w \qquad \begin{array}{c} \forall v \in V(G), \forall w \in V(H), \\ \forall w'' \in \operatorname{orb}_{\{w' \in V(H): w' < w\}}(w) \setminus \{w\} \end{array} \qquad (50)$$

for the SAT model.

Fig. 2. Construction example of the auxiliary graph for a part of a transitioned graph $(G, \mathcal{T})$. The newly added artificial vertices have color 2, which is drawn white and the original vertices have color 1, which is drawn black.

3.1 Finding All Automorphisms and Stabilizers

To add constraints (47)–(48) or (49)–(50) to our model we need to compute the automorphism group $\operatorname{Aut}(G, \mathcal{T})$, its orbits, the automorphism group $\operatorname{Aut}(H, \mathcal{S})$, its orbits, and the orbits $\operatorname{orb}_{\{w' \in V(H): w' < w\}}(w)$ of the stabilizers for each $w \in V(H)$.

The problem of computing a set of generators of the automorphism group of a simple graph is well studied. It is closely related to the famous graph isomorphism problem. Since no polynomial time algorithm is known for the graph isomorphism problem, which can be reduced to computing generators of the automorphism group of the graph, all proposed algorithms in literature require exponential time in general. Nevertheless, if we restrict the problem to graphs with bounded degree, like it is the case for the input graph H, which is always 4-regular, there are polynomial time algorithms, see [9]. On the other hand, there are efficient algorithms in practice, which can handle graphs with unbounded degree. See for example McKay and Piperno [10] where they solved the problem for graphs with several thousand vertices in reasonable time.

The algorithm of McKay and Piperno and also other algorithms in the literature working similarly get as an input a simple undirected graph $G = (V, E)$ with a vertex coloring $c : V \rightarrow \{1, \ldots m\}$ and compute a generator of $\operatorname{Aut}^c(G)$, which is the subgroup of $\operatorname{Aut}^c(G)$ which preserves the colors given by c, i.e.

$$\operatorname{Aut}^c(G) := \{f \in \operatorname{Aut}(G) \mid c(f(v)) = c(v) \quad \forall v \in V(G)\}.$$

Since we need to compute automorphism groups of transitioned multigraphs, we need to transform our graphs in such a way that we can apply McKay's algorithm to it. Let $(G, \mathcal{T})$ be a transitioned graph. We construct an auxiliary graph $\mathrm{Aux}(G, \mathcal{T})$ by inserting in each edge $e = v_1 v_2$ of G two vertices $w_e^{v_1}$ and $w_e^{v_2}$. This gives us immediately a simple graph. Furthermore, for each transition $t = (v, e_1, e_2) \in \mathcal{T}$ we add an edge between the vertices $w_{e_1}^v$ and $w_{e_2}^v$. We also define a coloring c on the auxiliary graph by coloring all original vertices with the color 1 and all artificially added vertices with the color 2. See Fig. 2 for an example on how to construct the auxiliary graph for a part of a given transitioned graph $(G, \mathcal{T})$.

Theorem 1.

$$\mathrm{Aut}(G, \mathcal{T}) = \{f|_{V(G)}\} \, [f \in \mathrm{Aut}^c(\mathrm{Aux}(G, \mathcal{T}))]$$

Proof. By adding the two artificial vertices with a second color between each edge we can associate with each automorphism in the auxiliary graph a vertex mapping and an edge mapping in the original graph. The edge mapping is defined by mapping an edge e_1 to an edge e_2 if the two artificial vertices on e_1 get mapped to the two artificial vertices on e_2 in the auxiliary graph. Furthermore, since there are edges between two added vertices $w_{e_1}^v$ and $w_{e_2}^v$ if and only if there is a transaction $(v, \{e_1, e_2\})$ in the original graph we also get that the mappings are transition-preserving. On the other hand, given a vertex mapping and an edge mapping as in the definition of an automorphism in a transitioned graph, we can use those to formulate an edge-preserving vertex mapping on the auxiliary graph.

Theorem 1 shows us that we can use the auxiliary graph $\mathrm{Aux}(G, \mathcal{T})$ to get the automorphism group of a transitioned graph $(G, \mathcal{T})$ by using an algorithm to compute $\mathrm{Aut}^c(\mathrm{Aux}(G, \mathcal{T}))$. What remains is how to compute the orbits and the orbits of the stabilizers which can be done with the Schreier-Sims algorithm [12]. To get the orbits of the stabilizers we may have to reorder our vertices (which in effect changes the needed stabilizers) according to the result of the Schreier-Sims algorithm. This is no problem for our model, since the order of the vertices is only relevant for the symmetry breaking and can therefore be adjusted.

4   Computational Results

To test our SAT model and compare it with the MIP model proposed in [8] we implemented both in C++ using Glucose 4.1 to solve the SAT model and Gurobi 8.1 to solve the MIP model. We also tested the impact of the symmetry breaking constraints for both models. To get the automorphism groups as described in Sect. 3.1 we used nauty 2.6 [10] and to get a strong generating set we used the implementation of the Schreier-Sims algorithm contained in the nauty program. All tests were performed on a single core of an Intel Xeon E5-2640 v4 processor with 2.40 GHz and 8 GB RAM.

We consider the instance sets S1, S2, and G1 from [8] together with a new instance set G2 of larger random graphs to also test the limits of the SAT

Table 1. Computation results for instance set S1.

$|V(C)|$	$|I|$	$|I_{\text{feas}}|$	$|I_{\text{inf}}|$	MIP			MIP$_{\text{sym}}$			SAT		SAT$_{\text{sym}}$	
				$t_{\text{feas}}[s]$	$t_{\text{inf}}[s]$	$|I_{\text{tl}}|$	$t_{\text{feas}}[s]$	$t_{\text{inf}}[s]$	$|I_{\text{tl}}|$	$t_{\text{feas}}[s]$	$t_{\text{inf}}[s]$	$t_{\text{feas}}[s]$	$t_{\text{inf}}[s]$
10	4	4	0	**< 1**	–	0	**< 1**	–	0	**< 1**	–	**< 1**	–
18	8	8	0	3	–	0	4	–	0	**< 1**	–	**< 1**	–
20	24	24	0	2	–	0	2	–	0	**< 1**	–	**< 1**	–
22	124	121	3	4	2035	0	5	727	0	**< 1**	11	**< 1**	**1**
24	620	604	16	8	3600	15	6	2955	2	**< 1**	26	**< 1**	**2**
26	5188	5124	64	12	3600	64	9	3600	58	**< 1**	78	**< 1**	**4**
28	4000	3970	30	19	3600	30	14	3600	30	**< 1**	166	**< 1**	**7**

Table 2. Computation results for instance set S2.

$|V(C)|$	$|I|$	$|I_{\text{feas}}|$	$|I_{\text{inf}}|$	MIP			MIP$_{\text{sym}}$			SAT		SAT$_{\text{sym}}$	
				$t_{\text{feas}}[s]$	$t_{\text{inf}}[s]$	$|I_{\text{tl}}|$	$t_{\text{feas}}[s]$	$t_{\text{inf}}[s]$	$|I_{\text{tl}}|$	$t_{\text{feas}}[s]$	$t_{\text{inf}}[s]$	$t_{\text{feas}}[s]$	$t_{\text{inf}}[s]$
18	98	15	83	2	194	0	1	9	0	**0.04**	0.12	**0.04**	**0.04**
20	1116	416	700	3	468	6	2	24	0	**0.05**	0.28	**0.05**	**0.06**
22	10694	4873	5821	4	1173	892	3	74	0	**0.06**	0.78	**0.06**	**0.08**

model. Set S1 consists of four random perfect matching contractions of all snarks with up to 26 vertices plus 1000 snarks with 28 vertices using the UD-K_5 as transitioned graph $(H, \mathcal{S})$. The set S2 consists of all PPM contractions of all snarks with up to 22 vertices. Furthermore, set G1 consists for each combination of $n \in \{9, \ldots, 15\}$ and $m \in \{5, \ldots, 7\}$ of ten instances, where each of those consists of a random 4-regular completely-transitioned graph G with n vertices and a random 4-regular completely-transitioned graph H with m vertices. The additional new instance set G2 is constructed the same way as G1 but with $n \in \{16, \ldots, 30\}$ and $m \in \{6, \ldots, 10\}$.

We compare the running times of four algorithms for the given instances, the original MIP model, the MIP model with the symmetry breaking constraints (47)–(48), which will be called MIP$_{\text{sym}}$, the SAT model, and the SAT model with the symmetry breaking constraints (49)–(50), which will be called SAT$_{\text{sym}}$.

Table 1 lists the computational results for instance set S1 for all four algorithms. The instances are grouped by the number of vertices $|V(C)|$ of the snark C used for the generation, one column per group. Column $|I|$ contains the numbers of instances, $|I_{\text{feas}}|$ the numbers of feasible instances, and $|I_{\text{inf}}|$ the numbers of infeasible instances. The time columns $t_{\text{feas}}[s]$ and $t_{\text{inf}}[s]$ list median running times of all feasible instances respectively the infeasible instances in seconds rounded to integer. Furthermore, for the MIP models columns I_{tl} contain the numbers of instances that could not be solved within the CPU-time limit of 3600 s. The best running times of the groups of feasible instances and infeasible instances are marked bold.

As we can see the SAT model outperforms the MIP model considerably and the symmetry breaking constraints improve the running times for the infeasible instances, especially for the SAT model but also for the MIP model.

To further compare the four models we applied a Wilcoxon signed-rank test for each pair of them using a p-value of 5%. The algorithm MIP_{sym} is significantly faster than MIP for the instance groups with $|V(C)| \geq 24$, for the infeasible but also for the feasible instances. The two SAT models are significantly faster than both MIP models for all instance groups except for $|V(C)| = 18$ and the infeasible instances of $|V(C)| = 22$ since those are too few to get a significant result. In fact the SAT models are faster on almost all instances except a few feasible instances. For the SAT model the variant without the symmetry breaking constraints is significantly faster on all feasible instance groups with $|V(C)| \geq 22$ although the difference in the values is only within hundredth of seconds. On the other hand for the infeasible instance groups with $|V(C)| \geq 24$ the model with the symmetry breaking constraints is significantly faster.

Table 2 shows the computational results for instance set S2. The columns are the same as in Table 1. The results are similar as for instance set S1, but this time MIP_{sym} can solve all instances within the time limit. Applying the Wilcoxon signed-rank test we get that MIP_{sym} is significantly faster than MIP except for the infeasible instance group with $|V(C)| = 18$. Both SAT models are significantly faster than the MIP models for all instance groups. This time SAT is not significantly faster than SAT_{sym} on the feasible instance groups, SAT_{sym} is even significantly faster than SAT for the feasible instance group with $|V(C)| = 22$. For the infeasible instances SAT_{sym} is significantly faster.

Table 3. Computation results for instance set G1.

$|V(G)|$	$|V(H)|$	$|I|$	$|I_{\text{feas}}|$	$|I_{\text{infl}}|$	MIP		MIP_{sym}		SAT	SAT_{sym}
					$t[s]$	$|I_{t1}|$	$t[s]$	$|I_{t1}|$	$t[s]$	$t[s]$
09	5	30	15	15	106	0	91	0	< 1	< 1
09	6	30	4	26	440	1	409	0	1	< 1
09	7	30	0	30	2059	11	2735	14	< 1	< 1
10	5	30	19	11	90	1	87	1	< 1	< 1
10	6	30	4	26	1939	12	1862	10	1	1
10	7	30	0	30	3600	16	3600	16	2	1
11	5	30	25	5	42	1	19	0	< 1	< 1
11	6	30	9	21	2777	14	3600	16	3	2
11	7	30	1	29	3600	22	3600	20	3	3
12	5	30	28	2	50	1	17	0	< 1	< 1
12	6	30	21	9	2204	13	2124	11	3	2
12	7	30	1	29	3600	30	3600	30	8	7
13	5	30	28	2	23	2	26	2	< 1	< 1
13	6	30	20	10	3600	17	2055	13	4	4
13	7	30	7	23	3600	30	3600	27	14	13
14	5	30	30	0	24	0	30	0	< 1	< 1
14	6	30	28	2	562	7	823	8	2	2
14	7	30	8	22	3600	29	3600	27	27	28
15	5	30	30	0	30	0	24	0	< 1	< 1
15	6	30	29	1	670	2	1475	11	3	2
15	7	30	18	12	3600	26	3600	27	27	30

Table 3 shows the computation results for the instance set $G1$. We group the instances by the number of vertices of the input graphs G and H. We do not distinguish between feasible and infeasible instance groups in this table, since the running time characteristics are similar for both types of instances. Columns $t[s]$ show the median running time for all instances of the instance group. Again both SAT models could solve all instances within one hour and outperform the MIP models. This time the differences between the models with symmetry breaking constraints and without are smaller, since the probability that a random graph has symmetries is small. Now the SAT models are on all instances faster than the MIP models. Between the MIP models there are only few instance groups where there is a significant difference in the running times in favor of both models. The situation between the two SAT models is similar although there are slightly more instance groups where SAT_{sym} is significantly faster.

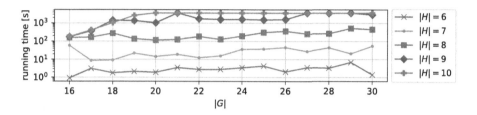

Fig. 3. Median running times of SAT_{sym} for instance set $G2$.

All instances in all three instance sets $S1$, $S2$, and $G1$ could be solved within the time limit of one hour by both SAT models. To also analyze the limits of our SAT models we also tested instance set $G2$. Figure 3 shows the median running times of the SAT_{sym} model for different sizes of $|G|$ and $|H|$. As we can see the running time heavily depends on the size of H and not so strongly on the size of G. For $|H| = 10$ and $|G| \geq 20$ we run into the time limit of one hour in most of the instances. Similarly, as for instance set $G1$ also in $G2$ the running times for SAT_{sym} and SAT are similar.

Using SAT_{sym} we also implemented the framework described at the end of Sect. 2. We use Boost's implementation of the Boyer-Myrvold planarity test to check for planar graphs. Furthermore, we use a simple SAT model for checking if a graph contains a K_5-minor and another SAT model for checking if it has a CCD. Since the bottleneck of this framework are the solving times for checking SUD-K_5-minor freeness, the running time improvements by the SAT model were crucial to check for all snarks with up to 32 vertices if they contain a planar contraction, a K_5-minor free contraction, a SUD-K_5-minor free contraction, or a CCD-containing contraction of a PPM. From the $1\,918\,812$ tested snarks we found $25\,248$ snarks that do not contain a planar contraction of a PPM, $19\,130$ snarks that do not contain a K_5-minor free contraction of a PPM, and $1\,095$ snarks that do not contain a SUD-K_5-minor free contraction of a PPM. The found snarks can be downloaded from https://www.ac.tuwien.ac.at/klocker-snark-collections/.

Up until now it was not known if there exist snarks that do not have a PPM whose contraction leads to planar/K_5-minor free/SUD-K_5-minor free graphs. With our implementation we could find many examples of snarks that have those properties. Nevertheless, all tested snarks always had a PPM whose contraction leads to a graph which has a CCD. Therefore, it remains an open question if there exists a snark that does not have a PPM whose contraction leads to a CCD-containing graph.

5   Conclusion and Future Work

In this work we proposed a SAT model for checking if a given transitioned graph $(G, \mathcal{T})$ has a Sup-$(H, \mathcal{S})$-transition minor. The model is based on the mathematical model developed in a previous work [8]. To improve the performance of the SAT model, but also of the MIP model we developed symmetry breaking constraints that are based on the automorphism groups of both input graphs restricted by the additional structure given through the transition systems. In our computational study we could verify that the SAT model outperforms the MIP model significantly and the symmetry breaking constraints could improve the running times especially for proving infeasibility. Using the SAT model in a framework we were able to find many snarks that do not have PPM whose contraction leads to SUD-K_5-minor free graphs.

In future work it may be interesting to consider a CP model for our problem to be able to use non-binary variables in the model for representing the mappings between the two input graphs. Furthermore, the framework for finding snarks that do not contain a SUD-K_5-minor free contraction of a PPM may be improved by adding symmetry breaking during the enumeration of the PPMs.

References

1. Aloul, F.A., Ramani, A., Markov, I.L., Sakallah, K.A.: Solving difficult SAT instances in the presence of symmetry. In: Proceedings of the 39th Annual Design Automation Conference, DAC 2002, pp. 731–736. ACM, New York (2002)
2. Fan, G., Zhang, C.-Q.: Circuit decompositions of Eulerian graphs. J. Comb. Theory Ser. B **78**(1), 1–23 (2000)
3. Fleischner, H.: Eulersche Linien und Kreisüberdeckungen, die vorgegebene Durchgänge in den Kanten vermeiden. J. Comb. Theory Ser. B **29**(2), 145–167 (1980)
4. Fleischner, H., Bagheri Gh., B., Zhang, C.-Q., Zhang, Z.: Cycle covers (III) - Compatible circuit decomposition and K5-transition minor. Technical report, Algorithms and Complexity Group, TU Wien (2018)
5. Jaeger, F.: A survey of the cycle double cover conjecture. In Alspach, B.R., Godsil, C.D. (eds.) Annals of Discrete Mathematics (27): Cycles in Graphs, volume 115 of North-Holland Mathematics Studies, pp. 1–12. North-Holland (1985)

6. Janota, M., Grigore, R., Manquinho, V.: On the Quest for an Acyclic Graph. arXiv:1708.01745 (2017)

7. Januschowski, T., Pfetsch, M.E.: Branch-cut-and-propagate for the maximum k-colorable subgraph problem with symmetry. In: Achterberg, T., Beck, J.C. (eds.) CPAIOR 2011. LNCS, vol. 6697, pp. 99–116. Springer, Heidelberg (2011). https://doi.org/10.1007/978-3-642-21311-3_11

8. Klocker, B., Fleischner, H., Raidl, G.: A Model for Finding Transition-Minors. Technical report, Algorithms and Complexity Group, TU Wien (2018)

9. Luks, E.M.: Isomorphism of graphs of bounded valence can be tested in polynomial time. J. Comput. Syst. Sci. **25**(1), 42–65 (1982)

10. McKay, B.D., Piperno, A.: Practical graph isomorphism, II. J. Symbolic Comput. **60**, 94–112 (2014)

11. Seymour, P.D.: Sums of circuits. Graph Theory Relat. Top. **1**, 341–355 (1979)

12. Sims, C.C.: Computational methods in the study of permutation groups. In: Leech, J. (ed.) Computational Problems in Abstract Algebra, pp. 169–183. Pergamon (1970)

13. Szekeres, G.: Polyhedral decompositions of cubic graphs. Bull. Australian Math. Soc. **8**(03), 367 (1973)

14. Tseitin, G.S.: On the complexity of derivation in propositional calculus. In: Siekmann, J.H., Wrightson, G. (eds.) Automation of Reasoning: 2: Classical Papers on Computational Logic 1967-1970. Symbolic Computation, pp. 466–483. Springer, Heidelberg (1983). https://doi.org/10.1007/978-3-642-81955-1_28

Barrier Covering in 2D Using Mobile Sensors with Circular Coverage Areas

Adil Erzin[1,2]([⊠]) [iD], Natalya Lagutkina[2] [iD], and Nika Ioramishvili[2] [iD]

[1] Sobolev Institute of Mathematics, SB RAS, Novosibirsk 630090, Russia
[2] Novosibirsk State University, Novosibirsk 630090, Russia
adilerzin@math.nsc.ru, {lagutnat,ioramishvili}@yandex.ru

Abstract. In the problem of barrier monitoring using mobile sensors with circular coverage areas, it is required to move the sensors onto some line (barrier) so that each barrier point belongs to the coverage area of at least one sensor. One of the criteria for the effectiveness of coverage is the minimum of the total length of the paths traveled by sensors. If we give up the requirement to move the sensors onto the barrier, then the problem (which is NP-hard) will not be easier. But at the same time, the value of the objective function can be reduced significantly. In this paper, we propose a new pseudo-polynomial algorithm which in the case of equal disks builds an optimal solution in the L_1 metric and a $\sqrt{2}$-approximate solution in the Euclidean metric. This algorithm is an efficient implementation of the dynamic programming method in which at the stage of preliminary calculations for each sensor it is possible to find a finite number of analytical functions equal to the minimal length of the path traveled by the sensor depending on the positions of the circle and the barrier. The conducted numerical experiment showed that if we remove the requirement to move the sensors onto the barrier, then the value of the objective function may decrease several times.

Keywords: Barrier monitoring · Mobile sensors · Covering

1 Introduction

Mobile sensor networks (MSNs) consists of mobile devices that collect information within a limited zone (sensor's *coverage area*). Covering a plane area is one of the main goals of MSNs. The problem of building a cover using mobile sensors is to move the sensors so that each point of the area is within the coverage zone of at least one sensor. Many articles are devoted both to the problems of covering plane areas [12,14,16,22], and also to cover the barrier, presented, as a rule, in the form of a straight line segment [2,6,9,15,18]. In this paper, the problem of monitoring (covering) a barrier in the form of a straight line segment

The research is partly supported by the program of fundamental scientific researches of the SB RAS (project 0314–2019–0014) and by the Russian Foundation for Basic Research (Projects 19–47–540007).

N. F. Matsatsinis et al. (Eds.): LION 13 2019, LNCS 11968, pp. 342–354, 2020.
https://doi.org/10.1007/978-3-030-38629-0_28

by mobile sensors with circular coverage areas is considered, therefore in the text we identify the sensor and the circle corresponding to its coverage area. Problems of efficient barrier coverage occur in many applications, including monitoring of extended objects (roads, pipelines, borders, etc.) using mobile devices, such as unmanned aerial vehicles (UAV) equipped with video cameras. The effectiveness is characterized by different criteria, the most popular of which are discussed in the next subsection.

1.1   Background

In the literature devoted to covering the barriers with mobile sensors, the barrier is usually a straight line segment and the following three criteria are considered: (MinSum) minimizing the total length of the sensor movement paths, (MinMax) minimizing the maximum length of the sensor movement path, and (MinNum) minimizing the number of sensors involved in barrier covering. These problems are considered in a one-dimensional formulation (1D) when initially the sensors are on the line containing the segment, and in a two-dimensional (2D) case when the sensors are initially arbitrarily distributed on a plane.

MinSum. Many mobile sensors, such as UAVs, have a limited supply of energy, which can be replenished only in certain places. Therefore, the most efficient scenario for UAV is to save energy or minimize the overall length of the paths traveled by the devices. It is this criterion that will interest us in this paper. In the 1D problem, the total movement length of all sensors involved in the covering, initially located on the line containing the barrier, is minimized [3]. It is known that the 1D problem is NP-hard in the general case, however, for the case of identical circles, a polynomial algorithm is proposed for constructing an optimal solution with $O(n^2)$–time complexity, where n is the number of sensors [8]. This result is improved in [1] where an algorithm for constructing an optimal coverage with $O(n \log n)$–time complexity is proposed. In [20] a weighted 1D problem is considered, in which each sensor has weight and it is required to minimize the total length of the paths traveled by the sensors involved in the covering multiplied by the weights of the sensors. The author proved that such a problem is NP-hard even in the case when all sensors are located on one side of the barrier being covered but have different radii. For this case, an FPTAS is proposed with time complexity $O(n^3/\varepsilon^2)$, where ε is a ratio. For the case when the sensors are initially located on both sides of the segment, an $O(n^5/\varepsilon^3)$–time complexity FPTAS has also proposed.

In the 2D formulation, the MinSum problem is naturally also NP-hard in the case of different circles. In the case of equal disks, a complexity status of the problem is unknown. The paper [7] considers a 2D problem of covering a line segment with equal disks, the initial positions of which are arbitrary. The authors proposed an $O(n^4)$–time complexity algorithm for constructing an optimal solution in the L_1 metric, which is a $\sqrt{2}$-approximate solution in the Euclidean metric. This result is improved in [10], where the complexity is reduced to $O(n^2)$. The paper [11] considers the same problem, where all circles touch each other in the cover. In this case an FPTAS was proposed, whose complexity is at most $O(n^3/\varepsilon^2)$. The problem of uniformly covering a circle with equal disks (with equal distances between the centers of neighboring disks) was considered in [4] and an $O(n^4/\varepsilon)$–time FPTAS was proposed to build an ε–approximate solution to the problem.

MinMax and MinNum. Knowing the maximum length of the UAV path allows determining how much fuel vehicle needs. The shorter the path, the less fuel is required. In this connection, problem MinMax arises. The MinMax problem is to move the sensors in such a way that the maximum path length of the sensors involved in the covering would be minimal. It is known that in the case of identical circles, the 1D MinMax problem is solved optimally with $O(n^2)$–time complexity [8]. This result improved in [5], in which the complexity of constructing a solution is reduced to $O(n \log n)$. The paper [21] discusses a weighted version of the 1D problem, in which each sensor has a weight and it is necessary to minimize the maximum path length traveled by sensor multiplied by its weight. In the case of equal circles, a $O(n^2 \log n \log(\log n))$–time algorithm is proposed. The 2D MinMax problem is considered in [9], which shows that the problem is NP-hard in the case of different circles. In the case of identical circles in [17] a $O(n^3 \log n)$–time algorithm is proposed that builds a solution to the problem. A similar problem was considered in [4], where it is necessary to cover a boundary of a simple polygon, and a $O(mn^{3.5} \log n)$–time algorithm proposed that builds an optimal solution, where m is the number of vertices of the polygon. The problem of covering the curve was considered in [13].

Another natural criterion is a minimization of the number of sensors in the cover (MinNum problem). In [19] it is proved that the MinNum problem in the case of various circles is NP-hard, but is polynomially solvable with $O(n^2)$–time complexity in the case of identical circles.

All the above results refer to the case when the sensors are moved *onto the barrier*. Such a requirement simplifies the problem, but the value of the objective function can be substantially larger than in the case when there is no need to move the sensors onto the barrier.

1.2   Our Contribution

In this paper, the problem MinSum of covering a barrier in the form of a straight line segment when it is not necessary to move the sensors onto the barrier is considered for the first time. Naturally, it can be assumed that the barrier in the Cartesian coordinate system is horizontal (located on the x-axis between the points $(0,0)$ and $(L,0)$). Then the sensors can be renumbered from left to right according to their abscissas, thus specifying their order. In the *order preserving covering* (OPC), if one sensor was to the left of another, this order will remain in the cover after the sensors are moved. For this case, an efficient dynamic programming algorithm is proposed, which builds an OPC [7,10]. It is known that in the case of equal circles whose centers move onto the barrier, there is an OPC that is optimal in the L_1 metric and it is a $\sqrt{2}$-approximate solution in the Euclidean metric. We have proved that this is also the case when the centers of circles are not required to be moved onto the barrier. This means that the dynamic programming algorithm proposed by us builds a solution no worse than a covering in which all sensors move onto the segment, and it is of interest to estimate the reduction of the objective function. To do this, we conducted a numerical experiment, which shows that if we abandon the requirement to move the sensors onto the segment, then the value of the objective function can decrease several times!

To effectively implement the dynamic programming algorithm, for each sensor (circle) k, we used preliminary calculations of the function $d_k(x, l)$ equals to the minimum length of the path for moving the circle k in the L_1 metric to cover the segment $[x, l]$. As a result, a constant number of variants of the relative position of the sensor k and the segment $[x, l]$ and a constant number of analytically defined functions $d_k(x, l)$ were found. Moreover, it was proved that the functions $d_k(x, l)$ are either linear or convex in x, which made it possible to reduce the complexity of the algorithm.

The rest of the paper is organized as follows. Section 2 presents the mathematical formulation of the problem. Section 3 is devoted to the description of the new dynamic programming algorithm and the justification of its complexity. Section 4 presents the results of a numerical experiment on comparing solutions built with the requirement of moving sensors onto a barrier and without this requirement. The final section presents the concluding considerations.

2   Problem Formulation

Let barrier is a straight line segment of length $L > 0$ on the x-axis between the points $(0, 0)$ and $(L, 0)$, and a set of circles (disks) S corresponding to the sensor's coverage zones ($|S| = n$) are given. Denote by $p_i = (x_i, y_i)$ the initial and by $\hat{p}_i = (\hat{x}_i, \hat{y}_i)$ the final positions of the sensor $i \in S$ correspondingly. Each sensor defines a circle with the center of the sensor location and radius r_i, which is sensor's coverage area. Moreover, without loss of generality, we assume that all sensors are located not lower than the x-axis, i.e. $y_i \geq 0$, $i \in S$.

Definition 1. *The function $\hat{p} : S \to R^2$ is called a cover if each point of the segment is within the coverage area of at least one sensor when the final positions of the sensors are $\hat{p}_i = (\hat{x}_i, \hat{y}_i)$, $i \in S$.*

In general, not all sensors need to be involved in the coverage. Let $B \subseteq S$ be a subset of sensors that, after moving, cover the barrier. Denote by $d(p_i, \hat{p}_i)$ the Euclidean distance between the points p_i and $\hat{p}_i$. The problem of barrier coverage by mobile sensors is to find the function $\hat{p}^*$, which is the solution to the problem

$$cost(\hat{p}^*) = \min_{\hat{p}} \sum_{i=1}^{n} d(p_i, \hat{p}_i). \tag{1}$$

Note that we do not require to move the sensors onto the barrier. As mentioned above, the problem (1) is NP-hard when using different circles. If the circles are equal, then the complexity status of the problem (1) is not known. However, for this case, polynomial algorithms for constructing a $\sqrt{2}$–approximate solution are known [7,10].

3   Algorithm

Let's renumber the circles from left to right according to the abscissas of their centers x_i, $i \in S$.

3.1   Properties of the Feasible Solutions

Definition 2. *The function $\hat{p}$ is called an* order preserving cover *(OPC) if $\hat{x}_i <$ $\hat{x}_j$ if and only if $i < j$ for all $i, j \in B$.*

It is known that in the case of putting the centers of equal disks on a segment, there is an optimal OPC in the L_1 metric [7]. It turns out this is true without the requirement to move the centers of circles onto the barrier.

Lemma 1. *If all circles are equal, then there is an optimal OPC in the L_1 metric.*

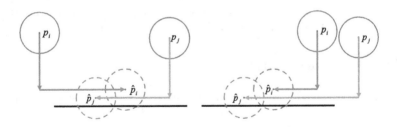

Fig. 1. Illustration to the proof of Lemma 1.

Proof. Suppose that all optimal covers do not preserve order and $\hat{p}$ is one of them. Consider a couple of such sensors $i, j \in S$ that $x_i < x_j$ and $\hat{x}_i > \hat{x}_j$. The sum of the displacement lengths of these two sensors in the L_1 metric is $w_1 = |y_i - \hat{y}_i| + |x_i - \hat{x}_i| + |y_j - \hat{y}_j| + |x_j - \hat{x}_j|$. Consider another cover $\hat{p}'$ in which the circles i and j preserve the order, namely $\hat{p}'_i = \hat{p}_j$ and $\hat{p}'_j = \hat{p}_i$. In this case, the sum of the sensor movements is $w_2 = |y_i - \hat{y}_j| + |x_i - \hat{x}_j| + |y_j - \hat{y}_i| + |x_j - \hat{x}_i|$. Remind that we assumed that $y_i \geq 0$, $i \in S$, therefore $y_i - \hat{y}_i \geq 0$ for all $i \in S$ and modules can be opened. Let's consider all possible cases of mutual initial and final arrangement of circles. Two main cases are possible (Fig. 1), and the remaining cases are reduced to these two. We have that $w_1 - w_2 = |x_i - \hat{x}_i| + |x_j - \hat{x}_j| - |x_i - \hat{x}_j| - |x_j - \hat{x}_i| \geq 0$, then $w_1 \geq w_2$. Changing the order of the circles for other pairs of circles that do not preserve the order, we obtain the optimal OPC in the L_1 metric. The lemma is proved.

Lemma 1 makes it possible to search for the optimal OPC in the L_1 metric, which can be built using the next dynamic programming algorithm.

3.2   Dynamic Programming Algorithm

The algorithm consists of one forward and one reverse recursion.

Forward Recursion. We write the recurrence relations of dynamic programming to build the OPC for different radii (which is optimal in the metric L_1 only if all disks are equal). Let $S_k(l)$ be the minimum of the sum of the lengths of the paths of the first $k = 1, \ldots, n$ circles covering the segment $[0, l]$, $0 \le l \le L$. Then for disk 1, the following expression holds

$$S_1(l) = \begin{cases} d_1(0, l), & 2r_1 \ge l; \\ +\infty, & 2r_1 < l. \end{cases}$$

The value of $d_1(0, l)$ is determined analytically in the L_1 metric depending on the initial position of the circle 1 and the segment $[0, l]$ to be covered.

$\sqrt{(x-x_k)^2 + y_k^2} \le r_k$ & $\sqrt{(l-x_k)^2 + y_k^2} \le r_k$		$l - x \le \sqrt{2}r_k$ $y_k \ge \frac{r_k}{\sqrt{2}}$				$l - x \ge \sqrt{2}r_k$	
		$x_k \le x + \frac{r_k}{\sqrt{2}}$	$x + \frac{r_k}{\sqrt{2}} \le x_k \le l - \frac{r_k}{\sqrt{2}}$	$x_k \ge l - \frac{r_k}{\sqrt{2}}$	$y_k \le \frac{r_k}{\sqrt{2}}$	$y_k \le \bar{y}_k$	$y_k \ge \bar{y}_k$
$x_k \le \bar{x}_k$	0	$F_{11}(x,l)$	$F_{12}(x,l)$ ↓	$F_{13}(x,l)$ ↙	$F_{14}(x,l)$ ⟶	$F_{15}(x,l)$ ↘
$x_k \ge \bar{x}_k$		↘	$F_{22}(x,l)$ ↓	↙	$F_{24}(x,l)$ ⟵	$F_{25}(x,l)$ ↙

Fig. 2. Functions $d_k(x, l)$ depending on the positions of the segment $[x, l]$ and the sensor k.

For an arbitrary circle $k = 1, \ldots, n$, the formula for calculating the minimum path length $d_k(x, l)$ in the L_1 metric to cover the segment $[x, l]$ is determined based on the table presented in Fig. 2. The arrows indicate the direction of movement of the circle to cover the segment (left, left-down, down, right-down or right). In this case, the direction to the right-down or left-down does not mean movement at a 45° angle. This means that the sensor moves both vertically and horizontally.

The functions $F_{ij}(x, l)$ depending on the position of the circle k and the segment $[x, l]$ are defined as follows:

- $F_{11}(x, l) = x - x_k + y_k$ (the center of the circle k moves to the point $(x + r_k/\sqrt{2}, r_k/\sqrt{2})$);
- $F_{12}(x, l) = y_k - \sqrt{r_k^2 - (l - x_k)^2}$ (the center of the circle k moves to the point $(x_k, \sqrt{r_k^2 - (l - x_k)^2})$);

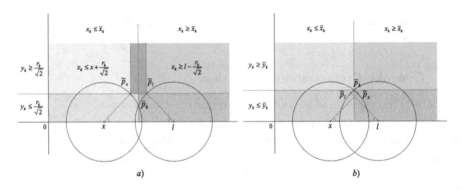

Fig. 3. (a) If $l - x \geq \sqrt{2}r_k$, then the disk k can move in six different ways depending on its initial location. (b) If $l - x \leq \sqrt{2}r_k$, then there are 4 options for moving the circle k. The line formed by the points $(x, 0)$ and $\widetilde{p}_x = (x + r_k/\sqrt{2}, r_k/\sqrt{2})$ and the line formed by the points $(l, 0)$ and $\widetilde{p}_l = (l - r_k/\sqrt{2}, r_k/\sqrt{2})$ intersect the x-axis at an angle of $45°$.

- $F_{13}(x, l) = x_k - l + y_k$ (the center of the circle k moves to the point $(l - r_k/\sqrt{2}, r_k/\sqrt{2})$);
- $F_{14}(x, l) = l - x_k - \sqrt{r_k^2 - y_k^2}$ (the center of the circle k moves to the point $(l - \sqrt{r_k^2 - y_k^2}, y_k)$);
- $F_{15}(x, l) = \overline{x}_k - x_k + y_k - \overline{y}_k$ (the center of the circle k moves to the point $(\overline{x}_k, \overline{y}_k)$);
- $F_{22}(x, l) = y_k - \sqrt{r_k^2 - (x - x_k)^2}$ (the center of the circle k moves to the point $(x_k, \sqrt{r_k^2 - (x - x_k)^2})$);
- $F_{24}(x, l) = x_k - x - \sqrt{r_k^2 - y_k^2}$ (the center of the circle k moves to the point $(x + \sqrt{r_k^2 - y_k^2}, y_k)$);
- $F_{25}(x, l) = x_k - \overline{x}_k + y_k - \overline{y}_k$ (the center of the circle k moves to the point $(\overline{x}_k, \overline{y}_k)$),

where $\overline{p}_k = (\overline{x}_k, \overline{y}_k)$ is the intersection point of two auxiliary circles of radius r_k with centers at the points $(x, 0)$ and $(l, 0)$ (Fig. 3). Obviously, $\overline{x}_k = (l + x)/2$, $\overline{y}_k = \sqrt{r_k^2 - (l - x)^2/4}$, $\overline{y}_k \geq 0$.

The region $O_k(x, l)$ where it is necessary to move the center of the circle k to cover the segment $[x, l]$ is the intersection of two auxiliary circles of radius r_k with centers at the points $(x, 0)$ and $(l, 0)$ correspondingly (see Figs. 3 and 4). To find $d_k(x, l)$, it suffices to find a point belonging to the domain $O_k(x, l)$, which is closest to the center p_k of the circle k. Obviously, such a point is on the boundary of the domain $O_k(x, l)$ and this is the point of tangency of the domain $O_k(x, l)$ with a circle centered at the point p_k and radius $d_k(x, l)$ in the L_1 metric. The circle in the L_1 metric is a square whose sides are inclined at an angle of $45°$ (see Fig. 4, in which the circles are shown with red dotted lines). Depending on the relative position of the circle k and the segment $[x, l]$, the point $\hat{p}_k$ where it is necessary to move the center of the circle k, and hence the function $d_k(x, l)$, is expressed differently. But the number of different situations is limited by a constant and they are all presented in the table in Fig. 2, and in Fig. 3. Additionally, Fig. 4 shows several cases for illustration. For example, if $l - x \leq \sqrt{2}r_k$, $y_k \geq r_k/\sqrt{2}$ and $x_k \leq x + r_k/\sqrt{2}$, then the circle k moves right-down to the point $(x + r_k/\sqrt{2}, r_k/\sqrt{2})$ (Fig. 4c).

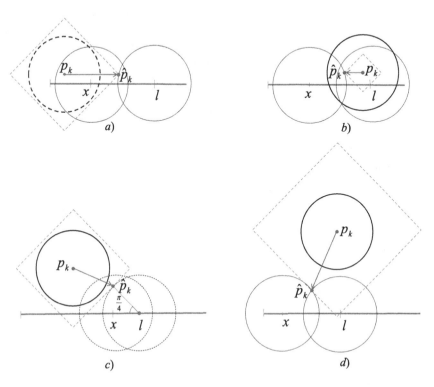

Fig. 4. The examples of moving a circle k. The red color shows a circle in the L_1 metric centered at p_k; the black circle is the initial position of the circle k; the arrows show the movement of the center of the circle to the endpoint $\hat{p}_k$. (Color figure online)

If the segment $[0, l]$ is covered by k first circles, then the following recurrence relations hold:

$$
S_k(l) = \begin{cases} \min\left\{S_{k-1}(l), \min_{x \in [0,l]}\{d_k(x, l) + S_{k-1}(x)\}\right\}, & 0 < l \leq \min\left\{2\sum_{i=1}^{k} r_i, L\right\}; \\ +\infty, & 2\sum_{i=1}^{k} r_i < l. \end{cases}
$$

The recurrence relations take place precisely in the OPC, because the circle k can cover the segment $[x, l]$, and the circles $1, \ldots, k - 1$ in the cover can be only to the left and cover the segment $[0, x]$.

At the last step n of a forward recursion, the optimal final position of the last circle n is determined. As a result of the reverse recursion, the whole barrier cover will be restored.

Reverse Recursion. If the circle n participates in the coverage, then the left boundary x of the segment $[x, L]$ covered by it, is known. Then the segment $[0, x]$ is covered by the first $n - 1$ circles. During the forward recursion, when finding $S_{n-1}(l)$, formulas were found for calculating this quantity for all values of the argument, and therefore, for $l = x$ too. Knowing the right border of the segment which is covered by the first $n-1$ circle, we also know the segment covered by the circle $n - 1$. If the circle n does not participate in the coverage, then we consider the case when circle $n - 1$ cover the segment $[0, L]$, and we know the segment which is covered by disk $n - 1$. If it participates in the coverage, then, similarly to the arguments above, we obtain the left boundary of the already covered segment $[x, l]$ and $[0, x]$ must be covered by circles $1, \ldots, n - 2$. Continuing this process, we will find the optimal coverage in $O(n)$ steps.

3.3   Complexity of the Algorithm

Lemma 2. *The functions $F_{ij}(x, l)$, $i = 1, 2$, $j = 1, 2, 3, 4, 5$, are either linear or convex in the variable x.*

Proof. The functions $F_{11}, F_{13}, F_{14}, F_{24}$ are linear. Let's show that $F_{12}, F_{15}, F_{22}, F_{25}$ are convex functions. To do this, we calculate the second derivatives. For example,

$$F''_{15} = \frac{1 + (l - x)^2}{4 \left(r_k^2 - \frac{(l-x)^2}{4} \right)^{3/2}}.$$

The numerator of this expression is greater than zero. Then it is enough to show that the denominator is not zero. If $\left(r_k^2 - (l - x)^2/4 \right)^{3/2} = 0$, then $l - x = \pm 2r_k$. It is possible that $l - x = 2r_k$, but in this case $\bar{y}_k$ in the function F_{15} vanishes, and this function becomes linear. Then $F''_{15} > 0$, except when $l - x = 2r_k$, and the convexity of the function follows. For the functions F_{12}, F_{22}, F_{25}, the convexity can be shown similarly. The lemma is proved.

Lemma 3. *For integers x, l and L the proposed algorithm builds an optimal OPC in L_1 metric with the pseudo-polynomial time complexity $O(nL \log L)$.*

Proof. The forward recursion of the algorithm is more time consuming. It consists of n steps, each of which calculates the value of $S_k(l)$ for any integer $l \in [0, L]$. It remains to show that the value of $\tilde{S}_k(l) = \min_{x \in [0, l]} \{ d_k(x, l) + S_{k-1}(x) \}$ is calculated with $O(\log L)$–time complexity. By Lemma 2, we see that the functions $F_{ij}(x, l)$, $i = 1, \ldots, 3$, $j = 0, \ldots, 9$, are either linear or convex in the variable x. Therefore, $S_k(l)$ and $\tilde{S}_k(l)$ are convex as sums of convex functions. Consequently, to find the minimum value of $\tilde{S}_k(l)$, it suffices to use, for example, the dichotomy method, whose complexity is $O(\log L)$. The lemma is proved.

4   Simulation

This section presents the results of a numerical experiment comparing the solutions constructed by the algorithm when the sensors are moved *onto the barrier* [10] and the solution constructed by the algorithm proposed in this paper. For this, both algorithms were implemented. The input data for instances were generated in a pseudo-random manner. Experiments were carried out for $L = 100, 1000, 10000$, $Y = 5, 20, 100, 300$ and $n = 10, 20, 30, 50, 75, 100, 250, 500$. It was assumed that the values $r_i \in [1, R]$, $i = 1, \ldots, n$ and L take integer values, and $y_i \in [0, Y]$, $R = Y/2$ for $L = 100$ and $R = L/10 + 4L/n \bmod L$ for $L = 1000, 10000$, $x_i \in [-L, 2L]$. For each n and Y, 75 instances were generated.

Table 1. The table cells indicate how many times the value of the objective function decreases on average unless the sensors are required to be moved onto a segment. Dashes mean that in these cases, the objective function was zero in all generated examples, and hence the gain is not upper limited.

n	$L = 100$	$L = 1000$	$L = 10000$
10	2,96	1,43	1,98
20	3,95	1,67	2,17
30	5,59	1,86	1,71
50	7,04	2,79	4,72
75	11,87	4,30	8,27
100	16,67	5,30	11,10
250	18,81	11,02	–
500	22,45	43,44	–

The results of the experiment show both the effectiveness of the proposed algorithm and the solution improvement opportunities in case of rejecting the requirement of moving the centers of the circles onto the barrier. Some results of the simulation are presented in Table 1 and Fig. 5. As can be seen from the the corresponding columns, the gain can be more than 40 times compared to the case when the sensors move onto the barrier. The greatest gain was when the ratio is $Y/R \approx 2$.

352   A. Erzin et al.

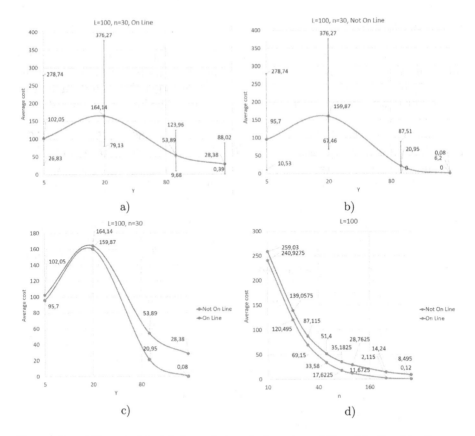

Fig. 5. $L = 100$, $n = 30$. (a) dependence of the value of the objective function on Y for the algorithm with the requirement to move the circles onto the segment; (b) dependence of the value of the objective function on Y for the proposed algorithm; (c) the values of the objective functions on Y; (d) the values of the objective functions depending on n.

5 Conclusion

Customarily, in the barrier coverage problems with mobile sensors, it is required to move the sensors *onto the barrier* so that each barrier point belongs to the coverage area (as usually, disk) of at least one sensor. We consider the effectiveness of coverage as a minimum of the total length of the paths traveled by the sensors (MinSum problem). If one gives up the requirement to move the sensors *onto the barrier*, then the problem (which is NP-hard in general) will not be easier. But at the same time, the value of the objective function can be significantly reduced. In this paper, we first propose a pseudo-polynomial algorithm for solving such a problem, which in the case of identical circles builds the optimal solution in the L_1 metric which is a $\sqrt{2}$-approximate solution in the Euclidean metric. This algorithm is an efficient implementation of the dynamic programming method in which at the stage of preliminary calculations it is possible to find the analytical forms of the functions for a finite number of variants of the relative position of the circles and the barrier. The conducted numerical experiment shows that if the requirement to move the sensors onto the barrier is removed, the value of the objective function can decrease several times.

As far as we know, this is the first algorithm that efficiently builds a barrier covering by mobile sensors with circular coverage zones without the requirement of moving the sensors onto the segment.

Acknowledgements. The authors express their sincere gratitude to Sergey Astrakov for the constructive suggestions that simplified the calculations.

References

1. Andrews, A.M., Wang, H.: Minimizing the aggregate movements for interval coverage. Algorithmica **78**(1), 47–85 (2017)
2. Bar-Noy, A., Rawitz, D., Terlecky, P.: Green barrier coverage with mobile sensors. In: Paschos, V.T., Widmayer, P. (eds.) CIAC 2015. LNCS, vol. 9079, pp. 33–46. Springer, Cham (2015). https://doi.org/10.1007/978-3-319-18173-8_2
3. Benkoczi, R., Friggstad, Z., Gaur, D., Thom, M.: Minimizing total sensor movement for barrier coverage by non-uniform sensors on a line. In: Bose, P., Gąsieniec, L.A., Römer, K., Wattenhofer, R. (eds.) ALGOSENSORS 2015. LNCS, vol. 9536, pp. 98–111. Springer, Cham (2015). https://doi.org/10.1007/978-3-319-28472-9_8
4. Bhattacharya, et al.: Optimal movement of mobile sensors for barrier coverage of a planar region. Theor. Comput. Sci. **410**(52), 5515–5528 (2009)
5. Chen, D., Gu, Y., Li, J., Wang, H.: Algorithms on minimizing the maximum sensor movement for barrier coverage of a linear domain. Discrete Comput. Geom. **50**(2), 374–408 (2013)
6. Chen, D.Z., Tan, X., Wang, H., Wu, G.: Optimal point movement for covering circular regions. Algorithmica **72**(2), 379–399 (2015)
7. Cherry, A., Gudmundsson, J., Mestre, J.: Barrier coverage with uniform radii in 2D. In: Fernández Anta, A., Jurdzinski, T., Mosteiro, M.A., Zhang, Y. (eds.) ALGO-SENSORS 2017. LNCS, vol. 10718, pp. 57–69. Springer, Cham (2017). https://doi.org/10.1007/978-3-319-72751-6_5
8. Czyzowicz, J., et al.: On minimizing the sum of sensor movements for barrier coverage of a line segment. In: Nikolaidis, I., Wu, K. (eds.) ADHOC-NOW 2010. LNCS, vol. 6288, pp. 29–42. Springer, Heidelberg (2010). https://doi.org/10.1007/978-3-642-14785-2_3
9. Dobrev, S., et al.: Complexity of barrier coverage with relocatable sensors in the plane. Theor. Comput. Sci. **579**, 64–73 (2015)
10. Erzin, A., Lagutkina, N.: Barrier coverage problem in 2D. In: Gilbert, S., Hughes, D., Krishnamachari, B. (eds.) ALGOSENSORS 2018. LNCS, vol. 11410, pp. 118–130. Springer, Cham (2019). https://doi.org/10.1007/978-3-030-14094-6_8
11. Erzin, A., Lagutkina, N.: FPTAS for barrier covering problem with equal circles in 2D. arXiv:1811.10161 (2018)
12. Han, C., Sun, L., Xiao, F., Guo, J.: An energy efficiency node scheduling model for spatial-temporal coverage optimization in 3D directional sensor networks. IEEE Access **4**, 4408–4419 (2016)

13. He, S., Gong, X., Zhang, J., Chen, J., Sun, Y.: Curve-based deployment for barrier coverage in wireless sensor networks. IEEE Trans. Wireless Commun. **13**(2), 724–735 (2014)
14. Gorain, B., Mandal, P.S.: Optimal covering with mobile sensors in an unbounded region. arXiv:1211.0157v1 (2012)
15. Kong, L., Zhu, Y., Wu, M.Y., Shu, W.: Mobile barrier coverage for dynamic objects in wireless sensor networks. In: Proceedings of the 9th IEEE International Conference on Mobile Ad Hoc and Sensor Systems, Las Vegas, Nevada (2012)
16. Li, X., Frey, H., Santoro, N., Stojmenovic, I.: Localized sensor self-deployment with coverage guarantee. ACM SIGMOBILE Mob. Comput. Commun. Rev. **12**(2), 50–52 (2008)
17. Li, J., Chen, J., Lai, T.: Energy-efficient intrusion detection with a barrier of probabilistic sensors. In: INFOCOM 2012 Proceedings IEEE, pp. 118–126. IEEE (2012)
18. Liu, B., Dousse, O., Wang, J., Saipulla, A.: Strong barrier coverage of wireless sensor networks. In: Proceedings of the ACM International Symposium on Mobile Ad Hoc Networking and Computing (MobiHoc) (2008)
19. Mehrandish, M., Narayanan, L., Opatrny, J.: Minimizing the number of sensors moved on line barriers. In: 2011 IEEE Wireless Communications and Networking Conference (WCNC), pp. 653–658. IEEE (2011)
20. Thom, M.: Investigation on two classes of covering problems. Ph.D. thesis, University of Lethbridge, Canada (2017)
21. Wang, H., Zhang, X.: Minimizing the maximum moving cost of interval coverage. In: Elbassioni, K., Makino, K. (eds.) ISAAC 2015. LNCS, vol. 9472, pp. 188–198. Springer, Heidelberg (2015). https://doi.org/10.1007/978-3-662-48971-0_17
22. Zou, Y., Chakrabarty, K.: A distributed coverage- and connectivity-centric technique for selecting active nodes in wireless sensor networks. IEEE Trans. Comput. **54**(8), 978–991 (2005)

Metaheuristics for Min-Power Bounded-Hops Symmetric Connectivity Problem

Roman Plotnikov[1(✉)] and Adil Erzin[1,2]

[1] Sobolev Institute of Mathematics, Novosibirsk, Russia
prv@math.nsc.ru
[2] Novosibirsk State University, Novosibirsk, Russia

Abstract. We consider a Min-Power Bounded-Hops Symmetric Connectivity problem that consists in the construction of communication spanning tree on a given graph, where the total energy consumption spent for the data transmission is minimized and the maximum number of edges in a path in the tree between any pair nodes is bounded by some predefined constant. We focus on the planar Euclidian case of this problem where the power cost necessary for the communication between two network elements is proportional to the squared distance between them. Since this is an NP-hard problem, we propose different heuristics based on the following metaheuristics: genetic local search, variable neighborhood search, and ant colony optimization. We perform a posteriori comparative analysis of the proposed algorithms and present the obtained results in this paper.

Keywords: Energy efficiency · Approximation algorithms ·
Symmetric connectivity · Bounded hops · Genetic local search ·
Variable neighborhood search · Ant colony optimization

1 Introduction

Due to the prevalence of wireless sensor networks (WSNs) in human life, the different optimization problems aimed to increase their efficiency remain actual. Since usually WSN consists of elements with the non-renewable power supply with restricted capacity, one of the most important issues related to the design of WSN is prolongation its lifetime by minimizing energy consumption of its elements per time unit. A significant amount of the sensor's energy is spent on the communication with other network elements. Therefore, modern sensors often can adjust their transmission ranges by changing the transmitter's power. Herewith, usually, the energy consumption of a network's element is assumed to be proportional to d^s, where $s \geq 2$ and d is the transmission range [1].

The research is supported by the Russian Science Foundation (project 18-71-00084).

The problem of determining the optimal power assignment in WSN is well-studied. The most general Range Assignment Problem, where the goal is to find a strongly connected subgraph in a given directed graph, has been considered in [2,3]. Its subproblem, Min-Power Symmetric Connectivity Problem (MPSCP), was first studied in [4]. The authors proved that the Minimum Spanning Tree (MST) is a 2-approximate solution to this problem. Also, they proposed a polynomial-time approximation scheme with a performance ratio of $1 + \ln 2 + \varepsilon \approx 1.69$ and a 15/8-approximation polynomial algorithm. In [5] a greedy heuristic, later called Incremental Power: Prim (IPP), was proposed. The IPP is similar to the Prim's algorithm for MST constructing. A Kruskal-like heuristic, later called Incremental Power: Kruskal, was studied in [6]. Both of these so-called incremental power heuristics have been proposed for the Minimum Power Asymmetric Broadcast Problem, but they are suitable for MPSCP too. It is proved in [7] that they both have an approximation ratio 2, and it was shown in the same paper that in practice these algorithms yield significantly more accurate solution than MST. Also, in a series of papers different heuristic algorithms have been proposed for MPSCP and the experimental studies have been done: local search procedures [7–9], methods based on iterative local search [10], hybrid genetic algorithm that uses a variable neighborhood descent as mutation [11], variable neighborhood search [12], and variable neighborhood decomposition search [13].

Another important property of WSN's efficiency is a message transmission delay, i.e. the minimum time necessary for transmitting a message from one sensor to another via the intermediate transit nodes. As a rule, the delay is proportional to the number of hops (edges) between two nodes in a shortest path that connects them in a network. In the general case, when the network is represented as a directed arc-weighted graph, and the goal is to find a strongly connected subgraph with minimum total power consumptions and bounded path length, the problem is called a Min-Power Bounded-Hops Strong Connectivity Problem. In [14] the approximation algorithms with guaranteed estimates have been proposed for the Euclidean case of this problem. The bi-criteria approximation algorithm for the general case (not necessarily Euclidean) has been proposed in [15]. The authors of [16] propose an improved constant factor approximation for the planar Euclidian case of the problem.

In this paper, we consider the symmetric case of Min-Power Bounded-Hops Strong Connectivity Problem, when the network is represented as an undirected edge-weighted graph. Such a problem is known as Min-Power Bounded-Hops Symmetric Connectivity Problem (MPBHSCP) [15]. We also assume that the sensors are positioned on Euclidian plane. The sensor's energy consumption for the data transmission in this case is assumed to be proportional to d^2, where d is the sensor's transmission range. This problem is still NP-hard in planar Euclidian case [17], and, therefore, the approximation heuristic algorithms that allow obtaining the near-optimal solution in a short time, are required for it.

A set of polynomial algorithms that construct the approximate solutions to MPBHSCP were proposed in [18]. In this paper, we suggest three metaheuristic approaches that aimed to improve the solutions obtained by the known constructive heuristics. Namely, we use a variable neighborhood search, a genetic local search, and an ant colony optimization that make use of different variants of local search procedure. This research was inspired by the papers where the different metaheuristics are successfully applied for the approximate solution of Bounded-Diameter Minimum Spanning Tree (BDMST) (e.g., see [19]), and for MPSCP (e.g., see [12]). We conducted an extensive numerical experiment to estimate the quality of our algorithms. For this purpose, we compared our algorithms with the best of the known constructive heuristics and, for the instances of small size, with CPLEX that was launched on the corresponding mixed integer linear programming (MILP) models. We present the results of the experiment in this paper. Note that, to the best of our knowledge, the metaheuristics of such kind were never applied to the MPBHSCP before.

The rest of the paper is organized as follows. In Sect. 2 the problem is formulated, in Sect. 3 descriptions of the proposed algorithms are given, Sect. 4 contains results and analysis of the experimental study, and Sect. 5 concludes the paper.

2   Problem Formulation

Mathematically, MPBHSCP can be formulated as follows. Given a connected edge-weighted undirected graph $G = (V, E)$ and an integer value $D \geq 2$, find such spanning tree T^* in G, which is the solution to the following problem:

$$W(T) = \sum_{i \in V} \max_{j \in V_i(T)} c_{ij} \to \min_{T}, \tag{1}$$

$$dist_T(u, v) \leq D \ \forall u, v \in V, \tag{2}$$

where $V_i(T)$ is the set of vertices adjacent to the vertex i in the tree T, $c_{ij} \geq 0$ is the weight of the edge $(i, j) \in E$, and $dist_T(u, v)$ is the number of edges in a path between the vertices $u \in V$ and $v \in V$ in T.

Obviously, in general case, MPBHSCP may even not have any feasible solution. In this paper, we consider a planar Euclidian case, where an edge weight equals the squared distance between the corresponding points and G is a complete graph. Therefore, a solution always exists.

Although any feasible solution of (1)–(2) is an undirected spanning tree with bounded diameter, we always can choose a center of this tree, i.e., a vertex (or two vertices if D is odd), such that a path from it to any other vertex in a tree contains not more than $\lfloor D/2 \rfloor$ edges. Therefore, it is convenient to consider a solution as a directed spanning tree rooted in one of its centers. Further, we assume that the centers and the root are predefined for each considered feasible spanning tree, and, therefore, we will handle with the following notations that are suitable for directed trees: v_0—a root of a tree $T = (V, E_T)$; $P_T(v)$—a parent vertex of $v \in V \setminus \{v_0\}$ in T; $L_T(v)$—a level (i.e., the number of edges in a path from v to the center) of $v \in V$ in T.

3   Heuristic Algorithms

In this section, we describe the heuristic algorithms for the approximation solution of MPBHSCP. Our methods are based on the following metaheuristics: variable neighborhood search (VNS), genetic local search (GLS), and ant colony optimization (ACO). All of our methods start with some initial *feasible* solution – a spanning tree with bounded diameter (or with a set of feasible solutions, as in GLS). We assume that at least one such solution was already constructed by some heuristic and the goal of our algorithms is to improve it preserving the feasibility in the best possible way.

Besides the obvious differences that are specific for the particular metaheuristics, our algorithms have common parts. Namely, they use the same variants of local search and random movement procedures. Therefore, we first describe these procedures.

3.1   Local Search

We suggest three types of neighborhood structure that are used in the local search procedures. The first neighborhood movement is called *LevelChange*. It consists in changing a parent node for a vertex in a such way that the level of a vertex changes and the diameter is feasible (at most D). The second procedure, *SameLevelParentChange*, consists in changing a parent for a vertex preserving its level. And the last one is *CenterChange*, which consists in changing of a center vertex of a tree. Note that none of these three variants of local movement can be replaced by a sequence of others.

The first two local movements, *LevelChange* and *SameLevelParentChange* are quite simple. In both cases one edge $e = (v, P_T(v))$ is removed from T, and then some vertex v_1, that is not a descendant of v in T, is chosen as a new parent of v. Herewith, some special conditions should be met: in the case of *LevelChange*, $L_T(v_1)$ should not be equal to $L_T(v) - 1$ and the diameter restriction should not be violated; in the case of *SameLevelParentChange*, the equality $L_T(v_1) = L_T(v) - 1$ should hold.

In the *CenterChange* movement one center $c \in V$ is chosen (it may be either a root or another center in a case of odd diameter), and then, some other non-center vertex $v \in V$ is chosen as a new center. In order to make v a new center instead of c the following steps are performed: (a) the children of c change their parent from c to v; (b) v is detached from its parent $v_p = P_T(v)$; (c) if c is a root then v becomes a root, otherwise it becomes a second center, and the root v_0 becomes a parent of v; (d) if $c \neq v_p$, then c becomes a child of v_p, otherwise, it becomes a child of v.

Our algorithms use these three variants of neighborhood movement as parts of one local search method based on variable neighborhood descent metaheuristic (VND) proposed in [20]. The idea of VND is to perform local search within more than one neighborhood structure. The pseudo-code of VND is given in Algorithm 1. In result, VND returns a local optimum for all considered neighborhood structures.

3.2   Random Movement

Besides the local search procedures, some of our metaheuristics (to be precise, GLS and VNS) involve an operator of randomized modification of a tree. For such random movement we suggest a procedure *RandomBranchReattaching*. In this procedure, some edge $(v, P_T(v))$ is chosen at random and removed from T. Then, v is connected with a non-descendant vertex u, which is chosen at random as well, if this operation keeps the feasibility of a tree. This process is repeated k times, where k is an external integer parameter provided by an upper-level metaheuristic.

3.3   Variable Neighborhood Search

Variable neighborhood search (VNS) is a metaheuristic developed by Hansen and Mladenovic [20], and it consists of two phases: randomized phase, or so-called *shaking*, when the current solution is changed in random or in half-random way, and deterministic phase, where VND is applied to the shaken solution. In our implementation, *RandomBranchReattaching* is used for the shaking phase, and the neighborhood movements *LevelChange*, *SameLevelParentChange*, and *CenterChange* are used in the local search phase. The pseudo-code is presented in Algorithm 2. The great advantage of this metaheuristic, comparing to others, is that it requires tuning of the only parameter $k_{\max}$. The algorithm starts with some feasible solution. As the first approximation for MPBHSCP, we use the best of the trees obtained by the heuristic algorithms that are proposed in [18]: MPCBTC, MPRTC, MPCBLSoC, MPCBRC, MPQBH, and MPIR.

Algorithm 1. Variable neighborhood descent

1: Select an initial solution T;
2: $k \leftarrow 0$;
3: Set the set of the local searches $(LS_l)_{l=1,2,3} \leftarrow \{LevelChange,\ SameLevelParentChange,\ CenterChange\}$;
4: *improved* $\leftarrow$ **true**;
5: **while** *improved* **do**
6:   *improved* $\leftarrow$ **false**, $l \leftarrow 1$;
7:   **while** $l \leq 3$ **do**
8:     $T' \leftarrow LS_l(T)$;
9:     **if** T' is better than T **then**
10:       $T \leftarrow T'$, $l \leftarrow 1$, *improved* $\leftarrow$ **true**;
11:     **else**
12:       $l \leftarrow l + 1$;
13:     **end if**
14:   **end while**
15: **end while**

Algorithm 2. Variable neighborhood search

1: Select an initial solution T;
2: $k \leftarrow 0$;
3: **while** the stopping criteria is not met **do**
4:     **while** $k \leq k_{\max}$ **do**
5:         Perform shaking: $T' \leftarrow RandomBranchReattaching(T, k)$;
6:         Apply the local search procedures to the shaken solution: $T'' \leftarrow VND(T')$;
7:         **if** T'' is better than T **then**
8:             $T \leftarrow T''$; $k \leftarrow 1$;
9:         **else**
10:            $k \leftarrow k + 1$;
11:        **end if**
12:    **end while**
13: **end while**

3.4   Genetic Local Search

Another approach suitable for the problem (1)–(2) is genetic local search algorithm. This metaheuristic deals with *population*—a set of feasible solutions. Before the algorithm starts, its first population should be generated. For the first population, we used all spanning trees constructed by 6 algorithms from [18]. Note that, since MPRTC is randomized, it may construct a set of different feasible solutions instead of only one solution, that yield other 5 deterministic heuristics. This allows us to generate the first population with the size that is larger than 6 and less than some predefined value. Each iteration of the algorithm consists of applying the following operators to the current population: (a) calculation of *fitness*, that expresses the quality of the solution; (b) *selection*, that chooses a subset of solutions from the population according to their fitness; (d) *crossover*, that creates a new solution (an offspring) from the selected pair of solutions; (e) *mutation*, that randomly modifies the offspring; (f) *local search*, that improves the offspring; (g) *join*, that selects the population of the next generation from the current population and the set of the offsprings. A brief description of the main steps of this algorithm is presented in Algorithm 3.

In our implementation of genetic local search, we take the value of $1/W(T)$ as a fitness of T. This corresponds to the rule that fitness has to be a positive value which is higher when the value of the objective function is closer to optimum. Within the selection procedure, a set of prospective parents of the next offspring is filled with solutions from the current population in the following way. Sequentially, two trees are taken from the current population in proportion to their fitness probability: the first tree of each pair is chosen randomly from the entire population, and the second tree is chosen from the remaining part of the population. Each pair should contain different trees, but the same tree may be included in many pairs.

For the crossover operator, a solution is represented as an array of integer values, that correspond to the vertex levels in a tree. In other words, we assume that the vertices are numbered, and for each number $i = 1, ..., n$, the value of i-th

Algorithm 3. Genetic local search

1: Generation of the first population;
2: Fitness calculation of the population;
3: **while** stop condition is not met **do**
4:    Selection;
5:    Crossover;
6:    Mutation;
7:    Local search by VND;
8:    Fitness calculation of the offspring;
9:    Join;
10:   $T \leftarrow$ the best tree among the current population;
11: **end while**

element in array is assigned to the level of i-th vertex in a tree. Given two integer arrays (let's call them the *parent* arrays), a new (*child*) array is generated in the following way. First of all, an offspring has to have a center. For that reason, one parent array is taken at random with probability of 0.5 and then the child array derives the elements assigned to 0 from this parent array. Each parent array has one or two such elements, depending on parity of D, and the child array should have the same number of elements assigned to 0. The values of all other elements in the child array derive the values at the same places from the parents, and each time the parent is chosen with probability of 0.5. Note that if the element that assigned to 0 is chosen to be derived by a child, then the corresponding element of a child is assigned to 1. This is done because the corresponding vertex cannot be a center of the offspring, since its center is already established.

The decoding of a tree from the integer array is performed in the following way. Let A be the array of integers that should be decoded to a tree T. At first, a such vertex v_0, that $A(v_0) = 0$, is assigned to the root of a tree, and, if another vertex v_1 with the same property exists, then it is assigned to the second center of a tree and v_0 is assigned to the parent of v_1. After that, for each other i-th element of an array its predecessor in T j is chosen in such way, that $A(j) < A(i)$ and the edge that connects i-th and j-th vertices, brings the minimum contribution to the value of the objective function.

The mutation procedure takes as an argument (an integer parameter) k—the maximum difference (number of different arcs in the initial tree and in the modified one). This parameter is taken randomly from the interval $[1, n/3]$, with probability proportional to its inverse value (i.e., smaller modifications are more possible). To perform a random movement for the mutation, we used the procedure *RandomBranchReattaching*. The mutation procedure is applied with probability PM (a parameter of the algorithm) to each offspring.

Additionally, our algorithm applies local search to improve the offsprings after the crossover operator. To do this, we used VND algorithm that performs local search within three neighborhood structures defined above: *LevelChange*, *SameLevelParentChange*, and *CenterChange*. This solution improvement procedure is applied with a predefined probability, as well as a randomized mutation.

At the *join* procedure a subset of solutions from the current population and the current offspring, which have the largest fitness values, are chosen to fill the population of the next generation.

Our version of GLS for MPBHSCP requires the following parameters:

- *PopSize*—the size of population;
- *OffspSize*—the size of offspring;
- *PM*—the probability of mutation;
- *PLS*—the probability of local search.

3.5   Ant Colony Optimization

As the third heuristic algorithm for the approximate solution of MPBHSCP, we propose an algorithm based on the *ant colony optimization metaheuristic* (ACO). A *path* of an ant corresponds to the solution to the problem. The path usually consists of the elements, each of which is chosen randomly with probability depending on *pheromone value* that stores information about the frequency of usage of a particular part of a path in the best-found solution. We designed our algorithm in a similar manner as it was done by Gruber et al. in [19]. To represent a feasible solution of MPBHSCP as a path we used the same vertex-level encoding that was used in the crossover operator of GLS, i.e., an array of n integers not greater than $\lfloor D/2 \rfloor$ corresponding to the vertex levels. As the pheromone values we used the matrix (τ_{il}) of size $n \times \lfloor D/2 \rfloor$, that is initially filled with equal non-negative real numbers $1/(n \cdot W(T_0))$, where T_0 is the initial solution. Our variant of the ACO algorithm consists of three phases: (a) paths construction, (b) solutions improvement, and (c) pheromone matrix updating. The main steps of the algorithm are briefly described in Algorithm 4.

Algorithm 4. Ant colony optimization

1: Generation of pheromone matrix;
2: **while** stop condition is not met **do**
3:     Construction of ant paths according to the pheromone matrix;
4:     Improvement of the solutions that are derived from the paths;
5:     Update of the pheromone matrix;
6: **end while**

In the paths construction phase, at first, the center of a corresponding tree should be defined. For that reason, we assign one or two (depending on parity of D) elements of ant path to 0. The vertices (or indices of ant path) that are assigned to the centers (with level 0) are chosen randomly with the probability $P_{i,0} = \tau_{i,0} / \sum_{j=1}^{n} \tau_{j,0}$. After that, for each vertex $i = 1, ..., n$, that has not been assigned to the center, its level is assigned randomly with the probability $P_{i,l} = \tau_{i,l} / \sum_{l'=1}^{\lfloor D/2 \rfloor} \tau_{j,l'}$, where $l = 1, ..., \lfloor D/2 \rfloor$.

After this construction, the paths are transformed into the spanning trees by the same decoding procedure that is used in the GLS. Each spanning tree is then improved by the VND procedure, that makes use of three neighborhood search types: *LevelChange*, *SameLevelParentChange*, and *CenterChange*. After that, the best solution found so far T_{best} is used for the updating of the pheromone matrix in the following way. For each $i = 1, ..., n$, and $l = 0, ..., \lfloor D/2 \rfloor$, $\tau_{i,l} = \tau_{i,l} + \rho/W(T_{best})$ if $l = L_{T_{best}}(i)$, and $\tau_{i,l} = \tau_{i,l}(1 - \rho)$, otherwise. ACO requires two parameters: *ColSize*—the number of ants in colony, and ρ—the pheromone decay coefficient.

4   Simulation

We have implemented all the described algorithms in C++ programming language and launched them on the Intel Core i5-4460 3.2 GHz processor with 8 GB RAM. In order to make our experiment results reproducible, we used as test instances the data sets that are given in Beasley's OR-Library for Euclidian Steiner Problem (http://people.brunel.ac.uk/~mastjjb/jeb/orlib). These test cases present the random uniformly distributed points in the unit square. We tested 3 variants of dimension: $n = 100$, 250, and 500. We also took different values of D for each dimension. Since all of our algorithms are partially probabilistic, we launched each algorithm 10 times on each instance, and calculated the average value of objective, the best value of objective, and the standard deviation. As a stop criteria the following condition was used: the best found solution is not changed during three iterations in a row.

We have performed the preliminary testing of each algorithm to determine such combination of its parameters that would provide in most cases the best result without consuming much time for the calculations. For VNS, $k_{\max}$ was chosen from the set $\{20, 30, 40\}$, and 30 appeared to be the best variant. For GLS, the pair $(PopSize, OffspSize) = (75, 40)$ appeared to be the best among the variants $\{(25, 15), (50, 20), (75, 40), (100, 50)\}$, and the pair $(PM, PLS) = (0.5, 0.5)$ appeared to be the best among the variants $\{(0.25, 025), (0.25, 05), (0.25, 075), (0.5, 05), (0.75, 0.25), (0.75, 075)\}$. As for ACO, we found out that $\rho = 0.2$ is the best choice among $\{0.005, 0.01, 0.05, 0.1, 0.2\}$ and $ColSize = 50$ is the best choice among $\{25, 50, 100\}$. We also tried to exclude one or more variants of local search in the VND subroutine of our algorithms, but, on average, this always deteriorated the results. Therefore, we decided to keep all the proposed variants of local search in each of our algorithms.

We have tested our algorithms on two groups of instances: small and large. The first group contains the instances when n is 20, 30, or 50, and D is 4, 5, and 6. For these instances, we used CPLEX with MILP models proposed in [18]. The results are presented in Table 1. In this table, in the first two columns, correspondingly, diameter of a tree D and problem size n are presented, the third column contains the numbers of instances in the OR-Library. In the fourth column, the objective function value on the best result of constructive heuristic, T_{CH}, is presented. Note that T_{CH} was passed to each metaheuristic algorithm as the initial solution. The fifth column contains the objective function value on the result obtained by CPLEX, and it is marked when the solution is proved to be optimal. The next three columns contain the best objective function values on solutions obtained by our algorithms (W_{best}), the following three columns contain and the average objective function values on solutions obtained by our algorithms (W_{av}), and the last four columns contain the average running time values of different algorithms measured in seconds. We bounded the calculation time for CPLEX by 1 h. The best objective function values among the launched algorithms are marked bold. It is seen that ACO and VNS often construct optimal or near-optimal solution, while GLS, on average, performs slightly worse than the best of other two metaheuristics. In four cases, when CPLEX was not able to find optimal solution within one hour, ACO constructed better solution than CPLEX. And all our algorithms outperform CPLEX on the instance 1 when $D = 4$ and $n = 50$.

Table 1. Comparison of the experiment's results obtained by different heuristics on small size instances.

D	n	nr	$W(T_{CH})$	$W(CPLEX)$	W_{best}			W_{av}			Time (in sec.)			
					ACO	GLS	VNS	ACO	GLS	VNS	CPLEX	ACO	GLS	VNS
4	20	1	1.33	1.28 (opt)	**1.28**	1.29	**1.28**	**1.28**	1.31	**1.28**	8.55	0.14	0.14	0.14
		2	1.32	1.24 (opt)	**1.24**	1.25	**1.24**	1.25	1.29	**1.24**	6.83	0.16	0.13	0.16
	30	1	1.92	1.51 (opt)	**1.51**	1.54	1.55	1.54	1.66	1.59	175.00	0.49	0.24	0.18
		2	2.11	1.58 (opt)	**1.58**	1.64	1.73	1.60	1.69	1.73	79.47	0.46	0.26	0.21
	50	1	3.10	2.85	**2.38**	2.45	2.43	2.52	2.85	2.43	3602.64	1.08	0.52	0.43
		2	3.27	2.28	**2.25**	2.44	2.84	2.46	2.59	2.84	3600.38	1.13	0.61	0.28
5	20	1	1.35	1.08 (opt)	1.09	1.09	1.29	1.09	1.14	1.29	39.94	0.19	0.13	0.11
		2	1.32	1.04 (opt)	1.24	1.06	1.24	1.24	1.14	1.24	17.25	0.17	0.13	0.10
	30	1	1.88	**1.15**	1.16	1.27	1.24	1.19	1.35	1.25	3606.57	0.48	0.25	0.17
		2	1.64	1.28 (opt)	1.30	1.33	1.42	1.31	1.40	1.48	2344.61	0.47	0.25	0.22
	50	1	2.67	**2.16**	2.20	2.20	2.44	2.27	2.34	2.44	3620.42	1.52	0.60	0.44
		2	2.57	2.35	**1.82**	1.98	1.94	2.00	2.15	1.97	3612.53	1.23	0.55	0.73
6	20	1	1.15	0.97 (opt)	**0.97**	**0.97**	1.03	0.99	1.03	1.03	12.41	0.25	0.14	0.12
		2	1.12	0.97 (opt)	**0.97**	**0.97**	1.03	0.98	1.06	1.03	15.14	0.23	0.13	0.09
	30	1	1.40	1.00 (opt)	**1.00**	1.03	**1.00**	1.01	1.10	**1.00**	222.75	0.62	0.28	0.32
		2	1.28	1.02 (opt)	**1.02**	1.08	**1.02**	1.03	1.12	**1.02**	127.82	0.64	0.29	0.19
	50	1	1.89	**1.53**	1.57	1.63	1.61	1.64	1.85	1.61	3614.82	1.70	0.53	0.37
		2	2.03	1.49	**1.41**	1.65	1.55	1.55	1.94	1.56	3611.65	1.97	0.54	0.40

The experiment results for the instances of larger size are presented in Table 2. The first three columns contain test instance properties: the tree diameter bound, D, the size of a problem, n, and the instance case number in the OR Library, nr. In the fourth column, the objective values on the best of constructive heuristics results are presented. In the other columns, the results of ACO, GLS, and VNS are presented: the objective values on the best found solutions, W_{best}, the average values of objective, W_{av}, the standard deviation of the set of objective values on the found solutions, W_{sd}, and the average running times. The best values among all algorithms are marked bold.

Table 2. Comparison of the experiment's results obtained by different heuristics on large size instances.

D	n	nr	W(T_CH)	Wbest ACO	GLS	VNS	Wav ACO	GLS	VNS	Wsd ACO	GLS	VNS	Time ACO	GLS	VNS
7	50	1	1.89	1.56	**1.47**	1.61	1.63	1.69	**1.61**	0.07	0.18	0	1.87	0.69	**0.36**
		2	1.77	**1.31**	1.42	1.34	**1.39**	1.55	1.42	0.06	0.12	0.04	2.22	**0.69**	0.75
		3	1.71	**1.24**	1.30	1.36	**1.33**	1.44	1.40	0.05	0.12	0.02	1.57	**0.67**	1.10
	100	1	2.07	**1.60**	1.74	1.66	1.96	1.80	**1.70**	0.14	0.05	0.04	5.93	3.26	**1.93**
		2	2.00	**1.48**	1.55	1.70	1.73	1.82	**1.70**	0.12	0.13	0.01	8.16	3.03	**1.88**
		3	2.35	2.35	1.99	**1.90**	2.35	2.03	**1.92**	0	0.04	0.04	**2.60**	3.15	3.04
	250	1	3.13	3.13	2.76	**2.60**	3.13	2.92	**2.62**	0	0.09	0.02	28.82	33.07	**21.16**
		2	3.30	3.30	2.80	**2.88**	3.30	2.98	**2.89**	0	0.17	0.01	28.73	32.36	**12.50**
		3	3.11	3.11	2.53	**2.43**	3.11	2.76	**2.49**	0	0.15	0.05	28.57	32.64	**30.41**
10	50	1	1.68	**1.14**	1.23	1.22	1.29	1.25	**1.24**	0.07	0.03	0.03	1.78	**0.76**	1.05
		2	1.18	**1.01**	1.13	1.06	1.16	1.18	**1.07**	0.05	0.02	0.00	0.99	**0.63**	0.73
		3	1.00	1.00	1.00	**0.88**	1.00	1.00	**0.88**	0	0	0	0.74	0.61	**0.41**
	100	1	1.73	1.18	1.34	**1.15**	1.36	1.38	**1.23**	0.07	0.07	0.05	9.73	**3.53**	8.66
		2	1.55	1.13	1.25	**1.07**	1.26	1.52	**1.09**	0.08	0.09	0.01	11.35	**3.33**	5.41
		3	1.88	**1.08**	1.29	1.21	1.33	1.49	**1.27**	0.14	0.23	0.04	12.53	**4.12**	5.45
	250	1	2.11	2.11	1.94	**1.75**	2.11	2.08	**1.84**	0	0.06	0.04	46.08	**38.13**	47.03
		2	2.30	1.99	1.84	**1.70**	2.22	2.14	**1.71**	0.11	0.17	0.03	54.10	**39.93**	47.24
		3	2.24	1.97	1.79	**1.74**	2.14	1.97	**1.80**	0.10	0.14	0.04	72.40	**38.87**	39.59
	500	1	2.57	2.57	2.13	**2.09**	2.57	2.47	**2.11**	0	0.18	0.03	277.90	255.59	**144.00**
15	50	1	1.07	1.01	0.98	**0.92**	1.06	1.06	**0.93**	0.02	0.03	0.01	0.89	**0.69**	1.25
		2	0.99	0.99	0.99	**0.88**	0.99	0.99	**0.89**	0	0	0.01	0.81	**0.67**	1.13
		3	0.89	0.84	0.89	**0.79**	0.88	0.89	**0.81**	0.01	0	0.02	0.92	**0.66**	1.51
	100	1	1.17	1.04	1.07	**0.97**	1.08	1.16	**0.97**	0.02	0.03	0	9.81	3.65	**3.51**
		2	1.14	1.02	0.97	**0.93**	1.06	0.99	**0.94**	0.04	0.05	0.01	8.44	4.35	**3.47**
		3	1.39	**0.91**	1.08	0.99	1.06	1.33	**1.01**	0.09	0.12	0.01	15.93	3.71	**2.88**
	250	1	2.05	1.33	1.34	1.26	1.46	1.46	**1.33**	0.07	0.07	0.07	140.88	**52.83**	93.75
		2	2.08	**1.28**	1.31	1.41	**1.39**	1.64	1.41	0.07	0.29	0.00	165.90	**45.20**	65.31
		3	1.71	1.28	1.23	**1.07**	1.38	1.62	**1.09**	0.06	0.16	0.02	128.30	41.15	**22.74**
	500	1	2.13	**1.64**	1.77	1.66	1.87	1.89	**1.68**	0.08	0.09	0.03	884.69	355.67	**271.71**
20	100	1	0.98	0.94	0.98	**0.83**	0.98	0.98	**0.84**	0.01	0	0.01	3.98	**3.11**	4.06
	250	1	1.17	1.17	1.17	**0.98**	1.17	1.17	**1.01**	0	0	0.02	53.33	39.26	**20.77**
	500	1	2.06	1.35	1.26	**1.11**	1.53	1.86	**1.14**	0.07	0.32	0.02	843.71	314.09	**268.58**
25	100	1	0.88	0.88	0.84	**0.80**	0.88	0.88	**0.82**	0	0.01	0.02	4.70	3.77	**2.93**
	250	1	0.99	0.99	0.97	**0.91**	0.99	0.98	**0.91**	0	0.00	0.00	58.07	45.87	**20.07**
	500	1	1.77	1.17	1.13	**1.01**	1.32	1.63	**1.03**	0.05	0.23	0.01	957.09	336.30	**220.94**

It is seen in the table that in more than in a half of all cases VNS works faster than other algorithms. But note that often, especially in small size cases, the difference in running time is not so significant. Besides, in the overwhelming majority of the cases, VNS constructs the best solution among all algorithms. Therefore, in general, the superiority of VNS is obvious. In some cases, especially when D is not too large, ACO yields the better solution than VNS and GLS. But often, even when the best solution found by ACO outperforms the best solution found by VNS, the average objective value of ACO remains inferior in quality than that of VNS: for example, see the cases $(D = 7, n = 100, \text{nr} = 1, 2)$ and $(D = 10, n = 100, \text{nr} = 3)$. Although GLS never appeared to be the best among all algorithms, it is worth to say that it often significantly improves T_{CH} and builds solutions that are very close to those constructed by ACO and VNS, in terms of the objective function. Moreover, in some cases GLS outperforms one of the other algorithms: for example, see the values of W_{best} in the cases $(D = 10, n = 250, \text{nr} = 2, 3)$, $(D = 15, n = 250, \text{nr} = 2)$, and $(D = 20, n = 500, \text{nr} = 3)$, and see the values of W_{av} in the cases $(D = 7, n = 100, \text{nr} = 3)$, $(D = 7, n = 250, \text{nr} = 2, 3)$, and $(D = 10, n = 250, \text{nr} = 2, 3)$.

In some cases, both algorithms ACO and GLS failed to improve the initial solution T_{CH}. This fact can be explained in the following way. These heuristics don't apply local search procedure directly to the initial solution, but

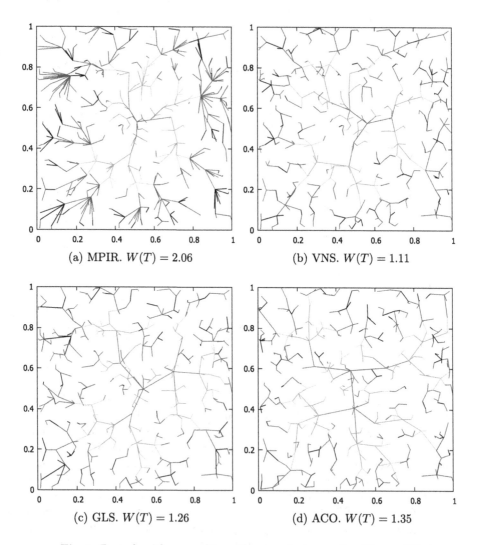

(a) MPIR. $W(T) = 2.06$ (b) VNS. $W(T) = 1.11$

(c) GLS. $W(T) = 1.26$ (d) ACO. $W(T) = 1.35$

Fig. 1. Best algorithms results on the same instance. $D = 20, n = 500$

they improve the derivative results of the initial solutions: mutated offspring or decoded ant path. Most probably, when solution space is rather large, there exists a risk that these two algorithms will explore only the solutions that are worse than initial solution, which may not lead to its improvement. It would be helpful to look into these cases deeper and to analyse the behaviour of the algorithms on them. Anyway, we believe that both algorithms ACO and GLS have a potential to be improved. It is also worth to say that in some cases the initial solution was significantly improved. For example, see the case $(D = 20, n = 500, nr = 1)$, where the initial solution was improved almost twice. In particular, this gives us the following negative result regarding the constructive heuristics from [18]: none of them provides an approximation with a guaranteed factor less than $2.06/1.11{\approx}1.856$.

As an illustration, we also present in Fig. 1 the best solutions that were obtained by different algorithms on the same instance when $D = 20, n = 500$. We chose this case, because of the big gap between the constructive heuristics and the metaheuristics results, that was discussed in the previous paragraph. For the convenience, the edges that remote from a center by an equal number of edges (or hops) are colored in the same color. This helps to easily verify that each tree is feasible, since the hops bound is never violated. Since the diameter bound is even in this case, there is the only center in all trees. In this case MPIR constructed the best solution among all constructive heuristics from [18]. The difference between the tree constructed by MPIR, and the new metaheuristic algorithms, is seen. In the solution obtained by the constructive heuristic MPIR, a part of a tree that lies far away from the center has a star-like structure, which is not desirable, because in this case a lot of rather long edges are connected with "star centers" (i.e., the vertices with high degree). Note that the trees constructed by VNS, GLS, and ACO, have no such star-like parts: the longer edges at the backbone of a tree allow to get rid of the need for the vertices with high degree.

5 Conclusion

In this paper, we considered the NP-hard Min-Power Bounded-Hops Symmetric Connectivity Problem. For its approximation solution, we proposed three different heuristic algorithms that are based on such known metaheuristics as variable neighborhood search, ant colony optimization, and genetic local search. To the best of our knowledge, this is the first application of such kind of heuristics to this problem. We implemented all the proposed algorithms and conducted the numerical experiment on different test instances that were generated on the data sets taken from the Beasley's OR-Library. The simulation showed that, in general, our methods allow to significantly improve the solution built by the best known polynomial constructive heuristics. Our algorithms appeared to be very efficient compared with optimal solution as well, as it follows from the experiments on small size instances when the problem was solved by CPLEX. In most cases, VNS based heuristic appeared to be more efficient than other methods both in terms of objective function and running time.

References

1. Rappaport, T.S.: Wireless Communications: Principles and Practices. Prentice Hall, Upper Saddle River (1996)
2. Clementi, A.E.F., Penna, P., Silvestri, R.: Hardness results for the power range assignment problem in packet radio networks. In: Hochbaum, D.S., Jansen, K., Rolim, J.D.P., Sinclair, A. (eds.) APPROX/RANDOM -1999. LNCS, vol. 1671, pp. 197–208. Springer, Heidelberg (1999). https://doi.org/10.1007/978-3-540-48413-4_21
3. Kirousis, L.M., Kranakis, E., Krizanc, D., Pelc, A.: Power consumption in packet radio networks. Theoret. Comput. Sci. **243**(1–2), 289–305 (2000)
4. Călinescu, G., Măndoiu, I.I., Zelikovsky, A.: Symmetric connectivity with minimum power consumption in radio networks. In: Baeza-Yates, R., Montanari, U., Santoro, N. (eds.) Foundations of Information Technology in the Era of Network and Mobile Computing. ITIFIP, vol. 96, pp. 119–130. Springer, Boston, MA (2002). https://doi.org/10.1007/978-0-387-35608-2_11
5. Cheng, X., Narahari, B., Simha, R., Cheng, M.X., Liu, D.: Strong minimum energy topology in wireless sensor networks: NP-completeness and heuristics. IEEE Trans. Mob. Comput. **2**(3), 248–256 (2003)
6. Chu, T., Nikolaidis, I.: Energy efficient broadcast in mobile ad hoc networks. In: Proceedings of AD-HOC Networks and Wireless (2002)
7. Park, J., Sahni, S.: Power assignment for symmetric communication in wireless networks. In: Proceedings of the 11th IEEE Symposium on Computers and Communications (ISCC), Washington, pp. 591–596. IEEE Computer Society, Los Alamitos (2006)
8. Althaus, E., Calinescu, G., Mandoiu, I.I., Prasad, S.K., Tchervenski, N., Zelikovsky, A.: Power efficient range assignment for symmetric connectivity in static ad hoc wireless networks. Wireless Netw. **12**(3), 287–299 (2006)
9. Erzin, A., Plotnikov, R., Shamardin, Y.: On some polynomially solvable cases and approximate algorithms in the optimal communication tree construction problem. J. Appl. Ind. Math. **7**, 142–152 (2013)
10. Wolf, S., Merz, P.: Iterated local search for minimum power symmetric connectivity in wireless networks. In: Cotta, C., Cowling, P. (eds.) EvoCOP 2009. LNCS, vol. 5482, pp. 192–203. Springer, Heidelberg (2009). https://doi.org/10.1007/978-3-642-01009-5_17
11. Erzin, A., Plotnikov, R.: Using VNS for the optimal synthesis of the communication tree in wireless sensor networks. Electron. Notes Discret. Math. **47**, 21–28 (2015)
12. Erzin, A., Mladenovic, N., Plotnikov, R.: Variable neighborhood search variants for Min-power symmetric connectivity problem. Comput. Oper. Res. **78**, 557–563 (2017)
13. Erzin, A., Plotnikov, R., Mladenovic, N.: VNDS for Min-power symmetric connectivity problem. Optim. Lett. **13**(8), 1897–1911 (2019). https://doi.org/10.1007/s11590-018-1324-0
14. Clementi, A.E.F., Ferreira, A., Penna, P., Perennes, S., Silvestri, R.: The minimum range assignment problem on linear radio networks. In: Paterson, M.S. (ed.) ESA 2000. LNCS, vol. 1879, pp. 143–154. Springer, Heidelberg (2000). https://doi.org/10.1007/3-540-45253-2_14
15. Calinescu, G., Kapoor, S., Sarwat, M.: Bounded-hops power assignment in ad hoc wireless networks. Discret. Appl. Math. **154**(9), 1358–1371 (2006)

16. Carmi, P., Chaitman-Yerushalmi, L., Trabelsi, O.: On the bounded-hop range assignment problem. In: Dehne, F., Sack, J.-R., Stege, U. (eds.) WADS 2015. LNCS, vol. 9214, pp. 140–151. Springer, Cham (2015). https://doi.org/10.1007/978-3-319-21840-3_12
17. Clementi, A.E.F., Penna, P., Silvestri, R.: On the power assignment problem in radio networks. Electronic Colloquium on Computational Complexity (ECCC) (054) (2000)
18. Plotnikov, R., Erzin, A.: Constructive heuristics for min-power bounded-hops symmetric connectivity problem. In: CCIS, vol. 1090, pp. 390–407 (2019). https://doi.org/10.1007/978-3-030-33394-2_31
19. Gruber, M., van Hemert, J., Raidl, G.R.: Neighbourhood searches for the bounded diameter minimum spanning tree problem embedded in a VNS, EA, and ACO. In: Proceedings of 8th Annual Conference on Genetic and Evolutionary Computation (GECCO 2006), Seattle, Washington, USA, 08–12 July, pp. 1187–1194 (2006). https://doi.org/10.1145/1143997.1144185
20. Hansen, P., Mladenovic, N.: Variable neighborhood search: principles and applications. Eur. J. Oper. Res. **130**, 449–467 (2001)

Optimization of Generalized Halton Sequences by Differential Evolution

Pavel Krömer[✉][iD], Jan Platoš[iD], and Václav Snášel[iD]

Department of Computer Science, VŠB Technical University of Ostrava,
Ostrava, Czech Republic
{pavel.kromer,jan.platos,vaclav.snasel}@vsb.cz

Abstract. Many practical applications such as multidimensional integration and quasi–Monte Carlo simulations rely on a uniform sampling of high–dimensional spaces. Halton sequences are d–dimensional quasirandom sequences that fill the d–dimensional hyperspace uniformly and can be generated with low computational costs. Generalized (scrambled) Halton sequences improve the properties of plain Halton sequences in higher dimensions by digit scrambling. Discrete nature–inspired optimization methods have been used to search for scrambling permutations of d–dimensional generalized Halton sequences that minimized the discrepancy of the generated point sets in the past. In this work, a continuous nature–inspired optimization method, the differential evolution, is used to optimize generalized Halton sequences.

1 Introduction

Halton sequences form a family of d–dimensional quasirandom sequences [5,7,9]. The Halton sequence (HS) consists of a set of d–dimensional points within the d–dimensional hypercube, $\mathcal{J}^d = [0,1)^d$, that fills the space in a uniform way [9]. The uniformity is expressed in terms of discrepancy, a measure that evaluates how the distribution of the points in the sequence deviates from an ideal uniform distribution of points in $\mathcal{J}^d$ [2,9]. The discrepancy also shows how evenly the sequence covers the space within the hypercube [3]. HSs are easy to implement and therefore popular for practical applications [2,9]. However, it is known that they suffer from high correlation between points in higher dimensions [2–4,6,18], which is in most cases an undesired property.

The optimization of generalized HSs constitutes a challenging combinatorial optimization problem that has been recently addressed by intelligent nature–inspired optimization methods, in particular, genetic algorithms [3,4,6]. The discrete nature of genetic algorithms corresponds with the discrete nature of combinatorial optimization. However, continuous optimization methods such as differential evolution and particle swarm optimization have demonstrated good ability to solve combinatorial optimization problems as well [11]. In this work, a method for the optimization of generalized HSs, based on an efficient continuous evolutionary optimization method, the differential evolution (DE), is proposed and evaluated.

© Springer Nature Switzerland AG 2020
N. F. Matsatsinis et al. (Eds.): LION 13 2019, LNCS 11968, pp. 370–382, 2020.
https://doi.org/10.1007/978-3-030-38629-0_30

The rest of this paper is organized in the following way. Halton sequences, generalized HSs, and the concept of discrepancy are formally described in Sect. 2. The optimization of generalized HSs is discussed in Sect. 3. The differential evolution is outlined in Sect. 4 and the proposed DE for the optimization of generalized HSs is detailed in Sect. 5. An experimental evaluation of the proposed approach is provided in Sect. 6. Finally, major conclusions are drawn and future work is outlined in Sect. 7.

2   Halton Sequences

The *Halton sequence* is a d–dimensional generalization of the one–dimensional *van der Corput sequence*. The van der Corput sequence in base b is for any integer, $n \geq 0$, defined by the radical inverse function [2,6,18]

$$\phi_b(n) = \frac{a_0}{b} + \frac{a_1}{b^2} + \ldots + \frac{a_k}{b^{k+1}} = \sum_{i=0}^{k} a_i b^{-i-1}, \tag{1}$$

where a_i is the i–th digit of the b–adic expansion of n in base b,

$$n = a_0 + a_1 b + \ldots + a_k b^k = \sum_{i=0}^{k} a_i b^i. \tag{2}$$

The HS is a d–dimensional sequence in $\mathcal{J}^d$, $\mathcal{X} = \langle x_1, x_2, \ldots, x_n \rangle$, with points defined by

$$x_n = (\phi_{b_1}(n), \phi_{b_2}(n), \ldots, \phi_{b_d}(n)), \tag{3}$$

and *pairwise co–prime* bases, $b_i > 1, i \in \{1, \ldots, d\}$. The simple definition and the ease of construction make HSs easier to implement than other low–discrepancy sequences [2,18]. On the other hand, they suffer from strong correlations between points in higher dimensions [6]. This problem is illustrated in Fig. 1, which shows 2D projections of the first 20 points of a 6D Halton sequence with bases corresponding to the first 6 primes (the most common choice of bases [18]). It can be seen that the 2D projections of higher dimensions (e.g., D4–D5, D4–D6, D5–D6) show highly correlated values.

There are several methods that can be used to disrupt the correlations between points in the Halton sequence [2]. *Generalized (scrambled) Halton sequence* is a variant of the sequence in which the digits of the points are in each dimension scrambled so that its good space filling properties are preserved and the correlations are minimized [18]. Generalized HS is defined using the scrambled radical inverse function

$$\phi_{\pi_b}(n) = \frac{\pi_b(a_0)}{b} + \frac{\pi_b(a_1)}{b^2} + \ldots + \frac{\pi_b(a_k)}{b^{k+1}}, \tag{4}$$

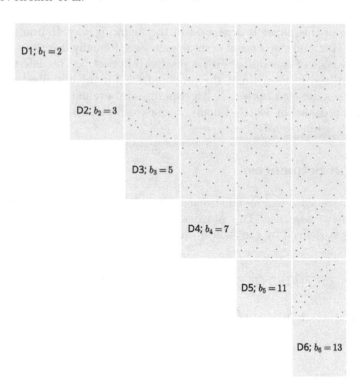

Fig. 1. 2D projections of the first 20 points of the 6D Halton sequence with the vector of bases $b = \langle 2, 3, 5, 7, 11, 13 \rangle$.

where π_b is a permutation of the digits $(0, 1, \ldots, b-1)$ that preserves the first digit fixed, i.e., $\pi_b(0) = 0$ [3,4,6,18]. The generalized Halton sequence in $\mathcal{J}^d$, $\hat{\mathcal{X}} = \langle \hat{x}_1, \hat{x}_2, \ldots, \hat{x}_n \rangle$, then consists of points defined by

$$\hat{x}_n = (\phi_{\pi_{b_1}}(n), \phi_{\pi_{b_2}}(n), \ldots, \phi_{\pi_{b_d}}(n)). \tag{5}$$

The vector of permutations, $\pi = \langle \pi_{b_1}, \pi_{b_2}, \ldots, \pi_{b_d} \rangle$, uniquely identifies the sequence and is called the *generating vector* of $\hat{\mathcal{X}}$ [6]. The use of different generating vectors leads to different sequences with different properties and the search for generating vectors that result in high–quality low–discrepancy sequences has been identified as a challenging combinatorial optimization problem [4].

2.1 Discrepancy

The *discrepancy* is a general concept that can be used to measure the uniformity of quasirandom sequences in $\mathcal{J}^d$ [2,9]. It reflects how much the distribution of a set of d–dimensional points, $\mathcal{X}$, deviates from the uniform distribution in $\mathcal{J}^d$ [4]. The discrepancy can be evaluated using a number of measures such as the star discrepancy, extreme discrepancy, L_2–star discrepancy, L_2–extreme discrepancy, and their modifications [3,4,6,18]. The L_2–star discrepancy of a sequence of n d–dimensional points, $\mathcal{X}$, is defined as [18]

$$[T_n^*(\mathcal{X})]^2 = \frac{1}{n^2} \sum_{k=1}^{n} \sum_{m=1}^{n} \prod_{i=1}^{d} (1 - \max(\boldsymbol{x}_k[i], \boldsymbol{x}_m[i]))$$

$$- \frac{2^{1-d}}{n} \sum_{k=1}^{n} \prod_{i=1}^{d} (1 - \boldsymbol{x}_k[i]^2) + 3^{-d}. \tag{6}$$

$T_n^*(\mathcal{X})$ is with computational complexity $O(dn^2)$ rather expensive to evaluate and Eq. (6) is ill–conditioned [18]. Nevertheless, it represents a natural optimization criterion that can be used to compare the properties of different generalized Halton sequences [6,18].

3 Optimization of Generalized Halton Sequences

The search for permutations that minimize the discrepancy of generalized Halton sequences has been addressed by a number of works recently [2–4,6,9,18]. Besides deterministic and constructive approaches [9,18], intelligent nature–inspired optimization methods have shown good ability to look for permutations that maximize modified L_2 [3,4] and L_2–star [6] discrepancies. Genetic algorithms (GAs) were used to search for generating vectors of generalized HSs with a low discrepancy in [3,4,6].

GAs form a family of nature–inspired search and optimization methods based on the idea of computational emulation of genetic evolution [8]. They evolve a population of encoded candidate problem solutions (chromosomes) by an iterative application of genetic operators. GAs use (most often) discrete encoding that can intuitively represent solutions of combinatorial optimization problems such as permutations and combinations. On the other hand, the basic genetic operators (e.g., one–point crossover, uniform mutation) assume that the values of genes (alleles) in the chromosomes are independent on each other. This is not the case of permutation problems and special genetic operators must be defined in order to avoid the creation of infeasible solutions [10].

De Rainville et al. [3,4] and Doerr and De Rainville [6] designed several GAs for the search for generating vectors of generalized HSs. They used a fixed–length vector of integer values to represent the generating vector. In [3,4], the chromosomes were translated to permutations using an integer–based version of the random key encoding [1]. In a more recent work, focusing on the generation of low–discrepancy *point sets*, the generating vectors were represented directly by permutation indices [6]. In all cases, the fixed 0s were removed from the genotype and special genetic operators that prevented the generation of infeasible solutions were proposed. The GAs looked either for all permutations simultaneously [3] or evolved one permutation (corresponding with one dimension of the sequence) at a time [4,6]. An example of all three encoding schemes is shown in Fig. 2.

In this work, a continuous nature–inspired optimization algorithm, the differential evolution, is used to search for the generating vector of a generalized

Simultaneous encoding of π_3, π_5, and π_7 by integer random keys [3]:

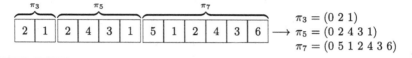

$$\pi_3 = (0\ 2\ 1)$$
$$\pi_5 = (0\ 2\ 4\ 3\ 1)$$
$$\pi_7 = (0\ 5\ 1\ 2\ 4\ 3\ 6)$$

Encoding of a single permutation at a time by integer random keys. Permutations π_3 and π_5 are fixed during the evolution of π_7 and not represented in the chromosome [4]:

$$\pi_3 = (0\ 2\ 1)$$
$$\pi_5 = (0\ 2\ 4\ 3\ 1)$$
$$\pi_7 = (0\ 5\ 1\ 2\ 4\ 3\ 6)$$

Simultaneous encoding of π_3, π_5, and π_7 by permutation indices [6]:

$$\pi_3 = (0\ 2\ 1)$$
$$\pi_5 = (0\ 2\ 4\ 3\ 1)$$
$$\pi_7 = (0\ 5\ 1\ 2\ 4\ 3\ 6)$$

Fig. 2. Example of the encoding of the generating vector $\pi = \langle \pi_3, \pi_5, \pi_7 \rangle$.

Halton sequence with a low L_2–star discrepancy. The DE is a successful evolutionary algorithm for continuous optimization driven by the idea of scaled vector differentials. This is in stringent contrast to the GAs that were designed to work primarily in discrete search spaces. As well as GAs, the DE represents a population–based stochastic search and optimization strategy. In comparison with the GAs, the differential evolution uses the real encoding of candidate solutions and different operations to evolve the population. It results in different search strategies and different directions found by the DE during the exploration of the search space associated with a particular problem. That makes it an interesting alternative to the GAs when solving various combinatorial optimization problems [11].

4   Differential Evolution

The DE is a population–based stochastic evolutionary optimization metaheuristic [14]. It is a population-based optimizer that evolves a population of real encoded vectors representing potential solutions to a given problem. The DE was introduced by Storn and Price in 1995 and it quickly became a popular alternative to the more traditional types of evolutionary algorithms. It evolves a population of candidate solutions by iterative modification of candidate solutions by the application of the differential mutation and crossover [14]. In each iteration, so–called trial vectors are created from the current population by the

differential mutation and further modified by various types of crossover operator. In the end, the trial vectors compete with existing candidate solutions for
survival in the population.

The DE starts with an initial population of M real-valued vectors. The vectors are initialized with real values either randomly or so, that they are evenly
spread over the problem space. The latter initialization leads to better results
of the optimization [14]. During the optimization, the DE generates new vectors that are scaled perturbations of existing population vectors. The algorithm
perturbs selected base vectors with the scaled difference of two (or more) other
population vectors in order to produce the trial vectors. The trial vectors compete with members of the current population with the same index called the
target vectors. If a trial vector represents a better solution than the corresponding target vector, it takes its place in the population [14].

The two most significant parameters of the DE are scaling factor and mutation probability [14]. The scaling factor $F \in [0, \infty]$ controls the rate at which the
population evolves and the crossover probability $C \in [0, 1]$ determines the ratio
of elements that are transferred to the trial vector from its opponent. The size
of the population and the choice of operators are other important parameters of
the optimization process.

The basic operations of the original DE can be summarized using the following formulae [14]: the random initialization of the ith vector with N parameters
is defined by

$$x_i[j] = \mathrm{rand}(b_j^L, b_j^U), \quad j \in \{1, \dots, N\}, \tag{7}$$

where b_j^L is the lower bound of j-th parameter, b_j^U is the upper bound of j-th
parameter, and $\mathrm{rand}(a, b)$ is a function generating a random number from the
range $[a, b]$. A simple form of the standard differential mutation is given by

$$v_i = v_{r_1} + F(v_{r_2} - v_{r_3}), \tag{8}$$

where F is the scaling factor and v_{r_1}, v_{r_2}, and v_{r_3} are three random vectors from
the population. The vector v_{r_1} is the base vector, v_{r_2} and v_{r_3} are the difference
vectors, and v_i is the trial vector. It is required that $i \neq r_1 \neq r_2 \neq r_3$. The
standard differential mutation is illustrated in Fig. 3.

The uniform (binomial) crossover that combines the target vector, x_i, with
the trial vector, v_i, is given by

$$v_i[j] = \begin{cases} v_i[j] & \text{if } (\mathrm{rand}(0, 1) < C) \text{ or } j = j_{\mathrm{rand}} \\ x_i[j], & \text{otherwise} \end{cases} \tag{9}$$

for each $j \in \{1, \dots, N\}$. The random index j_{rand} in Eq. (9) is randomly selected
as $j_{\mathrm{rand}} = \mathrm{rand}(1, N)$. The uniform crossover replaces the parameters in v_i by
the parameters from the target vector x_i with probability $1 - C$. The outline of
the DE according to [8, 14] is summarized in Algorithm 1.

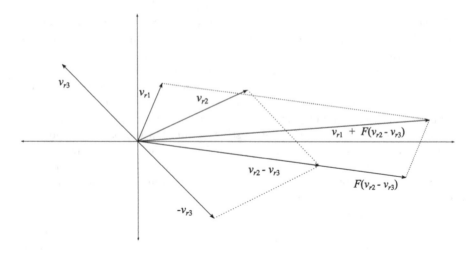

Fig. 3. Geometric interpretation of differential mutation.

1 Initialize the population P consisting of M vectors using eq. (7);
2 Evaluate the fitness function for all vectors in the population;
3 **while** *Termination criteria not satisfied* **do**
4 | Let G = current generation;
5 | **for** $i \in \{1, \ldots, M\}$ **do**
6 | Differential mutation. Create trial vector $\boldsymbol{v}_i$ according to eq. (8);
7 | Validate the range of coordinates of $\boldsymbol{v}_i$. Optionally adjust coordinates of $\boldsymbol{v}_i$ so, that it is valid solution to given problem;
8 | Perform uniform crossover. Select randomly one parameter j_{rand} in $\boldsymbol{v}_i$ and modify the trial vector using eq. (9);
9 | Evaluate the trial vector. Compute the fitness of $\boldsymbol{v}_i$;
10 | **if** *trial vector $\boldsymbol{v}_i$ represents a better solution than target vector $\boldsymbol{x}_i$* **then**
11 | | add $\boldsymbol{v}_i$ to P^{G+1}
12 | **else**
13 | | add $\boldsymbol{x}_i$ to P^{G+1}
14 | **end**
15 | **end**
16 **end**

Algorithm 1. An outline of the DE algorithm [8].

5 DE for Halton Sequence Optimization

In this work, a DE for the optimization of generalized Halton sequences is proposed. The DE looks for a generating vector, $\boldsymbol{\pi}$, that minimizes the discrepancy of the sequence and evolves the permutations associated with all dimensions of the sequence simultaneously. It is defined by the *encoding* of the generating vector and the *fitness function* expressing the discrepancy of the sequence.

The proposed DE uses an encoding that represents permutations by vectors of continuous numbers. This approach is widespread and has been used to model permutations for the DE [12,13,15], real–valued GAs (as the random key encoding) [16], and particle swarm optimization (as the smallest position value) [17]. In the random key encoding, the coordinates of the vector, x_i, are sorted from smallest to largest and their position changes are associated with permutation indices. This principle is illustrated in Eq. (10).

$$
\begin{aligned}
x_i = \begin{pmatrix} 0.2 & 0.3 & 0.1 & 0.5 & 0.4 \end{pmatrix} &\longrightarrow \begin{pmatrix} 0.2 & 0.3 & 0.1 & 0.5 & 0.4 \\ 1 & 2 & 3 & 4 & 5 \end{pmatrix} \xrightarrow[\text{value}]{\text{sort by}} \\
\begin{pmatrix} 0.1 & 0.2 & 0.3 & 0.4 & 0.5 \\ 3 & 1 & 2 & 5 & 4 \end{pmatrix} &\longrightarrow \begin{pmatrix} 3 & 1 & 2 & 5 & 4 \end{pmatrix} = \pi_i
\end{aligned} \tag{10}
$$

In the representation of the generating vector, π, the random key encoding is applied to each permutation, $\pi_{b_i} \in \pi$, independently. This approach is essentially a continuous version of the generating vector representation from [3]. An example encoding of a particular generating vector, $\pi = \langle (0\ 4\ 1\ 3\ 2\ 5), (0\ 2\ 3\ 5\ 4\ 1\ 6) \rangle$, is shown in Eq. (11).

$$
\overbrace{\boxed{0.2}\ \boxed{0.4}\ \boxed{0.3}\ \boxed{0.1}\ \boxed{0.9}}^{\pi_6}\ \overbrace{\boxed{0.5}\ \boxed{0.1}\ \boxed{0.2}\ \boxed{0.4}\ \boxed{0.3}\ \boxed{0.6}}^{\pi_7} \rightarrow \begin{array}{l} \pi_6 = (0\ 4\ 1\ 3\ 2\ 5) \\ \pi_7 = (0\ 2\ 3\ 5\ 4\ 1\ 6) \end{array} \tag{11}
$$

The fitness function, used in the proposed DE to evaluate the discrepancy of the generated sequence, is the L_2–star discrepancy defined by Eq. (6).

The encoding and the fitness function can be used with an arbitrary variant of the DE (in fact with an arbitrary optimization algorithm for d–dimensional continuous spaces). In this work, the traditional /DE/rand/1 differential evolution [8] was used to assess the ability of the DE to optimize generalized HSs.

6   Experimental Evaluation

A series of computational experiments was conducted in order to assess the ability of the proposed DE to optimize generalized HSs in terms of the L_2–star discrepancy. The DE was used to optimize the discrepancy of three different generalized HSs, in particular

- the 2–dimensional generalized HS, $\hat{\mathcal{X}}_1$, with the vector of bases, $b_1 = \langle 6, 7 \rangle$,
- the 3–dimensional generalized HS, $\hat{\mathcal{X}}_2$, with the vector of bases, $b_2 = \langle 3, 5, 7 \rangle$,
- the 5–dimensional generalized HS, $\hat{\mathcal{X}}_3$, with the vector of bases, $b_3 = \langle 4, 5, 7, 9, 11 \rangle$, and
- the 6–dimensional generalized HS, $\hat{\mathcal{X}}_4$, with the vector of bases, $b_4 = \langle 2, 3, 5, 7, 11, 13 \rangle$.

From the optimization point of view, they constitute 11, 12, 31, and 35–dimensional optimization problems, respectively. The sequences were selected so that the behaviour of the algorithm in both low and high–dimensional variants of this optimization problem could be observed. In all experiments, the DE was executed with the following parameters, selected on the basis of best practices, previous experience, and extensive trial–and–error runs: the DE was /DE/rand/1 with population size $M = 100$, scaling factor $F = 0.9$, crossover probability $C = 0.9$, and the maximum number of generations $G_{max} = 1,000$. The algorithm used the random key–based encoding of the generating vectors, illustrated in Eq. (11), and the L_2–star discrepancy of the first 1,000 points of the sequence as the fitness function. Due to the stochastic nature of the DE, all optimization runs were repeated 50 times independently.

The results of the optimization are summarized in Table 1. For each test case, it shows the discrepancy of the best, average, and worst sequence found by the DE. The table also shows the standard deviation, σ, of the discrepancies and the discrepancy of the corresponding plain HSs (i.e., sequences without digit scrambling). It can be immediately seen that the DE was able to find generating vectors that improved the discrepancy of the sequences. The discrepancies of the best found $\hat{\mathcal{X}}_1$, $\hat{\mathcal{X}}_2$, $\hat{\mathcal{X}}_3$, and $\hat{\mathcal{X}}_4$ were 2.62 times, 1.38 times, 1.35 times, and 1.09 times lower than the discrepancies of the corresponding plain HSs. The table also shows that the evolution of the generating vectors for $\hat{\mathcal{X}}_1$ always ended with the same result with the discrepancy 0.0009891. The fact that the DE converged in this case in all independent runs to the same solution is quite remarkable for a stochastic optimization method and suggests that it facilitates a robust search strategy. A further analysis of the search space (i.e. the evaluation of all possible 86,400 generating vectors of $\hat{\mathcal{X}}_1$) confirmed that the generating vector found by the DE, $\pi_{\hat{\mathcal{X}}_1}^{DE} = \langle (0\ 5\ 2\ 4\ 1\ 3), (0\ 3\ 5\ 1\ 2\ 4\ 6) \rangle$, is the global optimum. That clearly documents that the proposed DE was always able to solve this 11–dimensional version of the problem to optimality within 1,000 DE generations.

Table 1. Optimization results.

Test sequence	L_2–star discrepancy				
	Optimized generalized HS				No scrambling
	Best	Mean	σ	Worst	
$\hat{\mathcal{X}}_1$	0.0009891	0.0009891	0	0.0009891	0.00258861
$\hat{\mathcal{X}}_2$	0.0015005	0.0015010	0.12e-5	0.0015038	0.00207106
$\hat{\mathcal{X}}_3$	0.0015915	0.0016275	1.67e-5	0.0016621	0.00215402
$\hat{\mathcal{X}}_4$	0.0014899	0.0015041	0.85e-5	0.0015241	0.00162466

The optimization of $\hat{\mathcal{X}}_2$, $\hat{\mathcal{X}}_3$, and $\hat{\mathcal{X}}_4$ resulted after 1,000 DE generations in different generating vectors that led to sequences with different discrepancies. The optimization of all four test sequences is illustrated in Fig. 4. The figure shows the discrepancy of the best (minimum), average, and worst (maximum) best solution discovered by the independent optimization runs in each DE generation. It also displays the 95% confidence interval around the average best solution found by the proposed algorithm. The figure clearly illustrates that the proposed DE in all cases systematically minimized the discrepancy of the test sequences. It also shows that the DE needed only around 100 generations to find the globally best generating vector for the test sequence $\hat{\mathcal{X}}_1$ and that it did not improve significantly after the 100th generation in the case of the test sequence $\hat{\mathcal{X}}_2$. Finally, the figure suggests that the evolution of the generating vectors for the test sequences $\hat{\mathcal{X}}_3$ and $\hat{\mathcal{X}}_4$ might still continue and further improve the discrepancy of the sequence. Nevertheless, the 2D projections of the first 20 points of $\hat{\mathcal{X}}_2$ configured by the best generating vector found by the DE after 1,000 gen-

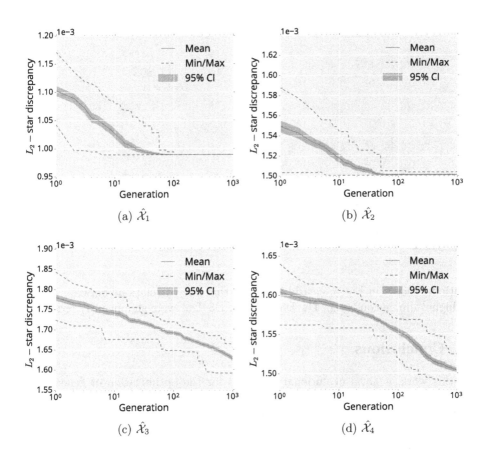

(a) $\hat{\mathcal{X}}_1$ (b) $\hat{\mathcal{X}}_2$

(c) $\hat{\mathcal{X}}_3$ (d) $\hat{\mathcal{X}}_4$

Fig. 4. The evolution of generating vectors for generalized Halton sequences. The graphs show the average, minimum, and maximum discrepancy of the best sequence found by all independent runs in each DE generation. They also show the 95% confidence interval around the mean best generating vector found in each DE generation.

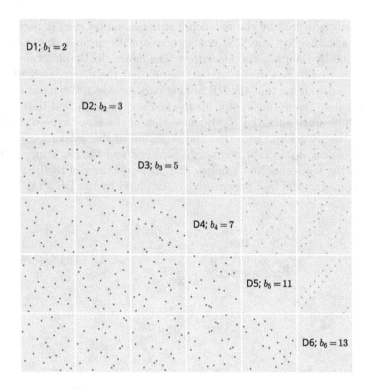

Fig. 5. 2D projections of the first 20 points of the 6D generalized Halton sequence, $\hat{\mathcal{X}}_4$, configured by the generating vector found by the DE (below main diagonal in blue) compared with the 2D projections of the first 20 points of the same unscrambled sequence (above main diagonal in red). (Color figure online)

erations, shown in Fig. 5, clearly illustrate that the correlations between points in higher dimensions (e.g., D4–D5, D4–D6, D5–D6) were significantly reduced.

7   Conclusions

In this work, a novel evolutionary method for the optimization of generalized Halton sequences was designed. The proposed approach takes advantage of the differential evolution, a successful and widely used nature–inspired search and optimization algorithm. The goal of the optimization was the minimization of the L_2–star discrepancy of multidimensional quasirandom sequences. The optimization problem was formulated as a search for a series of scrambling permutations that minimize the discrepancy of the multidimensional sequence. It was addressed by the differential evolution algorithm that used a variant of the random key encoding to represent the permutations. The algorithm was designed to optimize all permutations associated with all dimensions of the sequence simultaneously.

The proposed approach was evaluated in a series of computer experiments. The algorithm showed the ability to improve the discrepancy of 2, 3, 5, and 6–dimensional generalized Halton sequences. A search space analysis revealed that the proposed algorithm has in the case of the 2–dimensional sequence found in all independent optimization runs the best possible combination of the scrambling permutations, i.e. the global optimum. The proposed algorithm was able to improve the discrepancy of the point sets generated by higher–dimensional generalized Halton sequences as well. However, the global optimum was not found in the case of the 3, 5, and 6–dimensional Halton sequences. This can be attributed to a number of reasons including too short optimization time (i.e., not enough processed DE generations), imperfect fitness function (there are known problems associated with the standard L_2–star discrepancy measure [4]), and premature convergence of the standard /DE/rand/1 algorithm (e.g., loss of diversity in the population). Nevertheless, the initial success of the differential evolution has demonstrated a good ability of continuous optimization methods to address the combinatorial optimization problem associated with the optimization of generalized Halton sequences.

Future research in this area will follow several directions. Alternative continuous optimization algorithms including other DE variants, and, e.g., the PSO algorithm will be applied to Halton sequence optimization. Different discrepancy measures (e.g., the modified discrepancy [4]) will be evaluated as optimization criteria. Second, the ability of the differential evolution to optimize other types of low–discrepancy sequences such as the Sobol, Faure, and Niederreiter sequence will be investigated. Last but not least, different encodings of scrambling permutations will be evaluated in the context of sequence optimization.

Acknowledgement. This work was supported from ERDF in project "A Research Platform focused on Industry 4.0 and Robotics in Ostrava", reg. no. CZ.02.1.01/ 0.0/0.0/17_049/ 0008425, by the Technology Agency of the Czech Republic in the frame of the project no. TN01000024 "National Competence Center – Cybernetics and Artificial Intelligence", and by the projects SP2019/135 and SP2019/141 of the Student Grant System, VSB – Technical University of Ostrava.

References

1. Bean, J.C.: Genetic algorithms and random keys for sequencing and optimization. ORSA J. Comput. **6**(2), 154–160 (1994). https://doi.org/10.1287/ijoc.6.2.154
2. Chi, H., Mascagni, M., Warnock, T.: On the optimal Halton sequence. Math. Comput. Simul. **70**(1), 9–21 (2005). https://doi.org/10.1016/j.matcom.2005.03.004
3. De Rainville, F.M., Gagné, C., Teytaud, O., Laurendeau, D.: Optimizing low-discrepancy sequences with an evolutionary algorithm. In: Proceedings of the 11th Annual Conference on Genetic and Evolutionary Computation, GECCO 2009, pp. 1491–1498. ACM, New York (2009). https://doi.org/10.1145/1569901.1570101
4. De Rainville, F.M., Gagné, C., Teytaud, O., Laurendeau, D.: Evolutionary optimization of low-discrepancy sequences. ACM Trans. Model. Comput. Simul. **22**(2), 9:1–9:25 (2012). https://doi.org/10.1145/2133390.2133393

5. Dick, J., Pillichshammer, F.: Digital Nets and Sequences: Discrepancy Theory and Quasi-Monte Carlo Integration. Cambridge University Press, Cambridge (2010)

6. Doerr, C., De Rainville, F.M.: Constructing low star discrepancy point sets with genetic algorithms. In: Proceedings of the 15th Annual Conference on Genetic and Evolutionary Computation, GECCO 2013, pp. 789–796. ACM, New York (2013). https://doi.org/10.1145/2463372.2463469

7. Drmota, M., Tichy, R.F.: Sequences, Discrepancies and Applications. LNM, vol. 1651. Springer, Heidelberg (1997). https://doi.org/10.1007/BFb0093404

8. Engelbrecht, A.: Computational Intelligence: An Introduction, 2nd edn. Wiley, New York (2007)

9. Faure, H., Lemieux, C.: Generalized Halton sequences in 2008: a comparative study. ACM Trans. Model. Comput. Simul. **19**(4), 15:1–15:31 (2009). https://doi.org/10.1145/1596519.1596520

10. Krömer, P., Platoš, J., Nowaková, J., Snášel, V.: Optimal column subset selection for image classification by genetic algorithms. Ann. Oper. Res. **265**(2), 205–222 (2018). https://doi.org/10.1007/s10479-016-2331-0

11. Krömer, P., Platos, J., Snásel, V.: Traditional and self-adaptive differential evolution for the p-median problem. In: 2nd IEEE International Conference on Cybernetics, CYBCONF 2015, Gdynia, Poland, 24–26 June 2015, pp. 299–304 (2015). https://doi.org/10.1109/CYBConf.2015.7175950

12. Li, X., Yin, M.: An opposition-based differential evolution algorithm for permutation flow shop scheduling based on diversity measure. Adv. Eng. Softw. **55**, 10–31 (2013). https://doi.org/10.1016/j.advengsoft.2012.09.003

13. Ponsich, A., Tapia, M.G.C., Coello, C.A.C.: Solving permutation problems with differential evolution: an application to the jobshop scheduling problem. In: 2009 Ninth International Conference on Intelligent Systems Design and Applications, pp. 25–30, November 2009. https://doi.org/10.1109/ISDA.2009.49

14. Price, K.V., Storn, R.M., Lampinen, J.A.: Differential Evolution: A Practical Approach to Global Optimization. NCS. Springer, Heidelberg (2005). https://doi.org/10.1007/3-540-31306-0

15. Qian, B., Wang, L., Hu, R., Wang, W.L., Huang, D.X., Wang, X.: A hybrid differential evolution method for permutation flow-shop scheduling. Int. J. Adv. Manuf. Technol. **38**(7), 757–777 (2008). https://doi.org/10.1007/s00170-007-1115-8

16. Snyder, L.V., Daskin, M.S.: A random-key genetic algorithm for the generalized traveling salesman problem. Eur. J. Oper. Res. **174**(1), 38–53 (2006)

17. Tasgetiren, M.F., Liang, Y.C., Sevkli, M., Gencyilmaz, G.: A particle swarm optimization algorithm for makespan and total flowtime minimization in the permutation flowshop sequencing problem. Eur. J. Oper. Res. **177**(3), 1930–1947 (2007). https://doi.org/10.1016/j.ejor.2005.12.024

18. Vandewoestyne, B., Cools, R.: Good permutations for deterministic scrambled Halton sequences in terms of l2-discrepancy. J. Comput. Appl. Math. **189**(1), 341–361 (2006). https://doi.org/10.1016/j.cam.2005.05.022

Bayesian Optimization Approaches for Massively Multi-modal Problems

Ibai Roman[1]([⊠])(iD), Alexander Mendiburu[1](iD), Roberto Santana[1](iD),
and Jose A. Lozano[1,2](iD)

[1] University of the Basque Country (UPV/EHU), San Sebastian, Spain
{ibai.roman,alexander.mendiburu,roberto.santana,ja.lozano}@ehu.eus
[2] Basque Center for Applied Mathematics (BCAM), Bilbao, Spain

Abstract. The optimization of massively multi-modal functions is a challenging task, particularly for problems where the search space can lead the optimization process to local optima. While evolutionary algorithms have been extensively investigated for these optimization problems, Bayesian Optimization algorithms have not been explored to the same extent. In this paper, we study the behavior of Bayesian Optimization as part of a hybrid approach for solving several massively multi-modal functions. We use well-known benchmarks and metrics to evaluate how different variants of Bayesian Optimization deal with multi-modality.

Keywords: Bayesian Optimization · Multi-modal Optimization · Gaussian processes

1 Introduction

Massively multi-modal functions are characterized by having many optimal solutions, where some of them are local and some others are global. There may be numerous or no local optima, and only one global optimum or many global optima. Global optimization of this kind of functions is a difficult task since the optimization algorithm may get trapped in local optima, and due to their complexity, a high number of evaluations is expected. In addition, apart from finding the global optimum, a good covering of all the best solutions is required in many real-world problems [24].

This kind of functions have been extensively studied in Evolutionary Algorithms (EAs) [15]. Traditional methods to obtain a good optima covering in EAs include niching strategies [4,15,19]. In standard Genetic Algorithms (GAs), where the crossover operator produces an exploitation oriented behavior, niching methods try to reduce this effect by allowing further exploration.

On the other hand, model-based optimization algorithms take advantage of identifying the landscape of the problem to improve the search. They use all the information gathered from sampling the objective function by means of a Surrogate Model (SM). This type of algorithms have been successfully applied

© Springer Nature Switzerland AG 2020
N. F. Matsatsinis et al. (Eds.): LION 13 2019, LNCS 11968, pp. 383–397, 2020.
https://doi.org/10.1007/978-3-030-38629-0_31

to low-budget optimization tasks [8,26]. One of the most successful model-based optimization methods is Bayesian Optimization (BO) [1,11]. BO is a sequential optimization algorithm, where the sampling strategy is based on a probability distribution over all the possible objective functions. This probability distribution acts as a SM, and it is updated every time a solution is evaluated using the actual objective function.

Some of the work done in BO is closely related to the niching strategies proposed for EAs. There are some variants of BO algorithms that use sampling strategies that propose a batch of solutions to perform several evaluations at the same time [3]. Among them, the *Bayesian Optimization via Local Penalization* algorithm proposed in [5] avoids the concentration of samples, similar to sharing [4] and clearing [12] niching strategies investigated in EA literature. On the other hand, the clustering BO approaches [6,23], where the data is divided into clusters to reduce the complexity of the models, could benefit the discovery of new optima, resembling the clustering [25] techniques proposed for EA. Beyond the mentioned EC and BO approaches, methods that divide the search space have also been proposed in other fields [7,9].

In this paper, we face massively multi-modal problems bridging the work done in EA and BO literatures. We present preliminary results in this direction, proposing three algorithms that take BO beyond low-budget optimization: (1) Sequential BO with modeling of local optima. (2) Adaptive BO with clustering. (3) BO with batch sampling strategies. Furthermore, we evaluate their behavior using the benchmark presented in [10], and compare our results with those achieved by state-of-the-art algorithms in Evolutionary Computation (EC).

Our research aims to address a number of general questions that are significant for massively multi-modal problems. Among these questions are: Is it possible to get a good covering of multiple optima using a low-budget of function evaluations? Are model-based approaches, specifically BO, an efficient way to obtain a good coverage? Can we take advantage of the work done in EC to design more efficient BO techniques to improve the coverage?

The remainder of the paper is structured as follows: In Sect. 2, a brief introduction to BO and batch methods is presented. A reduced number of works significantly related to our approach are revised in Sect. 3. In Sect. 4, the different BO variants proposed in this paper are introduced. Section 5 presents the experimental benchmark and the numerical results of the experiments. Finally, Sect. 6 presents the conclusions of the paper.

2   Brief Introduction to Bayesian Optimization

BO [11] is a state-of-the-art global optimization technique suitable for low-budget optimization problems. In BO, the samples of the objective function are sequentially taken based on a sampling strategy. This next sample will be selected by optimizing an acquisition function $(u(\cdot))$ that provides a measure of the utility of each solution. A probability distribution over all the possible objective functions is used to determine this acquisition function, acting as a SM. As the analytic form of the objective function is generally unknown, BO treats the objective function like a random function, and places a prior belief about the space of possible objective functions. Every time the objective function is evaluated, this prior belief is updated with the likelihood of having those observations, generating a posterior distribution over functions.

Algorithm 1 illustrates the whole process of the BO algorithm [1].

Algorithm 1. BO algorithm

1: **procedure** BAYESIAN-OPTIMIZATION
2:   $y_1 = f(\mathbf{x}_1)$ ▷ Sample and evaluate a random point
3:   $D_{1:1} = \{(\mathbf{x}_1, y_1)\}$ ▷ Initialize the dataset
4:   $SM \leftarrow D_{1:1}$ ▷ Initialize the SM
5:   $t = 2$
6:   **repeat**
7:     $\mathbf{x}_t = \arg\max_{\mathbf{x} \in \mathbb{R}^d} u(\mathbf{x}|SM)$ ▷ Select the next point to evaluate
8:     $y_t = f(\mathbf{x}_t)$ ▷ Evaluate the objective function
9:     $D_{1:t} = D_{1:t-1} \cup \{(\mathbf{x}_t, y_t)\}$ ▷ Update the dataset
10:    $SM \leftarrow D_{1:t}$ ▷ Update the SM
11:    $t = t + 1$
12: **until** the stopping criterion is met
13: **end procedure**

The optimization process begins by sampling a point randomly, and evaluating it using the objective function. Typically, the SM is initialized at this point. Then, until the stopping criterion is met, the next point to evaluate is selected and evaluated, augmenting the dataset and updating the SM. At each step, except the first one, the point that maximizes the acquisition function is selected to be sampled.

2.1   Gaussian Processes

SMs are a key element in Algorithm 1, as the acquisition function will rely on them to select the next point. One of the most popular choices in BO is to use a GP as SM. A GP is a stochastic process, defined by a collection of random variables, any finite number of which has a multivariate Gaussian distribution [17]. What makes GPs interesting for BO is that the posterior distribution of a GP given some observations of the objective function is also a GP.

GPs can be completely defined by a mean function ($m(\mathbf{x})$) and a covariance function, which depends on a kernel ($k(\mathbf{x}, \mathbf{x}')$). Given that, the GP can be expressed as follows:

$$f(x) \sim GP(m(\mathbf{x}), k(\mathbf{x}, \mathbf{x}'))$$ (1)

Usually, a non-informative mean function is used, such as $m(\mathbf{x}) = 0$. However, in more sophisticated approaches, this function will depend on the data.

The kernel establishes the covariance between the objective function values of two different points. Although many kernels have been proposed in the literature, we use the Matern32 kernel, due to the performance shown in previous studies [18]. This kernel is expressed as follows:

$$k_{M32}(\mathbf{x}, \mathbf{x}') = \theta_0^2 \left(1 + \sqrt{3}\sqrt{r_{x,x'}^2}\right) exp\left(-\sqrt{3}\sqrt{r_{x,x'}^2}\right) + \theta_n$$

$$\text{where } r^2(\mathbf{x}, \mathbf{x}') = \sum_{d=1}^{n_d} \left(\frac{x_d - x_d'}{\theta_d}\right)^2 \tag{2}$$

where θ_d, θ_0 and θ_n are the length-scale, amplitude and noise parameters respectively. While the length-scale parameter expresses the relevance of each dimension d, the amplitude parameter scales all dimensions. The noise parameter allows the GP to adapt the random function when the observations of the objective function are noisy.

2.2 Acquisition Function

Once we have a way to approximate the value of the objective function through the SM, the acquisition function is placed to select the next point. It assigns a measure of utility for each point in the search space given the SM. This measure of utility balances the exploration versus exploitation trade-off.

In this paper we use GP-UCB [22] as acquisition function. This upper-confidence based algorithm is a combined strategy that balances the reduction of the uncertainty over the objective function with maximizing the expected reward:

$$GP\text{-}UCB(\mathbf{x}) = \mu(\mathbf{x}) + \sqrt{\beta_t}\ \sigma(\mathbf{x}) \tag{3}$$

where β_t is a constant specified depending on the context.

In the previous equation $\mu(\mathbf{x})$ and $\sigma(\mathbf{x})$ represent the expected value for the objective function at point $\mathbf{x}$ and its variance. At step $t + 1$, these values are given by the following equations:

$$\mu(\mathbf{x}) = m(\mathbf{x}) + k(\mathbf{X}_{1:t}, \mathbf{x})k(\mathbf{X}_{1:t}, \mathbf{X}_{1:t})^{-1}f(\mathbf{X}_{1:t})$$

$$\sigma^2(\mathbf{x}) = k(\mathbf{x}, \mathbf{x}) - k(\mathbf{X}_{1:t}, \mathbf{x})k(\mathbf{X}_{1:t}, \mathbf{X}_{1:t})^{-1}k(\mathbf{x}, \mathbf{X}_{1:t}) \tag{4}$$

where $k(\mathbf{X}_{1:t}, \mathbf{x})$ and $k(\mathbf{x}, \mathbf{X}_{1:t})$ are vectors with the value of the kernel function between $\mathbf{x}$ and all the previously evaluated solutions. $k(\mathbf{X}_{1:t}, \mathbf{X}_{1:t})$ is the matrix of the kernel values between all the pairs of the previously evaluated solutions.

3   Related Work

Our research is closely related to previous work in a number of areas. In EAs, research on multi-modal problems have been addressed mainly using niching methods [19,21]. However, most of these methods do not explicitly construct a model to guide the search. Among model based methods, some of the BO strategies can resemble the work done in EAs. Although they were not specifically designed to solve multi-modal problems, these BO methods can benefit from using a model.

In batch BO sampling strategies, instead of proposing one solution at each iteration, B solutions are simultaneously evaluated. Among these approaches, the *Batch Bayesian Optimization via Local Penalization* approach [5] tries to avoid the surrounding area of the already selected points applying a penalization to the utility of these selected solutions. This property can prevent the algorithm from getting stuck on local optima.

The clustering BO approaches [6,23] try to the reduce the complexity of the model by dividing the data in several clusters and using several models to create the acquisition function. Our intuition is that BO could take advantage of dividing the data into niches to encourage exploration.

Sharing [4] is one of the first attempts in niching methods. It consists in modifying the fitness value of each individual. When several individuals are occupying a niche, their fitness value is shared. Two individuals belong to the same niche when they are closer than a niche radius. The computational cost and the difficulty to set a good niche radius are the main downsides of this method. Besides, in clearing techniques [12], the neighboring solutions of the k best ones are "cleared", i.e., their fitness value is set to zero. Although it is closely related to sharing, it reduces the computational cost by only measuring the distances between the best solutions and the rest. On the contrary, it requires one extra parameter to adjust.

Other niching techniques use a clustering algorithm to divide the population into niches [25]. The fitness of each individual depends on the number of solutions in the niche and the distance to their centroids. This approach also reduces the distance computations required in sharing and encourages the individuals to leave the niche and explore new regions.

In research conducted on multi-modal functions in EAs, some of the state of art results are reported for Niching Evolutionary Algorithm 2 (NEA2) [14] and Niching migratory multi-swarm optimizer (NMMSO) [2]. NEA2 is a combination of clustering and local optimization techniques. Covariance Matrix Adaptation Evolution Strategy (CMA-ES) is used to find the local optima while Nearest-better Clustering allows a good covering of the optima. On the other hand, NMMSO uses concurrent swarms. Each swarm exploits its local mode and the algorithm splits or mergers swarms depending on characteristics of the optimization problem. While NEA2 incorporates a covariance matrix model as part of CMA-ES, these two algorithms do not learn a SM in order to predict the fitness value of a solution as BO does.

4 Dealing with Multi-modal Problems with BO Optimization

In the multi-modal optimization problems we address in this work, the number of evaluations that might be needed to cover the different optima is much higher than in low-budget optimization problems. Being BO an optimization technique well suited for low-budget optimization problems, devising efficient ways to learn a GP model with a large dataset is not straightforward. The reason is that the covariance matrix grows according to the number of evaluations. The inverse or the Cholesky decomposition of this matrix is required to compute the posterior distribution, and this computational cost may be prohibitive for massively multi-modal problems if the usual approach were used to learn the model.

In BO literature, some variants that attempt to reduce the computational cost of computing a model when the number of points is high have been proposed. These variants use various mechanisms, such as, "forgetting" older or less informative points, to handle a large number of points. Works that propose sparse GP approaches include the Subset of Data (SoD) approximations [16] and the Subset of Regressors (SoR) [20]. These methods have their own drawbacks, e.g., SoR is a very sophisticated and complex approach, that make them not directly applicable to the type of multi-modal problems we address.

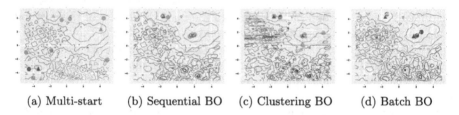

(a) Multi-start   (b) Sequential BO   (c) Clustering BO   (d) Batch BO

Fig. 1. BO approaches. The objective function is represented with a contour plot. Each figure describes the behavior of the proposed algorithms in a certain step. The yellow and red surface and contour plot refer to the acquisition function. The current step is represented with a triangle and a circle, where the circle represents the solution selected by the model, and the triangle represents the result of the local search starting from the selected solution. Previous steps are faded. Note that in batch_ls several solutions are selected and they are not faded. Finally, in the last figure, the centroids of the clusters are represented with big purple starts. (Color figure online)

Our proposal consists of combining BO with local search techniques. This way we can take full profit of the budget of available evaluations and guarantee that those solutions modeled by the covariance matrix are of better quality. In spite of including the local search, still a large dataset is needed to model the landscape of the objective function. Thus, we propose the following three algorithms that are able to deal with these datasets: The first algorithm is inspired in the SoD approach. The second algorithm avoids the unmanageable growth of the covariance matrix by splitting the current set of solutions into different clusters. Finally, the third method we propose is similar to the first but it is conceived for situations where the parallel evaluation of several points is desired. Their behavior is illustrated in Fig. 1.

4.1 Sequential BO with Modeling of Local Optima

In this approach, the "forgetting" SoD approach is combined with a local search algorithm (see Algorithm 2). Two major changes have been applied compared to Algorithm 1. First of all, a local search algorithm is used to exploit each solution proposed by the BO. Then, the SM is updated with the point suggested by the BO and the best solution achieved by the local search, provided that the local search is able to improve the solution proposed by the BO. The second major change is that the size of the dataset has been limited. When the dataset outsizes the maximum number of points (N), the oldest point is discarded, i.e. the most recently proposed points are those used for modeling. This approach might be the simplest solution when dealing with sparse GP models, but also is the computationally cheapest.

Algorithm 2. Sequential BO with modeling of local optima

1: **procedure** BO-LS(N)
2:     $y_1 = f(\mathbf{x}_1)$
3:     $D_{1:1} = \{(\mathbf{x}_1, y_1)\}$
4:     $SM \leftarrow D_{1:1}$
5:     $t = 2$
6:     **repeat**
7:         $\mathbf{x}_{bo} = \arg\max_{\mathbf{x} \in \mathbb{R}^d} u(\mathbf{x}|SM)$
8:         $y_{bo} = f(\mathbf{x}_{bo})$
9:         $(\mathbf{x}_{ls}, y_{ls}) = ls(f, \mathbf{x}_{bo})$ ▷ Run the local search
10:         $D_{1:t+1} = D_{1:t-1} \cup \{(\mathbf{x}_{bo}, y_{bo})\} \cup \{(\mathbf{x}_{ls}, y_{ls})\}$
11:         $SM \leftarrow D_{t-N-1:t+1}$ ▷ "Forget" old data
12:         $t = t + 2$
13:     **until** the stopping criterion is met
14: **end procedure**

4.2 Adaptive BO with Clustering

Our second approach is based on the previous one, but it aims to take more advantage of the knowledge acquired during the optimization process. Inspired in clustering niching techniques and the work done in [6] and [23], this approach divides the dataset into several subsets using clustering techniques. This way, to obtain the conditional distribution of the GP given some solution, only the closest solutions will be taken into account, reducing the size of the covariance matrix. It can be seen as a single SM that is composed by several GPs, each one with an exclusive subset of the dataset.

This algorithm differs from Algorithm 2 in the way the SM is updated and evaluated. Initially, only one subset of the dataset is used, and the posterior of the GP is calculated in the original manner. Every time a data point is sampled, it is added to the closest cluster. As can be seen in Algorithm 3, this updating process consists on updating the centroid and the GPs. When the maximum size of a cluster is reached, the dataset is split into two subsets using a clustering technique, and their centroids are calculated. Two new GPs are generated and updated, one for each subset, and the old cluster is removed. Note that this sparse GP technique can be incrementally applied in the BO process, as only one GP is updated at each step.

Algorithm 3. Update *ClusterSM*

1: **procedure** UPDATE-CLUSTERS($\mathbf{x}_t, y_t, \{\mathbf{c}\}, \{D\}, \{SM\}$)
2:     $\mathbf{c}_{closest}, D_{closest}, SM_{closest} = closest(\mathbf{x}_1, \{\mathbf{c}\})$
3:     $D_{closest} = D_{closest} \cup \{(\mathbf{x}_t, y_t)\}$
4:     **if** $size(D_{closest}) < N$ **then**
5:         $\mathbf{c}_{closest} = \text{centroid}(D_{closest})$
6:         $SM_{closest} \leftarrow D_{closest}$
7:     **else**
8:         $\{D_k\}_{k=1}^K = split\_dataset(D_{closest}, K)$
9:         **for** $k = 1$ to K **do**
10:             $\mathbf{c}_k = \text{centroid}(D_k)$
11:             $SM_k \leftarrow D_k$
12:         **end for**
13:         $\{\mathbf{c}\} = \{\mathbf{c}\} + \{\mathbf{c}_k\}_{k=1}^K - \{\mathbf{c}_{closest}\}$
14:         $\{D\} = \{D\} + \{D_k\}_{k=1}^K - \{D_{closest}\}$
15:         $\{SM\} = \{SM\} + \{SM_k\}_{k=1}^K - \{SM_{closest}\}$
16:     **end if**
17: **end procedure**

To evaluate the posterior distribution of this system given a certain solution, the distance to each centroid must be calculated. The *SM* associated to the closest data-subset is used to predict the outcome of the fitness.

4.3   BO with Batch Sampling Strategies

In order to expand the scope of the analysis of BO techniques for multi-modal problems, we also propose a modified version of Batch BO via Local Penalization [5]. This technique iteratively selects a batch of solutions by optimizing an acquisition function that penalizes the previous selections in the batch. Similar to the work that has been done in EAs to avoid the exploitation oriented nature of the selection operator, this penalization step might be suitable for multi-modal optimization problems, when parallel evaluations can be implemented.

The algorithm works as follows: First, a solution is selected according to a traditional acquisition function (we use UCB in this paper as suggested by the results of the original work). Then, according to Algorithm 4, a batch of solutions is obtained iteratively, selecting a solution and penalizing the acquisition function according to this selection.

Algorithm 4. Batch Selection via Local Penalization

1: **procedure** SELECT-BATCH-LP(t, K, SM)
2:     $\tilde{u}_{t,0}(\mathbf{x}) = u(\mathbf{x}|SM)$
3:     $\hat{M} = min_i\{y_i\}$
4:     $\hat{L} = max_{\mathbf{x} \in \mathbb{R}^d} ||\mu_\Delta(\mathbf{x})||$ ▷ Approximate L.
5:     **for** $k = 0$ **to** K **do**
6:         $\mathbf{x}_{t,k} = \arg\max_{\mathbf{x} \in \mathbb{R}^d} \tilde{u}_{t,k}(\mathbf{x})$
7:         $\tilde{u}_{t,k}(\mathbf{x}) = \tilde{u}_{t,0}(\mathbf{x}) \prod_{j=1}^{k} \varphi(\mathbf{x}, \mathbf{x}_{t,j}, \hat{L}, \hat{M})$
8:     **end for**
9: **end procedure**

To adapt the acquisition function, a penalization ball is applied for each solution already selected for the batch, as can be seen in Eq. 5.

$$\varphi(\mathbf{x}, \mathbf{x}_j, \hat{L}, \hat{M}) = \frac{1}{2} erfc(-z) \ where \ z = \frac{1}{\sqrt{2\sigma^2(\mathbf{x}_j)}}(\hat{L}||\mathbf{x}_j - \mathbf{x}|| + \hat{M} - \mu(\mathbf{x}_j)) \quad (5)$$

where $erfc$ is the complementary Gauss error function and L is the Lipschitz constant of the objective function (assuming that it is Lipschitz continuous). As suggested in [5], it can be approximated through the GP.

Finally, as in the first approach, it uses the SoD approach. In addition, we add the local search step to guarantee exploitation, where all the solutions in the batch are used as a starting point for the local search.

5 Experiments

The goal of this experimentation is to validate the BO approaches introduced in the previous section in the context of multi-modal optimization problems. We expect that BO will be able to model multi-modal landscapes and that will help to achieve a good covering of the optima. Moreover, the low-budget orientation of those algorithms may also help to achieve these goal with a limited number of function evaluations. The proposed clustering technique is supposed to have a better ability to model the landscape than the sequential one, and the batch approach may provide more diversity in the solutions due to the local penalization procedure. Although our main goal is investigate the benefits and limitations of the BO approaches, we will compare them to some of the best performing EAs since this is an area where highly multi-modal problems have been extensively investigated.

5.1 A Framework for Solving Low-Budget Multi-modal Optimization Problems

To test our proposals, we will use the benchmark proposed in *CEC Niching Methods for Multi-modal Optimization* competition [10], where 20 multi-modal optimization problems were introduced, along with 3 performance metrics. Table 1 describes the benchmark.

Table 1. Set of multi-modal functions as originally introduced in [10].

Id	Dim	# GO	Name	Characteristics
F_1	1	2	Five-Uneven-Peak Trap	Simple, deceptive
F_2	1	5	Equal Maxima	Simple
F_3	1	1	Uneven Maxima	Simple
F_4	2	4	Himmelblau	Simple, non-scalable, non-symmetric
F_5	2	2	Six-Hump Camel Back	Simple, non-scalable, non-symmetric
F_6	2, 3	18,81	Shubert	Scalable, # optima increase with d, unevenly dist. grouped optima
F_7	2, 3	36,216	Vincent	Scalable, # optima increase with d, unevenly dist. optima
F_8	2	12	Modified Rastrigin	Scalable, # optima independent from d, symmetric
F_9	2	6	Composition Function 1	Scalable, separable, non-symmetric
F_{10}	2	8	Composition Function 2	Scalable, separable, non-symmetric
F_{11}	2, 3, 5, 10	6	Composition Function 3	Scalable, non-separable, non-symmetr
F_{12}	2, 3, 5, 10	8	Composition Function 4	Scalable, non-separable, non-symmetr

To measure the performance of the optimization algorithms three performance metrics were introduced: Peak Ratio, Success Ratio, and Convergence speed.

- Peak Ratio (P_{ratio}) measures how many optima has been found over multiple runs in average: $P_{ratio} = \frac{\sum_{i=1}^{n_r} n_{g_i}}{n_g * n_r}$ where i is the run index and n_{g_i} is the number of global optima found at the end of process in each run. n_g refers to the number of known global optima, and n_r to the number of runs.
- Success Ratio (S_{ratio}) denotes the percentage of runs where all known global optima were found, out of all runs: $S_{ratio} = \frac{n_{sr}}{n_r}$ where n_{sr} is the number of successful runs.
- Convergence speed (C_{speed}) measures the number of function evaluations required to locate all known global optima, over multiple runs: $C_{speed} = \frac{\sum_{i=1}^{n_r} e_i}{n_r}$ where e_i is the number of evaluations used in each run to find all global optima. If all global optima is not found, the maximum function evaluations allowed is used.

To consider that a global optimum is found, the fitness value of the solution should be close to the best possible value, given an accuracy level. Five accuracy levels were considered to measure the performance metrics, as indicated in the benchmark: $\epsilon \in \{10^{-1}, 10^{-2}, 10^{-3}, 10^{-4}, 10^{-5}\}$. Apart from that, the distance between the solution and the optimum should be lower than a threshold that depends on the optimization problem.

5.2 Experimental Setup

We have conducted the experiments with the BO approaches introduced in Sect. 4 for all the functions in the benchmark and measured the performance metrics. Considering the stochastic nature of the algorithms, we conducted several runs of each algorithm configuration for each optimization problem. We were only able to perform 10 runs with our BO algorithms, while the EAs shown in the benchmark are averaged across 50 runs. All the results presented in this section have been measured with $\epsilon = 10^{-3}$ as accuracy level.

For all the BO approaches we have used the following configuration. First of all, regarding GPs, the mean function was set to $m(\mathbf{x}) = \theta_\mu$, where θ_μ is the mean of all fitness values obtained so far. On the other hand, for the kernel function, the Matern32 was employed, and for its parameter selection, the likelihood function was optimized every step with a grid search. Moreover, to limit the size of the covariance matrix, N was set to 500. UCB was used as the acquisition function adjusting the β constant as suggested in [22]. Finally, to maximize this acquisition function, a stochastic version of DIRECT [7] was applied. Being DIRECT a deterministic algorithm, repeated solutions may be selected to sample. In consequence, we decided to add a random shift to the starting point, in order to produce stochastic solutions.

On the other hand, for the batch approach B was set to 10, and in the clustering BO algorithm, k-means was used to create the clusters with $K = 2$.

We used a bounded version of the Powell's conjugate direction method [13] to guide the local search. This algorithm does not need derivatives, which makes it suitable for our optimization problem. When an out-of-bounds solution is required by this algorithm, the worst possible value is returned.

In addition to the three variants of BO investigated, a random multi-start optimization algorithm has been added as a reference. It also uses Powell's method as the local search algorithm, starting from a random solution at each iteration.

5.3 Comparison Between the Different BO Variants

Figure 2 shows the evolution of the peak ratio over function evaluations. In the figure, for illustration proposes, some functions with different number of dimensions are shown. In the lower axis, the number of evaluations are shown in percentage, while the P_{ratio} is illustrated in the vertical axis.

As can be seen in Fig. 2a, there are some functions in the benchmark were all the optima are found in the first 10% of the evaluations for all BO approaches. In Fig. 2b, for example, it takes a little bit longer to achieve the best possible

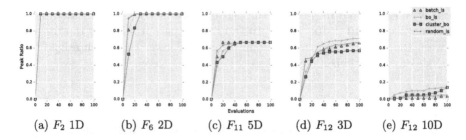

(a) F_2 1D   (b) F_6 2D   (c) F_{11} 5D   (d) F_{12} 3D   (e) F_{12} 10D

Fig. 2. Evolution of the peak ratio over function evaluations (%) for the proposed algorithms and the random multi-start algorithm in five optimization problems.

result, but the algorithms are able to solve this problem with a low-budget. On the contrary, in Fig. 2c, all the approaches find always the same optima but are not able to find more difficult ones in the second part of the optimization.

Figure 2d and e correspond to F_{12} function, with 3 and 10 dimensions respectively. Here, it can be seen that the performance of all methods decreases in the higher dimensional problems. In this problem, bo_ls performs better than the other BO approaches. Regarding the C_{speed}, cluster_bo takes more evaluations to find all global optima in F_6 2D.

5.4   Comparison to State-of-the-Art EAs for Multi-modal Problems

Here we compare BO approaches to NEA2 [14] and NMMSO [2] algorithms. Both algorithms won the *CEC Niching Methods for Multi-modal Optimization* competition [10] in 2013 and 2015 respectively.

Table 2 shows the peak-ratio of both algorithms, taken from the CEC 2015 competition, along with the peak-ratios computed from the results of all the algorithms we evaluate in our experimentation. The first five functions were omitted as all approaches were able to obtain the best possible score.

The sequential and the batch BO methods show competitive results in low dimensional functions compared to NEA2 and NMMSO. Our methods improve the results of NEA2 and NMMSO for F_6 2D, F_6 3D, F_9 2D and F_{10} 2D functions. However, for these functions, the random multi-start algorithm achieves similarly good performance. Although the results of the sequential approach are similar to the random multi-start algorithm overall, the batch BO method is able to beat the random algorithm in 5 functions while the opposite occurs only once. The comparison among the BO approaches is favorable to the batch approach, beating the other proposed approaches in F_{12} 5D function, and obtaining the best result in F_6 3D overall. The cluster BO approach is able to get good results in F_7 2D and F_{10} 11D, but most of the comparisons are favorable to the other approaches. In functions with more dimensions, NEA2 clearly outperforms the rest of the algorithms.

Table 2. Peak Ratio for the EAs, the random multi-start algorithm and the proposed algorithms for the functions presented in the benchmark. Best result for each function is written in bold. As all the algorithms get the best possible results in the first five functions their results are not shown. $\epsilon = 10^{-3}$ was used as accuracy level. *cluster_bo* was not able to finish the experimentation on time for the F_6 3D and F_7 3D problems.

	NEA2	NMMSO	random_ls	bo_ls	batch_ls	cluster_bo
F_6 2D	0.958	0.992	**1.000**	**1.000**	**1.000**	**1.000**
F_7 2D	0.918	**1.000**	0.889	0.864	0.892	0.917
F_6 3D	0.240	0.922	0.985	0.973	**0.989**	–
F_7 3D	0.584	**0.978**	0.571	0.782	0.722	–
F_8 2D	**1.000**	**1.000**	**1.000**	**1.000**	**1.000**	**1.000**
F_9 2D	0.967	0.990	**1.000**	**1.000**	**1.000**	**1.000**
F_{10} 2D	0.843	0.995	**1.000**	**1.000**	**1.000**	**1.000**
F_{11} 2D	0.960	**0.983**	0.684	0.767	0.684	0.684
F_{11} 3D	**0.810**	0.723	0.667	0.667	0.667	0.667
F_{12} 3D	**0.720**	0.642	0.662	0.713	0.662	0.569
F_{11} 5D	**0.673**	0.660	0.667	0.667	0.667	0.667
F_{12} 5D	**0.695**	0.470	0.412	0.375	0.425	0.338
F_{11} 10D	**0.667**	0.650	0.367	0.333	0.400	0.433
F_{12} 10D	**0.667**	0.457	0.050	0.138	0.037	0.138
F_{12} 20D	**0.360**	0.172	0.000	0.000	0.000	0.000

5.5 Discussion

Among our BO approaches, bo_ls and batch_ls seem to be the most competitive proposals. They show good results for the lowest dimension (2D) problems, beating in some functions the state-of-the-art algorithms. However, for higher dimensions, particularly in F_{11} and F_{12}, their performance decreases. Having only one hyperparameter for all dimensions instead of many, may cause this dimensionality problem. It would be interesting to study in depth the characteristics of such parameters.

It is somehow surprising the good behavior of the baseline random multi-start approach, being able to identify all the optima in four problems, and outperforming the best algorithms. Being such low dimension, one could think that the function is easy to solve, but this is not the case (looking at NEA2 and NMMSO).

6 Conclusions

In this paper we have investigated the suitability of BO methods for massively multi-modal optimization problems traditionally addressed with EAs.

We have proposed three new approaches to deal with these problems. These approaches incorporate ideas and insights of the previous work done in the EC filed. Although BO is usually used in low-budget optimization problems, we were able to achieve competitive results in optimization problems with low dimensions by modeling the search space.

Acknowledgments. This research has been partially supported by the Basque Government (ELKARTEK programs), and Spanish Ministry of Economy and Competitiveness MINECO (project TIN2016-78365-R). Jose A. Lozano is also supported by BERC 2018-2021 (Basque Government), and Severo Ochoa Program SEV-2017-0718 (Spanish Ministry of Economy, Industry and Competitiveness).

References

1. Brochu, E., Cora, V.M., de Freitas, N.: A tutorial on Bayesian optimization of expensive cost functions, with application to active user modeling and hierarchical reinforcement learning, December 2010. arXiv:1012.2599
2. Fieldsend, J.E.: Running up those hills: multi-modal search with the niching migratory multi-swarm optimiser. In: 2014 IEEE Congress on Evolutionary Computation (CEC), pp. 2593–2600. IEEE (2014)
3. Ginsbourger, D., Riche, R.L., Carraro, L.: A multi-points criterion for deterministic parallel global optimization based on Gaussian processes. Technical report, CCSD, March 2008. https://hal.archives-ouvertes.fr/hal-00260579/document
4. Goldberg, D.E., Richardson, J.: Genetic algorithms with sharing for multi-modal function optimisation. In: Proceedings of the of Second International Conference on Genetic Algorithms and Their Applications, pp. 41–49 (1987)
5. González, J., Dai, Z., Hennig, P., Lawrence, N.D.: Batch Bayesian optimization via local penalization (2015). arXiv:1505.08052
6. Guhaniyogi, R., Li, C., Savitsky, T.D., Srivastava, S.: A divide-and-conquer Bayesian approach to large-scale kriging, December 2017. arXiv:1712.09767
7. Jones, D.R., Perttunen, C.D., Stuckman, B.E.: Lipschitzian optimization without the Lipschitz constant. J. Optim. Theory Appl. **79**(1), 157–181 (1993). https://doi.org/10.1007/BF00941892
8. Jones, D.R.: A taxonomy of global optimization methods based on response surfaces. J. Global Optim. **21**(4), 345–383 (2001). https://doi.org/10.1023/A:1012771025575
9. Kvasov, D.E., Sergeyev, Y.D.: Deterministic approaches for solving practical black-box global optimization problems. Adv. Eng. Softw. **80**, 58–66 (2015). https://doi.org/10.1016/j.advengsoft.2014.09.014. http://www.sciencedirect.com/science/article/pii/S096599781400163X
10. Li, X., Engelbrecht, A., Epitropakis, M.G.: Benchmark functions for CEC 2013 special session and competition on niching methods for multimodal function optimization. Technical report, Royal Melbourne Institute of Technology, March 2013. http://goanna.cs.rmit.edu.au/xiaodong/cec13-niching/
11. Mockus, J.: The Bayesian approach to local optimization. In: Mockus, J. (ed.) Bayesian Approach to Global Optimization. Mathematics and Its Applications, vol. 37, pp. 125–156. Springer, Dordrecht (1989). https://doi.org/10.1007/978-94-009-0909-0_7

12. Petrowski, A.: A clearing procedure as a niching method for genetic algorithms. In: Proceedings of IEEE International Conference on Evolutionary Computation, pp. 798–803 (1996). https://doi.org/10.1109/ICEC.1996.542703
13. Powell, M.J.D.: An efficient method for finding the minimum of a function of several variables without calculating derivatives. Comput. J. **7**(2), 155–162 (1964). https://doi.org/10.1093/comjnl/7.2.155. http://comjnl.oxfordjournals.org/content/7/2/155
14. Preuss, M.: Niching the CMA-ES via nearest-better clustering. In: Proceedings of the 12th Annual Conference Companion on Genetic and Evolutionary Computation, GECCO 2010, pp. 1711–1718. ACM, New York (2010). https://doi.org/10.1145/1830761.1830793
15. Preuss, M.: Multimodal Optimization by Means of Evolutionary Algorithms. Natural Computing Series. Springer, Cham (2015). https://doi.org/10.1007/978-3-319-07407-8
16. Quiñonero-Candela, J., Rasmussen, C.E.: A unifying view of sparse approximate Gaussian process regression. J. Mach. Learn. Res. **6**(Dec), 1939–1959 (2005). http://www.jmlr.org/papers/v6/quinonero-candela05a
17. Rasmussen, C.E., Williams, C.K.: Gaussian Processes for Machine Learning. MIT Press, Cambridge (2006)
18. Roman, I., Santana, R., Mendiburu, A., Lozano, J.A.: Dynamic kernel selection criteria for Bayesian optimization. In: BayesOpt 2014: NIPS Workshop on Bayesian Optimization, Montreal (2014). http://bayesopt.github.io/papers/paper13.pdf
19. Sareni, B., Krahenbuhl, L.: Fitness sharing and niching methods revisited. IEEE Trans. Evol. Comput. **2**(3), 97–106 (1998)
20. Silverman, B.W.: Some aspects of the spline smoothing approach to non-parametric regression curve fitting. J. Roy. Stat. Soc. Series B (Method.) **47**(1), 1–52 (1985)
21. Singh, G., Deb, K.: Comparison of multi-modal optimization algorithms based on evolutionary algorithms. In: Proceedings of the 8th Annual Conference on Genetic and Evolutionary Computation, GECCO 2006, pp. 1305–1312. ACM, New York (2006). https://doi.org/10.1145/1143997.1144200
22. Srinivas, N., Krause, A., Kakade, S., Seeger, M.: Gaussian process optimization in the bandit setting: no regret and experimental design. In: Proceedings of the 27th International Conference on Machine Learning (ICML 2010), Haifa, 21–24 June 2010, pp. 1015–1022 (2010). http://www.icml2010.org/papers/422.pdf
23. Wang, H., van Stein, B., Emmerich, M., Bäck, T.: Time complexity reduction in efficient global optimization using cluster kriging. In: Proceedings of the Genetic and Evolutionary Computation Conference, GECCO 2017, pp. 889–896. ACM, New York (2017). https://doi.org/10.1145/3071178.3071321
24. Wong, K.C., Leung, K.S., Wong, M.H.: Protein structure prediction on a lattice model via multimodal optimization techniques. In: Proceedings of the 12th Annual Conference on Genetic and Evolutionary Computation, GECCO 2010, pp. 155–162. ACM, New York (2010). https://doi.org/10.1145/1830483.1830513
25. Yin, X., Germay, N.: Investigations on solving the load flow problem by genetic algorithms. Electr. Power Syst. Res. **22**(3), 151–163 (1991). https://doi.org/10.1016/0378-7796(91)90001-4. http://www.sciencedirect.com/science/article/pii/0378779691900014
26. Zhigljavsky, A., Žilinskas, A.: Methods based on statistical models of multimodal functions. In: Zhigljavsky, A., Žilinskas, A. (eds.) Stochastic Global Optimization. Springer Optimization and Its Applications, vol. 9, pp. 149–244. Springer, Boston (2008). https://doi.org/10.1007/978-0-387-74740-8_4

Author Index

Printed in the United States
By Bookmasters